ADVANCES IN MACROFUNGI
Industrial Avenues and Prospects

Series: Progress in Mycological Research

- Fungi from Different Environments (2009)
- Systematics and Evolution of fungi (2011)
- Fungi from Different Substrates (2015)
- Fungi: Applications and Management Strategies (2016)
- Advances in Macrofungi: Diversity, Ecology and Biotechnology (2019)
- Advances in Macrofungi: Pharmaceuticals and Cosmeceuticals (2021)

Series: Progress in Mycological Research

ADVANCES IN
MACROFUNGI
Industrial Avenues and Prospects

Editors

Kandikere R. Sridhar

Department of Biosciences
Mangalore University, Mangalore, India

Sunil K. Deshmukh

Nano Biotechnology Centre
The Energy and Resources Institute, New Delhi, India

CRC Press
Taylor & Francis Group
Boca Raton   London   New York

CRC Press is an imprint of the
Taylor & Francis Group, an **informa** business

A SCIENCE PUBLISHERS BOOK

First edition published 2021
by CRC Press
6000 Broken Sound Parkway NW, Suite 300, Boca Raton, FL 33487-2742

and by CRC Press
4 Park Square, Milton Park, Abingdon, Oxon OX14 4RN

© 2021 Taylor & Francis Group, LLC

CRC Press is an imprint of Taylor & Francis Group, an Informa business

ISBN: 978-0-367-56205-2 (hbk)
ISBN: 978-0-367-56209-0 (pbk)
ISBN: 978-1-003-09681-8 (ebk)

Typeset in Times New Roman
by Radiant Productions

Preface

Macrofungi are a vital component of our ecosystem and serve as a unique research tool in the fields of mycology, biotechnology, nanotechnology and bioremediation. In addition to well-known nutritional benefits of macrofungi, they are potential producers of value-added bioactive compounds and metabolites. Many mushrooms have three-fold benefits in nutrition, bioactive components and serve as mutualists with tree species (boletes and truffles). Some macrofungi have a mutualistic association with termites (*Termitomyces*) and serve as a precious food source, while some are pathogens to insects (*Cordyceps*), known for their value-added drugs and pharmaceuticals. Owing to the multifarious benefits of macrofungi, several challenges lie ahead to explore the diversity of macrofungi in different ecosystems as well as under various geographic conditions (e.g., documentation, cataloguing, conservation, domestication and utilisation).

Mushrooms are an important source of agricultural earnings, thus cultivated or gathered (wild mushrooms) all over the world. The global turnover of edible as well as medicinal mushrooms is about $ 63 billion per annum (Zied and Pardo-Giménez, 2017). Although 200 species are in cultivation on an experimental basis, 60 are cultivated commercially, while only 10 are elevated to industrial production (Chang, 2006). Biochemical profiles of wild and cultivated macrofungi provide evidence of their resourcefulness and functions beyond mere biological and nutritional value. Such insights stimulate further research on industrial applications and environmental bioremediation. Recent advances in biotechnology, nanotechnology and bioinformatics have further broadened our knowledge on macrofungi towards nutraceuticals, cosmeceuticals, immunoceuticals, enzymes, alcoholic beverages, perfumes, pigments and so on. Besides, macrofungi will be a potential source to produce unanticipated ecofriendly merchandise, especially microbial fuel cells, wearable materials, substitutes to plastics, nanomaterials and packaging and building materials (Meyer et al., 2020). Macrofungi also serve environmental remediation, such as waste management, degradation of xenobiotics and biopesticides.

Macrofungi serve as indispensable components of basic and applied research in different disciplines. In recent past, there have been remarkable developments in applied aspects of macrofungi in various fields, such as food, agriculture, health, environment and industry. The major aspects of macrofungi dealt in this book include methods of mushroom cultivation, mushroom market, industrial applications, biomedical applications, functional aspects of their enzymes, exploitation of pigments produced by them and significance in environmental bioremediation. Several contributors have documented their expertise, and offered fascinating chapters on industrial applications and potential prospects of macrofungi.

This book addresses: (1) commercial cultivation techniques, sustainable pest management during cultivation with examples of mushroom industry and markets; (2) bioprospects, biotechnological and biomedical applications of mushrooms; (3) industrial and medicinal applications of truffles and *Pleurotus*; (4) production of alcoholic beverages (beer, wine and spirits), their quality control and production of flavouring agents by macrofungi; (5) production of enzymes (ligninolytic, cellulolytic and proteases) by macrofungi including white rot fungi with emphasis on usefulness in paper and textile industries, nanotechnology, and bioremediation (degradation of effluents, textile/paper dyes and xenobiotics); (6) production, applications of industrially valued (food, textile and wood), health-protective, antimicrobial and cytotoxic pigments by macrofungi; (7) applications of macrofungi in environmental remediation and production of other agriculturally vital products (e.g., pesticides). This book also reflects other applications of macrofungi, such as microbial fuel cell generator, wearable products like fabrics, substitutes to plastics, packaging materials and building insulation materials.

This contribution draws attention towards applied values of macrofungi. It will be a precious resource in applied disciplines of mycology (biotechnology, enzymology, food science, cosmetic biology, applied biology and medicine) to the graduates, post-graduates and researchers. Our appreciation goes to the friendly gesture of the contributors and reviewers towards submission, assessment and revision of the chapters on time. We express our gratitude to Dr. Dietmar Schlosser for sketching an attentive overview on the applied aspects of macrofungi and the contents of this book. The CRC Press has extended co-operation in spite of the current pandemic by straightforward official formalities to present this book on time.

Mangalore, India   **Kandikere R. Sridhar**
New Delhi, India   **Sunil K. Deshmukh**
December 24, 2020

References

Chang, S.T. (2006). Int. J. Med. Mush., 8: 297–314.

Meyer, V., Basenko, E.Y., Benz, J.P. et al. (2020). Fungal Biol. Biotechnol., 7: 5. 10.1186/s40694-020-00095-z.

Zied, D.C. and Pardo-Giménez, A. (2017). Edible and Medicinal Mushrooms: Technology and Applications. Wiley-Blackwell, England.

Contents

Enzymes

Pigments

Bioremediation

List of Contributors

Paul Agastian

Research Department of Plant Biology and Biotechnology, Loyola College, Nungambakkam, Chennai, India.

Naif Abdullah Al-Dhabi

Department of Botany and Microbiology, College of Science, King Saud University, P.O. Box 2455, Riyadh 11451, Saudi Arabia.

Ramzi Abdel Alsaheb

Al-khwarizmi College of Engineering, University of Baghdad, Baghdad, Iraq.

Mariadhas Valan Arasu

Department of Botany and Microbiology, College of Science, King Saud University, P.O. Box 2455, Riyadh 11451, Saudi Arabia.
Xavier Research Foundation, St. Xavier's College, Palayamkottai, Thirunelveli, India.

Jaime Carrasco

Department of Plant Sciences, University of Oxford, South Parks Road, Oxford OX1 3RB, United Kingdom.

Ekta Chaudhary

Department of Microbiology, Punjab University, Chandigarh-160014, India.

Ghoson Daba

Chemistry of Natural and Microbial Products Department, Pharmaceutical Industries Researches Division, National Research Centre, El Buhouth St., Dokki, 12311, Giza, Egypt.

Rosy Agnes De Souza

Mycological Laboratory, Department of Botany, Goa University, Taleigao Plateau, Goa-403206, India.

Laurent Dufossé

Chimie et Biotechnologie des Produits Naturels (ChemBioProLab) & ESIROI. Agroalimentaire, Université de la Réunion, 15 Avenue René Cassin, CS 92003, F-97744 Saint-Denis CEDEX, France.

Veeramuthu Duraipandiyan

Department of Botany and Microbiology, College of Science, King Saud University, P.O. Box 2455, Riyadh 11451, Saudi Arabia.
Entomology Research Institute, Loyola College, Chennai, India.

Hesham Ali El-Enshasy

Institute of Bioproduct Development (IBD), Universiti Teknologi Malaysia (UTM), Johor Bahru, Johor, Malaysia.
School of Chemical and Energy Engineering, Faculty of Engineering, Universiti Teknologi Malaysia (UTM), Johor Bahru, Johor, Malaysia.
City of Scientific Research and Technology Application, New Burg Al-Arab, Alexandria, Egypt.

Waill Elkhateeb

Chemistry of Natural and Microbial Products Department, Pharmaceutical Industries Researches Division, National Research Centre, El Buhouth St., Dokki, 12311, Giza, Egypt.

Marwa Elnahas

Chemistry of Natural and Microbial Products Department, Pharmaceutical Industries Researches Division, National Research Centre, El Buhouth St., Dokki, 12311, Giza, Egypt.

Elsayed Ahmed Elsayed

Chemistry of Natural and Microbial Products Department, National Research Centre, Cairo, Egypt.

Shin-Yee Fung

Medicinal Mushroom Research Group (MMRG), Department of Molecular Medicine, Faculty of Medicine, University of Malaya, 50603 Kuala Lumpur, Malaysia.

Francisco J Gea

Centro de Investigación, Experimentación y Servicios del Champiñón, Quintanar del Rey, Cuenca, Spain.

Siti Zulaiha Hanapi

Institute of Bio-product Development (IBD), Universiti Teknologi Malaysia (UTM), Johor Bahru, Johor, Malaysia.

Ting Ho

Global Agro-innovation (HK) Limited, Hong Kong.

Shin Hyun-Jae

Department of Biochemical and Polymer Engineering, Chosun University, Gwangju, Republic of Korea.

Savarimuthu Ignacimuthu

Xavier Research Foundation, St. Xavier's College, Palayamkottai, Thirunelveli, India.

Kab-yel Jang

Mushroom Research Division, National Institute of Horticultural and Herbal Science, Rural Development Administration, Eumseong, Republic of Korea.

Nandkumar Mukund Kamat

Mycological Laboratory, Department of Botany, Goa University, Taleigao Plateau, Goa-403206, India.

Wong-sik Kong

Department of Biochemical and Polymer Engineering, Chosun University, Gwangju, Republic of Korea.

Ameer Khusro

Research Department of Plant Biology and Biotechnology, Loyola College, Nungambakkam, Chennai, India.

Ajay C Lagashetti

National Fungal Culture Collection of India (NFCCI), Biodiversity and Palaeobiology Group, Agarkar Research Institute, G.G. Agarkar Road, Pune-411004, India.

Hariprasath Lakshmanan

Department of Biochemistry, Karpagam Academy of Higher Education, Coimbatore-641021, India.

R Lekshmi

Department of Botany and Biotechnology, MSM College, Kayamkulam, Kerala, India.

María J Navarro

Centro de Investigación, Experimentación y Servicios del Champiñón, Quintanar del Rey, Cuenca, Spain.

Leopold M Nyochembeng

Department of Biological and Environmental Sciences, CCS-Bonner Wing Room 218 B Alabama A & M University, PO Box-1208 Normal, AL -35762, United States.

Soosaimanickam Maria Packiam

Department of Botany and Biotechnology, MSM College, Kayamkulam, Kerala, India.

Arturo Pardo-Gimenez

Centro de Investigación, Experimentación y Servicios del Champiñón, Quintanar del Rey, Cuenca, Spain.

Riikka Räisänen

HELSUS Helsinki Institute of Sustainability Science, Craft Studies, P.O. Box 8 (Siltavuorenpenger 10), 00014 University of Helsinki, Finland.

Deepak K Rahi

Department of Microbiology, Panjab University, Chandigarh-160014, India

Sonu Rahi

Department of Botany, Government Girls College, A.P.S. University, Rewa-486003, India.

R Rajakrishnan

Department of Botany and Microbiology, College of Science, King Saud University, P.O. Box 2455, Riyadh-11451, Saudi Arabia.

Madaiah Rajashekhar

Department of Biosciences, Mangalore University, Mangalagangotri, Mangalore, Karnataka, India.

Jegadeesh Raman

Department of Biochemical and Polymer Engineering, Chosun University, Gwangju, Republic of Korea.
Mushroom Research Division, National Institute of Horticultural and Herbal Science, Rural Development Administration, Eumseong, Republic of Korea.

Venugopalan Ravikrishnan

Department of Biosciences, Mangalore University, Mangalagangotri, Mangalore, Karnataka, India.

Muhammad Fazril Razif

Medicinal Mushroom Research Group (MMRG), Department of Molecular Medicine, Faculty of Medicine, University of Malaya, 50603 Kuala Lumpur, Malaysia.

Seri C Robinson

Department of Wood Science and Engineering, 119 Richardson Hall, Oregon State University, Corvallis, OR 97331, USA.

Dietmar Schlosser

Department of Environmental Microbiology, Helmholtz Centre for Environmental Research - UFZ, Permoserstraße 15, 04318 Leipzig, Germany.

Paras N Singh

National Fungal Culture Collection of India (NFCCI), Biodiversity and Palaeobiology Group, Agharkar Research Institute, GG Agarkar Road, Pune-411004, India.

Sanjay K Singh

National Fungal Culture Collection of India (NFCCI), Biodiversity and Palaeobiology Group, Agharkar Research Institute, GG Agarkar Road, Pune-411004, India.

Kandikere R Sridhar

Department of Biosciences, Mangalore University, Mangalagangotri, Mangalore, Karnataka, India.

Centre for Environmental Studies, Yenepoya (deemed to be) University, Mangalore, Karnataka, India.

Dalia Sukmawati

Biology Department, Laboratory of microbiology, 9th Floor Hasyim Ashari Building, Faculty of Mathematics and Natural Sciences, Universitas Negeri Jakarta, Jakarta, Indonesia.

Malika Suthar

National Fungal Culture Collection of India (NFCCI), Biodiversity and Palaeobiology Group, Agharkar Research Institute, GG Agarkar Road, Pune-411004, India.

Chon-Seng Tan

LiGNO Research Initiative, LiGNOBiotech Sdn. Bhd., Balakong Jaya-43300, Selangor, Malaysia.

Paul Thomas

Mycorrhizal Systems Ltd, Lancashire, PR25 2SD, UK.
University of Stirling, Stirling, FK9 4LA, UK.

Sonja Veljović

Institute of General and Physical Chemistry, P.O. Box 551, 11001 Belgrade, Serbia.

Ponnuswamy Vijayaraghavan

Bioprocess Engineering Division, Smykon Biotech Pvt LtD, Nagercoil, Kanyakumari, India.

Jovana Vunduk

Research Associate, Institute for Food Technology and Biochemistry, Faculty of Agriculture, Nemanjina 6, 11080, University of Belgrade, Belgrade, Serbia.

Racha Wehbe

Laboratory of Cancer Biology and Molecular Immunology, Faculty of Science I, Lebanese, University, Hadath, Lebanon.

Diego C Zied

Universidade Estadual Paulista, Faculdade de Ciências Agrárias e Tecnológicas, Dracena, São Paulo, Brazil.

1

Past, Present and Future of Macrofungal Applications
Old Friends, Yet Unknown Strangers

Dietmar Schlosser

1. INTRODUCTION

Fungi existed long before the first humans appeared. They evolved from phagotrophic ancestors and branched off from animals more than one billion years ago (Berbee et al., 2020) (https://www.nationalgeographic.com/science/2020/01/oldest-fungus-fossils-found-earth-history/; accessed December 14, 2020). One particular group of true fungi (Kingdom Fungi) referred to as macrofungi (or more colloquially and frequently simply denoted as mushrooms) may be defined as fungi forming spore-bearing structures during sexual reproduction and which are visible to the naked eye and may emerge above or below the ground (Mueller et al., 2007). Macrofungi include Ascomycetes, Basidiomycetes and also a few former Zygomycetes (members of which are now being assigned to the phyla Mucoromycota and Zoopagomycota; Spatafora et al., 2016) and under certain conditions may also grow in the form of asexual reproduction stages (anamorphs).

The so far oldest molecularly identified remains of fungi were found in 810–715 million-year (Ma)-old dolomitic shale from the Democratic Republic of Congo,

Department of Environmental Microbiology, Helmholtz Centre for Environmental Research - UFZ, Permoserstraße 15, 04318 Leipzig, Germany.
Email: dietmar.schlosser@ufz.de

whereas various older Precambrian fossils suggesting fungal filament fragments, spores and lichen-like structures are difficult to distinguish from fossil prokaryotes and hence remain ambiguous (Bonneville et al., 2020). Starting from osmotrophic ancestors living in freshwater habitats, early fungal colonisation of land is currently thought to have occurred sometime between the Ordovician (443–485 Ma) to not later than about 800 Ma ago (Berbee et al., 2020; Bonneville et al., 2020). Fossil records from the early Devonian Age (407–397 Ma) suggest the presence of early members of Ascomycota and Mucoromycota (Berbee et al., 2020; Bonneville et al., 2020). Woody plants representing new fungal substrates also evolved during the Devonian period, driving the subsequent expansion of Ascomycete and Basidiomycete-decay fungi during the following geological eras (Floudas et al., 2012; Berbee et al., 2020). Macrofungi-related fungal taxa and the evolution of corresponding members can hence be traced back to the Paleozoic and continued during Mesozoic times (Floudas et al., 2012; Berbee et al., 2020; Bonneville et al., 2020).

Published numbers of known macrofungal species range from about 14,000 (Hawksworth, 2019) to approximately 22,000 (Mueller et al., 2007), with a recent estimate of between 220,000–380,000 species for all (including the so far unknown) Macromycete species on earth (Hawksworth, 2019). Such figures suggest that only up to (and more probably less than) 10 per cent of the globe's macrofungi may be known today, as based on a total predicted number of fungal species lying between 2.2–3.8 million (Hawksworth and Lücking, 2017; Hawksworth, 2019).

The use of macrofungi for different purposes by humans has a very long history, with particular presumable applications seemingly already existing in the Neolithic period and proven record for other types of uses dating back to ancient times (refer to the next sub-section for more details). The diverse lifestyles of macrofungi and their multifaceted capabilities to cope with the biotic and abiotic factors of their respective habitats gathered during their long-lasting evolutionary history have formed the basis for their current biotechnological usability by humans. This chapter aims to provide a brief sketch of historical aspects and the current state in different important fields of Macromycete applications. It further advocates for the continued research into Macromycetes and their lifestyles in so far underexplored environments.

2. Macromycetes and Humans: Old and Current Industrial Avenues

The evolution of fungal sexual reproduction strategies involving the formation of fruit bodies, which are most prominent in basidiomycota and ascomycota, resulted in the most likely oldest form of utilisation of fungi by mankind. The collection and eating of mushroom fruit bodies by humans dates back to early human history. Edible species were found associated with people living 13,000 years ago in Chile and mushroom consumption has been evidenced since ancient Chinese, Greek and Roman times (Boa, 2004). Today the commercial production and exploitation of easily cultivatable edible mushrooms for food represents one of the economically most important forms of utilisation of macrofungi, even if limited to only a handful of genera (Hawksworth, 2019) (Carrasco et al., Chap. 2). Globally, a much higher number of Macromycetes in the range of several hundred (and perhaps even more) species can reasonably be assumed to be currently collected for nutrition purposes from the wild for both consumption by local communities and trade (Boa, 2004; Hawksworth, 2019) (Fung and Tan, Chap. 3). In this context a clear need to extend ethnomycological surveys in remote areas of many parts of the world has been emphasised (Hawksworth, 2019) and it is quite likely that the full potential beyond the use for food purposes of such underexplored resources still remains to be discovered (Ravikrishnan et al., Chap. 9).

Macrofungi were also used for various medicinal purposes since a long time and various genera are well known to produce different bioactive compounds (Hawksworth, 2019; Hyde et al., 2019). These include polysaccharides, peptides, amino acids, phenolics, triterpenoids, sterols, steroids, dietary fibres, fatty acids and corresponding esters and vitamins, and form the basis for various health-promoting functions of mushrooms (e.g., immune-stimmulatory, anticancer, anti-inflammatory, antiviral, antioxidant, antibacterial, antifungal, antihypotensive, and antidiabetic) (Hyde et al., 2019; Lu et al., 2020). The famous mummified Ice Man 'Ötzi', who lived in the Alps about 5,300 years ago and was found in the receding ice of the Val Senales glacier (South Tyrol, northern Italy), carried a part of a fruit body of the birch fungus (*Piptoporus betulinus*), which is known to have anthelmintic properties and was perhaps used for medicinal purposes (Capasso, 1998). Highly prized delicacies, such as truffles, are ectomycorrhizal fungi and therefore much more difficult to cultivate than saprotrophic decomposer fungi, which may easily be cultivated on agricultural residues (Hyde et al., 2019). Truffles provide their mycorrhizal host plants with macronutrients (potassium, phosphorus, nitrogen, sulphur) and micronutrients (iron, copper, zinc, chloride) in exchange for carbohydrates mainly via an intercellular hyphal network between plant root cells, which is referred to as Hartig net (Allen et al., 2003). These fungi are known for their complex chemical-based ecological interactions with other organisms, which involve various volatiles and bioactive compounds and therefore also offer medicinal and industrial values beyond those solely related to nutrition (Splivallo et al., 2011) (Thomas et al., Chap. 6; Elsayed et al., Chap. 10). Contrary to truffles, *Pleurotus* species involving the oyster mushroom (*Pleurotus ostreatus*) together with other species of the genus (Raman et al., Chap. 7) can easily be cultivated on lignocellulosic residues from agriculture and forestry. These fungi belong to the commercially most important mushrooms produced and have also been implicated with various health-promoting effects in addition to their nutritional value (Hyde et al., 2019; Lu et al., 2020). Due to their impressive biodegradation capabilities, *Pleurotus* species are further attractive biocatalysts for mycoremediation purposes (Hyde et al., 2019) (see also below). Medicinal fungi may further be employed in the production of alcoholic beverages and due to the excretion of bioactive compounds improve certain functional beverage properties (Veljović et al., 2019) (Vunduk and Veljović, Chap. 8). Fungal bioactive compounds are also attractive for cosmetic industry (Hyde et al., 2019; Lu et al., 2020) (Fung and Razif, Chap. 4).

Arising from growing concern related to potential harmful effects of synthetic colourants on both human and environmental health, the demand for natural colourants in the food, cosmetic and textile industries is rapidly increasing (Kalra et al., 2020). Pigments from Basidiomycete mushrooms were used for the dyeing of wool and silk in ancient times (Hernández et al., 2019). Especially filamentous fungi (in particular ascomycetous and basidiomycetous mushrooms), which can be grown in fermenters, and also lichens (symbiotic associations of fungi with green algae and/ or cyanobacteria) produce a wide range of pigments of different chemical classes, such as, e.g., melanins, anthraquinones, hydroxyanthraquinones, azaphilones, carotenoids, oxopolyene, quinones and naphthoquinones (Hyde et al., 2019; Kalra et al., 2020). Fungal pigments are usually reported as secondary metabolites with either yet unknown or known biological functions, such as acting as enzyme cofactors (flavins), or preventing from photooxidative (carotenoids) and other environmental stress (melanins) (Gmoser et al., 2017; Kalra et al., 2020). Pigments from Macromycetes hence represent a very attractive but still underexplored resource for a wide range of applications (Suthar et al., Chap. 13; Lagshetti et al., Chap. 14; De Souza et al., Chap. 15).

Fungi can attack a wide range of organic environmental pollutants, resulting in the formation of organic biotransformation products or in mineralisation to CO_2. Only a limited number of organic pollutants with mostly rather simple and only rarely more complex structures are utilised as fungal growth substrates, whereas the vast majority of such pollutants are cometabolised in the presence of carbon-and energy-delivering cosubstrates (Harms et al., 2017; Schlosser, 2020). The ecological background of this type of biochemical attack may relate to fungal defence against and detoxification of natural toxic compounds present in many fungal environments, e.g., in lignocellulosic plant material utilised by saprotrophic fungi, or in plants as defence compounds against pytopathogenic fungi (Harms et al., 2017; Schlosser, 2020). The cometabolic mineralisation of lignin to CO_2 and H_2O by Basidiomycetes, causing the so-called 'white rot' decay type of wood, aims to access lignocellulosic polysaccharides that serve as fungal carbon and energy sources. The corresponding fungal degraders and their special ligninolytic enzymes have evolved from non-ligninolytic ancestors following the evolution of wood lignin in plants (Floudas et al., 2012; Eastwood, 2014; Ayuso-Fernández et al., 2019). The lignin-degrading machinery enables white rot Basidiomycetes to mineralise a very broad range of environmental pollutants. To a considerably lesser extent, mineralisation of environmental pollutants is also known from brown rot decay Basidiomycetes, which employ extracellularly produced hydroxyl radicals as very unspecific and highly reactive oxidants to attack organic compounds. Cometabolic biotransformations of environmental pollutants to organic products predominate in other Macromycete groups (Harms et al., 2017; Schlosser, 2020). Fungal cometabolism is generally much less compound-specific than growth on organic pollutants as frequently found in bacteria. Moreover, cometabolic degraders do not depend on the utilisation of pollutants as growth substrates, thereby providing advantages under conditions of poor bioavailability of pollutants or very low pollutant concentrations. Such characteristics render macrofungi attractive for mycoremediation purposes by targeting organic environmental pollutants (Harms et al., 2011; Harms et al., 2017; Schlosser, 2020). Beyond that, fungi are also reported to exhibit pesticidal properties (Hyde et al., 2019), which may be employed in mycoremediation schemes (Nyochembeng, Chap. 16).

The extraordinary capabilities of Macromycetes to decompose complex natural organic matter in their diverse habitats depend on the effective arrays of extracellular enzymes, which initiate the breakdown of macromolecules. Such enzymes include oxidoreductases (oxidases like laccase and various peroxidases; among them are the powerful ligninolytic peroxidases) and many different hydrolases (e.g., amylases, cellulases, lipases, phytases, proteases, tannases and xylanases), which possess a very wide range of applicability for both biodegradative and biosynthesis processes (Harms et al., 2011; Hyde et al., 2019; Schlosser, 2020). Examples of this include the degradation of fat in wastewater treatment; applications in animal feed, pulp and paper, and detergent industries; for food and leather processing; in textile and pharmaceutical industries; treatment of contaminated water and biorefinery applications (Hyde et al., 2019) to mention a few (Al-Dhabi et al., Chap. 11; Rahi et al., Chap. 12).

Moreover, beyond main traditional and current industrial uses, macrofungi are also attractive for niche applications and based on their great metabolic and adaptive versatility with regard to diverse habitats, hold promise for possible new applications in future (Elkhateeb et al., Chap. 5). The fascinating natural networks of fungal mycelia have attracted humans since a long time. Meanwhile, mushrooms are also being used as objects of art (https://v-meer.de/; accessed December 22, 2020) and fungal mycelia are considered as composite construction (Jones et al., 2020) and packaging materials (https://interestingengineering.com/ikea-moves-mushrooms-replace-current-packaging/; accessed December 22, 2020) (Elkhateeb et al., Chap. 5; Suthar et al., Chap. 13).

3. Outlook

Humans have been benefiting from macrofungi since several thousands of years and for obvious reasons, they were focusing on clearly visible species sharing the same (i.e., mainly terrestrial) environments. Most terrestrial macrofungi are saprotrophs or mycorrhizal symbionts and others are known as plant or fungal pathogens (Mueller et al., 2007). Nevertheless, there is much more to discover as current estimates of the so far unknown macrofungi exceed the number of known species by far (Hawksworth and Lücking, 2017; Hawksworth, 2019; also refer to the introduction). Fungi have further conquered so far underexplored habitats beyond terrestrial ones, such as diverse freshwater and marine environments (Berbee et al., 2020; Lücking et al., 2020). Moreover, macrofungi form mutualistic symbioses not only with plants (e.g., with chlorophyte algae in lichens) but also with various animals, but this is less investigated as compared to fungal-plant interactions (Tang et al., 2015; Naranjo-Ortiz and Gabaldón, 2019; Berbee et al., 2020). Taken together, yet undiscovered, the potential novel characteristics and capabilities of macrofungi which could be expected due to possible fungal adaptations to (or interactions in) so far underexplored environments or lifestyles, may improve the existing or inspire new biotechnological applications of macrofungi. For instance, pigment production by cold-adapted fungi of the cryosphere has been identified as a novel source of pigments with potential use in textile and food industries, and for biomedical applications (Sajjad et al., 2020). In addition to the use of wild-type of strains, engineering of proteins or whole biosynthetic gene clusters, CRISPR/Cas9 genome editing, adaptive evolution techniques and the heterologous expression of desired gene products are expected to boost the development of diverse future biotechnological applications of Macromycetes (Kun et al., 2019; Skellam, 2019; Song et al., 2019; Stanzione et al., 2020). Technologically advanced fungal biotechnologies would provide a valuable contribution to the transition from the globe's petroleum-based economy into a bio-based circular economy in future (Meyer et al., 2020).

Acknowledgement

I am thankful for the support of this work by the Helmholtz Association of German Research Centres in the frame of the integrated project 'Controlling Chemicals' Fate' of the Chemicals In The Environment (CITE) research programme (conducted at the Helmholtz Centre for Environmental Research - UFZ).

References

Allen, M.F., Swenson, W., Querejeta, J.I., Egerton-Warburton, L.M. and Treseder, K.K. (2003). Ecology of mycorrhizae: A conceptual framework for complex interactions among plants and fungi. Ann. Rev. Phytopathol., 41(1): 271–303. 10.1146/annurev.phyto.41.052002.095518.

Ayuso-Fernández, I., Rencoret, J., Gutiérrez, A., Ruiz-Dueñas, F.J. and Martínez, A.T. (2019). Peroxidase evolution in white-rot fungi follows wood lignin evolution in plants. Proc. Nat. Acad. Sci., 116(36): 17900–17905. doi:10.1073/pnas.1905040116.

Berbee, M.L., Strullu-Derrien, C., Delaux, P.-M., Strother, P.K., Kenrick, P. et al. (2020). Genomic and fossil windows into the secret lives of the most ancient fungi. Nat. Rev. Microbiol., 18(12): 717–730. 10.1038/s41579-020-0426-8.

Boa, E. (2004). Wild Edible Fungi: A Global Overview of their use and Importance to People, Food and Agriculture Organisation of the United Nations, Rome.

Bonneville, S., Delpomdor, F., Préat, A., Chevalier, C., Araki, T. et al. (2020). Molecular identification of fungi microfossils in a neoproterozoic shale rock. Sci. Adv., 6(4): eaax7599. 10.1126/sciadv. aax7599.

Capasso, L. (1998). 5300 years ago, the ice man used natural laxatives and antibiotics. The Lancet, 352(9143): 1864. 10.1016/S0140-6736(05)79939-6.

Eastwood, D.C. (2014). Evolution of fungal wood decay. pp. 93–112. *In*: Deterioration and Protection of Sustainable Biomaterials, vol. # 1158, American Chemical Society.

Floudas, D., Binder, M., Riley, R., Barry, K., Blanchette, R.A. et al. (2012). The paleozoic origin of enzymatic lignin decomposition reconstructed from 31 fungal genomes. Science, 336(6089): 1715–1719. 10.1126/science.1221748.

Gmoser, R., Ferreira, J.A., Lennartsson, P.R. and Taherzadeh, M.J. (2017). Filamentous Ascomycetes fungi as a source of natural pigments. Fungal Biol. Biotechnol., 4(1): 4. 10.1186/s40694-017-0033-2.

Harms, H., Schlosser, D. and Wick, L.Y. (2011). Untapped potential: Exploiting fungi in bioremediation of hazardous chemicals. Nat. Rev Microbiol., 9(3): 177–192. 10.1038/nrmicro2519.

Harms, H., Wick, L.Y. and Schlosser, D. (2017). The fungal community in organically polluted systems. pp. 459–469. *In*: Dighton, J. and White, J.F. (eds.). The Fungal Community: Its Organisation and Role in the Ecosystem, 4th ed., CRC Press, Boca Raton, USA.

Hawksworth, D. (2019). The macrofungal resource: Extent, current utilisation, future prospects and challenges. pp. 1–9. *In*: Sridhar, K.R. and Deshmukh, S.K. (eds.). Advances in Macrofungi: Diversity, Ecology and Biotechnology, CRC Press, Boca Raton, USA.

Hawksworth, D.L. and Lücking, R. (2017). Fungal diversity revisited: 2.2 to 3.8 million species. Microbiol Spectrum, 5(4): FUNK-0052-2016. 10.1128/microbiolspec.FUNK-0052-2016.

Hernández, V.A., Galleguillos, F., Thibaut, R. and Müller, A. (2019). Fungal dyes for textile applications: Testing of industrial conditions for wool fabrics dyeing. J. Text. Inst., 110(1): 61–66. 10.1080/00405000.2018.1460037.

Hyde, K.D., Xu, J., Rapior, S., Jeewon, R., Lumyong, S. et al. (2019). The amazing potential of fungi: 50 ways we can exploit fungi industrially. Fungal Divers, 97(1): 1–136. 10.1007/s13225-019-00430-9.

Jones, M., Mautner, A., Luenco, S., Bismarck, A. and John, S. (2020). Engineered mycelium composite construction materials from fungal biorefineries: A critical review. Mat. Des., 187: 108397. https://doi.org/10.1016/j.matdes.2019.108397.

Kalra, R., Conlan, X.A. and Goel, M. (2020). Fungi as a potential source of pigments: Harnessing filamentous fungi. Front. Chem., 8: 369–369. 10.3389/fchem.2020.00369.

Kun, R.S., Gomes, A.C.S., Hilden, K.S., Salazar Cerezo, S., Makela, M.R. (2019). Developments and opportunities in fungal strain engineering for the production of novel enzymes and enzyme cocktails for plant biomass degradation. Biotechnol. Adv., 37(6): 107361. 10.1016/j.biotechadv.2019.02.017.

Lu, H., Lou, H., Hu, J., Liu, Z. and Chen, Q. (2020). Macrofungi: A review of cultivation strategies, bioactivity, and application of mushrooms. Compr. Rev. Food Sci. Food Saf., 19(5): 2333–2356. https://doi.org/10.1111/1541-4337.12602.

Lücking, R., Aime, M.C., Robbertse, B., Miller, A.N., Ariyawansa, H.A. et al. (2020). Unambiguous identification of fungi: Where do we stand and how accurate and precise is fungal DNA barcoding? IMA Fungus, 11: 14–14. 10.1186/s43008-020-00033-z.

Meyer, V., Basenko, E.Y., Benz, J.P., Braus, G.H., Caddick, M.X. et al. (2020). Growing a circular economy with fungal biotechnology: A white paper. Fungal Biol. Biotechnol., 7: 5. 10.1186/s40694-020-00095-z.

Mueller, G.M., Schmit, J.P., Leacock, P.R., Buyck, B., Cifuentes, J. et al. (2007). Global diversity and distribution of macrofungi. Biodiver. Conser., 16(1): 37–48. 10.1007/s10531-006-9108-8.

Naranjo-Ortiz, M.A. and Gabaldón, T. (2019). Fungal evolution: Major ecological adaptations and evolutionary transitions. Biol. Rev., Cambridge Phil. Soc., 94(4): 1443–1476. 10.1111/brv.12510.

Sajjad, W., Din, G., Rafiq, M., Iqbal, A., Khan, S. et al. (2020). Pigment production by cold-adapted bacteria and fungi: Colourful tale of cryosphere with wide range applications. Extremophiles: Life Under Extreme Conditions, 24(4): 447–473. 10.1007/s00792-020-01180-2.

Schlosser, D. (2020). Fungal attack on environmental pollutants representing poor microbial growth substrates. pp. 33–57. *In*: Nevalainen, H. (ed.). Grand Challenges in Fungal Bbiotechnology, Springer International Publishing.

Skellam, E. (2019). Strategies for engineering natural product biosynthesis in fungi. Tr. Biotechnol., 37(4): 416–427. 10.1016/j.tibtech.2018.09.003.

Song, R., Zhai, Q., Sun, L., Huang, E., Zhang, Y. et al. (2019). Crispr/cas9 genome editing technology in filamentous fungi: Progress and perspective. Appl. Microbiol. Biotechnol., 103(17): 6919–6932. 10.1007/s00253-019-10007-w.

Spatafora, J.W., Chang, Y., Benny, G.L., Lazarus, K., Smith, M.E. et al. (2016). A phylum-level phylogenetic classification of zygomycete fungi based on genome-scale data. Mycologia, 108(5): 1028–1046. 10.3852/16-042.

Splivallo, R., Ottonello, S., Mello, A. and Karlovsky, P. (2011). Truffle volatiles: From chemical ecology to aroma biosynthesis. New Phytol., 189(3): 688–699. 10.1111/j.1469-8137.2010.03523.x.

Stanzione, I., Pezzella, C., Giardina, P., Sannia, G. and Piscitelli, A. (2020). Beyond natural laccases: Extension of their potential applications by protein engineering. Appl. Microbiol. Biotechnol., 104(3): 915–924. 10.1007/s00253-019-10147-z.

Tang, X., Mi, F., Zhang, Y., He, X., Cao, Y. et al. (2015). Diversity, population genetics and evolution of macrofungi associated with animals. Mycology, 6(2): 94–109. 10.1080/21501203.2015.1043968.

Veljović, S., Nikićević, N. and Nikšić, M. (2019). *Ganoderma lucidum* as raw material for alcohol beverage production. pp. 161–197. *In*: Grumezescu, A.M. and Holban, A.M. (eds.). Alcoholic Beverages, Woodhead Publishing.

Cultivation and Market

Commercial Cultivation Techniques of Mushrooms

Jaime Carrasco,[1,]* Diego C Zied,[2] María J Navarro,[3]
Francisco J Gea[3] and Arturo Pardo-Giménez[3]

1. INTRODUCTION

Mushrooms are tasty and worldwide appreciated as a food item that grows in the wild and has been consumed by humans since ages (Kaliyaperumal et al., 2018). Initially in Asia and later in Europe and America, different species were domesticated and a number of agricultural techniques were developed for the cultivation of mushrooms. Currently, Asia, especially China, is the largest producer and consumer of cultivated mushrooms, followed by Europe and America (Royse et al., 2017). There is a sustained growing trend both in demand and production of cultivated mushrooms, for instance, to meet the requirements of protein alternatives other than meat, and for their nutritional and medicinal properties (Kaliyaperumal et al., 2018). In order to cultivate mushrooms, the first approach consists of the selection and isolation of active mycelium in the growing media under strict hygienic conditions in the laboratory (Moreaux, 2017). The goal is production of the so-called mushroom spawn in different carriers, such as cereal grains, to inoculate the selective substrates with multiple points of inoculation (Zied et al., 2018). The selective substrate is formulated with mixtures from different lignocellulosic agricultural wastes, which are selected on the basis of proximity, continuous availability, adequate formulation and price. A number of phases are needed to produce the selective substrate with

[1] Department of Plant Sciences, University of Oxford, South Parks Road, Oxford OX1 3RB, United Kingdom.
[2] Universidade Estadual Paulista, Faculdade de Ciências Agrárias e Tecnológicas, Dracena, São Paulo, Brazil.
[3] Centro de Investigación, Experimentación y Servicios del Champiñón, Quintanar del Rey, Cuenca, Spain.
* Corresponding author: carraco.jaime@gmail.com

the nutritional requirements that the button mushrooms (*Agaricus bisporus*) and oyster mushrooms (*Pleurotus* spp.) require (Pardo et al., 2007; Carrasco et al., 2018). Different species of mushrooms demand singular environmental conditions and operations for the management of the crop for a profitable production of mushrooms. This must be performed indoors to control the parameters like temperature, relative humidity, aeration, water supply and light (Bellettini et al., 2019). Besides, mushroom crops are subjected to production losses owing to biotic and abiotic causes (Preston et al., 2018). Among the biotic disorders, bacterial and fungal diseases are the most common which create a dramatic yield drop if not controlled, and pests, such as diptera, mites and nematodes which can be devastating due to their myceliophagous behaviour or as vectors for dispersion of other crop diseases (Gea and Navarro, 2017). In this chapter, initially we present a picture of the worldwide situation regarding the production of mushrooms and subsequently build a sequence of sections, as a source of guidelines for mushroom production, based on spawn production (generation of mycelium inoculum as initial phase), production of selective substrates (formulation of substrates to cover specifically the nutritional requirements of the cultivated species), crop management (requirements for commercial cultivation including, controlled environment and operations) and finally an approach to the most important diseases and pests and the actions to implement integrated pest management programmes (IPM).

2. Global Situation in Mushroom Production

Mushrooms currently represent an important source of agricultural income as they are cultivated or collected (in natural environment) in all parts of the world. Edible, medicinal and wild species generate a turnover of $ 63 billion dollars per year (Royse et al., 2017). Given the high diversity of the fungal kingdom, the great potential of this activity is notorious, especially if we consider that there are an estimated 140,000 species that produce fruit bodies, among which only 14,000 are reported, 7,000 are edible and 2,000 are considered as prime edible mushrooms (Hawksworth, 2001). Even considering these numbers, only 200 species have been cultivated experimentally, 60 cultivated commercially and 10 cultivated on an industrial scale (Chang, 2006).

Currently the most cultivated mushroom in the world is the *Lentinula edodes* (shiitake), followed by *Pleurotus* spp. (oyster mushroom), *Auricularia* spp. (wood-ear mushroom), *Agaricus bisporus* (button mushroom) and *Flammulina velutipes* (enoki) (Royse et al., 2017). In the last four decades, there has been a great increase in mushroom production in China, leading to a shift in the position of *A. bisporus* from the most cultivated mushroom in the world to the fourth in place. However, *A. bisporus* mushroom has the most advanced cultivation technology, in addition to being cultivated commercially in all the countries that produce mushrooms. In the Americas, unlike *L. edodes*, for instance, which covers 23 countries, *A. bisporus* is cultivated commercially only in the USA, Canada, Brazil, Chile, Colombia and Peru (Sánchez et al., 2018).

China is by far the main mushroom-producing country, responsible for 30.4 billion kg of a total of 34.8 billion kg of mushrooms produced worldwide.

According to Royse et al. (2017), the most cultivated mushrooms in China are shiitake, wood-ear, oyster and button mushrooms produced at a total of 7, 6.9, 6, 2.37 billion kg per year, respectively. Currently China is responsible for 87 per cent of the total mushroom production. The Asian continent also has other important countries that collaborate with global production of mushrooms, such as India (130 million kg per year), Iran (81.4 million kg per year), Japan (65.75 million kg per year), Turkey (46.14 million kg per year), Indonesia (31 million kg per year) and South Korea (22.7 million kg per year) (Sharma et al., 2017; FAO, 2018).

The second largest mushroom producing country is the US with an annual production of 421.9 million kg per year (NASS, 2018). Currently, America is responsible for the production of 666.24 million kg per year. Other important countries that collaborate in the American mushrooms production task are Canada (133.93 million kg per year), Mexico (63.76 million kg per year), Brazil (15.69 million kg per year), Chile (12 million kg per year), Colombia (8 million kg per year), and Argentina (4.5 million kg per year) (Sánchez et al., 2018).

The third largest mushroom producing country is the Netherlands with an annual production of 300 million kg per year. It is very close to Poland, producing 280.23 million kg per year. In the European Union, Spain produces 166.25 million kg per year, France 83.01 million kg per year, Germany 73.23 million kg per year, Italy 70.67 million kg per year and Ireland 65.3 million kg per year (NASS, 2018). The United Kingdom has a production of 98.5 million kg per year (NASS, 2018). In addition to all the countries previously mentioned, two countries should also be mentioned and these are Australia which produces 75 million kg per year and South Africa, which has an annual production of 21.11 million kg (Devochkina et al., 2019; NASS, 2018). Table 1 shows the annual production of the largest countries responsible for mushroom cultivation in the world.

Table 1. Largest mushroom producing countries.

Ranking	Country	Mushroom production (kg)
1	China	34,800,000,000
2	United State	421,900,000
3	Netherlands	300,000,000
4	Poland	280,000,000
5	Spain	166,250,000
6	Canada	133,930,000
7	India	130,000,000
8	United Kingdom	98,500,000
9	France	83,010,000
10	Iran	81,406,000
11	Australia	75,000,000
12	Germany	73,230,000
13	Italy	70,670,000
14	Japan	65,750,000
15	Ireland	65,300,000

3. Spawn Production

One of the most important steps in the production of cultivated mushrooms consists of the production of spawn. Currently, several commercial companies (Sylvan®, Amycel®, Lambert®, Mycelia® and others) have hybrid materials that provide a greater yield and maintain a stable production cycle. The mycelium for cultivation can be produced under laboratory conditions through fragments of basidiome and spores (Moreaux, 2017). In this text, the procedures for the production of mycelium from fragments of basidiomes will be addressed, followed by the following steps— preparation of primary culture, secondary culture, tertiary culture and mycelium or mushroom spawn understood as the mushroom seed (Fig. 1).

Specialised human resources, equipment and infrastructure are required in all these stages. All substrate-preparation procedures, culture medium and substrates are carried out in clean rooms, using an autoclave (121°C, 1 atm.), sterile tools for microbiology and laminar flow hood. Despite the use of asepsis and sterilisation techniques, contamination is always present, but at low levels (≤ 2 per cent). To prevent contamination, a continuous evaluation of contaminant agents is required. Materials which change in colour and odour should be discarded immediately (Minhoni and Zied, 2008).

3.1 Culture Media

The culture media are used to prepare the primary and secondary culture in a petri dish. The culture media most used in the isolation of mushrooms are PDA (potato-destrose-agar), grains of wheat-agar and malt-dextrose-agar (Oei, 2003). Culture media similar to the growing substrate are also used and are more recommended as they promote adaptation of the fungus to the growing substrate, thus avoiding delay in colonisation. We recommend the CA (compost-agar) medium for mushrooms to be grown on pasteurised substrates and the SA (substrate-agar) medium for mushrooms grown on sterile substrates. Initially, a sample of mushroom cultivation substrate is boiled for 90 min in 900 ml of distilled water. For SA, the sample can be 200 g of substrate [composed of 80 per cent sawdust + 20 per cent bran (wheat and/or rice and/or soy) + 2 pert cent calcitic limestone]. For CA, the sample can be 60 g of compost (collected at the end of phase II of composting, dehydrated at 60–62°C for 72 hours). Filter through a common fine mesh sieve and if necessary, cotton muslin cloth. Add water to the filtrate to make up 1000 ml with distilled water. The filtrate is placed in Duran flasks and autoclaved at 121°C for 20 min. After 12 to 24 hours, agar (20 g/l) is added and autoclaved again at 121°C for 20 min. Once the growing medium is cool (45°C aprox.), pour it into Petri dishes (Zied et al., 2018).

3.2 Biological Carrier Production

The biological carrier is used to prepare the tertiary culture and spawn in glass bottles and in plastic bags with filter. Cereal grains, sawdust, wooden plugs have been used and these must be new and of high quality. We recommend the GC (grain-carrier) for mushrooms grown on pasteurised substrates and the SC (sawdust-carrier) for mushrooms grown on sterile substrates. The GC production is based on grains of

Fig. 1. Sequence of steps necessary for spawn production, the primary culture being prepared from the cloning of mushrooms; the secondary culture through the transfer of fragments from the primary culture; the tertiary culture receiving a fraction in the 'pizza' broth from the secondary culture and finally, the production of spawn from portions of the tertiary culture (*Note*: In the image, the person responsible for the procedures is not wearing gloves, as we have always chosen to use 70 per cent alcohol for constant hand hygiene. It is important to mention that no ring, watch or bracelet should be used during the process).

triticale, sorghum, wheat and others. Initially, the dried grains are immersed and boiled in water for 20–40 min, depending on the type of grain. Excess water is drained for approximately 40–50 min. Then the grains are mixed with gypsum

Fig. 2. On the *left*, dehydrated wheat grains; in the *centre*, the same grains after cooking; on the *right*, the grains added with gypsum and limestone, ready for use in tertiary culture and spawn.

(4 per cent) and calcitic limestone (2 per cent) (Fig. 2). The humidity of the substrate, at the end of the process, should be 55–60 per cent.

An SC formulation may contain 80 per cent eucalyptus sawdust, 10 per cent wheat bran, 10 per cent rice bran and 2 per cent calcitic limestone. The sawdust is sieved and added to the other materials. Homogenise and add water until it reaches 70 per cent humidity. Finally, the substrate is placed in glass bottles and plastic bags and autoclaved at 121°C (1 atm) for 3–4 hours (Fig. 3).

3.3 Primary Culture

The procedure consists of transferring a fragment of the internal pseudo-tissue from a young and healthy basidiome to the culture medium. The base of the stalk is removed with a sharp instrument and any substrate debris is removed from the basidiome, using a clean and dry brush. Break the mushroom in half with the hands, being careful not to touch the inside of the mushroom. With the aid of tweezers or a platinum loop, a 2–5 mm fragment is removed from the pseudo-tissue and placed in the centre of Petri dishes containing the culture medium and the edges are closed with plastic film.

3.4 Secondary Culture

The growing medium and the procedures for preparing the secondary culture are the same as that in primary culture, except for the inoculum, which are now primary culture discs that will be placed in the centre of Petri dishes. To maintain and guarantee the viability, vigour and high productivity of a strain, it must be periodically re-isolated from healthy basidiomes.

3.5 Tertiary Culture

The inoculation of tertiary culture is done after cooling the substrate in the bottles/ plastic bags to room temperature, using segments of the secondary matrix. Initially, the secondary matrix is divided into eight 'pizza' segments and with the aid of tweezers, one segment per bottle is transferred.

For substrates with an irregular structure, that is based on sawdust, straw and others, the secondary culture segment is placed inverted on the substrate (Fig. 4A).

Fig. 3. A: Glass bottle with a twistable lid, with two filter leaves on the bottom of the lid, containing grains of wheat; B: high density polyethylene plastic bag (HDPE), with Tyvek® type filter on the upper left, containing wheat grains (Zied et al., 2018); C: spawn with sawdust substrate, in HDPE bag with cotton filter on the upper left (Zied et al., 2018); D: tertiary culture containing substrate based on sawdust; and E: three commercial packages of different sizes, suitable for the production of spawn; yellow arrows indicate the location of the filter (SACO2®) (*Note* in image 3A, the technician is using a watch, as the bottle is just been filled with grains and which will be closed and autoclaved).

For substrates with a regular structure, based on grains, the inoculation is initiated by placing a secondary culture segment, with the colonised face facing up, at the bottom of an empty, previously sterilised bottle (Fig. 4B). Then the grains from one vial are poured into another containing the inoculum. The process is continued sequentially, in which the empty flask initially receives the inoculum segment and then the substrate of another flask.

Fig. 4. A: Addition of a fragment of pizza (secondary culture) in the tertiary culture on the substrate based on sawdust; B: addition of a pizza fragment (secondary culture) in the tertiary culture at the bottom of the empty bottle, with subsequent filling with the grains (*Note* since we have always chosen to use 70 per cent alcohol for constant hand hygiene, it is not required that the staff handle the process by wearing globes. Any accessories, such as watch, ring or bracelet must be taken off to prevent contamination).

3.6 Spawn Production

The substrate and the procedures for preparing the spawn are the same as that of tertiary culture, except for the inoculum and sometimes the bottles (or packaging) for the substrate. Inoculation is performed, the spawn is prepared by the addiction of fractions of a tertiary culture in the substrate accommodated in plastic bags type PP (Polypropylene) or HDPE (High Density Polyethylene), at a dose of 1.2–2 per cent (weight of fresh substrate) using a sterile metal spoon. Ready-made packaging, with a filter, can be purchased commercially (Fig. 3E). However, bags without the filter are less expensive and the filter is attached to the spawn production laboratory (Fig. 3B, C).

For each cultivation cycle, a new spawn must be used, that is, freshly prepared for the cycle in question. In addition, the preparation of this must always be with a new tertiary culture, meaning freshly prepared from a secondary culture. This procedure avoids possible genetic changes in the inoculum that could compromise productivity. Cell degeneration in fungi can occur due to nutrient and/or oxigen deficiency, changes in pH, accumulation of toxic metabolites, infections (bacteria, viruses and other fungi) and so on (Oei, 2003).

4. Nutrition and Substrate Design

4.1 Requirements for Growth

Fungi benefit from an adaptable and flexible metabolism and their versatile biochemical capabilities are used in a variety of ways during morphogenesis (Moore et al., 2008). Different factors, such as environmental pH and the availability of oxygen, influence the metabolic processes of fungi together with the ability of a fungal species to utilise certain substances as nutrient influences. Nutritional requirements of mushrooms can be grouped as follows (Chang and Miles, 2004).

Water: Since a high percentage of mushroom composition consists of water, this is the substance which is taken from the substrate in the largest quantity.

Carbon: Lignin and polysaccharides, cellulose and hemicellulose, are the most important carbon compounds used by fungi. These complex molecules are the fundamental part of the plant cell wall and they are digested by extracellular enzymes excreted from the hyphae.

Nitrogen: The synthesis of essential compounds requires nitrogen which is obtained by fungi from a variety of sources. For instance, critical components for the biological functions of fungi with nitrogen are amino acids, proteins or nucleic acids (organic nitrogen) and other inorganic compounds like nitrate or ammonium and nitrogen. A large part of the nitrogen needed by some mushrooms (secondary decomposers) can be supplied by the nitrogen found in the N-rich lignin-humus complex obtained after composting.

Minerals: Other essential elements are required in lower amounts than carbon and nitrogen. These are necessary for all kinds of physiological processes and among them, we find sulphur, phosphorus, potassium, calcium, magnesium and the so-called trace elements as iron, zinc, manganese, copper or molybdenum, needed only in small concentrations.

Vitamins: Most fungi have relatively simple nutritional requirements, but some do require one or more vitamins. Vitamins most commonly required by fungi are thiamine (B_1) and biotin (B_7, H). Other vitamins for which certain fungi have a natural requirement include nicotinic acid (B_3), pantothenic acid (B_5), and 4-aminobenzoic acid (B_{10}).

Overall, fungal development requires certain environmental variables. Among them are temperature, light, relative humidity, moisture content of substrate, aeration (oxygen and carbon dioxide concentrations) and gravity (Chang and Miles, 2004; Moore et al., 2008). In the case of these physical environmental factors, growth invariably occurs over a range of values. In addition, other factors include the quality genetics of the strain, the age of the mycelium and the cultural conditions employed that impact the crop performance (Chang and Miles, 2004).

Finally, in addition to the detailed genetic, chemical and nutritional or physical and environmental factors, it is certain microbiological issues also count. Among them, for instance, those microorganisms associated in the process of composting (Cao et al., 2019; Carrasco and Preston, 2020), the bacteria species associated with the substrates employed for the cropping of *Pleurotus ostreatus* (Cho et al., 2003; Velázquez-Cedeño et al., 2008; Vieira et al., 2019), or the specific bacterial strains with a stimulating role in the process of fructification of *Agaricus bisporus* (Pardo et al., 2002; Carrasco and Preston, 2020).

To achieve maximum yield in commercial production of edible mushrooms, it is necessary to know the nature of the mushrooms, ecological habitats in wood and straw (primary decomposers) or dung (secondary decomposers), requirements of the preparation of selective substrate materials and the appropriate management of physical, chemical and biological parameters and, ultimately, proper management of mushroom beds (Chang, 2008). In respect to ecological habitats, edible mushrooms are able to colonise and degrade a large variety of lignocellulosic substrates and other wastes, which are produced primarily through the activities of the agricultural, forest, and food-processing industries (Sánchez, 2010). They have two kinds of saprophytic lifestyles: (1) firstly, most of them are primary decomposers that can be cultivated on pasteurised or sterilised lignocellulosic substrates; (2) secondly, those that can be cultivated on compost formulated from various agricultural wastes including manure since they are leaf-litter secondary decomposers (Savoie et al., 2013).

Figure 5 presents a simplified flow diagram for the commercial production of edible mushrooms. The production of tailor-made substrates for the cultivation of two of the most cultivated species worldwide (*Agaricus bisporus* and *Pleurotus ostreatus*) is briefly described.

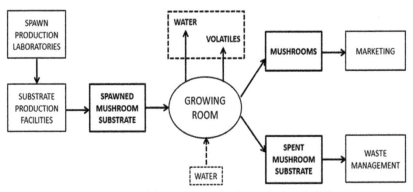

Fig. 5. Flow diagram summarising main stages for the production of edible mushrooms.

4.2 Production of Agaricus bisporus Substrate as Example of Secondary Decomposer

Secondary decomposers, such as button mushroom (*Agaricus bisporus*) and almond mushroom (*Agaricus subrufescens*) require prior action of other microorganisms that partially decompose the organic matter to generate a substrate which can be effectively digested by the fungal enzymatic systems to absorb the needed nutrients for the mycelial growth and mushroom fruiting. This pre-semi-decomposition process is called composting (Carrasco et al., 2018; Carrasco and Preston, 2020).

Commonly, wheat straw is selected as the preferred raw material for compost preparation as it is the base material and main source of carbon. Depending on local availability, other materials, such as cereal straws (rice, oat, barley, rye), corn cobs, coarse hay, maize stalks and leaves, sugarcane bagasse, soybean stalks, or elephant grass may be used. Horse manure, chicken manure and other nitrogen-containing substances as cottonseed meal, sunflower agricultural wastes, cotton hulls, peanut meal, malt sprouts, brewers' grains, urea, ammonium nitrate or ammonium sulphate are added to the base material, mainly as a nitrogen source (Savoie and Mata, 2016; Buth, 2017). Pardo et al. (2007) evaluated composts based on waste products from grape-growing and wine-making (vine shoots, grape stalks and grape pomace). Gypsum is also added to the compost to improve its texture, making the compost less greasy and facilitating aeration, while lowering the pH (Buth, 2017). Table 2 presents an example of mushroom compost formulation based on wheat straw and chicken manure for *A. bisporus* cultivation.

Table 2. Example of mushroom compost formulation for *Agaricus bisporus* cultivation (adjustment in the mixture of raw materials: C/N 30.0; Nitrogen 17.3 g/kg).

Ingredients	Fresh weight (g)	Moisture (g kg⁻¹)	Dry weight (g)	Nitrogen (g kg⁻¹)	Total nitrogen (g)	Ash (g kg⁻¹)	Organic matter (g kg⁻¹)	Total carbon (g)
Wheat straw	1000.0	100	900.0	5.5	5.0	70.0	930.0	485.5
Chicken manure	450.0	180	369.0	36.0	13.3	175.0	825.0	176.6
Urea	8.5	0	8.5	460.0	3.9	0.0	1000.0	1.7
Gypsum	5.0	200	4.0	0.0	0.0	1000	0.0	0.0
TOTAL	1463.5		1281.5		22.1			663.7

The compost for mushroom culture is a homogeneous and selective nutritive substrate for the production prepared in a two-phase standardised process (Buth, 2017; Carrasco and Preston, 2020).

Phase I: The procedure consists of initial hydration and mixing of the straw with manure and/or other activators with optimal balance between nitrogen and carbohydrate content in the mixture. The material is disposed in piles or aerated floors that are mixed by tumbling at regular intervals. It is an aerobic, biological and chemical process, which takes place at temperatures between 50–80°C to facilitate primary decomposition of the raw materials. The objective is to initiate microbial activity, to homogenise the solid substrate physically as well as chemically to increase its uniformity and to increase its water content. Sufficient levels of nitrogen must be achieved for rapid microbial growth and biodegradation in an available form for *A. bisporus* and achieve an optimal material for its mechanical properties, which allow a good gas exchange and has good water-holding capacity.

Phase II: This is a complementary fermentation process fundamentally oriented to pasteurisation and conditioning of the compost mass. It occurs in tunnels or inside the culture rooms (tray system), where the temperature and the supply of air through forced ventilation are regulated. It is also an aerobic process that takes place at

Fig. 6. Phase I composting facilities of Champinter (Villamalea, Spain).

temperatures between 45–60°C. The objective is to develop a specific microbial flora for obtaining a selective final product, based on microbiological and chemical basis, for the subsequent development of *A. bisporus* mycelium, while eliminating volatile ammonia.

Due to the increasing demand from the society for eco-friendly production techniques, several methods of indoor composting are developed. During indoor composting, the odours can be trapped before air is vented out and the climatic parameters of the process are controlled at all times, achieving sustainable development by saving time, energy, raw materials and space, by improving sanitary control and by favouring the stabilisation of compost and the quality of fungi (Savoie and Mata, 2016). Non-composted substrates (NCS) with and without sterilisation have been also evaluated with a limited practical application (Savoie and Mata, 2016). Other lines of research for alternatives involve the use of thermophilic fungi, with reduced composting systems.

In Fig. 7, the different phases of the process of making substrates for cultivation of *A. bisporus* are presented in a flow chart. The so-called Phase III (compost fully colonised by the crop mycelium) appears as optional. Phase III consists of the spawn-running (incubation) in tunnels and has certain advantages over the use of Phase II compost. For instance, a greater number of crop cycles per growing room per year, reduction of infections from pathogens and flies, possibility of simultaneous filling of the rooms with compost and casing, possibility of supplementation and watering at the time of filling and better compost temperature controls during incubation in tunnels (Buth, 2017).

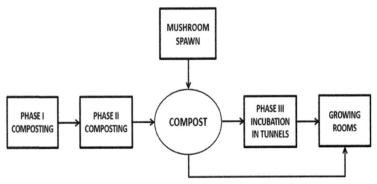

Fig. 7. General composting phases in *Agaricus* spp. cultivation.

4.3 *Pleurotus ostreatus* Substrate Production as Example of Primary Decomposer

Primary decomposers, such as oyster mushroom (*Pleurotus ostreatus*) and shiitake (*Lentinula edodes*), can use a diverse variety of organic materials as substrates. The oyster mushroom is one of the most suitable fungi for producing protein-rich food from various agro-wastes without composting (Upadhyay, 2011).

Different authors have compiled and characterised a wide range of materials used as a growing substrate for *Pleurotus*: cereal straws (wheat, barley, rye, rice and others), diverse types of sawdust, crop residues for industrial use or for spontaneous plants (rapeseed, flax, cotton, sunflower stalks, tobacco leaves, vine shoots, grape scrap, and reeds), agro-industrial by-products (cottonseed hulls, sunflower seed hulls, groundnut shells, rice husks, grapeseed meal, grape pomace and olive mill waste), hay (fescue, fleo, grama, vallico and alfalfa), legume straws (beans, lentils, vetch and peas), flour and bran (wheat, rye, barley, rice and oats) (Poppe, 2004; Upadhyay, 2011; Rodríguez and Pecchia, 2017).

The materials eligible to be used in the preparation of substrates for the cultivation of *Pleurotus* must possess, as a starting point, good availability in quantity and continuously, adequate physico-chemical characteristics, consistency in their physico-chemical composition, advantageous acquisition price, easy and close location and ease for transportation and handling.

Unlike white mushrooms (*A. bisporus*), *Pleurotus* spp. do not require a substrate with chemical selectivity because they can grow in nutrient media with a C/N ratio between 30 and 300. However, they need to grow in a specific biological environment with accompanying flora to protect and promote growth. For all the aforementioned, it is understood that with such a versatile C/N ratio, a broad range of vegetable by-products or combinations of two or more of them are usable for the cultivation of *Pleurotus* spp. (Sánchez, 2010). Therefore, the design of substrate mixtures is often done on the basis of the availability of the raw materials, chemical composition of the raw materials and previous experience of the mushroom growers (Rodríguez and Pecchia, 2017).

In the preparation of substrates for cultivation of *P. ostreatus*, the different methods used share some general preliminary operations, such as adjusting the particle size by crushing, grinding or chopping, and adjusting the moisture content. A particle size of 2–5 cm and a moisture content of around 70–75 per cent are recommended as a general rule. The three main post-heat treatments applied to substrates are pasteurisation by immersion in hot water at 65–85°C for 10–40 min (Oei and van Nieuwenhuijzen, 2005); pasteurisation by injecting steam and air at 65°C in the substrate for one hour or aerobic treatment in tunnels at 60–80°C for 6–12 hours; and sterilisation in autoclave at 121°C for 1–2 hours (Oei and van Nieuwenhuijzen, 2005).

Other proposed treatments are aerobic fermentation, consisting of conventional pasteurisation followed by conditioning thermophilic fermentation, fermentation in the natural environment in piles, cold anaerobic fermentation, immersion in alkaline water and so-called chemical sterilisation by soaking in fungicide solutions (Rodríguez and Pecchia, 2017).

5. Crop Management

Mushrooms' survival and multiplication are related to several intrinsic and extrinsic factors, which may act individually or have interactive effects (Bellettini et al., 2019). The intrinsic factors are basically associated with the preparation of the substrates (formulation, sources of nitrogen, chemical composition, water activity, C/N ratio, pH, moisture, particle size and density of inoculum). Regarding extrinsic factors, crop management is associated with the control of various environmental cues in each of the different stages of the crop cycle.

5.1 Temperature

The temperature range is narrower for fruiting than for mycelial growth and the range for the optimum temperature is likewise narrower for fruiting than for growth (Chang and Miles, 2004). Thus, in the case of *Agaricus bisporus* the range for mycelial growth is between 3–32°C, with an optimum between 22–25°C while for fruiting, the range would be between 9–22°C, with an optimum between 15–17°C. In the case of *Pleurotus ostreatus*, the range for mycelial growth is between 7–37°C, with an optimum between 26–28°C, with different intervals for fruiting depending on different strains (Chang and Miles, 2004).

5.2 Relative Humidity

Evaporation, together with the action of the mycelium, maintains the transport of nutrients and water from the mycelium to the mushrooms for its growth. Evaporation can be regulated by modifying relative humidity, although the rate of evaporation is also affected by temperature and speed of air in the room and the substrate activity as interdependent factors. The various stages of cultivation require different levels of relative humidity. For instance, in the case of *A. bisporus*, the recommended RH values are > 95 per cent until the pin heading, RH 90–95 per cent for the pin heading and RH 80–85 per cent for the growth of fruit bodies.

5.3 Aeration

In general, the requirements for fruiting are more demanding than for vegetative growth. A generalisation can be made that fruit bodies of higher fungi typically form the best under conditions of good aeration (Chang and Miles, 2004). Instead, mycelium can tolerate very high CO_2 concentrations. Normally in the incubation phase, CO_2 concentration can reach up to 20,000 or even 30,000 ppm, while induction is usually carried out with intense ventilation by reducing the CO_2 concentration below 1000 ppm in many cases.

5.4 Light

Many fungi are apparently uninfluenced in reproduction by light in the visible range. However, there are some fungi that do not fruit without light and require light for normal development of the fruit body (Chang and Miles, 2004).

5.5 Stages and Management of Agaricus bisporus Growing Cycle

Spawn-Run: The spawned compost, in which the mycelium is evenly mixed in a ratio of 0.8–1 per cent (w/w), is filled in polythene bags or directly on the trays/shelves of the growing room, compacted and levelled. In the cropping room, compost is maintained at an approximate temperature of 25–27°C (air temperature 22–24°C), relative humidity of 95 per cent and high CO_2 concentration (above 10000 ppm, without fresh air) for spawn run. Process takes 12–14 days (Dhar, 2017). As mentioned above, spawn-running can also be done in tunnels (*see* Phase III).

Casing: Fructification of *A. bisporus* and in general of secondary decomposers, occurs in the casing layer. It's a layer of 3–5cm thick, usually based on soil or peat, used as a top covering of spawn run compost, to induce the transition from vegetative to reproductive growth (Pardo-Giménez et al., 2017).

Primordia Formation, Development and Harvesting: After the application of the casing layer, we find a period of approximately one week, called pre-fruiting (case run), which takes place in conditions similar to spawn-run. Once the stage is complete, when the mycelium appears on the surface of the covering layer, the fruiting is induced, lowering the temperature, relative humidity and introducing fresh air to reduce CO_2. In this way, primordial formation is induced (Fig. 8), so that the first harvest occurs approximately 35 days after inoculation.

Fig. 8. Fructification of *Agaricus bisporus* on a peat-based casing layer.

As a rule, during the entire cropping period, the air temperature of 15–17°C and 85 per cent relative humidity is maintained in the cropping room, with CO_2 concentration held at 800–1000 ppm (Dhar, 2017). Slight variations may be recommended for the different strains found on the market. The temperature of the compost tends to stay 1–2°C above the air temperature to promote slow evaporation from the casing layer. Mushrooms are harvested every day at their optimal commercial stage of development, when buttons are of 4–5 cm in diameter, though stout and hard (Dhar, 2017).

Watering: Mushroom gets water from compost and casing during case run and fruit body formation. Optimal mushroom growth is dependent on continuous water supply on the casing layer. The most common practice for watering is manual sprinkling or the use of automatic sprinklers. The use of drip irrigation enables the addition of water to the casing soil during the entire cropping cycle, thus allowing careful management of moisture balance in the compost and casing layers (Navarro et al., 2020).

After the crop cycle, by normally picking three flushes in successive weeks, it is required to apply cook-out (if available), emptying of the growing room and

Fig. 9. Dutch-style growing room of Mercajúcar (Villalgordo del Júcar, Spain).

managing the spent mushroom substrate. Currently, the Dutch system of growing rooms with computerised environmental control has gained more presence in professional farms worldwide (Dhar, 2017) (Fig. 9).

5.6 Stages in Management of Pleurotus ostreatus Growing Cycle

Spawn-run (incubation): After pasteurisation, the substrate is inoculated at a ratio of 2–5 per cent and usually filled into black or transparent polypropylene bags with pre-punched holes for future pin formation. Incubation is done at 24–30°C, in total darkness, and high CO_2 concentration (up to 20000 ppm). Length of incubation typically ranges from 12–21 days (Upadhyay, 2011; Dhar, 2017; Rodriguez and Pecchia, 2017).

Primordia Formation, Development and Harvesting: Primordia formation is induced by changing the environmental conditions once the substrate is completely colonised with the mycelium. Air temperature is lowered to 10–22°C (cold shock), depending on the strain; relative humidity is adjusted to 85–90 per cent and fresh air ventilation is forced to decrease the carbon dioxide concentration below 600–1000 ppm (Upadhyay, 2011; Rodriguez and Pecchia, 2017). In general, the photoperiod of mycelia stimulation to promote mushroom fruit body formation should be enough to read a newspaper (200–640 lux; 8–12 h/day) at a temperature compatible with the mushroom necessities (Bellettini et al., 2019). According to other authors, light requirement for primordia formation and fruit body development ranges between 200–1500 lux, 12 h/day (Upadhyay, 2011; Dhar, 2017). Primordia formation takes three to five days and the later fruit body development, between four to seven days. Clusters of mature oyster mushrooms are harvested from the holes at maximum size before wrinkling begins in the external edges (Fig. 10). The diameter of pileus is about 5–7.5 cm but the length comprising pileus and stipe is about 7.5–10 cm (Dhar, 2017). Flushing break is approximately 10 days.

As in the case of *A. bisporus*, once the cultivation cycle has finished, normally after harvesting, in this case two to three flushes, the cultivation room is emptied and the agricultural waste, spent mushroom substrate, is managed. The cultivation is appropriate and affordable in farm-glasshouses specifically designed for *Pleurotus* production, perfectly isolated and equipped with sophisticated air conditioning. In Spain, a greenhouse of this type is currently used in the Champignon Research, Experimentation and Services Center (CIES for research work (Fig. 11).

Fig. 10. Clusters of *Pleurotus ostreatus* from wheat straw-based substrate in bags.

Fig. 11. Glasshouse for *Pleurotus* spp. and other edible mushrooms cultivation (Quintanar del Rey, Spain).

5.7 Uses of Spent Mushroom Substrates

Spent mushroom substrate (SMS), agricultural waste after mushroom cultivation, is a low-cost feedstock with huge potential for valorisation. Research focused towards the application of SMS for sustainable and circular processes has found possible uses in bioremediation (purification of air, water, soils and degradation of pesticides) as a substrate for cultivating other crops (greenhouse flowers and vegetables, fruit and vegetables in the field among others), as a general amendment and fertiliser for soils, in nurseries and landscaping, as animal feed or in aquaculture, in pests and disease management, as energy feedstock (alternative fuel and production of biogas), in vermiculture, as source of degradative enzymes, in further edible mushrooms production and additionally other diverse uses (Oei et al., 2008; Rinker, 2017; Grimm and Wösten, 2018; Hanafi et al., 2018; Jasińska, 2018).

6. Biotic Disorders and Control in Mushroom Crops

Cultivated mushrooms are susceptible to a variety of pests and diseases that may affect the mushroom yield and quality. The nature of the crop, while cultivated under controlled environmental conditions indoors, facilitates the implementation of integrated pest management (IPM) programmes that allow the application of pest exclusion techniques, cultural control measures through the correct management of temperature and humidity and sanitation in the growing rooms. Usually, these types of measures have been accompanied by the preventive use of phyto-sanitary products. However, in recent years, due to legal shifts towards a regulatory landscape that targets food safety for consumers and the protection of the environment, there is an increasing restrictive scheme for the use of chemicals. In this context, biological control methods for pests and diseases are drawing growing interest, looking for alternatives for chemical agents.

6.1 Mushroom Diseases

6.1.1 Bacterial Diseases

Cultivated mushrooms can be affected by various bacterial diseases, such as brown blotch disease, soft rot, internal stipe necrosis, drippy gills and ginger blotch (González et al., 2009, 2012). The most common and harmful of all of them is brown blotch disease (*Pseudomonas tolaasii*) characterised by superficial, shining brown-stained lesions, irregular and often sunken, on the mushroom caps (Navarro et al., 2018). *Pseudomonas tolaasii* is a pathogen for button mushroom and speciality mushrooms, such as *Pleurotus ostreatus* and *Flammulina velutipes* or *Pleurotus eryngii* (González et al., 2009; Han et al., 2012; Navarro et al., 2018). External brownish lesions in cap tissues affect only the outer layers, 2–3 mm below the surface of the cap (Soler-Rivas et al., 1999). Symptoms can be detected from the beginning of the harvest period to the final stages of the crop cycle with higher incidence during spring and autumn, in which the temperature amplitude is greater. Brown blotch renders crops unmarketable by strongly affecting the aesthetic quality of the caps pre-harvest and reducing the shelf-life of the mushrooms post-harvest (Taparia et al., 2020). *Pseudomonas tolaasii*, considered as the main causal agent of the disease, may be present in the composts and materials used in the casing layer and is easily carried by sciarid and phorid flies, mites, dust particles, pickers and their tools. Conditions of temperature and environmental relative humidity within the farms are directly related to the appearance of brown blotch. In many cases, external weather conditions are also correlated. The development of the disease is induced by high relative air humidity and inadequate ventilation, which entails condensation on the cap surface (Navarro et al., 2018). Modifying the environmental conditions by increasing the temperature before irrigation and moderating the manoeuvres of the drying process help control bacterial blotch disease (Navarro et al., 2018). To fight this harmful disease, farmers must implement strict hygiene and control the environmental conditions in the growing facilities.

6.1.2 Fungal Diseases

Dry bubble: Dry bubble, caused by the fungus *Lecanicillium fungicola* (Zare and Gams, 2008) causes serious losses in commercial button-mushroom farms worldwide. The observed symptoms include detorted mushrooms (undifferentiated spherical masses known as bubbles), curved and/or cracked stipes called blowout and spotty caps (Berendsen et al., 2010). The primary source of *L. fungicola* in mushroom crops is the casing material, especially peat. Infected mushrooms, contaminated equipment and diptera could also be important sources of dry bubble infection (Berendsen et al., 2010). The air in ventilation systems can play an important role in dispersing the disease inside the mushroom farm or even transporting it to other installations since conidia can adhere to specks of dust or other debris. Methods of control include the application of strict hygiene and fungicides. An ocular inspection prior to any operation, including watering or harvesting to remove infected mushrooms, is highly advisable to prevent the dispersion of the disease (Gea and Navarro, 2017). The range of fungicides available to control the dry bubble is scarce. Prochloraz is the most effective but reduced sensitivity to this fungicide has been reported (Berendsen et al., 2010).

Wet Bubble: The wet bubble is a disease caused by the mycoparasite *Mycogone perniciosa* that parasitises only the fruit bodies of *A. bisporus*, while the vegetative mycelium remains immune. The mycoparasite sporulates heavily on *Agaricus* basidiomes, producing small, thin-walled phialoconidia called aleuriospores on *Verticillium*-like conidiophores or chlamydospores. The pathogen conditions the morphogenesis of the fruit body, driving round-shape masses of tissue without signs of differentiation into stipe and cap called wet bubbles (Gea et al., 2010). Wet bubbles are initially white and fluffy and then turn to brown when aging. They may exude amber-coloured droplets on their surface containing bacteria and spores (Fig. 12). They eventually rot, releasing an unpleasant smell.

   The mycoparasite may spread by the splashing of water, flies and operators (tools, hands and clothes) and contaminated casing material is considered the main initial source of infection. Therefore, casing sanitation is a requirement to eliminate the primary sources of inoculum. Conidia can also be transported by air (Gea and Navarro, 2017). Prochloraz and chlorothalonil are the most effective fungicides for the control of the wet bubble (Gea et al., 2010).

Cobweb: The cobweb, caused by several species from the mycopathogenic genus *Cladobotryum* (*C. dendroides*, *C. mycophilum*, *C. varium*) is one of the most serious diseases affecting mushroom cultures worldwide (Carrasco et al., 2016, 2017a). Cobweb has been reported as a harmful agent in different species of cultivated edible mushrooms: *Agaricus bisporus*, *Lentinula edodes*, *Pleurotus eryngii* and *P. ostreatus* (Carrasco et al., 2016; Gea et al., 2017, 2018, 2019). Its occurrence generally reduces the production and quality and thus leads to economic losses due to cap spotting; the impact on crop surface colonised by the parasite and the necessity for early crop

Fig. 12. Undifferentiated mass of *Agaricus bisporus* tissue infected by *Mycogone perniciosa* (wet bubble). Arrows pointing amber-coloured droplets on the infected tissue.

Fig. 13. Fluffy white mycelium of *Cladobotryum mycophilum* colonising the surface of cultivated *Pleurotus ostreatus* (cobweb disease).

termination when the disease becomes epidemic. The parasitic white fluffy mycelium grows over the surface of the casing layer, primordia and adult fruit bodies which generally discolour and finally rot (Fig. 13). Cobweb mycelium becomes dense and sometimes develops a rosy red or yellow hue.

Another manifestation of the disease is the appearance of spots on the carpophores. Two types of cap spotting may be observed: brown dark spots and light brown to grey spots (Carrasco et al., 2017a). The main sources of inoculum may be fragments of mycelium transported by air or contaminated casing material. The chief means of dispersion of conidia is air. In addition, water splashes during irrigation may contribute to dispersion. Mushroom flies and pickers seem to be less important for dispersion since the spores are not sticky. As soon as a punctual outbreak of cobweb is located over the casing or carpophores, it has to be treated before sporulation by covering it with thick damp paper to avoid conidia release and disease spread (Carrasco et al., 2017a). Cobweb appears more often at the end of the crop cycle during autumn and winter cycles. Methods to control cobweb include: strict hygiene and environmental control measures to prevent dispersion of the conidia and the application of fungicides, mainly prochloraz and metrafenone (Carrasco et al., 2017b).

***Green Mould Disease*:** Different species belonging to the genera *Trichoderma* (*T. harzianum, T. atroviride, T. koningii, T. aggressivum, T. viride, T. citrinoviride, T. deliquescens, T. pleurotum* or *T. pleuroticola*) have been described to cause economic losses in cultivated mushroom worldwide (Kim et al., 2016; O'Brien et al., 2017). *Trichoderma* produces a vast quantity of tiny conidia that vary in their shades of green. They form into chains of conidia, which can be easily scattered by the movement of air when they are on debris or by insects and mites, particularly pygmephorid mites or by workers, containers and so on. Their optimal temperature varies from one species to another, from 22 to 28°C. Some species of *Trichoderma* grow particularly well at pH lower than 6, especially if the nitrogen level is low. Thus, a C/N ratio of 22–23 favours the growth of *Trichoderma* in compost (Gea and Navarro, 2017).

Nowadays, in button mushroom crops, the most harmful species is *T. aggressivum*, with two subspecies (*T. aggressivum* f. *aggressivum* and *T. aggressivum* f. *europaeum*) (O'Brien et al., 2017). The green mould disease (*T. pleuroti* (previously *T. pleurotum*) and *T. pleuroticola*) also is a serious problem in *Pleurotus ostreatus* crop (Park et al., 2006; Komon-Zelazowska et al., 2007). The *T. harzianum, T. atroviride* and *T. citrinoviride* are also antagonistic against the mycelial growth of shiitake, causing green mould in *Lentinula edodes* crop (Kim et al., 2016).

Typical symptoms of the disease are green sporulation areas on the surface of the cultivation substrate that is exposed to green mould infection, mostly during spawn run (Innocenti et al., 2019). The disease is characterised by the presence of dense white mycelia of fast-growing colonies on casing material or compost that changes colour into green after extensive sporulation. In areas colonised by *T. aggressivum*, the mushroom fruit body formation gets retarded and fruit bodies may be of poor quality due to damage or discolouration (Fig. 14).

Fig. 14. Spotting mushroom caps due to green mould disease (*Trichoderma* spp.).

It is necessary to take extra precautions during the critical period, such as compost spawning and packaging. The addition of nutritional supplements with a high content of carbohydrates at spawning may favour the development of *Trichoderma*. By contrast, a compost completely colonised by the *Agaricus* mycelium is not affected by the pathogen. The application of strict hygiene measures is a must to reduce green mould impact and prevent outbreaks when *Trichoderma* is detected in the compost or in those mushroom farms that have previously suffered the disease.

6.2 Mushroom Pests

***Diptera*:** Several species of diptera are among the most serious arthropod pest problems affecting the cultivation of mushrooms throughout the world, as much as button mushroom (Erler and Polat, 2008; Shamshad, 2010) as shiitake (Shin et al., 2012) and oyster mushroom crops (Zhang et al., 2016). Mushroom yield losses are either directly due to the larvae of flies feeding on mycelia or else due to other pests and diseases vectored by the mushroom flies (Fletcher and Gaze, 2008) (Fig. 15). Mushroom flies have been reported as pest and disease carriers while transporting mites and/or spores of different mycoparasite species (Shamshad et al., 2009; Keum et al., 2015; Navarro et al., 2019).

During the pasteurisation process, the compost is exposed to elevated temperatures enough to kill pests; however, growing substrates can be a source of fly larvae if the process is not conducted efficiently (Jess et al., 2007). Besides, phorid and sciarid diptera fly into the mushroom farms during the beginning of the cycle through ventilation holes. The emergence of the first generation of flies developed inside the growing medium and those coming from eggs laid during the first days (Lewandoski et al., 2012; Jess et al., 2017), starts with the first flush. Preventing flies from accessing the farm during the early stages of the mushroom crop (physical barriers) and an early interruption of the growth cycle would prevent the spread of the pests.

Traditionally, control of diptera is based almost exclusively on good cropping management and the use of conventional pesticides. However, the continued use of

Fig. 15. Phorid fly on dry bubble infected mushroom (distorted mushrooms and spotted caps). The sticky conidia are attached to the fly legs which act as a vector for the dispersion of the disease.

chemicals may lead to reduction in mushroom production and quality losses, and even the presence of residues in the mushrooms once harvested (Navarro et al., 2017) and also the occurrence of resistances in the flies. Moreover, the actual regulatory restrictions on the use of insecticides makes it necessary to find alternatives for controlling these pests, such as entomopathogenic nematodes (EPNs), mites, bacteria and plant extracts (Jess and Bingham, 2004; Shamshad et al., 2009; Andreadis et al., 2016).

Phorids: The predominant species in recent literature is *Megaselia halterata* (Wood). Females of *M. halterata* are attracted by the volatiles produced by actively growing mycelia (Tibbles et al., 2005); their larvae are obligate mycobionts with acephalic head and mouth parts adapted to feeding on mycelium; thus, directly reduce its growth (Lewandoski et al., 2012). Phorids spend less lifetime than sciarid at the stage of larvae; therefore their impact on mushroom losses is lower. Temperature range of 24–25°C and 15–20 days is needed for *M. halterata* to complete its development from oviposition to adult eclosion (Lewandoski et al., 2012; Barzegar et al., 2016).

Regarding new control agents, mainly EPNs and plant extracts (essentials oils and hot-water plant extracts) have been studied against phorid pest. However, it has been reported that application of EPNs does not mean reduction of phorid flies to acceptable level (Jess and Bingham, 2004; Navarro et al., 2014). Then further efforts should target the detection and development of more efficient entomopathogenic nematodes against *M. halterata*. On the other hand, applications of some hot-water extracts may be potential alternatives to conventional pesticides for the control of this pest (Erler et al., 2009).

Sciarids: Most authors that have studied mushroom flies describe sciarids (Fig. 16) as the major mushroom arthropod pest (Jess et al., 2007; Shamshad, 2010; Eui and Seo, 2016). *Lycoriella auripila* (Winnertz) is considered one of the major species (Fletcher and Gaze, 2008). Female flies are strongly attracted to odours emanating from the compost after the pasteurisation and conditioning phases. Larvae of this fly are general feeders, but they prefer to feed on developing mycelium, thereby damaging primordial and mature fruit bodies by tunnelling into them. Since sciarids

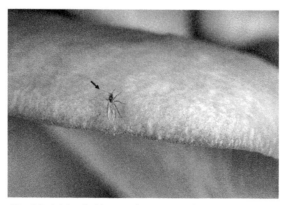

Fig. 16. Adult sciarid on mushroom cap (*Pleurotus ostreatus*).

spend most of their life-time as larvae, the developmental time of sciarid flies depends on the temperature and fungus varieties; on button mushroom, at 25°C, an average of 21 days was needed for *L. auripila* to complete its development from oviposition to adult eclosion (Shirvani-Farsani et al., 2013).

Regarding new control agents, many authors defend the use of *Steinernema feltiae* for the control of sciarids, with effectiveness ranging between 66–95 per cent, depending on the doses and the timing of application (Shamshad et al., 2008, 2010). As an alternative to conventional pesticides, the application of essential oils from aromatic plants can be employed for the control of this pest (Choi et al., 2006).

Cecids: Cecids are an occasional pest of the mushroom crops. Species of the genera *Mycophila* and *Heteropeza* have been described (Fletcher and Gaze, 2008). They could reproduce by a unique process, called paedogenesis, in which a mature (mother) larva will give birth to 10–40 daughter larvae without becoming an adult and mating. At 24°C, the process spends 18 days; a logarithmic increase of the cecids population. Cecid larvae feed on mycelium and the yield of the crop could be decreased. But this pest also influences the quality of the marketable mushrooms because larvae accumulate at the junction of the stipe and gills. Peat is considered as the main source of contamination, so its pasteurisation is proposed as the more important method of control (Fletcher and Gaze, 2008).

6.3 Mites

Pygmephorids: These mites feed on moulds (*Monilia, Humicola, Penicillium* and others), but some of them feed on *Trichoderma* spp. and also the mycelium of the cultivated mushroom. They usually are phoretic on tools and cloths or flies (Navarro et al., 2019). The red pepper mites, *Bakerdania mesembrinae* (Canestrini) and *B. sellnicki* (Krczal) feed principally from *Trichoderma* spp. and other molds. In addition, these mites can transport *Trichoderma* spores as a disease vector for dispersion. They tend to congregate on top of the carpophores, causing 'red patches'. They do not cause direct damage to the mushroom crop (*Pleurotus* and *Agaricus* spp.) but may cause losses in the marketable yield. The pygmy mite, *Microdispus lambi* (Krczal) is a myceliophagus mite, which causes substantial economic losses, even more than 70 per cent to the production, in button mushroom crops. When density is low, the mite may easily attack, unnoticed by growers, but it quickly increases in numbers and moves to the surface of the growing substrate and on to the mushroom caps (Navarro et al., 2010).

Tarsonemids: *Tarsonemus myceliophagus* Hussey mite feeds on several moulds and the mycelium of edible fungi. It can remove the connetion between vegetative mycelium and fruit bodies, causing yield and quality losses (Fletcher and Gaze, 2008).

Acaridaes: *Tyrophagus putrescentiae* (Schrank) is a myceliophagus mite that affects the commercial quality of the mushrooms, causing black holes in cap and stipe, where bacterial putrefaction can be detected. The hipopus of this mite is phoretic on diptera.

There are no registered chemical treatments for the control of mushroom mites. Right pasteurisation of the compost and an integrated pest management programme is required to minimise the spread of the mites and moulds (Fletcher and Gaze, 2008).

6.4 Nematodes

Saprophytic Nematodes (Rhabditidae): They are relatively usual in commercial mushroom crops, pointing at poor composting. In the presence of nematodes, the compost looks wet and with black necrotic areas; their action fragments the crop mycelium and consequently impacts negatively on mushroom yield. The nematodes and their associated bacteria also reduce the quality of fresh mushrooms (Rinker, 2017).

Parasitic (Mycophagus) Nematodes (Ditylenchus myceliophagus and Aphelenchoides composticola): They can be occasionally detected as a mushroom pest. These nematodes feed exclusively on the mushroom mycelium and destroy it. The affected area lack production and the compost becomes sticky. They ultimately group together, forming clumps or swarms when the availability of nutrients drops. This behaviour permits nematode phoretic dispersion to other locations (Fletcher and Gaze, 2008).

Unfortunately, there is no chemical treatment to control nematode pests. In order to cope with this biological threat, the application of correct pasteurisation with adequate temperatures is a requirement to ensure nematodes free substrate. Insects, equipment and workers can also disperse the nematodes; therefore, to prevent dispersion, a complete sanitation programme and correct hygiene practices must be implemented on the farm.

Concluding Remarks

Mushrooms are a nutritious income for human diet that can be efficiently cultivated around the world. The production is not subjected to external climate conditions since the mushrooms are cultivated indoors under controlled environment. Besides, due to their powerful enzymatic machinery, mushrooms are able to degrade multiple lignocellulosic substrates that are generated from local agricultural wastes, favouring the circular economy. The initial inoculum or spawn is produced under strict cleanliness standards in lab conditions to prevent contamination. It is inoculated in selective substrates that subsequently are colonised by the crop mycelium which is induced to fructification by forced environmental cues. Crop production is sensitive in a number of biotic disorders that can provoke drastic yield losses when inappropriately controlled. In this chapter we have compiled an overview of the global production and described a series of biotechnological stages required for commercial production of the crop as a source of guidelines for technicians and growers. Considering the current sustained human population growth, mushroom cultivation is an opportunity to provide food in a sustainable manner and fight hunger.

Acknowledgements

The pictures illustrating this chapter have been generously provided by Jaime Carrasco (University of Oxford), Diego C. Zied (*Universidade Estadual Paulista, Brazil*), Francisco J. Gea and Arturo Pardo (*Centro de Investigación, Experimentación y Servicios del Champiñón*, Spain).

References

Andreadis, S.S., Cloonan, K.R., Bellicanta, G.S., Paley, K., Pecchia, J. and Jenkins, N.E. (2016). Efficacy of *Beauveria bassiana* formulations against the fungus gnat *Lycoriella ingenua*. Biol. Cont., 103: 165–171.

Barzegar, S., Zamani, A.A., Abbasi, S., Vafaei Shooshtari, R. and Shirvani Farsani, N. (2016). Temperature-dependent development modelling of the phorid fly *Megaselia halterata* (Wood) (Diptera: Phoridae). Neotrop. Entomol., 45: 507–517.

Bellettini, M.B., Fiorda, F.A., Maieves, H.A., Teixeira, G.L., Ávila, S., Hornung, P.S., MaccariJúnior, A. and Ribani, R.H. (2019). Factors affecting mushroom *Pleurotus* spp. Saudi J. Biol. Sci., 26(4): 633–646.

Berendsen, R.L., Baars, J.J.P., Kalkhove, S.I.C., Lugones, L.G., Wösten, H.A.B. and Bakker, P.A.H.M. (2010). *Lecanicillium fungicola*: Causal agent of dry bubble disease in white-button mushroom. Mol. Plant Pathol., 11: 585–595.

Buth, J. (2017). Compost as a food base for *Agaricus bisporus*. pp. 129–147. *In*: Zied, D.C. and Pardo-Giménez, A. (eds.). Edible and Medicinal Mushrooms: Technology and Applications, Wiley-Blackwell, West Sussex, England.

Cao, G., Song, T., Shen, Y., Jin, Q., Feng, W., Fan, L. and Cai, W. (2019). Diversity of bacterial and fungal communities in wheat straw compost for *Agaricus bisporus* cultivation. Hort. Science, 54(1): 100–109.

Carrasco, J., Navarro, M.J., Santos, M., Diánez, F. and Gea, F.J. (2016). Incidence, identification and pathogenicity of *Cladobotryum mycophilum*, causal agent of cobweb disease on *Agaricus bisporus* mushroom crops in Spain. Ann. Appl. Biol., 168: 214–224.

Carrasco, J., Navarro, M.J. and Gea, F.J. (2017a). Cobweb, a serious pathology in mushroom crops: A review. Span. J. Agric. Res., 15(2): e10R01.

Carrasco, J., Navarro, M.J., Santos, M. and Gea, F.J. (2017b). Effect of five fungicides with different modes of action on cobweb disease (*Cladobotryum mycophilum*) and mushroom yield. Ann. Appl. Biol., 171: 62–69.

Carrasco, J., Tello, M.L., Perez, M. and Preston, G. (2018). Biotechnological requirements for the commercial cultivation of macrofungi: substrate and casing layer. pp. 159–175. *In*: Singh, B., Lallawmsanga, Passari A. (eds.). Biology of Macrofungi, Fungal Biology, Springer, Cham.

Carrasco, J. and Preston, G.M. (2020). Growing edible mushrooms: A conversation between bacteria and fungi. Environ. Microbiol., 22(3): 858–872.

Chang, S.T. (2006). The world mushroom industry: trends and technological development. Int. J. Med. Mush., 8(4): 297–314.

Chang, S.T. (2008). Mushrooms and mushroom cultivation. pp. 3552–3555. *In*: Considine, G.D. and Kulik, P.H. (eds.). Van Nostrand's Scientific Encyclopedia, 10th ed., John Wiley & Sons, Inc., Hoboken, New Jersey, USA.

Chang, S.T. and Miles, P.G. (2004). Mushrooms: Cultivation, Nutritional Value, Medicinal Effect, and Environmental Impact, 2nd ed., CRC Press. Boca Raton, Florida, USA, pp. 451.

Cho, Y.S., Kim, J.S., Crowley, D.E. and Cho, B.G. (2003). Growth promotion of the edible fungus *Pleurotus ostreatus* by fluorescent pseudomonads. FEMS Microbiol. Lett., 218(2): 271–276.

Choi, W.S., Park, B.S., Lee, Y.H., Jang, D.Y., Yoon, H.Y. and Lee, S.E. (2006). Fumigant toxicities of essential oils and monoterpenes against *Lycoriellamali* adults. Crop Prot., 25: 398–401.

Devochkina, N., Nurmetov, R. and Razin, A. (2019). Economic assessment of the development potential of mushroom production in the Russian Federation. pp. 108–115. *In*: IOP Conference Series: Earth and Environmental Science, Vol. 395: 012076. 10.1088/1755-1315/395/1/012076.

Dhar, B.L. (2017). Mushroom farm design and technology of cultivation. pp. 271–308. *In*: Zied, D.C. and Pardo-Giménez, A. (eds.). Edible and Medicinal Mushrooms: Technology and Applications, Wiley-Blackwell, West Sussex, England.

Erler, F. and Polat, E. (2008). Mushroom cultivation in Turkey as related to pest and pathogen management. Israel J. Pl. Sci., 56: 303–308.

Erler, F., Polat, E., Demir, H., Cetinc, H. and Erdemira, T. (2009). Control of the mushroom phorid fly, *Megaselia halterata* (Wood), with plant extracts. Pest Manag. Sci., 65: 144–149.

Eui, L.B. and Seo, G.S. (2016). Occurrence and control of mushroom flies during *Agaricus bisporus* cultivation in Chungnam, Korea. J. Mush., 14: 100–104.

FAO. (2018). Food and Agriculture Organisation Corporate Statistical Database, Crops, Mushrooms and Truffes, Production Quantity, Rome.

Fletcher, J.T. and Gaze, R.H. (2008). Mushroom Pest and Diseases Control. Manson Publishing Ltd., London, UK, pp. 192.

Gea, F.J., Tello, J.C. and Navarro, M.J. (2010). Efficacy and effects on yield of different fungicides for control of wet bubble disease of mushroom caused by the mycoparasite *Mycogone perniciosa*. Crop Prot., 29(9): 1021–1025.

Gea, F.J. and Navarro, M.J. (2017). Mushroom diseases and control. pp. 239–259. *In*: Zied, D.C. and Pardo-Giménez, A. (eds.). Edible and Medicinal Mushrooms: Technology and Applications, John Wiley & Sons Ltd., Chichester, West Sussex, UK.

Gea, F.J., Carrasco, J., Suz, L.M. and Navarro, M.J. (2017). Characterisation and pathogenicity of *Cladobotryum mycophilum* in Spanish *Pleurotus eryngii* mushroom crops and their sensitivity to fungicides. Eur. J. Plant Pathol., 147: 129–139.

Gea, F.J., Navarro, M.J. and Suz, L.M. (2018). First report of cobweb disease caused by *Cladobotryum dendroides* on shiitake mushroom (*Lentinula edodes*) in Spain. Plant Dis., 102: 1030.

Gea, F.J., Navarro, M.J. and L.M. Suz. (2019). Cobweb disease on oyster culinary-medicinal mushroom (*Pleurotus ostreatus*) caused by the mycoparasite *Cladobotryum mycophilum*. J. Plant Pathol., 101: 349–354.

González, A., Gea, F.J., Navarro, M.J. and Fernández, A.M. (2012). Identification and RAPD-typing of *Ewingella americana* on cultivated mushroom in Castilla-La Mancha, Spain. Eur. J. Plant Pathol., 133: 517–522.

González, A.J., González-Varela, G. and Gea, F.J. (2009). Brown blotch caused by *Pseudomonas tolaasii* on cultivated *Pleurotus eryngii* in Spain. Pl. Dis., 93: 667.

Grimm, D. and Wösten, H.A. (2018). Mushroom cultivation in the circular economy. Appl. Microbiol. Biotechnol., 102(18): 7795–7803.

Han, H.-S., Jhune, C.S., Cheong, J.C., Oh, J.A., Kong, W.S., Cha, J.S. and Lee, C.J. (2012). Occurrence of black rot of cultivated mushrooms (*Flammulina velutipes*) caused by *Pseudomonas tolaasii* in Korea. Eur. J. Pl. Pathol., 133: 527–535.

Hanafi, F.H.M., Rezania, S., Taib, S.M., Din, M.F.M., Yamauchi, M., Sakamoto, M., Hara, H., Park, J. and Ebrahimi, S.S. (2018). Environmentally sustainable applications of agro-based spent mushroom substrate (SMS): An overview. J. Mater. Cycles Waste Manag., 20: 1383–1396.

Hawksworth, D.L. (2001). Mushrooms: The extent of the unexplored potential. Int. J. Med. Mush., 3(4). 10.1615/IntJMedMushr.v3.i4.50.

Innocenti, G., Montanari, M., Righini, H. and Roberti, R. (2019). *Trichoderma* species associated with the green mould disease of *Pleurotus ostreatus* and their sensitivity to prochloraz. Pl. Pathol., 68: 392–398.

Jasińska, A. (2018). Spent mushroom compost (SMC)—retrieved added value product closing loop in agricultural production. Acta Agraria Debreceniensis, 150: 185–202.

Jess, S. and Bingham, J.F.W. (2004). Biological control of sciarid and phorid pests of mushrooms, with predatory mites of the genus *Hypoaspis* (Acari: Hypoaspidae) and the entomopathogenic nematode *Steinernema feltiae*. Bull. Entomol. Res., 94: 159–167.

Jess, S., Murchie, A.K. and Bingham, J.F.W. (2007). Potential sources of sciarid and phorid infestations and implications for centralised phases I and II mushroom compost production. Crop Prot., 26: 455–464.

Jess, S., Kirbas, J.M., Gordon, A.W. and Murchie, A.K. (2017). Potential for use of garlic oil to control *Lycoriella ingenua* (Diptera: Sciaridae) and *Megaselia halterata* (Diptera: Phoridae) in commercial mushroom production. Crop Prot., 102: 1–9.

Kaliyaperumal, M., Kezo, K. and Gunaseelan, S. (2018). A global overview of edible mushrooms. pp. 15–56. *In*: Singh, B. and Lallawmsanga, Passari A. (eds.). Biology of Macrofungi, Fungal Biology, Springer, Cham.

Keum, E., Kang, M. and Jung, C. (2015). New record of *Arctos eiuscetratus* (Sellnick, 1940) (Mesostigmata: Ascidae) phoretic to sciarid fly from mushroom culture in Korea. Kor. J. Environ. Biol., 33: 209–214.

Kim, J.Y., Kwon, H.W., Yun, Y.H. and Kim, S.H. (2016). Identification and characterisation of Trichoderma species damaging shiitake mushroom bed-logs infested by Camptomyiapest. J. Microbiol. Biotechnol., 26(5): 909–917.

Komon-Zelazowska, M., Bisset, J., Zafari, D., Hatvani, L., Manczinger, L., Woo, S., Lorito, M., Kredics, L., Kubicek, C.P. and Druzhinina, I.S. (2007). Genetically closely related but phenotypically divergent *Trichoderma* species cause green mold disease in oyster mushroom farm worldwide. Appl. Environ. Microbiol., 73: 7415–7426.

Lewandoski, M., Kozak, M. and Sznyk-Basalyga, A. (2012). Biology and morphometry of *Megaselia halterata*, an important insect pest of mushrooms. Bull. Insectol., 65: 1–8.

Minhoni, M.T.A. and Zied, D.C. (2008). Tecnologias de Produção de Inoculantes para o Cultivo de Cogumelos Comestíveis. pp. 42–50. *In*: IV International Symposium on Mushrooms, 4th International Symposium on Mushrooms, Documentos # 266 Embrapa, Simpósio Nacional sobre Cogumelos Comestíveis, Caxias do Sul., Brazil.

Moore, D., Gange, A.C., Gange, E.G. and Boddy, L. (2008). Fruit bodies: Their production and development in relation to environment. pp. 79–103. *In*: Boddy, L., Frankland, J.C. and van West, P. (eds.). Ecology of Saprotrophic Basidiomycetes. The British Mycological Society, Elsevier Ltd., Oxford, UK.

Moreaux, K. (2017). Spawn production. pp. 89–128. *In*: Zied, D.C. and Pardo-Giménez, A. (eds.). Edible and Medicinal Mushrooms: Technology and Applications, Wiley-Blackwell, West Sussex, England.

NASS. (2018). Mushrooms, National Agricultural Statistics Service (NASS), Agricultural Statistics Board, United States Department of Agriculture (USDA), ISSN: 1949–1530.

Navarro, M.J., Gea, F.J. and Escudero, A. (2010). Abundance and distribution of *Microdispuslambi* (Acari: Microdispidae) in Spanish mushroom crops. Exp. Appl. Acarol., 50: 309–316.

Navarro, M.J. and Gea, F.J. (2014). Entomopathogenic nematodes for the control of phorid and sciarid flies in mushroom crops. Pesc. Agropec. Bras., 49: 11–17.

Navarro, M.J., Merino, L.L. and Gea, F.J. (2017). Evaluation of residue risk and toxicity of different treatments with diazinon insecticide applied to mushroom crops. J. Environ. Sci. Health. B., 52: 218–221.

Navarro, M.J., Gea, F.J. and González, A.J. (2018). Identification, incidence and control of bacterial blotch disease in mushroom crops by management of environmental conditions. Sci. Hort., 229: 10–18.

Navarro, M.J., López-Serrano, F.R., Escudero-Colomar, L.A. and Gea, F.J. (2019). Phoretic relationship between the myceliophagus mite *Microdispuslambi* (Acari: Microdispidae) and mushroom flies in Spanish crops. Ann. Appl. Biol., 174: 277–283.

Navarro, M.J., Gea, F.J., Pardo-Giménez, A., Martínez, A., Raz, D., Levanon, D. and Danay, O. (2020). Agronomical valuation of a drip irrigation system in a commercial mushroom farm. Sci. Hort., 265: 109234.

O'Brien, M., Kavanagh, K. and Grogan, H. (2017). Detection of *Trichoderma aggressivum* in bulk phase III substrate and the effect of *T. aggressivum* inoculum, supplementation and substrate-mixing on *Agaricus bisporus* yields. Eur. J. Pl. Pathol., 147(1): 199–209.

Oei, P. (2003). Mushroom Cultivation, 3rd ed., Backhuys Publishers, Leiden, Netherlands, pp. 429.

Oei, P. and van Nieuwenhuijzen, B. (2005). Small-scale Mushroom Cultivation: Oyster, Shiitake and Wood Ear Mushrooms (Agrodok 40), Agromisa Foundation and CTA, Wageningen, Netherlands, pp. 86.

Oei, P., Zeng, H., Liao, J., Dai, J., Chen, M. and Cheng, Y. (2008). Alternative uses of spent mushroom compost. pp. 231–245. *In*: Lelley, J.I. and Buswell, J.A. (eds.). Mushroom Biology and Mushroom Products, GAMU GmbH, Bonn, Germany.

Pardo, A., De Juan, J.A. and Pardo, J.E. (2002). Bacterial activity in different types of casing during mushroom cultivation (*Agaricus bisporus* (Lange) Imbach). Acta Alimentaria, 31(4): 327–342.

Pardo, A., Perona, M.A. and Pardo, J. (2007). Indoor composting of vine by-products to produce substrates for mushroom cultivation. Spanish J. Agric. Res., 5(3): 417–424.

Pardo-Giménez, A., Pardo, J.E. and Zied, D.C. (2017). Casing materials and techniques in *Agaricus bisporus* cultivation. pp. 149–174. *In*: Zied, D.C. and Pardo-Giménez, A. (eds.). Edible and Medicinal Mushrooms: Technology and Applications, Wiley-Blackwell, West Sussex, England.

Park, M.S., Bae, K.S. and Yu, S.H. (2006). The new species of *Trichoderma* associated with green mold of oyster mushroom cultivation in Korea. Mycobiology, 34: 11–13.

Poppe, J. (2004). Agricultural wastes as substrates for oyster mushroom. pp. 75–85. *In*: Mushroom Growers' Handbook 1. Oyster Mushroom Cultivation, Mush World, Seoul, Korea.

Preston, G.M., Carrasco, J., Gea, F.J. and Navarro, M.J. (2018). Biological control of microbial pathogens in edible mushrooms. pp. 305–317. *In*: Singh, B. and Lallawmsanga, Passari A. (eds.). Biology of Macrofungi, Fungal Biology, Springer Cham.

Rinker, D.L. (2017). Spent mushroom substrate uses. pp. 427–454. *In*: Zied, D.C. and Pardo-Giménez, A. (eds.). Edible and Medicinal Mushrooms: Technology and Applications, Wiley-Blackwell, West Sussex, England.

Rodríguez, A.E. and Pecchia, J. (2017). Cultivation of *Pleurotus ostreatus*. pp. 339–360. *In*: Zied, D.C. and Pardo-Giménez, A. (eds.). Edible and Medicinal Mushrooms: Technology and Applications, Wiley-Blackwell, West Sussex, England.

Royse, D.J., Baars, J. and Tan, Q. (2017). Current overview of mushroom production in the world. pp. 5–13. *In*: Zied, D.C. and Pardo-Giménez, A. (eds.). Edible and Medicinal Mushrooms: Technology and Applications, Wiley-Blackwell, West Sussex, England.

Sánchez, C. (2010). Cultivation of *Pleurotus ostreatus* and other edible mushrooms. Appl. Microbiol. Biotechnol., 85: 1321–1337.

Sánchez, J.E., Zied, D. and Albertó, E. (2018). Edible mushroom production in the Americas. pp. 2–11. *In*: 9th International Conference on Mushroom Biology and Mushroom Products, Shanghai, China.

Savoie, J.-M., Foulogne-Oriol, M., Barroso, G. and Callac, P. (2013). Genetics and genomics of cultivated mushrooms, application to breeding of *Agarics*. pp 3–33. *In*: Esser, K. and Kempken, F. (eds.). The Mycota XI, Agricultural Applications, 2nd ed., Springer, Heidelberg, Germany.

Savoie, J.-M. and Mata, G. (2016). Growing *Agaricus bisporus* as a contribution to sustainable agricultural development. pp. 69–91. *In*: Petre, M. (ed.). Mushroom Biotechnology: Developments and Applications, Academic Press, Cambridge, USA.

Shamshad, A., Clift, A.D. and Mansfield, S. (2008). Toxicity of six commercially formulated insecticides and biopesticides to third instar larvae of mushroom sciarid, *Lycoriella ingenua* Dufour (Diptera: Sciaridae), in New South Wales, Australia. Aust. J. Entomol., 47: 256–260.

Shamshad, A., Clift, A.D. and Mansfield, S. (2009). The effect of tibia morphology on vector competency of mushroom sciarid flies. J. Appl. Entomol., 133: 484–90.

Shamshad, A. (2010). The development of integrated pest management for the control of mushroom sciaridflies, *Lycoriella ingenua* (Dufour) and *Bradysia ocellaris* (Comstock) in cultivated mushrooms. Pest Manag. Sci., 66: 1063–1074.

Sharma, V.P., Annepu, S.K., Gautam, Y., Singh, M. and Kamal, S. (2017). Status of mushroom production in India. Mush. Res., 26(2): 111–120.

Shin, S.W., Lee, H.S. and Lee, S. (2012). Dark winged fungus gnats (Diptera: Sciaridae) collected from shiitake mushroom in Korea. J. Asia-Pac. Entomol., 15: 174–181.

Shirvani-Farsani, N., Zamani, A.A., Abbasi. S. and Kheradmand, K. (2013). Effect of temperature and button mushroom varieties on life history of *Lycoriella auripila* (Diptera: Sciaridae). J. Econ. Entomol., 106: 115–123.

Soler-Rivas, C., Jolivet, S., Arpin, N., Olivier, J.M. and Wichers, H.J. (1999). Biochemical and physiological aspects of brown blotch disease of *Agaricus bisporus*. FEMS Microbiol. Rev., 23: 591–614.

Taparia, T., Krijger, M., Hodgetts, J., Hendriks, M., Elphinstone, J.G. and van der Wolf, J. (2020). Six multiplex TaqManTM-qPCR assays for quantitative diagnostics of *Pseudomonas* species causative of bacterial blotch diseases of mushrooms. Front. Microbiol., 11: 989. 10.3389/fmicb.2020.00989.

Tibbles, L.L., Chandler, D., Mead, A., Jervis, M. and Boddy, L. (2005). Evaluation of the behavioural response of the flies *Megaselia halterata* and *Lycoriella castanescens* to different mushroom cultivation materials. Entomol. Exp. Appl., 116: 73–81.

Upadhyay, R.C. 2011. Oyster mushroom cultivation. pp. 129–138. *In*: Singh, M., Vijay, B., Kamal, S. and Wakchaure, G.C. (eds.). Mushrooms: Cultivation, Marketing and Consumption, Directorate of Mushroom Research (Indian Council of Agricultural Research), Solan, India.

Velázquez-Cedeño, M., Farnet, A.M., Mata, G. and Savoie, J.-M. (2008). Role of *Bacillus* spp. in antagonism between *Pleurotus ostreatus* and *Trichoderma harzianum* in heat-treated wheat-straw substrates. Biores. Technol., 99: 6966–6973.

Vieira, F.R., Pecchia, J.A., Segato, F. and Polikarpov, I. (2019). Exploring oyster mushroom (*Pleurotus ostreatus*) substrate preparation by varying phase I composting time: Changes in bacterial communities and physicochemical composition of biomass impacting mushroom yields. J. Appl. Microbiol., 126(3): 931–944.

Zare, R. and Gams, W. (2008). A revision of the *Verticillium fungicola* species complex and its affinity with the genus *Lecanicillium*. Mycol. Res., 112: 811–824.

Zhang, Z., Li, X., Chen, L., Wang, L. and Lei, C. (2016). Morphology, distribution and abundance of antennal sensilla of the oyster fly, *Coboldia fuscipes* (Meigen) (Diptera: Scatopsidae). Rev. Bras. Entomol., 60: 8–14.

Zied, D.C., Pardo-Giménez, A., Pardo Gonzalez, J.E. and Minhohi, M.T.A. (2018). Métodos y técnicas para producción de inóculo de hongos comestibles y medicinales. pp. 127–140. *In*: Patronato de desarrollo provincial (Org.), Avances en la tecnología de la producción comercial del champiñón y otros hongos cultivados 5. 1ed.Cuenca: Diputación Provincial de Cuenca, v. 5.

3

Mushroom Market in Malaysia

Shin-Yee Fung[1,2,3,*] and *Chon-Seng Tan*[4]

1. INTRODUCTION

Malaysia has been recognised as one of the 12 known 'mega-diversity' countries in the world. The Malaysian government has taken the necessary steps to enhance effective management, promote conservation and maintain the development of forest biodiversity (Forestry Department of Peninsular Malaysia). Malaysia is endowed with over 15,000 species of flowering plants, 1,500 species of terrestrial vertebrates and about 150,000 species of invertebrates (Dawson, 1997). In such an environment, microorganisms abound and more than 4,000 species of known fungi can be found.

Mushrooms are macrofungi that have been known since time immemorial. They are placed in a kingdom of their own, known as *Myceteae* due to their unique characteristics, which distinguishes them from being a plant or an animal (Miles and Chang, 1997). They can be either epigeous (above ground) or hypogeous (underground) with fruit bodies large enough to be seen with the naked eye and to be picked by hand (Chang and Miles, 1992). In a broad sense, mushrooms can be either Ascomycetes or Basidiomycetes with diverse morphological shapes. According to their characteristics, they can be divided into four categories: (1) edible mushrooms: fleshy and edible [e.g., *Agaricus bisporus* (button mushroom), *Volvariella volvacea*

[1] Medicinal Mushroom Research Group (MMRG), Department of Molecular Medicine, University of Malaya, 50603 Kuala Lumpur, Malaysia.
[2] Centre for Natural Products Research and Drug Discovery (CENAR), University of Malaya, 50603 Kuala Lumpur, Malaysia.
[3] University of Malaya Centre for Proteomics Research (UMCPR), University of Malaya, 50603 Kuala Lumpur, Malaysia.
[4] LiGNO Research Initiative, LiGNO Biotech Sdn. Bhd., Balakong Jaya 43300, Selangor, Malaysia.
* Corresponding author: syfung@ummc.edu.my

(straw mushroom), *Lentinula edodes* (shiitake)]; (2) medicinal mushrooms: possess medicinal applications [e.g., *Ganoderma lucidum* (lingzhi mushroom), *Inonotus obliquus* (chaga mushroom), *Fomes fomentarius* (hoof mushroom)]; (3) poisonous mushrooms: identified as poisonous [e.g., *Amanita phalloides* (death cap), *Cortinarius orellanus* (fools webcap), *Clitocybe acromelalga*]; and (4) other mushrooms: less well defined (Chang, 2008). Mushrooms are known for not only their high protein content but also provide profound health-promoting benefits but with the exception of poisonous species.

Recent scientific studies confirm their medical efficacy as well as identify their many bioactive molecules. Extensive studies have been performed on the bioactive compounds responsible for their antioxidative, anticancer, antidiabetic, anti-allergic, immunomodulating, anti-cholesterolemic, antimicrobial and hepatoprotective effects (Yu et al., 2009; Zhang et al., 2011a, b; Chang and Wasser, 2012; Finimundy et al., 2013). Numerous bioactive compounds have been extracted from their fruit bodies, mycelium and culture medium such as polysaccharides, proteins, glycoproteins, unsaturated fatty acids, terpenoids, tocopherols, phenolic, flavonoids, carotenoids, alkaloids, tocopherols, ergosterols and lectins (Valverde et al., 2015; Ma et al., 2018). Although over 2000 edible fleshy mushroom species are available worldwide, only about 100 are cultivated (Leifa et al., 2006). The cultivation technologies range from conventional sawdust logs to high-tech methods where even the environment for the mushroom's growth is highly regulated to ensure their quality.

2. Mushroom Industry in Malaysia

In Malaysia, mushrooms have been cultivated since the late 1960's. Prior to this era, Malaysia relied on imported dried and canned mushroom products, such as button (*Agaricus bisporus*), shiitake (*Lentinula edodes*) and wood ear (*Auricularia polytricha*), all of which were imported mainly from mainland China.

Malaysia, which is geographically situated in the equator and blessed with tropical climate, experiences hot and humid weather conditions, with a high annual rainfall. Temperatures range between 23–35°C and humidity is between 80–90 per cent. This climatic condition can be tolerated only by certain species of mushrooms. Species which have a difficulty in acclimatising to this weather condition require control rooms designed to ensure that the environment is within acceptable limits to ensure growth of mushroom species of human interest.

In recent years with increasing awareness, mushrooms have now been declared as an industrial crop under the 4th National Agricultural Policy (2010–2015) by the Ministry of Agriculture. As one of the seven high-valued industrial crops in Malaysia, mushrooms will be commercially grown under the National Agro-Food Policy (Dasar Agro-Makanan Negara, 2011–2020). The export of mushrooms in Malaysia has grown at a rate of 19 per cent from RM12 million a year in 2000 to RM67 million in 2010. The main destination for export is to Singapore. Approximately 49 per cent of mushrooms are exported as fresh variety [i.e., button (*Agaricus bisporus*) and oyster (*Pleurotus ostreatus*)], while 51 per cent is in processed form [i.e., Ling zhi (*Ganoderma lucidum*), wood ear (*Auricularia auricula-judae*) and straw mushroom (*Volvariella volvacea*)] (National Agro-Food Policy, Dasar Agro-Makanan Negara,

2011–2020). Three varieties [straw (*Volvariella volvacae*), abalone (*Pleurotus cystidiosus*) and chestnut (*Agrocybe* spp.)] are seen as potentially viable species for commercial use.

In 2007, Vikineswary and co-workers reported that only seven varieties [white and grey oyster (*Pleurotus ostreatus*), abalone (*Pleurotus cystidiosus*), Ling zhi (*Ganoderma lucidum*), button (*Agaricus bisporus*), shiitake (*Lentinus edodes*), wood ear (*Auricularia polytricha*) and monkey head (*Hericium erinaceus*)] out of 20 different species of mushrooms in Malaysia are cultivated for commercial reasons. However, the National Agro-Food Policy (Dasar Agro-Makanan Negara 2011–2020) has documented more species that possess commercial value. Many new species are being cultivated and this pose a severe challenge on how to maintain the viability of the industry. Suitable technologies are required to ensure the viability of the mushroom industry. The identification of species for commercial use would require technological input from the private sector via the industry-academia-government collaborations to improve and advance the existing research and development. Stringent quality control and improvement of high-quality strains are also important to ensure success of this industry. At present, grey oyster mushroom (*Pleurotus ostreatus*) is the dominant commercial variety as it is easy to grow and can be cultivated in lowlands (Table 1 and Fig. 1). There are approximately 320 farmers engaged in the cultivation of this species in Peninsular Malaysia. Most of them operate under a small-scale agriculture model and produce 50–500 kg with not more than 10 farms producing tons of fresh mushrooms per day (Vikineswary, 2007). All of them still practice growing mushrooms in open-air mushroom houses. Mushroom growing is a highly technical industry and the number of growers have shrunk of late, from about 2,000 growers in the 1980s to about 300 growers today. The distribution of mushroom growers in Malaysia is shown in Fig. 2.

The expansion of the mushroom industry into a fully automated industry is possible only when the return of investment is ensured. On one hand, it will be forward looking to improve on technical grounds to increase production. However, the cost for this advancement does not equal the returns as the local markets are not at par in terms of demand. In addition, the import of dried mushrooms [such as shiitake (*Lentinus edodes*), black and white jelly (*Auricularia* sp.) and fresh mushrooms like king oyster (*P. eryngii*) and enoki (*F. velutipes*)] from China and Korea is more economical than growing them locally.

In the year 2009, a local small to medium enterprise known as LiGNO™ Biotech Sdn. Bhd. was established to mass cultivate a local Malaysian medicinal

Table 1. Production of grey oyster mushroom per day in comparison to other species of mushrooms in Malaysia.

Type of mushroom	kg/day
Grey Oyster	25000
White Oyster	600
Button	500
Shiitake	400
Black Jelly	100

Fig. 1. Mushroom houses situated in lowland area with indoor view of the cultivation of grey oyster mushroom in Malaysia.

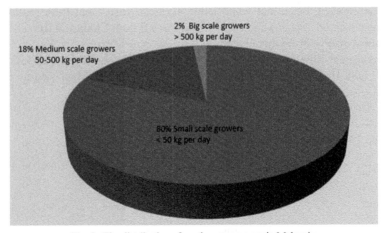

Fig. 2. The distribution of mushroom growers in Malaysia.

Fig. 3. *Lignosus rhinocerus* (*top*) and its cultivation of TM02® in a controlled environment (*bottom*).

mushroom—the tiger milk mushroom (*Lignosus rhinocerus*) (Fig. 3). The TM02® strain is cultivated by using a proprietary solid-state fermentation technique, where the cultivation medium is specially formulated with rice, water and other food-based materials. The culture is grown in a controlled environment for up to six months until the formation of the sclerotium.

Numerous attempts were made to cultivate the sclerotium with high nutritional and bioactive values. Developing a suitable environment and condition for harvest of this specific species is important. There are various methods of cultivating mushrooms: (a) submerged fermentation and (b) solid-state fermentation (Elisashvili, 2012; Letti et al., 2018). The solid-state fermentation technology employed so far provides a higher yield and allows the formation of sclerotium (Letti et al., 2018). Furthermore, the controlled environment allows for the production of nutrient and bioactive-filled sclerotia.

The TM02® cultivar has been registered under Section 28 of the Protection of New Plant Varieties (PNPV) Act 2004 (Act 634) by the Plant Varieties Board of the Malaysian government bearing the registration number: PBR0182 as a recognised cultivated variety, distinct from the wild type (Fung and Tan, 2017). This is one step ahead for a local Malaysian cultivator who not only is the first in the world to successfully cultivate a local species but has made it into a commodity that is widely sought after for various uses in the nutraceutical industry.

Besides LiGNO Biotech Sdn. Bhd., there are other success stories of mushroom cultivation in Malaysia where a variety of innovative concepts have been adopted in their business. The most notable ones are Ganoexcel Malaysia and DXN who specialise in Reishi (*Ganoderma lingzhi*) and *Ganoderma lucidum*. Biofact Life Sdn. Bhd., incorporated in 2005, specialises in production of *Cordyceps* under its traditional name *Dong Chong Xia Cao* through laboratory cultivation. Another grower based in Teluk Gong, Port Klang, Malaysia is Champ-Fungi. There are a diversified group of companies which are active in the cultivation of button mushrooms (initial endeavour), Shitake and related 'Asian' mushrooms. They cultivate, pack, ship and distribute all throughout Southeast Asia, from farm to the shelf or the plate in less than twenty-four hours. They utilise imported substrate with inoculants from EC and induce the fruit in a cold room. Hokto Malaysia Sdn. Bhd. is a Japanese company based in Negeri Sembilan, Malaysia which employs the state-of-the-art facility that allows tight control of temperature, humidity and light exposure throughout the mushrooms' growing cycle. Much of the operation in this cultivation facility is fully or partially automated, using conveyor belts among growing rooms, cooling chambers and harvesting and packing facility. Their main produce include bunapi, bunashimeiji and maitake.

The prospects for the mushroom industry in Malaysia are promising as the demand is steadily increasing for both fresh and processed mushroom produce. The demand is seen to increase in tandem with the population rise and consumption per capita (1 kg as at 2008 to 2.4 kg in 2020). Production of mushrooms has also increased as much as 16 per cent, from 15,000 metric tonnes in 2010 to 67,000 metric tonnes in 2020. It is expected that by the end of 2020, the export value of Malaysian mushroom will reach RM300 million, a 16 per cent increase annually (National Agro-Food Policy: Dasar Agro-Makanan Negara, 2011–2020).

2.1 Challenges to the Malaysian Mushroom Industry

There are many issues and challenges faced by this mushroom industry due to its newness. The current issues and challenges can be divided into two categories: (1) production or cultivation technology and (2) marketing. In Malaysia which happens to be a tropical country, the environmental temperature favours the oyster mushroom cultivation. It is of utmost importance for mushroom growers to understand environmental issues and challenges prior to getting involved in its industrial cultivation. Mushrooms that thrive in a cooler environment, such as shiitake and button mushroom, will find this tropical climate challenging. For instance, the hot Malaysian weather can cause the cultivation medium to dry easily, leading to a general decline in the mushroom species at cultivation plants. Advancement in technology enabling controlled environment for mushroom cultivation has been proven effective in some countries.

Pest contamination is yet another major problem faced by mushroom growers with which they need to reconcile especially in a hot and humid local environment. Green mold is the most prevalent problem for mushrooms cultivated in exposed plastic bags. Green mold is an umbrella term that refers to a number of species of fungi with spores which take on a green tint. The more common species usually belong to three genera: *Aspergillus, Cladosporium* and *Penicillium*. The mold is able to penetrate the dead spaces and can be difficult to eradicate. As the temperature and humidity being greater within the bag than the surrounding environment, it encourages the mold to propagate. Mold contamination is usually noticeable on the part of the substrate situated at the plug end of the bag.

Insects are another group of pests that can significantly affect mushroom production in Malaysian weather. Ants and mites are the two main foes of mushrooms grown in the *al fresco* area of tropical Malaysia. These two insects create problems by eating the substrate used for mushroom growth and thereby significantly reduce the mushroom production output. Another pest, an infamous rodent, the mouse can also pose a major challenge to mushroom cultivators. Mice will eat the spawn containing the immature mushrooms grown in the spawn bags and/or destroy them while searching for grains.

Lastly, and most crucial challenge faced by mushroom cultivators/growers in Malaysia is the quality of the seed culture and raw materials. Most of the cultivators/ growers depend on private suppliers for seeds for which they have to pay an exorbitant price. The cost of raw material, for instance, rubber wood sawdust has been increasing annually and small-scale cultivators/growers find it economically exasperating. The limited availability of rubber wood sawdust has also caused a conspicuous competition and led to a demand for an unrealistic price. The obstacles listed have universally increased the production cost for mushroom cultivators/ growers and indirectly handicapped Malaysian mushroom growers from competing with other countries, which are able to produce mushrooms at a significantly lower cost.

3. Waste Management in the Mushroom Industry

Waste management is always a serious issue in any industry as it comes at a cost to the industry, especially as it impacts the environment. However, if careful planning can turn these wastes into useful by-products, rendering the industry capable of zero waste, then the industry would appear more appealing to the investors.

Mushroom production is regarded as the largest solid-substrate fermentation industry in the world (Moore and Chiu, 2001). Approximately 5 kg of spent mushroom substrates (SMS) are left abandoned or discarded from 1 kg of mushroom production. Spent mushroom substrates are commonly made from renewable agriculture residues, such as sawdust, sugarcane bagasse, oil-palm fruit bunches, wheat straw-bedded horse manure, hay, poultry manure, ground corncobs, cottonseed meal, cocoa shells, gypsum and other substances (Jordan et al., 2008).

In Malaysia, the production of 100 tonnes of fresh mushroom per annum generates about 440 tonnes of spent mushroom substrate. This figure is expected to rise due to an increase in the demand for mushrooms. Thus, an efficient and practical method is needed to ensure that the spent mushroom substrates are disposed off via a constructive and utilitarian approach. Currently, small-scale local mushroom cultivators/growers are discarding their spent mushroom substrates by spreading the spent substrates on open land so that it gets buried and the spent substrates is composted with animal-manure or land filling.

It is interesting to note that spent mushroom substrates are useful materials, which can be put to proper use, such as for composting and to be used in horticulture. The spent mushroom substrates contain high nutrients and are generally non-toxic to plants. It is therefore beneficial to use them as a good source for manure to be used for different crops (Beyer, 1996). However, it is interesting to note that only the large-scale mushroom growers have sufficient volume of spent mushroom substrates for composting and turn into bio-fertiliser. Spent mushroom substrates have a variety of practical utilities. Their nutritional composition makes them suitable as soil mix for gardens and potted plants. They can also be used to improve the soil condition (generally known as soil amendment) in wetland areas for improving the growth of lowland plants, such as grass. These useful substrates can also be applied to treat polluted groundwater by removing the pollutants or converting them into harmless products through bioremediation of contaminated water.

An innovative use of spent mushroom substrates is to make them soil stabilisers or conditioners. Soil stabilisation a general term used for any physical, chemical, mechanical, biological or combined method of coverting natural soil to meet an engineering purpose. Soil properties are altered and enhanced to increase its suitability for construction purposes. Improvements include increasing the weight-bearing capabilities, tensile strength and overall performance of *in situ* subsoil, sand and waste material for strengthening the road pavements.

Other uses of spent mushroom substrates include their use as plant-disease controller and as bedding for animals. Rinker (2017) has written an indepth article in a book entitled, *Spent Mushroom Substrate Uses*, in which he outlines the characteristics of spent substrate and details the uses of these substrates (Rinker, 2017). Bakar and colleagues have discussed yet another method in managing spent mushroom substrates (Bakar et al., 2011). They have suggested that the spent mushroom substrates can be used as a vermicompost medium either on its own or through different combinations of agricultural and vegetable farm-waste. Vermicomposting refers to employment of the decomposition process in which various species of worms (red wigglers, white worms and other earthworms) are used to create a mixture of decomposing vegetable or food waste, bedding material and vermicast. In vermiculture, the earthworms help to aerate, mix, grind and fragment the substrate via enzymatic digestion and microbial decomposition of substrate in their intestine (Hand et al., 1988). The decomposition time of spent mushroom substrates through vermiculture can be reduced to a few weeks as compared to regular composting materials.

4. Conclusion

It is irrefutable that mushroom is a promising food item but unfortunately a relatively untapped source of material with immense potential in nutritional, medicinal and industrial applications. Malaysia, being one of the 12 megadiversity countries in the world, is actively promoting the growth of its mushroom industry. The vast potential for mushrooms as edible, pharmaceutical, nutritional and nutraceutical materials can be a good source of income for interested entrepreneurs. Growing mushrooms for commercialisation should be directed at not just providing a sufficient supply for the market, but ensuring that the mushroom quality is assured. Whilst it is impossible for small growers to compete with the mass quantities produced by larger growers, it is crucial that the produce retains the health properties associated with mushroom consumption. The Government of Malaysia has been carrying out various activities and exercises to encourage new cultivators/growers with the aim to continue to invest in the development of mushroom industry in the country.

References

Bakar, A.A., Mahmood, N.Z., Teixeira, J.A., Abdullah, N. and Jamaludin, A.A. (2011). Vermicomposting of sewage sludge by *Lumbricus rubellus* using spent mushroom compost as feed material. Biotechnol. Bioproc. Eng., 16: 1036–1043.

Beyer, M. (1996). The impact of the mushroom industry on the environment. Mushroom News, 44(11): 6–13.

Chang, S.-T. and Miles, P.G. (1992). Mushroom biology: A new discipline. The Mycologist, 6: 64–65.

Chang, S.-T. and Wasser, S.P. (2012). The role of culinary-medicinal mushrooms on human welfare with a pyramid model for human health. Int. J. Med. Mushr., 14(2): 95–134.

Dasar Agro-Makanan Negara 2011–2020. (2011). Kementerian Pertanian dan Industri Asas Tani, Percetakan Watan Sdn. Bhd.

Dawson, G.W.H. (1997). Endau Rompin, a Malaysian Heritage, Malayan Nature Society.

Elisashvili, V. (2012). Submerged cultivation of medicinal mushrooms: Bioprocesses and products. Int. J. Med. Mushr., 14: 211–239. 10.1615/IntJMedMushr.v14.i3.10.

Finimundy, T., Gambato, G., Fontana, R., Camassola, M., Salvador, M. et al. (2013). Aqueous extracts of *Lentinula edodes* and *Pleurotus sajor-caju* exhibit high antioxidant capability and promising *in vitro* antitumor activity. Nutr. Res., 33(1): 76–84.

Fung, S.Y. and Tan, C.S. (2017). The bioactivity of tiger milk mushroom: Malaysia's prized medicinal mushroom. pp. 111–133. *In*: Agrawal, D., Tsay, H.S., Shyur, L.F., Wu, Y.C. and Wang, S.Y. (eds.). Medicinal Plants and Fungi: Recent Advances in Research and Development. Medicinal and Aromatic Plants of the World, vol. 4, Springer, Singapore. 10.1007/978-981-10-5978-0_5.

Hand, P., Hayes, W.A., Satchell, J.E., Frankland, J.C., Edwards, C.A. and Neuhauser, E.F. (1988). The vermicomposting of cow slurry. Earthworms Waste Environ. Manage., 31: 49–63.

Jordan, S.N., Mullen, G.J. and Murphy, M.C. (2008). Composition variability of spent mushroom compost in Ireland. Biores. Technol., 99: 411–418.

Leifa, F., Pan, H., Soccol, A.T., Pandey, A. and Soccol, C.R. (2006). Advances in mushroom research in the last decade. Food Technol. Biotechnol., 44: 303–311.

Letti, L., Vítola, F., Pereira, G., Karp, S., Medeiros, A.B.P. et al. (2018). Solid-state fermentation for the production of mushrooms. pp. 285–318. *In*: Pandey, A., Larroche, C. and Soccol, C.R. (eds.). Current Developments in Biotechnology and Bioengineering. Current Advances in Solid-state Fermentation, Elsevier, B.V.

Ma, G., Yang, W., Zhao, L., Pei, F., Fang, D. and Hu, Q. (2018). A critical review on the health promoting effects of mushrooms nutraceuticals. Food Sci. Hum. Wellness, 7(2): 125–133.

Miles, P.G. and Chang, S.T. (1997). Mushroom biology: Concise basics and current developments. World Scientific, p. 216.

Moore, D. and Chiu, S.W. (2001). Filamentous fungi as food. pp. 223–251. *In*: Pointing, S.B. and Hyde, K.D. (eds.). Bio-exploitation of Filamentous Fungi. Fungal Diversity Research Series # 6, Fungal Diversity Press, Hong Kong.

Rinker, D.L. (2017). Spent mushroom substrate uses. pp. 427–454. *In*: Diego, C.Z. and Pardo-Giménez, A. (eds.). Edible and Medicinal Mushrooms: Technology and Applications, John Wiley and Sons Ltd. 10.1002/9781119149446.ch20.

Valverde, M.E., Hernández-Pérez, T. and Paredes-López, O. (2015). Edible mushrooms: Improving human health and promoting quality life. Int. J. Microbiol., 2015: 376387. 10.1155/2015/376387.

Vikineswary, S., Norlidah, A., Normah, I., Tan, Y.H., Fauzi, D. and Jones, E.B.G. (2007). Edible and medicinal mushroom. pp. 287–305. *In*: Malaysian Fungal Diversity, NRE Malaysia.

Yu, S., Weaver, V., Martin, K. and Cantorna, M.T. (2009). The effects of whole mushrooms during inflammation. BMC Immunology, 10(1): 12. 10.1186/1471-2172-10-12.

Zhang, L., Fan, C., Liu, S., Zang, Z., Jiao, L. and Zhang, L. (2011a). Chemical composition and antitumor activity of polysaccharide from *Inonotus obliquus*. J. Med. Pl. Res., 5(7): 1251–1260.

Zhang, Y., Li, S., Wang, X., Zhang, L. and Cheung, P.C. (2011b). Advances in lentinan: Isolation, structure, chain conformation and bioactivities. Food Hydrocolloids, 25(2): 196–206.

Industrial Applications

Bioprospecting Macrofungi for Biotechnological and Biomedical Applications

Shin-Yee Fung* and Muhammad Fazril Razif*

1. INTRODUCTION

Mushrooms (also called *mushrom, mushrum, muscheron, mousheroms, mussheron* or *musserouns*) (Ramsbottom, 1954) are ubiquitous and diverse in nature. Basidiomycetes, a large and diverse phylum in the fungal kingdom, includes jelly and shelf fungi, mushrooms, puffballs, stinkhorns, certain yeasts, rusts and smuts. This phylum, composed of numerous mushroom species that comprise a diverse range of metabolites, which have nutraceutical, therapeutic and industrial significance. These compounds are valuable for use in many industries with several applications, including human nutrition and medication. There have been increasing scientific explorations towards the therapeutic, nutritional and industrial applications of mushrooms worldwide. Hyde et al. (2019) described 50 ways by which fungi could be exploited for use in a variety of ways, including to combat human and plant diseases, enhance crops and forestry, used in food and beverages industry, in planet conservation and in production of other commodities. These ingenious uses have positioned fungi in a prominent spot for continuing research and development. Being nature's most accomplished chemist, fungi contain a plethora of nutraceutical compounds, such as proteins, carbohydrates, fats, essential amino acids, vitamins, minerals, nucleic acid and nucleotide derivatives. From these, therapeutic metabolites are often attributed to their polysaccharides, terpenoids, glycoproteins, polyketides, alkaloids, flavonoids, saponins, tannins and anthraquinones.

Medicinal Mushroom Research Group (MMRG), Department of Molecular Medicine, Faculty of Medicine, University of Malaya, 50603 Kuala Lumpur, Malaysia.
* Corresponding authors: syfung@ummc.edu.my; fazril.razif@um.edu.my

Polysaccharides of mushrooms have different compositions with most belonging to the β-d-glucans group which has β-(1–3) linkages in the main chain of glucan and β-(1–6) in the branch points, most of which have high molecular weight. For example, a heteroglycan (molecular weight ranging from 5.82×10^5 Da to 3.74×10^6 Da) from *Tremella fuciformis* (known as 'Yiner' or 'Baimuer' in China) was shown to have immunomodulatory, antitumor, antioxidation, antiaging, hypoglycemic, hypolipidemic and neuroprotective effects (Yang et al., 2019). Polysaccharopeptides from *Trametes versicolor* (also known as *Coriolus versicolor* and *Polystictus versicolor*) have been commercialised and are recognised as 'biological response modifiers' (BRM) and as useful adjuncts to conventional therapy for cancer and other diseases. Polysaccharopeptide Krestin (PSK) and polysaccharopeptide (PSP) are products obtained from the extraction of *T. versicolor* mycelia. Both products have similar physiological activities, but are structurally different. The PSK (Japan) and PSP (China) are produced from CM-101 and Cov-1 strains of *T. versicolor*, respectively (Cui and Chisti, 2003). The PSK as well as PSP have a molar mass of approximately 100 kDa (Yang et al., 1992; Ng, 1998) and these polysaccharopeptides appear to be resistant to enzymatic proteolysis (Hotta et al., 1981).

Terpenoids are naturally occurring bioactive metabolites that are produced by many higher fungi. They are formed from C5 isoprene units, leading to their characteristic branched chain structure and are divided into families on the basis of the number of isoprene units from which they are formed. They appear as monoterpenoids (C10), sesquiterpenoid (C15), diterpenoids (C20), sesterterpenoids (C25), triterpenoids (C30) and carotenoids (C40) with interesting biological activities, such as anticancer (Arpha et al., 2012; Wang et al., 2012; Kim et al., 2013; Wang et al., 2013a, b), antimalarial (Isaka et al., 2013), anticholinesterase (Lee et al., 2011), antiviral (Mothana et al., 2003), antibacterial (Shibata et al., 1998; Arpha et al., 2012) and anti-inflammatory activities (Kamo et al., 2004). Sterols, essential eukaryotic constituents, are biosynthesised through triterpenoids (Dewick, 2002). Duru and Chayan (2015) reviewed a total of 285 terpenoids consisting of 5 mono-, 70 sesqui-, 44 di-, 166 tri-terpenoids from a variety of mushroom species linking them to their biological activities with different potencies.

It is interesting to note that polysaccharides and proteins play diverse roles in activating the innate and adaptive immunity. For instance, LZ-8 protein from Reishi (*G. lucidum*) can activate murine macrophages and T lymphocytes but its polysaccharides only activate the former. Hence, many bioactive proteins and peptides from mushrooms, including lectins, fungal immunomodulatory proteins (FIP), ribosome inactivating proteins (RIP), antimicrobial/antifungal proteins, ribonucleases, and laccases (Xu et al., 2011) have been investigated for their role in antitumor, antiviral, antimicrobial, antioxidative and immunomodulatory activities. Lectins, for instance, are non-immune glycoproteins that bind specifically to cell-surface carbohydrates, with the ability to initiate cell agglutination and has been reported to promote antiproliferation, antitumor and immunomodulatory activities (Liu et al., 2006; Zhang et al., 2009; Pusparajah et al., 2016; Cheong et al., 2019). FIPs from mushrooms, some of which are also in glycated forms (Sheu et al., 2009), play important synergistic roles in antitumor and immunomodulation. Organic extracts of mushrooms usually contain polyketides, alkaloids, flavonoids, saponins, tannins and anthraquinones. These compounds have been shown to possess antioxidant capability amongst other bioactivities. Considering the industrial applications of mushrooms, this chapter addresses cultivation of mushrooms, applications in foods, beverages, nutraceuticals, cosmeceuticals, biomedical and other biotechnological pursuits.

2. Wild vs. Cultivated Mushrooms

The increasing use of mushrooms as food and for medicinal purposes has coerced its intentional cultivation. *Auricularia auricula-judae, Flammulina velutipes, Lentinula edodes, Tremella fuciformis* and *Volvariella volvacea* were first cultivated in China, whilst the *Agaricus bisporus* was the major commercially cultivated mushroom in France (Chang and Miles, 2004). The development of mushroom cultivation technology has spread from France to other European countries, North America and other parts of the world towards increased mushroom production to meet the consumer demand.

Much of the nutrition and therapeutic benefits of mushrooms have been documented in scientific journals, commercial write-ups, on various websites and social media platforms earning mushroom the title of 'super food' to help achieve optimal health for life. Xue and O'Brien (2003) have described the use of mushrooms as diet therapy to sustain or to improve health and/or to treat illness in the imperial court of China 2,000 years ago. The term 'immunomodulatory food' was also coined in reference to mushrooms for their nutritional qualities and tonic effects acting as nutraceuticals (Chang and Buswell, 1996). The cultivation of selected mushrooms (*Agaricus bisporus, Agaricus brasiliensis, Ganoderma lucidum, Lentinula edodes, Pleurotus pulmonarius* var. *stechangii* and *Volvariella volvacea*) have been described at length by Chang and Wasser (2017). The authors also ascertained that in recent years, many other species have been cultivated on a larger scale.

In Malaysia, for instance, the once uncultivated medicinal mushroom, *Lignosus rhinocerus* or locally known as tiger milk mushroom or '*cendawan susu harimau*' was successfully mass cultivated by a local group of scientists and a local company LiGNO Biotech Sdn. Bhd. by using a proprietary solid-state fermentation (SSF) technology with rice-based media. It is a specially formulated culture medium consisting of water and other food-based materials. The cultivation, using standard, sterile and hygienic protocol, is done in a controlled environment and harvested in optimum condition to produce a consistent supply of the *L. rhinocerus* sclerotia.

This specific cultivation has successfully addressed problems related to irregularities in supply, coupled with the inconsistency in quality and nutritional (medicinal) content of the sclerotium (which is highly dependent on the harvesting conditions) along with scarcity of the tiger milk mushroom due to changing weather patterns and growth environment (Fung and Tan, 2017). The cultivated *L. rhinocerus* TM02® are authenticated by using standard genetic markers, targeting the internal transcribed spacer (ITS) regions of the ribosomal RNA to ensure that it is identical to the wild type (Tan et al., 2010). The Medicinal Mushroom Research Group (MMRG) from University of Malaya has also collaborated with the local small and medium-sized enterprise (SME), LiGNO Biotech Sdn. Bhd. to perform safety studies to ascertain the innocuous nature of the cultivated species (Lee et al., 2011, 2013).

Fig. 1. Cultivation facility of a Malaysian SME: LiGNO Biotech Sdn. Bhd. is the largest mass cultivator using proprietary SSF technology in an indoor, controlled system for medicinal mushrooms in Malaysia; (*top*) LiGNO Biotech Sdn. Bhd.; (*middle*) Indoor controlled system; (*below*) Visitor Atrium for education on Tiger milk mushroom.

The proprietary SSF cultivation technology by this Malaysian SME has created an avenue for other medicinal mushrooms to be cultivated. Other prized medicinal mushrooms, such as *Antrodia camphorata, Inonotus obliquus* (Chaga), *Ophiocordyceps sinensis* and *Phellinus linteus* (sanghwang) are also being mass cultivated, using sterile and hygienic protocols in a controlled environment and harvested in an optimum condition. This technique of cultivation has enabled consistency in all its produce and given rise to consistent amounts of biomolecules, leading to consistencies in bioactive potential. This route of cultivation-commercialisation has taken mushroom cultivation to a whole new level and an emerging commodity, not only for economic purposes, but also to improve academia-industry collaborations to enhance the research, growth and development of fungal culture in Malaysia.

3. The Multifaceted Applications of Mushrooms

The worldwide production of mushrooms has been increasing by 10–20 per cent year-on-year. The major cultivated species of mushrooms accounting for 90 per cent of worldwide production include *Agaricus bisporus, Flammulina velutipes, Ganoderma lucidum, Lentinula edodes* and *Pleurotus eryngii.* There are numerous bioactive compounds that can be found in mushrooms, hence it is not surprising that various mushroom products, including pharmaceuticals, functional foods, cosmeceutical and nutraceuticals are currently available in the market.

3.1 Foods and Beverages

Mushroom is a common staple in vegetarian, vegan and meat-based products due to its taste and nutritional content that is rich in essential amino acids, vitamins, minerals and trace elements. Mushrooms are also high in fibre, low in fat and starch, and considered to be ideal ingredients in a healthy diet. Among the 3,000 edible species of mushrooms, only ~ 100 species are cultivated commercially and only ~ 25 are widely accepted as food (Valverde et al., 2015). Carbohydrates constitute the bulk of the nutritional composition of mushrooms, accounting for 50–65 per cent (dry weight) of the fruit body. Depending on the species of mushroom, the carbohydrate constituents can include free sugars, fructose, glucose, glycogen, hemicellulose, mannitol, raffinose, sucrose and xylose (Wannet et al., 2000; Singh and Singh, 2002). With regards to protein content, the general composition is dependent on the species of mushroom, the size of its pileus, the substratum and harvest time. Mushrooms generally have higher protein content than most vegetables, with the protein content varying between 14–42 per cent (dry weight). One of the benefits of mushroom consumption is that they can contain all the essential amino acids required by human diets (Bach et al., 2017).

The fat content in fresh mushrooms varies, ranging between 1.75–15.5 per cent (per 100 g dry weight). Among these, unsaturated fatty acids are predominantly found, with the main fatty acid in the fruiting bodies of all mushrooms being cis-linoleic acid (18:2), ranging between 22.39–65.29 per cent. The other fatty acids include cis-oleic, palmitic and stearic acids (Pedneault et al., 2006; Yilmaz et al., 2006). Besides fats, the major mineral components found in mushrooms are Ca, K, Mg, Na, P and Se with trace amounts of Cu, Cd, Fe, Mo and Zn. It is interesting to note that the mineral contents in mushrooms can be altered by fortifying the substratum used in their cultivation stage.

In recent years, mushrooms have been incorporated into various food products and marketed as functional foods. A mix of *Auricularia auricula* flour with wheat dough was found to enhance the water-holding capacity of the dough (Yuan et al., 2017). Certain mushrooms, including *Schizophyllum commune* possess enzymes (e.g., lactate dehydrogenase) and proteases that can be used in cheese production and milk clotting. The latter is also used for the production of protein hydrolysates (flavour enhancers). In addition, mushrooms have been found to contain amylase, which can be utilised in the production of fermented food products (Fukuda et al., 1999; Okamura et al., 2000). Some species of mushrooms have been used in the production of alcoholic beverages owing to the presence of enzyme alcohol dehydrogenase (Okamura-Matsui et al., 2003). *Agaricus blazei, Pleurotus ostreatus* and *Tricholoma matsutake* were found to produce high concentrations of alcohol (2648 mM, 12.2 per cent).

Mushroom polysaccharides have drawn attention due to their bioactive properties. The oral intake of mushroom polysaccharide is safe as well as efficacious. Mushroom polysaccharides have the potential to be used as prebiotics to enhance the overall intestinal health. Chou et al. (2013) found that mushroom polysaccharides from *Flammulina velutipes, Lentinula edodes* and *Pleurotus eryngii* have a protective effect on probiotics in yoghurt. These polysaccharidess (0.1–0.5 per cent) were found to be effective in maintaining probiotics survival with synergistic action complementing peptides and amino acids in yoghurt. Furthermore, administration of mushroom polysaccharides extracted from *Ganoderma lucidum, Hericium erinaceus* and *Lentinula edodes* improved gut microbiota and prevented intestinal diseases, besides showing insulin resistance (Nie et al., 2017).

Song et al. (2020) recently described the development of *Cordyceps* coffee, which contains cordycepin and β-glucan that improve the quality and nutritional functionality of coffee. Green coffee beans were mixed with extracts from three medicinal mushrooms, *Cordyceps militaris, Phellinus linteus* and *Inonotus obliquus.* The *Cordyceps* coffee (37.34 mg GAE/g) was found to have a higher total polyphenol content than *Cordyceps* alone (23.17 mg GAE/g). Cordycepin, β-glucan, flavonoids and polyphenols were also detected in the *Cordyceps* coffee, contributing to an additional enrichment of the product. Cordycepin is known to possess antitumor, immunomodulation, anti-inflammatory and antiobesity effects (Wu et al., 2006; Wu et al., 2007; Wong et al., 2010; Patel and Ingalhalli, 2013; Li et al., 2018).

3.2 Cosmeceuticals and Nutraceuticals

Cosmeceuticals are products with bioactive ingredients that produce pharmaceutical action on skin (Reed, 1964). On the other hand, Stephen L. Defelice (1989), the person who coined the term nutraceuticals, defines it as *"any substance that is a food or part of a food and provides medicinal or health benefits, including the prevention and treatment of disease."* The recent linkage between body-/skin-health and nutrition has resulted in increased research to examine the molecular mechanisms of the action of mushrooms and its extracts when applied topically or consumed orally (cosmenutraceuticals) (Hyde et al., 2010). Various studies have shown that mushrooms are rich in glycoproteins, antioxidants and enzymes that are beneficial for overall health.

With regards to cosmeceuticals, studies have revealed that mushrooms possess bioactive metabolites with anticollagenase, antielastase, antihyaluronidase and antityrosinase activities. For instance, the antioxidant activity of polysaccharides derived from black fungus has resulted in a decrease in skin lipofuscin via increasing hydroxyproline (HYP) content (Peng et al., 2010). Polysaccharides from *Tremella fuciformis* has been shown to reduce transepidermal water loss (TEWL) and enhance type I collagen synthesis (Wen et al., 2016). It has been reported that polysaccharides from *Grifola frondosa* can inhibit skin melanin biosynthesis (Kim et al., 2007). Ganodermanondiol from *Ganoderma lucidum* was found to inhibit the activity and expression of cellular tyrosinase and the expression of tyrosinase-related protein-1 (TRP-1) and TRP-2 that cause melanin production. Kim et al. (2014) reported that MMP levels were reduced upon treatment with *T. matsutake* (*Tricholomataceae*) mycelium extract, inhibiting elastase activity. A study by Kawagishi et al. (2002) discovered a novel hydroquinone, (E)-2-(4-hydroxy-3-methyl-2-butenyl)-hydroquinone from the mushroom *Piptoporus betulinus* that acts as an MMP inhibitor. Hydroquinone acts as a skin depigmentation agent by inhibiting melanin synthesis. Liao et al. (2014) revealed that the water extracts of *Auricularia fuscosuccinea* possessed moisture-retention capacity equally as potent as sodium hyaluronate. Hence, many cosmetic products are formulated with mushroom extracts in the hope of delaying skin-aging and improving skin health. However, scientific data on the effectiveness of these cosmeceutical formulations has yet to be made.

The mushroom nutraceutical market has grown tremendously over the past several years due to the increasing characterisation of their bioactive compounds. Mushroom nutraceuticals typically contain extracts or fractions derived from the mycelium or fruiting bodies to be consumed as a capsule and tablet or to drink (cosme-nutraceuticals). Most mushroom nutraceuticals are made from edible mushroom species, hence toxicity has not been an issue. Numerous researches have revealed that mushrooms are able to produce antioxidant, antimicrobial, hepatoprotective, hypoglycemic, immunomodulating and prebiotic compounds. Mushroom compounds that are classified as BRMs, include β-glucans, heteropolysaccharides, flavonoids, phenols, sterols and terpenes (Giavasis, 2013; Mizuno and Nishitani, 2013). The term is derived based on their ability to exert numerous biological effects in the human body, either directly or indirectly (Giavasis and Biliaderis, 2006; Sullivan et al., 2006). However, despite these discoveries, concerns regarding the application of these molecules as nutraceuticals remain, as issues such as dosage (i.e., appropriate concentration), bioavailability, toxicity, stability and molecular diversity of these compounds remain unanswered.

There are a large number of *in vitro* and *in vivo* studies using various human cell lines (normal cancer and others) and rodent models (delivered orally or via intravenous injection) that explore and verify the bioactivities of these mushroom compounds. However, only a handful of clinical studies have been conducted. These studies vary in the source and starting material used, either as extract, fraction or purified active substance. Due to this dissimilarity, it is generally understood that only research on pure compounds can deduce the true bioactivity of specific molecules and their mode of action. Nonetheless, from a holistic standpoint, because mushrooms contain several bioactive ingredients, some researchers believe that the synergistic effects of these compounds should not be disregarded and that the compounds should be consumed as a collective, rather than individually. Similarly, the formulation of these nutraceuticals may alter the bioactivity of the mushroom compounds, depending on the ingredients incorporated. Consequently, clinical trials should be performed on the consumption of these nutraceuticals to fully understand their potential therapeutic effects, providing scientific validation of their health claims. Also, toxicity studies in humans must be performed to determine the maximum allowable dose that can be consumed and to avoid any unforeseen side effects even if taken only as a supplement.

Table 1. Cosmetic products containing mushroom currently available in the market.

Mushroom	Commercial product name
Sparassis crispa (Cauliflower fungus)	Acwell Betaglution Ultra Moisture Toner Some By Mi Galactomyces Pure Vitamin C Glow Serum
Agaricus bisporus (Common)	Zelens Provitamin D Treatment Drops it Cosmetics Bye Bye Redness Sensitive Skin Moisturiser
Inonotus obliquus (Chaga)	Blithe Tundra Chaga Pressed Serum Dr Dennis Gross B$_3$ Adaptive Superfoods Stress Rescue Super Serum
Tricholoma matsutake (Matsutake)	Dr. Jart+ V7 Serum The Face Shop Yehwadam Heaven Grade Ginseng Collection
Ganoderma lucidum (Lingzhi)	Aveeno Positively Ageless® Collection CV Skin Labs Calming Moisture
Phellinus linteus (Sanghwang)	Clinelle Purifying Gel Moisturiser Nature Republic Collagen Dream 70 Collection
Tremella fuciformis (Snow fungus)	Tarte Face Tape Foundation The Inkey List Snow Mushroom
Cordyceps sinensis	Bobbi Brown Intensive Skin Serum Foundation SPF40 Estee Lauder Re-Nutriv Ultimate Diamond Transformative Energy Creme
Lentinus edodes (Shiitake)	Murad Invisiblur Perfecting Shield Broad Spectrum SPF 30 Yves Saint Laurent Blanc Pur Couture Brightening Cream
Grifola frondosa (Hen-of-the-wood)	Helena Rubinstein Collagenist Re-Plump Day Cream Dry Skin SPF 15 Lancôme Blanc Expert Skin Tone Brightening Emulsion
Trametes versicolor (Turkey tail)	Clinique Even Better Clinical™ Radical Dark Spot Corrector + Interrupter Dermalogica Invisible Physical Defence Sunscreen Spf 30
Pleurotus ferulae (White Elf)	CLIV Ginseng Berry Premium Ampoule PureHeal's Ginseng Berry 80 Eye Essence
Agaricus blazei (Almond)	May Coop Raw Activator Neogen Blueberry Real Fresh Foam Cleanser
Hericium erinaceum (Lion's Mane)	The Plant Base Time Stop Collagen Ampoule Yuripibu Cucu Black Truffle Serum
Schizophyllum commune (Split Gills)	it Cosmetics Secret Sauce Clinically Advanced Miraculous Anti-aging Moisturiser OSSOLA Skincare the Turmeric Emulsion

Some mushroom-based nutraceuticals include 'Nature's Answer, Reishi, 1,000 mg Capsules' (120 mg *G. Lucidum* fruit body extract with 10 per cent polysaccharides; 880 mg *G. lucidum* whole mushroom), which is a nutritional supplement for liver health and healthy function of the immune system; the 'Lingzhi Creacked Spores Plus Capsules' (240 mg Cracked Spores *G. Lucidum* and 60 mg Sporophore *G. Lucidum* concentrate) taken twice daily for health and general well-being; the 'Zhou Nutrition, Mushroom 8-Plex Powder' (250 mg *I. obliquus*, 250 mg *G. frondosa*, 250 mg *G. lucidum*, 250 mg *A. blazei*, 250 mg *L. edodes*, 250 mg, *C. versicolor*, 250 mg, *C. militaris*, 250 mg *H. erinaceus*), which claims to support immune health, brain power, energy and endurance; and the 'KIKI Health Organic Multi-Mushroom 8 Extract Blend' (*C. militaris* extract, *H. erinaceus* extract, *I. obliquus* extract, *G. lucidum* extract, *L. edodes* extract, *A. brasiliensis* extract,

G. frondosa extract, *A. auricula-judae* extract), which contains 50 per cent β-glucan rich polysaccharides extracted by using enzymatic preparations and temperature control to produce substantial quantity of plant sterols and triterpenes and to preserve the bio-effectiveness of the active compounds.

3.3 *Biotechnological Applications*

Explorations into the application of mushrooms and their by-products in various industrial sectors have also increased over the past decade. With the need to manufacture eco-friendly, biodegradable packaging to reduce environmental pollution and improve the waste management, a novel type of packaging made from mushrooms has been developed. The products include Myco Foam and Ecocradle. Myco Foam was developed by Ecovative (USA) in 2006. It involves the incorporation of sterilised agricultural waste, such as straw, corn husks or lentil pods mixed with mushroom mycelium. The mycelium subsequently forms a network of fibres entwined within and around the substrate composite, forming a compact structure. The resulting material is subsequently dried to halt the mycelium from further growth. Similarly, Ecocradle was developed by Ecovative Design (USA) in 2010, using a similar concept. The packaging generally takes 30 days to compost. Both Myco Foam and Ecocradle are seen as alternatives or replacements for foam packaging materials made of expanded polystyrene (EPS), polypropylene (EPP) and polyethylene (EPE).

Research into using natural compounds for crop protection is ongoing in the hope of reducing the harmful effects of pesticides on non-target organisms and the environment. To improve pest management, researchers have been developing pesticides using mushroom lectins, a component of the fungal defence mechanism against parasites. Lectins are univalent or polyvalent carbohydrate-binding proteins first discovered by Ford (1910) and found to be associated with the fungal defence system. Mushroom lectins have also been found to act as a storage protein, typically found in the fruiting body. Some lectins are stage-specific and promote mycelial differentiation in fruiting bodies. Lectin, derived from *Xerocomellus chrysenteron* called XCL, was found to be effective against the aphids *Acyrthosiphon pisum* and *Myzuspersicae* (Jaber et al., 2007). Lectin from *Xerocomus chrysenteron* was found to possess insecticidal activity against dipteran *Drosophila melanogaster* and hemipteran *A. pisum* (Trigueros et al., 2003). Lectins from *Agrocybe cylindracea, Boletus edulis, Ganoderma cepense* and *Xylaria hypoxylon* have been recently demonstrated to possess anti-nematode activity against *Dictylenchus dispsaci* and *Heterodera glycines* (Zhao et al., 2009).

Peroxidase [including lignin peroxidase (LiP), manganese peroxidase (MnP) and versatile peroxidase (VP)], oxidoreductases and other oxidases derived from mushrooms are currently investigated for their potential application in waste management. Mushrooms are excellent decomposers as they are able to degrade lignocellulose biomass typically associated with wood and plant materials (Andlar et al., 2018). Lignocellulose is a natural polymer, composed of cellulose, hemicellulose and lignin. One such application is the use of biodegradable portions of agricultural and municipal waste as substrates to grow mushrooms for consumption of animal feed via solid state fermentation. Mushrooms are also used for mycoremediation, with the purpose of partially or completely removing contaminants, such as heavy metals and dyes from polluted soil (Kulshreshtha et al., 2014). Their cost-effectiveness and ability to enhance soil health make them excellent candidates for this purpose. Examples of mushrooms that have the ability to degrade or absorb contaminants and pollutants include *Pleurotus ostreatus* for green polyethylene and plastics (Da Luz et al., 2013), *Jelly* sp. and *Schizophyllum commune* for malachite green (Yogita et al., 2011), *Pleurotus palmonarius* for crude oil (Chiu et al., 2009), *Pleurotus chrysosporium* for styrene (Lee et al., 2006), *Agaricus bisporus* for cadmium biosorption (Vimala and Das, 2009) and *Fomes fasciatus* for copper biosorption (Sutherland and Venkobachar, 2013).

Mushroom supplementation is also exploited for use in the agricultural industry. For example, enhancement of serum lysozyme activity and phagocytic activity in kelp grouper fish via diet enrichment with *Phellinus linteus* extract has increased their resistance against certain pathogens (Harikrishnan et al., 2011). In addition, enhancement of glutathione peroxidase, phenoloxidase and superoxide dismutase activities in white shrimp was achieved by using feed supplemented with *Hericium erinaceum* (Campa-Córdova et al., 2002). Extract from *Lentinula edodes* has been used to enhance the immune system of chickens via direct activation of macrophages and lymphocytes (Ullah et al., 2017).

3.4 *Biomedical Applications*

Mushrooms have been used in the development of numerous biomedical and health-related products. Mushroom polysaccharides, including lentinan, polysaccharide K (Kerstin) and polysaccharide peptide (PSP) are currently available on the pharmaceutical market. Lentinan, an immunomodulating glucan derived from the fruiting bodies of *Lentinus edodes*, is the first mushroom polysaccharide that has been applied in clinical therapy. It has been used in treating patients with gastric cancers and to enhance impairment of natural killer cell activity after cardiopulmonary bypass (Wang et al., 2017). Lentinan-based drugs are available as capsules, injections and tablets and are clinically used for the treatment of various cancers, including colorectal cancer, gastric cancer and lung cancer (Maekawa et al., 1990; Tanabe et al., 1990; Ogawa et al., 1994; Mio et al., 1994; Fujimoto et al., 2006). Mushroom polysaccharides have also been used for the treatment of gastric ulcers and inflammation (He et al., 2017). As a commercial product, PSP from the strains of 'COV-1' (China) and PSK from the strain 'CM-101' (Japan) have been approved as adjuvants in cancer therapy (Habtemarium, 2020). A study examining adjuvant immunochemotherapy with oral Tegafur/Uracil plus PSK in patients with stage II or III colorectal cancer by Oshwada et al. (2004) found that patients with stage II and III colorectal cancer exhibited reduced recurrence and increased survival in stage III. In another study, *Astragalus* polysaccharides (APS) independently induced apoptosis and strengthened the proapoptotic effect of Adriamycin (0.1 µg/ml) on gastric cancer cells, suggesting that APS may act as a chemotherapeutic sensitiser (Song et al., 2020). Other fungal β-glucans, such as schizophyllan from *Schizophyllum commune*, pleuran from *Pleurotus chrysosporium ostreatus*, and ganoderan from *Ganoderma lucidum* have also been extensively studied due to their ability to stimulate the immune system, with antioxidant, anticancer and antitumor properties. *P. umbellatus* polysaccharide (PUPS), known for its anti-tumor and anti-inflammatory activities, has been approved by the Chinese Food and Drug Administration (CFDA) to be used alone or in combination with a variety of clinical drugs for the treatment of Hepatitis B, lung and liver cancers (Guo et al., 2019). Clinical trials pertaining to Hepatitis B treatment have revealed that combined therapies for PUPS with acyclovir, interferon or Hepatitis B vaccine are better than when treated with either drugs alone. In 2000, *G. lucidum* polysaccharide (GLPS) was developed into a drug and approved by the Chinese FDA for the treatment of atrophic myotonica, dermatomyositis, muscular dystrophy, neurosis, polymyositis and refractory myopathy. Results of a study by Gao et al. (2009) where 34 advance-stage cancer patients were treated with 1800 mg Ganopoly® (product containing *G. lucidum* polysaccharides) three times daily orally before meals for 12 weeks, found significant increase in (i) mean plasma concentrations of interleukin (IL-2), IL-6, and interferon (IFN)-γ, (ii) mean absolute number of CD56+ cells and (iii) mean natural killer cell activity. It is also recognised for its ability to reduce adverse reactions and improve the quality of life for cancer patients during chemotherapy.

Another compound, cordycepin, a derivative of the nucleoside adenosine, was initially extracted from the *Cordyceps* species but is now produced synthetically. The two main effects of cordycepin are inhibition of polyadenylation and activation of AMP-activated kinase pathway (Wong et al., 2010). *Cordyceps sinensis* (Cs-4), an experimentally developed strain has been used extensively in China with the approval of the Chinese FDA (Zhu et al., 1998). Products derived from this strain have been clinically tested against many diseases.

Mushroom chitin is being developed for application in wound healing. Wound dressings generally possess good barrier properties and are permeable to oxygen. However, there is a need for enhanced wound dressing options that are more biocompatible and can actively accelerate healing as issues such as irritation still persists with conventional dressings. Chitin, a linear macromolecule composed of *N*-acetylglucosamine and its derivative chitosan, is a great polymer to be used to synthesise nanofibres which mimic the skin's natural extracellular matrix (Croisier and Jérôme, 2013). The benefits of mushroom chitin are that its structure contains more pliable branched β-glucan and does not require additional treatments (i.e., acid treatment, demineralisation) prior to its use. Sacchachitin, derived from *G. tsugae*, was the first commercialised, fungi-derived wound-dressing material whereby human trials revealed improvement in healing of chronic wounds, increased proliferation of keratinocytes and fibroblast, and matrix metalloproteinases (MMPs) activity (Hung et al., 2001).

Mushroom lectins are also being developed for application in diagnostic technologies. Fungal lectins are able to bind with high specificity to epitopes present on human glycoconjugates. Mushroom lectins can be used to determine blood type due to the specificity of carbohydrate structures present on the cell surface of erythrocytes. Lectin from *T. clypeatus* can agglutinate to human type-A erythrocytes; *C. atramentarius* lectin was found to agglutinate with type-O erythrocytes; while *L. squarrosulus* lectin agglutinated to both type-B and -AB erythrocytes (Gorakshakar and Ghosh, 2016). Also, changes in glycosylation are associated with certain pathological conditions, such as inflammation and autoimmune disorders. For example, lectin PVL from *Lacrymaria velutina* has been used to detect IgG in rheumatoid arthritis patients via ELISA-based assays (Tsuchiya et al., 1993) or affinity biosensor technology (Liljeblad et al., 2000). They are now included in lectin microarrays for high-sensitivity profiling of glycomes.

The mechanisms of action of many bioactive compounds found in mushrooms are still being elucidated as numerous biochemical pathways remain to be studied. There is an urgent need for translational research and clinical trials to determine the effectiveness of these compounds, as many products extrapolate *in vitro* information to imply therapeutic importance. Through many show potent pharmacological actions, the mechanism behind their effects on a molecular level in humans remain unclear.

4. Conclusion and Future Prospects

A paradigm shift towards the utilisation of mushrooms and their bioactive compounds as components for wellness is fast becoming a reality. Mushrooms are deemed as healthy 'super foods' for what they do not contribute to the diet. They are deemed wholesome, nutritious for the diet as they do not contain cholesterol, gluten and are low in fat, sugars, sodium and calories. These are not the only reasons why mushrooms should be on everyone's list of food to eat for they also consist of bioactive compounds, such as protein and essential amino acids, polysaccharides and other small yet significant biomolecules. Not only are mushrooms good for the palate and our body, abundant studies have shown that they have limitless potential in the field of cosmetics, biotechnology and assist the environment as part of the ecosystem as decomposers, cleansers and allies of the environment and play an essential role in microforestry. Mushrooms with key biomolecules can and will some day be the key to a subtler, but no less important, revolution through which we can fulfil the admonition of Hippocrates to 'let food be thy medicine'. Paul Edward Stamets, a mycologist and an advocate of medicinal fungi and mycoremediation, summarised it well in a TED talk he gave in 2008 entitled *6 Ways Mushrooms can Save the world* (https://www.ted.com/talks/paul_stamets_6_ways_mushrooms_can_save_ the_world/up-next) where he shared about a revolution in how we view mushrooms. In the words of Daniel Christian Wahl, in his interview with P.E. Stamets in 2010: "...(we) need to shed all remnants of mycophobia and embrace these important allies in healing ourselves and the soils, forests, streams as we go about regenerating the Earth and her people" (adapted from https://medium.com/age-of-awareness/allies-of-the-regeneration-paul-stamets-speaks-for-the-fungi-1c701b1863d7).

References

Andlar, M., Rezic, T., Mardetko, N., Kracher, D., Ludwig, R. and Šantek, B. (2018). Lignocellulose degradation: An overview of fungi and fungal enzymes involved in lignocellulose degradation. Eng. Life Sci., 18(11): 1–11. 10.1002/elsc.201800039.

Arpha, K., Phosri, C., Suwannasai, N., Mongkolthanaruk, W. and Sodngam, S. (2012). Astraodoric acids A-D: New lanostane triterpenes from edible mushroom *Astraeus odoratus* and their anti-*Mycobacterium tuberculosis* H37Ra and cytotoxic activity. J. Agric. Food Chem., 60: 9834–9841.

Bach, F., Helm, C.V., Bellettini, M.B., Maciel, G.M., Windson, C. and Haminiuk, I. (2017). Edible mushrooms: A potential source of essential amino acids, glucans and minerals. Int. J. Food Sci. Technol., 52(11): 10.1111/ijfs.13522.

Campa-Córdova, A.I., Hernández-Saavedra, N.Y. and Ascencio, F. (2002). Superoxide dismutase as modulator of immune function in American white shrimp (*Litopenaeus vannamei*). Comp. Biochem. Physiol. Toxicol. Pharmacol., 133(4): 557–565.

Chang, S. and Wasser, S. (2017). The Cultivation and Environmental Impact of Mushrooms, Oxford Research Encyclopedia of Environmental Science (retrieved on July 18, 2020: https://oxfordre. com/environmentalscience/view/10.1093/acrefore/9780199389414.001.0001/acrefore-9780199389414-e-231).

Chang, S.T. and Buswell, J.A. (1996). Mushroom nutriceuticals. World J. Microbiol. Biotechnol., 12: 473–476.

Chang, S.T. and Miles, P.G. (2004). Mushroom: Cultivation, Nutritional Value, Medicinal Effect and Environmental Impact (2nd ed.), CRC Press, Boca Raton, FL.

Cheong, P.C.H., Yong, Y.S., Fatima, A., Ng, S.T., Tan, C.S., Kong, B.H., Tan, N.H., Rajarajeswaran, J. and Fung, S.Y. (2019). Cloning, over-expression, purification and modelling of a lectin (Rhinocelectin) with anti-proliferative activity from tiger milk mushroom, *Lignosus rhinocerus*. IUBMB Life, 71(10): 1579–1594.

Chiu, S.W., Gao, T., Chan, C.S. and Ho, C.K. (2009). Removal of spilled petroleum in industrial soils by spent compost of mushroom *Pleurotus pulmonarius*. Chemosphere, 75(6): 837–842.

Chou, W.-T., Sheih, I.-C. and Fang, T.J. (2013). The applications of polysaccharides from various mushroom wastes as prebiotics in different systems. J. Food Sci., 78: M1041–M1048.

Croisier, F. and Jérôme, C. (2013). Chitosan-based biomaterials for tissue engineering. Eur. Polym. J., 49(4): 780–792.

Cui, J. and Chisti, Y. (2003). Polysaccharopeptides of *Coriolus versicolor*: Physiological activity, uses, and production. Biotech. Adv., 21: 109–122.

Da Luz, J.M.R., Paes, S.A., Nunes, M.D., da Silva, M.C.S. and Kasuya, M.C.M. (2013). Degradation of oxo-biodegradable plastic by *Pleurotus ostreatus*. PLoS ONE, 8(8): 69386.

DeFelice, S. (1989). The NutraCeutical Revolution: Fueling a Powerful, New International Market, The Foundation for Innovation in Medicine (retrieved on May 1, 2020: https://fimdefelice.org/library/ the-nutraceutical-revolution-fueling-a-powerful-new-international-market/).

Dewick, P.M. (2002). Medicinal Natural Products, John Wiley & Sons, West Sussex, UK, p. 507.

Duru, M.E. and Çayan, G.T. (2015). Biologically active terpenoids from mushroom origin: A Review. Rec. Nat. Prod., 9: 456–483.

Ford, W.W. (1910). The distribution of haemolysis agglutinins and poisons in fungi, especially the amanitas, the entolomas, the lactarius and the inocybes. J. Pharmacol. Exp. Ther., 2: 285–318.

Fujimoto, K., Tomonaga, M. and Goto, S. (2006). A case of recurrent ovarian cancer successfully treated with adoptive immunotherapy and lentinan. Anticanc. Res., 26(6A): 4015–4018.

Fukuda, S., Okamura, T., Tanaka, M., Sera, M., Takeno, T. and Ohsugi, M. (1999). Screening of lactate dehydrogenase and curd rennet of microorganisms. Bull. Mukogawa Women's Univ. Natl. Sci., 47: 51–56.

Fung, S.Y. and Tan, C.S. (2017). The bioactivity of tiger milk mushroom: Malaysia's prized medicinal mushroom. pp. 111−134. *In*: Agrawal, D.C., Tsay, H.-S., Shyur, L.-F., Wu, Y.-C. and Wang, S.-Y. (eds.). Medicinal Plants and Fungi: Recent Advances in Research and Development, Springer Nature, Singapore.

Gao, Y., Zhou, S., Jiang, W., Huang, M. and Dai, X. (2003). Effects of ganopoly (a *Ganoderma lucidum* polysaccharide extract) on the immune functions in advanced-stage cancer patients. Immunol. Invest., 32(3): 201–215.

Giavasis, I. and Biliaderis, C. (2006). Microbial polysaccharides. pp. 167–214. *In*: Biliaderis, C. and Izydorczyk, M. (eds.). Functional Food Carbohydrates. CRC Press, Boca Raton.

Giavasis, I. (2014). Bioactive fungal polysaccharides as potential functional ingredients in food and nutraceuticals. Curr. Opin. Biotechnol., 26C: 162–173.

Gorakshakar, A.C. and Ghosh, K. (2016). Use of lectins in immunohematology. Asian J. Trans. Sci., 10(1): 12–21.

Guo, Z., Zang, Y. and Zhang, L. (2019). The efficacy of *Polyporus umbellatus* polysaccharide in treating hepatitis B in China. Prog. Mol. Biol. Transl. Sci., 163: 329–360.

Habtemariam, S. (2020). *Trametes versicolor* (Synn. *Coriolus versicolor*) polysaccharides in cancer therapy: Targets and efficacy. Biomedicines, 8(5): 135.

Harikrishnan, R., Balasundaram, C. and Heo, M.S. (2011). Diet enriched with mushroom *Phellinus linteus* extract enhances the growth, innate immune response, and disease resistance of kelp grouper, *Epinephelus bruneus* against vibriosis. Fish Shellfish Immunol., 30(1): 128–34.

He, X., Wang, X., Fang, J., Chang, Y., Ning, N., Guo, H., Huang, L., Huang, X. and Zhao, Z. (2017). Polysaccharides in *Grifola frondosa* mushroom and their health promoting properties: A review. Int. J. Biol. Macromol., 910–921.

Hotta, T., Enomoto, A., Yoshikumi, C., Ohara, M. and Ueno, S. (1981). Protein-bound Polysaccharides, US Patent # 4,271,151

Hung, W.S., Fang, C.L., Su, C.H., Lai, W.F., Chang, Y.C. and Tsai, Y.H. (2001). Cytotoxicity and immunogenicity of sacchachitin and its mechanism of action on skin wound healing. J. Biomed. Mater. Res., 56(1): 93–100.

Hyde, K.D., Bahkali, A.H. and Moslem, M.A. (2010). Fungi—An usual source of cosmetics. Fungal Diversity, 42: 1–9.

Hyde, K.D., Xu, J., Rapior, S., Jeewon, R., Lumyong, S. et al. (2019). The amazing potential of fungi: 50 ways we can exploit fungi industrially. Fungal Diversity, 97: 1–136.

Isaka, M., P. Chinthanom, S. Kongthong, K. Srichomthong and R. Choeyklin. (2013). Lanostane triterpenes from cultures of the basidiomycete *Ganoderma orbiforme* BCC 22324. Phytochemistry, 87: 133–139.

Jaber, K., Francis, F., Paquereau, L., Fournier, D. and Haubruge, E. (2007). Effect of a fungal lectin from *Xerocomus chrysenteron* (XCL) on the biological parameters of aphids. Comm. Agric. Appl. Biol. Sci., 72(3): 629–638.

Kamo, T., Imura, Y., Hagio, T., Makabe, H., Shibata, H. and Hirota, M. (2004). Anti-inflammatory cyathane diterpenoids from *Sarcodon scabrosus*. Biosci. Biotech. Biochem., 68: 1362–1365.

Kawagishi, H., Hamajima, K. and Inoue, Y. (2002). Novel hydroquinone as a matrix metalloproteinase inhibitor from the mushroom, *Piptoporus betulinus*. Biosci. Biotechnol. Biochem., 66(12): 2748–2750.

Kim, K.H., Moon, E., Choi, S.U., Kim, S.Y. and Lee, K.R. (2013). Lanostane triterpenoids from the mushroom *Naematoloma fasciculare*. J. Nat. Prod., 76: 845–851.

Kim, S.W., Hwang, H.J., Lee, B.C. and Yun, J.W. (2007). Submerged production and characterization of *Grifola frondosa* polysaccharides—A new application to cosmeceuticals. Food Technol. Biotechnol., 45: 295–305.

Kim, S.Y., Go, K.C., Song, Y.S., Jeong, Y.S., Kim, E.J. and Kim, B.J. (2014). Extract of the mycelium of *T. matsutake* inhibits elastase activity and TPA-induced MMP-1 expression in human fibroblasts. Int. J. Mol. Med., 34: 1613–1621.

Kulshreshtha, S., Mathur, N. and Bhatnagar, P. (2014). Mushroom as a product and their role in mycoremediation. AMB Express, 4: 29.

Lee, I.S., Ahn, B.R., Choi, J.S., Hattori, M., Min, B.S. and Bae, K.H. (2011). Selective cholinesterase inhibition by lanostane triterpenes from fruiting bodies of *Ganoderma lucidum*. Bioorg. Med. Chem. Lett., 21: 6603–6607.

Lee, J.W., Lee, S.M., Hong, E.J., Jeung, E.B., Kang, H.Y., Kim, M.K., Choi, I.G. (2006). Estrogenic reduction of styrene monomer degraded by *Phanerochaete chrysosporium* KFRI 20742. J. Microbiol., 44(2): 177–84.

Lee, S.S., Tan, N.H., Fung, S.Y., Pailoor, J. and Sim, S.M. (2011). Evaluation of the sub-acute toxicity of the sclerotium of *Lignosus rhinocerus* (Cooke), the tiger milk mushroom. J. Ethnopharmacol., 138(1): 192–200.

Lee, S.S., Enchang, F.K., Tan, N.H., Fung, S.Y. and Pailoor, J. (2013). Preclinical toxicological evaluations of the sclerotium of *Lignosus rhinocerus* (Cooke), the tiger milk mushroom. J. Ethnopharmacol., 147(1): 157–163.

Li, Y., Li, Y., Wang, X., Xu, H., Wang, C., An, Y., Luan, W., Wang, X., Li, S., Liu, M., Tang, X. and Yu, L. (2018). Cordycepin modulates body weight by reducing prolactin via an adenosine A1 Receptor. Curr. Pharm. Des., 24(27): 3240–3249.

Liao, W.C., Hsueh, C.Y. and Chan, C.F. (2014). Antioxidative activity, moisture retention, film formation, and viscosity stability of *Auricularia fuscosuccinea*, white strain water extract. Biosci. Biotechnol. Biochem., 78: 1029–1036.

Liljeblad, M., Rydén, I., Ohlson, S., Lundblad, A. and Påhlsson, P. (2001). A lectin immunosensor technique for determination of alpha(1)-acid glycoprotein fucosylation. Anal. Biochem., 288(2): 216–224.

Liu, Q., Wang, H. and Ng, T.B. (2006). First report of a xylose-specific lectin with potent hemagglutinating, antiproliferative and anti-mitogenic activities from a wild ascomycete mushroom. Biochim. Biophys. Acta, 1760(12): 1914–1919.

Maekawa, S., Saku, M., Kinugasa, T., Ikejiri, K. and Yakabe, S. (1990). A case report of advanced gastric cancer remarkably responding to mitomycin C, aclacinomycin A, SF-SP and lentinan combination therapy. Gan to Kagaku Ryoho., 17: 137–140. 2105085 [in Japanese].

Mio, H. and Terabe, K. (1994). Postoperative immunochemotherapy for gastric carcinoma with peritoneal dissemination—The effects with the combination of CDDP, 5-FU and lentinan. Canc. Chemother., 21(4): 531–534.

Mizuno, M. and Nishitani, Y. (2013). Immunomodulating compounds in Basidiomycetes. J. Clin. Biochem. Nutr., 52(3): 202–207.

Mothana, R.A.A., Ali, N.A.A., Jansen, R., Wegner, U., Mentel, R. and Lindequist, U. (2003). Antiviral lanostanoid triterpenes from the fungus *Ganoderma pfeifferi*. Fitoterapia, 74: 177–180.

Ng, T.B. (1998). A review of research on the protein-bound polysaccharide (polysaccharopeptide, PSP) from the mushroom *Coriolus versicolor* (Basidiomycetes: Plyporaceae). Gen. Pharmacol., 30: 1–4.

Nie, Y., Lin, Q. and Luo, F. (2017). Effects of non-starch polysaccharides on inflammatory bowel disease. Int. J. Mol. Sci., 18(7): 1372.

Ogawa, K., Watanabe, T., Katsube, T., Miura, K., Hirai, M., Wakasugi, S., Yagawa, H., Kajiwara, T., Suga, T. and Hamuro, J. (1994). Study on intratumor administration of lentinan—Primary changes in cancerous tissues. Gan To Kagaku Ryoho., 21(13): 2101–2104.

Oshwada, S., Ikeya, T., Yokomori, T., Kusaba, T., Roppongi, T., Takahashi, T., Nakamura, S., Kakinuma, S., Iwazaki, S., Ishikawa, H., Kawate, S., Nakajima, T. and Morishita, Y. (2004). Adjuvant immunochemotherapy with oral Tegafur/Uracil plus PSK in patients with stage II or III colorectal cancer: a randomised controlled study. Br. J. Canc., 90(5): 1003–1010.

Okamura, T., Ogata, T., Toyoda, M., Tanaka, M., Minamimoto, N., Takeno, T., Noda, H., Fukuda, S., and Ohsugi, M. (2000). Production of sake by mushroom fermentation. Mushroom Sci. Biotech., 8: 109–114.

Okamura-Matsui, T., Tomoda, T., Fukuda, S. and Ohsugi, M. (2003). Discovery of alcohol dehydrogenase from mushrooms and application to alcoholic beverages. J. Mol. Catalysis B: Enzymatic, 23: 133–144. 10.1016/S1381-1177(03)00079-1.

Patel, K.J. and Ingalhalli, R.S. (2013). *Cordyceps militaris*: An important medicinal mushroom. J. Pharmacog. Phytochem., 2(1): 315–319.

Pedneault, K.P., Gosselia, A. and Tweddell, R.J. (2006). Fatty acid composition of lipids from mushrooms belonging to the family Boletaceae. Mycol. Res., 110: 1179–1183.

Peng, X.B., Li, Q., Ou, L.N., Jiang, L.F. and Zeng, K. (2010). GC-MS, FT-IR analysis of black fungus polysaccharides and its inhibition against skin aging in mice. Int. J. Biol. Macromol., 47(2): 304–307.

Pushparajah, V., Fatima, A., Chong, C.H., Gambule, T.Z. and Chan, C.J. (2016). Characterisation of a new fungal immunomodulatory protein from tiger milk mushroom, *Lignosus rhinocerotis*. Nat. Sci. Rep., 6: 30010. 10.1038/srep30010.

Ramsbottom, J. (1954). Mushrooms and Toadstools: A Study of the Activities of Fungi. Collins, London.

Reed, R.E. (1964). The definition of cosmeceuticals. J. Cosmet. Sci., 13: 103–110.

Sheu, F., Chien, P.J., Hsieh, K.Y., Chin, K.L., Huang, W.T., Tsao, C.Y., Chen, Y.F., Cheng, H.C. and Chang, H.H. (2009). Purification, cloning and functional characterisation of a novel immunomodulatory protein from *Antrodia camphorata* (bitter mushroom) that exhibits TLR2-dependent NF-kappa B activation and M1 polarisation within murine macrophages. J. Agric. Food Chem., 57(10): 4130–4141.

Shibata, H., Irie, A. and Morita, Y. (1998). New antibacterial diterpenoids from the *Sarcodon scabrosus*. Biosci. Biotech. Biochem., 62: 2450–2452.

Singh, N.B. and Singh, P. (2002). Biochemical composition of *Agaricus bisporus*. J. Ind. Bot. Soc., 81: 235–237.

Song, H.N. (2020). Functional *Cordyceps* coffee containing cordycepin and β-Glucan. Prev. Nutr. Food Sci., 25(2): 184–193.

Song, J., Chen, Y., He, D., Tan, W., Lv, F., Liang, B., Xia, T. and Li, J. (2020). Astragalus polysaccharide promotes adriamycin-induced apoptosis in gastric cancer cells. Canc. Manage. Res., 12: 2405–2414.

Sullivan, R., Smith, J.E. and Rowan, N.J. (2006). Medicinal mushrooms and cancer therapy: Translating a traditional practice into Western medicine. Perspect. Biol. Med., 49: 159–170.

Sutherland, C. and Enkobachar, C. (2013). Equilibrium modelling of cu (ii) biosorption onto untreated and treated forest macro-fungus *Fomes fasciatus*. Int. J. Pl. Anim. Environ. Sci., 3(1): 193–203.

Tan, C.S., Ng, S.T., Vikineswary, S., Lo, F.P. and Tee, C.S. (2010). Genetic markers for identification of a Malaysian medicinal mushroom, *Lignosus rhinocerus* (Cendawan Susu Rimau). Acta Hortic., 859: 161–167.

Tanabe, H., Imai, N. and Takechi, K. (1990). Studies on usefulness of postoperative adjuvant chemotherapy with lentinan in patients with gastrointestinal cancer. Nippon Gan Chiryo Gakki Shi., 25(8): 1657–1667. 2230448.

Trigueros, V., Lougarre, A., Ali-Ahmed, D., Rahbé, Y., Guillot, J., Chavant, L., Fournier, D. and Paquereau, L. (2003). *Xerocomus chrysenteron* lectin: Identification of a new pesticidal protein. Biochim. Biophys. Acta, 1621(3): 292–298.

Tsuchiya, N., Endo, T., Matsuta, K., Yoshinoya, S., Takeuchi, F., Nagano, Y., Shiota, M., Furukawa, K., Kochibe, N. and Ito, K. (1993). Detection of glycosylation abnormality in rheumatoid IgG using N-acetylglucosamine-specific *Psathyrella velutina* lectin. J. Immunol., 151: 1137–1146.

Ullah, M.I., Akhtar, M., Awais, M.M., Anwar, M.I. and Khaliq, K. (2018). Evaluation of immunostimulatory and immunotherapeutic effects of tropical mushroom (*Lentinus edodes*) against eimeriasis in chicken. Trop. Anim. Health Prod., 50(1): 97–104.

Valverde, M.E., Hernández-Pérez, T. and Paredes-López, O. (2015). Edible mushrooms: Improving human health and promoting quality life. Int. J. Microbiol., 376387. 10.1155/2015/376387.

Vimala, R. and Das, N. (2009). Biosorption of cadmium (II) and lead (II) from aqueous solutions using mushrooms: A comparative study. J. Hazard. Mater., 168(1): 376–382.

Wang, H., Cai, Y., Zheng, Y., Bai, Q., Xie, D. and Yu, J. (2017). Efficacy of biological response modifier lentinan with chemotherapy for advanced cancer: A meta-analysis. Canc. Med., 6(10): 2222–2233.

Wang, S., Bao, L., Zhao, F., Wang, Q., Li, S., Ren, J., Li, L., Wen, H., Guo, L. and Liu, H. (2013a). Isolation, identification, and bioactivity of monoterpenoids and sesquiterpenoids from the mycelia of edible mushroom *Pleurotus cornucopiae*. J. Agric. Food Chem., 61: 5122–5129.

Wang, S.J., Bao, L., Han, J.J., Wang, Q.X., Yang, X.L., Wen, H.A., Guo, L.D., Li, S.J., Zhao, F. and Liu, H.W. (2013b). Pleurospiroketals A-E, perhydrobenzannulated 5,5-spiroketal sesquiterpenes from the edible mushroom *Pleurotus cornucopiae*. J. Nat. Prod., 76: 45–50.

Wang, S.J., Li, Y.X., Bao, L., Han, J.J., Yang, X.J., Li, H.R., Wang, Y.Q., Li, S.J. and Liu, H.W. (2012). Eryngiolide A, a cytotoxic macrocyclic diterpenoid with an unusual cyclododecane core skeleton produced by the edible mushroom *Pleurotus eryngii*. Organ. Lett., 14: 3672–3675.

Wannet, W.J.B., Hermans, J.H.M., Vander Drift, C. and Op den Camp, H.J.M. (2000). HPCL detection of soluble carbohydrates involved in mannitol and trehalose metabolism in the edible mushroom, *Agaricus bisporus*. J. Agaric. Food Chem., 48: 287–291.

Wen, L., Gao, Q., Ma, C.-W., Ge, Y., You, L., Liu, R.H., Fu, X. and Liu, D. (2016). Effect of polysaccharides from *Tremella fuciformis* on UV-induced photoaging. J. Func. Foods, 20: 400–410.

Wong, J.H., Ng, T.B., Cheung, R.C., Ye, X.J., Wang, H.X., Lam, S.K., Lin, P., Chan, Y.S., Fang, E.F., Ngai, P.H., Xia, L.X., Ye, X.Y., Jiang, Y. and Liu, F. (2010). Proteins with antifungal properties and other medicinal applications from plants and mushrooms. Appl. Microbiol. Biotechnol., 87(4): 1221–1235.

Wong, Y.Y., Moon, A., Duffin, R., Barthet-Barateig, A., Meijer, H.A., Clemens, M.J. and de Moor, C.H. (2010). Cordycepin inhibits protein synthesis and cell adhesion through effects on signal transduction. J. Biol. Chem., 285(4): 2610–2621.

Wu, J., Zhang, Q. and Leung, P. (2007). Inhibitory effects of ethylacetate extract of *Cordyceps sinensis* mycelium on various cancer cells in culture and B16 melanoma in C57BL/6 mice. Phytomedicines, 14: 43–49.

Wu, Y., Sun, H., Qin, F., Pan, Y. and Sun, C. (2006). Effect of various extracts and a polysaccharide from the edible mycelia of *Cordyceps sinensis* on cellular and humoral immune response against ovalbumin in mice. Phytother. Res., 20: 646–652.

Xu, X., Yan, H., Chen, J. and Zhang, X. (2011). Bioactive proteins from mushrooms. Biotechnol. Adv., 29: 667–674.

Xue, C.C. and O'Brien, K.A. (2003). Modalities of Chinese medicine. pp. 21–46. *In*: Leung, P.C., Xue, C.C. and Cheng, Y.C. (eds.). Comprehensive Guide to Chinese Medicine, World Scientific, Singapore.

Yang, D., Liu, Y. and Zhang, L. (2019). *Tremella* polysaccharide: The molecular mechanisms of its drug action. Prog. Mol. Biol. Trans. Sci., 163: 383–421.

Yang, Q.Y., Jong, S.C., Li, X.Y., Zhou, J.X., Chen, R.T. and Xu, L.Z. (1992). Antitumor and immunomodulating activities of the polysaccharide-Peptide (PSP) of *Coriolus versicolor*. J. Immunol. Immunopharmacol., 12: 29–34.

Yilmaz, N.M., Solamaz, I. and El-mastas, M. (2006). Fatty acid composition in some wild edible mushrooms growing in the Middle Black region of Turkey. Food Chem., 99: 168–174.

Yogita, R., Simanta, S., Aparna, S. and Kamlesh, S. (2011). Biodegradation of malachite green by wild mushroom of Chhattisgrah. J. Exp. Sci., 2(10): 69–72.

Yuan, B., Zhao, L., Yang, W., McClements, D.J. and Hu, Q. (2017). Enrichment of bread with nutraceutical-rich mushrooms: Impact of *Auricularia auricula* (mushroom) flour upon quality attributes of wheat dough and bread. J. Food Sci., 82(9): 2041–2050.

Zhao, S., Guo, Y.X., Liu, Q.H., Wang, H.X. and Ng, T.B. (2009). Lectins but not antifungal proteins exhibit anti-nematode activity. Environ. Toxicol. Pharmacol., 28(2): 265–268.

Zhang, G.Q., Sun, J., Wang, H.X. and Ng, T.B. (2009). A novel lectin with antiproliferative activity from the medicinal mushroom *Pholiota adiposa*. Acta Biochim. Pol., 56(3): 415–421.

Zhong, L., Yan, P., Lam, W.C., Yao, L. and Bian, Z. (2019). *Coriolus versicolor* and *Ganoderma lucidum* related natural products as an adjunct therapy for cancers: A systematic review and meta-analysis of randomised controlled trials. Front. Pharmacol., 10: 703. 10.3389/fphar.2019.00703.

Zhu, J.S., Halpern, G.M. and Jones, K. (1998). The scientific rediscovery of an ancient Chinese herbal medicine: *Cordyceps sinensis* Part I. J. Altern. Comple. Med., 4: 289–303.

5

Infrequent Current and Potential Applications of Mushrooms

Waill Elkhateeb,* Marwa Elnahas and *Ghoson Daba*

1. INTRODUCTION

Mushrooms had reserved their position centuries ago as food and medicine. They are rich in nutrients and in biologically active compounds that belong to different chemical classes (Elkhateeb and Daba, 2019). Capabilities of different members of mushrooms have encouraged researchers to investigate further applications of these macrofungi in fields other than food and pharmaceutical industries. Specially, owing to the current shortage in global resources, contamination caused by plastic components and the incredible increase in population worldwide needs alternatives through macrofungi. Mushrooms have prestigious enzymatic machinery allowing their application in different industries; some mushrooms produce highly stable dyes that can be used in textiles and other dye-related products (Velíšek and Cejpek, 2011) and some mushrooms can be used as microbial fuel cell (MFC). In this chapter, some of the unusual current and potential applications of mushrooms are described, focusing on mushrooms as MFC, mushroom species in fabric production and footwear products, as natural substitute to some plastics and as an alternative insulation material in building, infrastructure construction, packaging, foams and green building materials, like bricks.

Chemistry of Natural and Microbial Products Department, Pharmaceutical Industries Researches Division, National Research Centre, El Buhouth St., Dokki, 12311, Giza, Egypt.
* Corresponding author: Waillahmed@yahoo.com

2. Mushroom as Microbial Fuel Cell Generator

Production of renewable energy is an essential requirement to support the human society. The traditional sources of energy have several disadvantages, such as reduction of fossil fuels, global warming, energy supply and security risk (Min et al., 2012). Active microorganisms can produce bioelectricity from renewable sources (Pant et al., 2010; Wang et al., 2012). The MFC is a bioreactor that converts chemical energy of organic compounds into bioelectrical energy. The MFCs are bio-electrochemical devices that convert biodegradable organic substrate into electricity through metabolic activities of microorganisms. They are the most efficient candidates for utilising complex fuels since microorganisms are capable of converting various organic compounds into carbon dioxide (CO_2), water and energy (Bullen, 2006; Shukla et al., 2004). The MFC technology has been known since two decades ago. It gains great importance due to its potential role in clean energy production in the form of electricity (Logan, 2008). Different microorganisms including bacteria, fungi and algae show their capacity to generate electric current from a wide variety of substrates, ranging from simple sugars to complicated toxic products, such as wastewaters (Toczyłowska-Mamińska et al., 2018). It is more common that microorganisms used to generate the electric current are placed in the anode compartment, where they are able to transfer the electrons released due to the oxidation of organic substrates. However, the protons developed in the oxidation process move to the cathode through a proton-selective membrane to combine with oxygen in order to form water molecules. Hence, the spontaneous movement of electrons from anode to cathode results in the production of electric current.

Electrogenic bacteria can produce electricity and these bacteria exist in all environments and ecosystems. The most popular electrogenic strains nowadays are *Shewanella* sp. or *Geobacter* sp. which were discovered in the late 1990s (Fredrickson et al., 2008; Lovley et al., 2011). It was interesting to find that electrogens directly transfer the electrons to the anode through transporting proteins on the outer membrane, such as 'cytochrome c', or through some membrane appendages named nanowires (Sanchez et al., 2015). *Pseudomonas aeruginosa*, another electrogenic bacteria, have been reported to produce endogenous mediators by themselves, such as pyocyanin. These mediators can shuttle the electrons towards the anode (Ali et al., 2018). Although many know about the electrical effects accompanied by the decomposition of organic matter by yeast and bacteria, but little is known about fungi-dependent MFCs (Potter, 1911). This is related to the lack of fungal electrogens and the lower power produced by MFCs depending on a single fungus. Nevertheless, over the last decade, investigations related to fungi-based MFCs have been reported to show promising electrogenic potential of fungi (Potter, 1911).

Recent studies suggest that the electron transfer mechanism of fungi may be direct, as in the case of bacteria, where the redox enzymes are found in the fungal cell membrane. Lactate dehydrogenase or ferricyanide reductase is among these redox enzymes (Prasad et al., 2007). It has been also found that some electroactive compounds (short-lived) may also be included, where they act like mediators (Ducommun et al., 2010).

Most of the fungal MFCs well-studied systems are yeast-based systems. Here, cytochrome c plays the main role in direct electron transfer (Wilkinson et al., 2006). Many other fungi also have redox-active enzymes. Fungi belong to the white-rot group as wood degraders exhibit 'an extracellular oxidative ligninolytic enzymatic system' to degrade lignin (Martínez et al., 2005). Many other enzymes are found in this system and these help in the degradation of several xenobiotic compounds. Another group of enzymes is also found in white-rot fungi named laccases. These are a group of N-glycosylated blue oxidases with four copper atoms found within active sites (Leonowicz et al., 2001; Wesenberg et al., 2003). The importance of these enzymes retains in its ability to catalyse the oxidation of aromatic amines and phenolic compounds by consuming atmospheric oxygen, which acts here as the electron acceptor (Leonowicz et al., 2001). Laccase has gained great importance in enzymatic fuel cells as it has been applied as a cathodic biocatalyst because of its high oxidation/reduction potential (Schaetzle et al., 2009; Sadhasivam et al., 2008).

Unlike bacteria, fungi can be used in the MFC systems in two modes. Firstly, it is used in the anode (electron transfer will be through fungal redoxactive proteins or through any chemical mediators that support electron transport). Secondly, it is used in the cathode (where fungi produce enzymes that catalyse the reduction of a terminal electron acceptor, which is mainly oxygen) (Sekrecka-Belniak et al., 2018). Many other fungi reveal their important roles in MFC system, such as *Ganoderma lucidum* BCRC 36123 (Fig. 1a) that has been inoculated in the cathode in order to enhance power production (Lai et al., 2017). *Gloeophyllum*, also known as brown-rot fungi (Fig. 1b), produces pyranose-2-oxidase (P_2O), which exhibits strong specific activity on glucose (Bakar et al., 2012; Nakanishi et al., 1987). On the other hand, *Rhizopus* is a laccase-producing fungi. Laccase is an oxidoreductase that belongs to the copper-containing enzyme family demonstrating specific affinity for oxygen as its electron acceptor (Oliveira et al., 1997; Thurston, 1994; Giardina et al., 2010; Rodgers et al., 2010). Therefore, Bakar et al. (2012) reported that *Gloeophyllum-Rhizopus* microbial fuel cell generates electricity from catalysed glucose/oxygen redox reactions resulting from metabolic activities of respective fungi. Today, we can feel that the power production in MFCs employing a single strain of fungi is more or less comparable to that employing a single bacterial strain.

3. Wearable Mushroom Mycelium

A few years ago, mushrooms were only known as edible vegetables used as food, but now its derivatives made from agricultural waste, like stalks or seed husks with fungal mycelium act as a self-assembling glue that could be used for several purposes (Miles and Chang, 2004). Because of the increasing demand for 'green' materials and productive processes, extensive investigation has been carried out on the so-called biocomposite and bio-based materials (Dicker et al., 2014).

The world footwear production was about 23 billion pairs in 2015 (World Footwear, 2016). It is common that after any footwear purchase, it is worn and used till it gets torn or damaged or is no more wanted by the consumer for other reasons. This used or unwanted footwear, besides other products, such as clothing, is required

Fig. 1. (a) *Ganoderma lucidum* (collected by Cwwood. Locality: Virginia, Willow Oaks, Richmond, USA); (b) *Gloeophyllum* sp. (collected by Kari Pihlaviita. Locality: Tali, Helsinki, Finland, hosted by: *https://mycoportal.org*); (c) *Pleurotus eryngii* (collected by Stefan Mintoff. Locality: Malta, Mgarr, Fort Chambray); (d) *Pleurotus citrinopileatus* (collected by observer 26. Locality: Lasalle County, USA, hosted by: *https://mycoportal.org*); (e) *Agaricus bisporus* (collected by Alan_Rockefeller. Locality: Manchester State Park, California, USA, hosted by: *https://mycoportal.org*); (f) *Pleurotus ostreatus* (collected by Patricia R. Miller. Locality: Mississippi, Lafayette, Puskus Lake, USA, hosted by: *https://mycoportal.org*).

to visit some place in order to get rid of it. Nowadays, disposal and recycling methods are not effectively applied to these types of waste.

The biological metabolism employs various natural resources to produce outputs that can be decomposed safely (Jacques et al., 2010). The inputs used for the production are obtained from the earth. However, outputs comprise of what comes from the product and returns back to the environment (Ashton, 2018). Most inputs of footwear contain toxic and unsafe materials. These include PVC that can form dioxins, which, upon accumulation in human bodies and the environment, disturb the hormone levels (Cao et al., 2014). Moreover, shoe's leather is tanned with chromium, which is a toxic heavy metal; further, the rubber soles contain another toxic heavy metal, that is, lead. These heavy metals can enter the air and soil and finally end their lives (McDonough and Braungart, 2010). Instead of spending more time and effort to clean up these solid wastes and chemicals, Staikos et al. (2006) suggested a 'proactive approach' to replace the conventional material inputs with biodegradable footwear.

Most mushrooms are composed of filaments called hyphae that join together to create a mat-like structure which is the mycelium (Hanson, 2008). A wearable mushroom mycelium composite can be used as a shoe sole that is completely biodegrade at the end of its life (Fig. 2a). Mycelium composites have been positively used as compostable styrofoam substitute to protect computers and other fragile shipments (Holt et al., 2012). These material attributes potentially allow mycelium composites to provide support to the wearer's foot by cushioning against the hard ground. The mycelium (mushroom roots) can be full-grown in a mould to form different shapes for different items and grows quickly into a dense material. When it reaches the desired density and shape, the material is dehydrated that curbs further growth; afterwards it is useful as a packaging material. Mushroom mycelium-based biofoam offers great possibility in application as an alternative insulation material in building and infrastructure construction, or as an alternative lightweight backfill material for geoengineering and others (Arifin and Yusuf, 2013; Farmer, 2013; Ashok et al., 2016; Elkhateeb and Daba, 2019). In a recent study, four mushrooms—king oyster, reishi, oyster and yellow oyster—were tested along with two fabric levels (with or without a natural fabric mat). Mycelial growth within the composite and around the substrates was confirmed (Silverman et al., 2020).

The studies reported that some mushrooms show an important fabric production, such as reishi (*Ganoderma lucidum*) (Haneef et al., 2017). Also, it was found that incorporating some mushroom species, such as king oyster (*Pleurotus eryngii*) (Fig. 1c) and yellow oyster (*Pleurotus citrinopileatus*) (Fig. 1d) together with chicken feather waste leads to the production of a new composite. This composite shows a high compressive strength. That is because the fungal mycelium grows into the feather waste fibres, then interweaves with the fibre, and finally embeds the

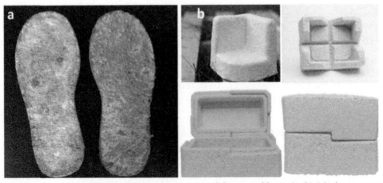

Fig. 2. (a) Mycelium shoe soles (hosted by: *https://www.delawarepublic.org*); (b) Mushroom mycelium packaging material (hosted by: *https://ecovativedesign.com*).

feathers into composite material. Consuming the waste products to produce natural composite fibres will save money on waste disposal as well as produce composite at a reasonable price (Väisänen et al., 2017). The compressive strength of the produced mycelium composite makes it very useful in shoe-sole applications. Additionally, this composite is made from natural, safe and biodegradable materials, which are eco-friendly and protect our environment (air and soil) from hazardous materials (McDonough and Braungart, 2010).

4. Mushroom Contribution in Plastic Replacement

Synthetic plastic and their non-renewable input materials cause environmental pollution. The plastic has properties that allows plastic to accumulate in the environment. This may last forever; because of which it needs to be recycled properly to counter its harmful effects (Andrady and Neal, 2009; Ashok et al., 2016). The plastic contains many compounds that make it harmful, such as plasticisers and colourants, which increase its harmful effect upon disposal (Andrady and Neal, 2009). The idea of developing bioplastic compounds from biodegradable and renewable resources attracted plenty of attention recently in order to decrease the environmental pollution (Flieger et al., 2003).

The process of producing bio-based packaging material is not simple; however, it requires many steps in order to design and manufacture the packaging material (Weber et al., 2002). There are many properties that should be achieved in order to obtain an ideal packaging material. These include permeability, resistance to various chemical substances, transparency, mechanical properties and sealing. Eventually, bioplastic shelf-life and disposal methods must be taken in to consideration (Brown, 1992; Weber et al., 2002; Kumar and Yaakob, 2011).

The composites formed by filamentous fungi are an analogue to green polymers. It is composed of biodegradable material (Zeller and Zocher, 2012; Elkhateeb and Daba, 2019). It was reported that some filamentous fungi enhance the particles of the composite residues to aggregates due to the effect of the cell-wall chitin (Islam et al., 2017; Yang et al., 2017). Plus, oxidative extracellular enzymes produced by the fungal hypha enhance the substrate degradation (Hernández et al., 2017), promoting cell growth, which in turn increases the mycelial density (Haneef et al., 2017). This rise in mycelial density is the key factor in the mechanical and physical properties of the composite (Zeller and Zocher, 2012). The produced composite has the advantage of lack of toxic material and also is not flammable (Arifin and Yusuf, 2013; Tudryn et al., 2018; Hemmati and Garmabi, 2013).

The wood-decaying fungi have the capacity to degrade lignin, cellulose or hemicelluloses. After that, the substrate matrix is pushed and penetrated by fungal hyphae. In turn, a tight net gets formed and its size increases gradually. Over time, the fungal biomass will partially replace the substrate and cement the substrate itself. Finally, the result will be a strong biocomposite material (Schwarze et al., 2013).

In many literatures, several filamentous fungi show their ability to produce biocomposites. Examples of these fungi are *Pleurotus eryngii* (Fig. 1c), *Agaricus bisporus* (Fig. 1e) (Hemmati and Garmabi, 2013), *Pleurotus ostreatus* (Hidayat and Tachibana, 2012), *Aspergillus nidulans* (Zhao et al., 2005) and *Phanerochaete chrysosporium*. Within these species, the white rot fungi are characterised by the production of oxidative enzymes, which are able to degrade lignin (Hernández et al., 2017). Moreover, the composites formed by white rot fungi are very useful in package production (Fig. 2b) (Zeller and Zocher, 2012).

Some fungal composites may show plastic behaviour. The one produced by *Pleurotus eryngii* (Fig. 1c) has been found to exhibit plastic behaviour after complete colonisation of the substrate for 15 days. The materials with plastic behaviour result in more durability as well as higher capacity to absorb energy than the elastic composites (Horvath, 1997). Hence, this composite will be very useful for products with higher strength and more tenacity. The final composite product is further shaped to make various products, including packaging materials (Fig. 2b), mycelium-based foam, green building material like bricks or even any new design objects (Schiffman, 2013; Santhosh et al., 2018; Girometta et al., 2019).

It should be noted that in order to make optimum use of mushroom mycelia, many studies have focused on reducing or inhibiting formation of fruiting bodies. This will contribute in maintaining the stability of mycelium-based material as well as lower the cost. Different strategies were developed for such purposes where heat killing and treatment with fungicide were the commonly used methods (Haneef et al., 2017). Recently a study has suggested the use of lithium chloride or CHIR99021 trihydrochloride, which is glycogen synthase kinase-3 (GSK-3) inhibitor for regulating mycelium growth and inhibiting formation of the fruiting body (Chang et al., 2019).

5. Mushrooms in Biodegradation

Contamination of soil, water and air by hazardous substances are the major environmental problems in today's world. Extensive agricultural practices have also contributed to the generation of enormous quantities of waste, which is being disposed off in non-cost-effective and environmentally unfriendly ways. There are various ways in which fungi can help decrease waste and reduce pollution.

Consumption of mushrooms has become a tradition among many people due to their richness in flavour, proteins and medicinal importance. But their ability to degrade hazardous substances, like plastic (*Pleurotus ostreatus*; Fig. 1f) (Da Luz et al., 2013), dyes and other wastes by secreting various enzymes or by absorption (*Agaricus bisporus*; Fig. 1e) and adsorption of colours from waste substances has made them more attractive for use in the field of bioremediation. Mushrooms serve as good decomposers as they degrade cellulose and lignin of plants for their growth and development. Hence, they maintain the soil health by performing the role of hyper accumulators (Gupta et al., 2018).

Mycoremediation tools describe the ability of mushrooms and their enzymes to degrade a wide variety of environmental pollutants and converting industrial and agro-industrial wastes into useful products. Biosorption is the second important process of removal of metals/pollutants from the environment by mushrooms (the most important is *Agaricus bisporus*; Fig. 1e) (Nagy et al., 2013). Biosorption represents an alternative to the remediation of industrial wastes as well as recovery of metals present in the effluent. Current studies concerned with conversion of industrial or agro-industrial sludge into other useful forms is drawing greater attention. The most potent mushroom to possess the bioconverting ability is *Pleurotus citrinopileatus* (Fig. 1c) (Kulshreshtha et al., 2010; Kulshreshtha et al., 2013; Kulshreshtha et al., 2014).

Mushrooms have an enzymatic machinery for the degradation of waste/pollutants and therefore, can be applied in degradation of a wide variety of pollutants (Purnomo et al., 2013; Kulshreshtha et al., 2013). Mushrooms can produce extracellular peroxidases, ligninase (lignin peroxidase, manganese-dependent peroxidase and laccase), cellulases, pectinases, xylanases and oxidases (Nyanhongo et al., 2007). Basidiomycetous mushrooms are becoming more popular today for remediation purposes because they are not only tools of bioremediation but also provide mycelium or fruit bodies as a source of protein. Many studies have highlighted the role of mushrooms in bioremediation of wastes by the process of biodegradation, biosorption and bioconversion (Akinyele et al., 2012; Kumhomkul and Panich-pat, 2013; Lamrood and Ralegankar, 2013). Mycoremediation through mushroom cultivation will solve the world's major problems, especially of waste accumulation and production of protein food source. There is need for further research to present the potential of mushrooms as bioremediation tools and their safety aspects for consumption as food product.

6. Conclusion and Future Prospects

Mushrooms are a rich source of nutritional and bioactive compounds; hence were used as food and medicine centuries ago. Shortage in global materials and resources, besides the increase in world population, has encouraged research in eco-friendly alternatives, which are also relatively cheap. Capabilities of macrofungi in general, and mushrooms in particular, have supported their use in further applications in fields other than food and pharmaceutical industries. Some mushrooms are able to produce highly stable dyes which can be used in textiles and other dye-related products; some mushrooms can be used as cathodes or anodes in microbial fuel cell (MFC). Mushroom mycelial growth contributes to production of bioplastics and as an alternative insulation material in building, infrastructure construction, packaging, foams and green building materials, like bricks. Moreover, mushrooms have potent enzymatic machinery that allows them to contribute in bioremediation processes.

Further studies are required to explore the additional applications of these magnificent macrofungi. Developing methods for cultivation and increasing the yield of mushrooms are most welcome. On the other hand, it is very important to encourage and improve at the same time the production of fully-degradable wearable components. Moreover, creating awareness among customers for acceptance of such products is of critical importance. Furthermore, exploring new applications for mycelium-based materials will contribute in finding alternatives to the worldwide limited resources.

References

Akinyele, J.B., Fakoya, S. and Adetuyi, C.F. (2012). Anti-growth factors associated with *Pleurotus ostreatus* in a submerged liquid fermentation. Malay. J Microbiol., 8: 135–140.

Ali, J., Sohail, A., Wang, L., Rizwan Haider, M., Mulk, S. and Pan, G. (2018). Electro-microbiology as a promising approach towards renewable energy and environmental sustainability. Energies, 11: 1822.

Andrady, A.L. and Neal, M.A. (2009). Applications and societal benefits of plastics. Phil. Trans. Royal Soc. B: Biol. Sci., 364: 1977–1984.

Arifin, Y. and Yusuf, Y. (2013). Mycelium fibres as new resource for environmental sustainability. Procedia Eng., 53: 504–508.

Ashok, A., Rejeesh, C.R. and Renjith, R. (2016). Biodegradable polymers for sustainable packaging applications: A review. Int. J. Bionics Biomat., 2(2): 1–11.

Ashton, E.G. (2018). Analysis of footwear development from the design perspective: Reduction in solid waste generation. Strat. Des. Res. J., 11: 2–8.

Bakar, A., Syahirah, A., Othman, R., Yahya, M.Z., Othman, R., Din, N. and Suhaimi, N.M. (2012). Bioenergy from *Gloeophyllum-Rhizopus* fungal biofuel cell. Adv. Mat. Res., 512: 1461–1465.

Brown, W.E. (1992). Plastics in Food Packaging: Properties: Design and Fabrication, Ist ed., CRC Press.

Bullen, T.A. (2006). Biofuel cells and their development. Biosen. Bioelec., 21: 2015–2045.

Cao, H., Wool, R.P., Bonanno, P., Dan, Q., Kramer, J. and Lipschitz, S. (2014). Development and evaluation of apparel and footwear made from renewable bio-based materials. Int. J. Fashion Des. Technol. Educ., 7: 21–30.

Chang, J., Chan, P.L., Xie, Y., Ma, K.L., Cheung, M.K. and Kwan, H.S. (2019). Modified recipe to inhibit fruiting body formation for living fungal biomaterial manufacture. PloS ONE, 14(5): p.e0209812.

Da Luz, J.M., Paes, S.A., Nunes, M.D., da Silva, M.C. and Kasuya, M.C. (2013). Degradation of oxo-biodegradable plastic by *Pleurotus ostreatus*. PLoS ONE, 8(8): 69386.

Dicker, M.P., Duckworth, P.F., Baker, A.B., Francois, G., Hazzard, M.K. and Weaver, P.M. (2014). Green composites: A review of material attributes and complementary applications. Compos. Part A Appl. Sci. Manuf. 56: 280–289.

Ducommun, R., Favre, M.F., Carrard, D. and Fischer, F. (2010). Outward electron transfer by Saccharomyces cerevisiae monitored with a bi-cathodic microbial fuel cell-type activity sensor. Yeast, 27: 139–148.

Elkhateeb, W.A. and Daba, G.M. (2019). The amazing potential of fungi in human life. ARC J. Pharmaceut. Sci., 5(3): 12–16.

Farmer, N. (2013). Trends in Packaging of Food, Beverages and other Fast-moving Consumer Goods (FMCG): Markets. Materials and Technologies, Woodhead Publishing, Elsevier.

Flieger, M., Kantorova, M., Prell, A., Řezanka, T. and Votruba, J. (2003). Biodegradable plastics from renewable sources. Folia Microbiol., 48: 27.

Fredrickson, J.K., Romine, M.F., Beliaev, A.S., Auchtung, J.M., Driscoll, M.E., Gardner, T.S., Nealson, K.H., Osterman, A.L., Pinchuk, G. and Reed, J.L. (2008). Towards environmental systems biology of Shewanella. Nat. Rev. Microbiol., 6: 592–603.

Giardina, P., Faraco, V., Pezzella, C., Piscitelli, A., Vanhulle, S. and Sannia, G. (2010). Laccases: A never-ending story. Cell. Mol. Life Sci., 67: 369–385.

Girometta, C., Picco, A.M., Baiguera, R.M., Dondi, D., Babbini, S., Cartabia, M., Pellegrini, M. and Savino, E. (2019). Physico-mechanical and thermodynamic properties of mycelium-based biocomposites: A review. Sustainability, 11: 281.

Gupta, S., Annepu, S.K., Summuna, B., Gupta, M. and Nair, S.A. (2018). Role of mushroom fungi in decolourization of industrial dyes and degradation of agrochemicals. *In*: Biology of Macrofungi, Cham, Springer, 177–190.

Haneef, M., Ceseracciu, L., Canale, C., Bayer, I.S., Heredia-Guerrero, J.A. and Athanassiou, A. (2017). Advanced materials from fungal mycelium: Fabrication and tuning of physical properties. Scientific Reports, 7(1): 1–11.

Haneef, M., Ceseracciu, L., Canale, C., Bayer, I.S., Heredia-Guerrero, J.A. and Athanassiou, A. (2017). Advanced materials from fungal mycelium: fabrication and tuning of physical properties. Sci. Rep., 7: 41292. 10.1038/srep41292.

Hanson, J.R. (2008). The Chemistry of Fungi, Royal Society of Chemistry. Cambridge.

Hemmati, F. and Garmabi, H. (2013). A study on fire retardancy and durability performance of bagasse fibre/polypropylene composite for outdoor applications. J. Thermoplastic Composite Mat., 26: 1041–1056.

Hernández, C., Da Silva, A.-M.F., Ziarelli, F., Perraud-Gaime, I., Gutiérrez-Rivera, B., García-Pérez, J.A. and Alarcón, E. (2017). Laccase induction by synthetic dyes in Pycnoporus sanguineus and their possible use for sugar cane bagasse delignification. Appl. Microbiol. Biotechnol., 101: 1189–1201.

Hidayat, A. and Tachibana, S. (2012). Characterisation of polylactic acid (PLA)/kenaf composite degradation by immobilised mycelia of *Pleurotus ostreatus*. Int. Biodeterior. Biodegrad., 71: 50–54.

Holt, G.A., Mcintyre, G., Flagg, D., Bayer, E., Wanjura, J.D. and Pelletier, M.G. (2012). Fungal mycelium and cotton plant materials in the manufacture of biodegradable molded packaging material: Evaluation study of select blends of cotton by-products. J. Biobased Mat. Bioenergy, 6: 431–439.

Horvath, J.S. (1997). The compressible inclusion function of EPS geofoam. Geotextiles and Geomembranes, 15: 77–120.

Islam, M., Tudryn, G., Bucinell, R., Schadler, L. and Picu, R. (2017). Morphology and mechanics of fungal mycelium. Sci. Rep., 7: 1–12.

Jacques, J.J., Agogino, A.M. and Guimarães, L.B. (2010). Sustainable product development initiatives in the footwear industry based on the cradle to cradle concept. International Design Engineering Technical Conferences and Computers and Information in Engineering Conference, 10.1115/ DETC2010-29061.

Kulshreshtha, S., Mathur, N. and Bhatnagar, P. (2013). Mycoremediation of paper, pulp and cardboard industrial wastes and pollutants. pp. 77–116. *In*: Goltapeh, E.M., Danesh, Y.R. and Varma, A. (eds.). Fungi as Bioremediators: Soil Biology, Springer, Berlin, Heidelberg.

Kulshreshtha, S., Mathur, N. and Bhatnagar, P. (2014). Mushroom as a product and their role in mycoremediation. AMB Express, 4(1): 29.

Kulshreshtha, S., Mathur, N., Bhatnagar, P. and Jain, B.L. (2010). Bioremediation of industrial wastes through mushroom cultivation. J. Environ. Biol., 31: 441–444.

Kumar, M. and Yaakob, Z. (2011). Biobased materials in food packaging applications. pp. 121–159. *In*: Handbook of Bioplastics and Biocomposites Engineering Applications, John Wiley and Sons.

Kumhomkul, T. and Panich-pat, T. (2013). Lead accumulation in the straw mushroom, *Volvariella volvacea*, from lead contaminated rice straw and stubble. Bull. Environ. Contam. Toxicol., 91: 231–234.

Lai, C.-Y., Wu, C.-H., Meng, C.-T. and Lin, C.-W. (2017). Decolourisation of azo dye and generation of electricity by microbial fuel cell with laccase-producing white-rot fungus on cathode. Appl. Energy, 188: 392–398.

Lamrood, P.Y. and Ralegankar, S.D. (2013). Biosorption of Cu, Zn, Fe, Cd, Pb and Ni by non-treated biomass of some edible mushrooms. Asian. J. Exp. Biol. Sci., 4: 190–195.

Leonowicz, A., Cho, N., Luterek, J., Wilkolazka, A., Wojtas-Wasilewska, M., Matuszewska, A., Hofrichter, M., Wesenberg, D. and Rogalski, J. (2001). Fungal laccase: Properties and activity on lignin. J. Basic Microbiol., 41: 185–227.

Logan, B.E. (2008). Microbial Fuel Cells, John Wiley & Sons.

Lovley, D., Ueki, T., Zhang, T., Malvankar, N.S., Shrestha, P.M., Flanagan, K.A., Aklujkar, M., Butler, J.E., Giloteaux, L. and Rotaru, A.-E. (2011). Geobacter: The microbe electric's physiology, ecology, and practical applications. Adv. Microb. Physiol., 59: 1–100.

MamińskaMartínez, Á.T., Speranza, M., Ruiz-Dueñas, F.J., Ferreira, P., Camarero, S., Guillén, F., Martínez, M.J., Gutiérrez Suárez, A. and Río Andrade, J.C. (2005). Biodegradation of Lignocellulosics: Microbial, Chemical, and Enzymatic Aspects of the Fungal Attack of Lignin, 8(3): 195–204.

McDonough, W. and Braungart, M. (2010). Cradle to Cradle: Remaking the Way We Make Things, Northpoint Press.

Miles, P.G. and Chang, S.T. (2004). Mushrooms: Cultivation, Nutritional Value, Medicinal Effect, and Environmental Impact, 2nd ed., CRC Press.

Min, B., Poulsen, F.W., Thygesen, A. and Angelidaki, I. (2012). Electric power generation by a submersible microbial fuel cell equipped with a membrane electrode assembly. Bioresource Technol., 118: 412–417.

Nakanishi, H., Kita, H. and Kitai, S.T. (1987). Intracellular study of rat *Substantia nigra pars* reticulata neurons in an *in vitro* slice preparation: Electrical membrane properties and response characteristics to subthalamic stimulation. Brain Res., 437: 45–55.

Nyanhongo, G.S., Gübitz, G., Sukyai, P., Leitner, C., Haltrich, D. and Ludwig, R. (2007). Oxidoreductases from *Trametes* spp. in biotechnology: A wealth of catalytic activity. Food Technol. Biotechnol., 45: 250–268.

Oliveira, P., Karmali, A. and Clemente, A. (1996). One-step purification of glucose 2-oxidase from *Coriolus versicolor.* Int. J. Biochromatogr., 1: 273–283.

Pant, D., Van Bogaert, G., Diels, L. and Vanbroekhoven, K. (2010). A review of the substrates used in Microbial Fuel Cells (MFCs) for sustainable energy production. Bioresource Technol., 101: 1533–1543.

Potter, M.C. (1911). Electrical effects accompanying the decomposition of organic compounds. Proc. Royal Soc. London, Ser. B., 84: 260–276.

Prasad, D., Arun, S., Murugesan, M., Padmanaban, S., Satyanarayanan, R., Berchmans, S. and Yegnaraman, V. (2007). Direct electron transfer with yeast cells and construction of a mediatorless microbial fuel cell. Biosensors Bioelectronics., 22: 2604–2610.

Purnomo, A.S., Mori, T., Putra, S.R. and Kondo, R. (2013). Biotransformation of heptachlor and heptachlor epoxide by white-rot fungus *Pleurotus ostreatus.* Int. Biodeterior. Biodegrad., 82: 40–44.

Rodgers, C.J., Blanford, S.F., Giddens, S.R., Skamnioti, P., Armstrong, F.A. and Gurr, S.J. (2010). Designer laccases: A vogue for high potential fungal enzymes. Tr. Biotechnol., 28: 63–72.

Sadhasivam, S., Savitha, S., Swaminathan, K. and Lin, F.-H. (2008). Production, purification and characterisation of mid-redox potential laccase from a newly isolated *Trichoderma harzianum* WL1. Proc. Biochem., 43: 736–742.

Sanchez, D.V., Jacobs, D., Gregory, K., Huang, J., Hu, Y., Vidic, R. and Yun, M. (2015). Changes in carbon electrode morphology affect microbial fuel cell performance with *Shewanella oneidensis* MR-1. Energies, 8: 1817–1829.

Santhosh, B., Bhavana, D. and Rakesh, M. (2018). Mycelium composites: An emerging green building material. Int. Res. J. Eng. Technol., 5: 3066–3068.

Schaetzle, O., Barrière, F. and Schröder, U. (2009). An improved microbial fuel cell with laccase as the oxygen reduction catalyst. Ener. Environ. Sci., 2: 96–99.

Schiffman, R. (2013). Packing materials grown from mushrooms. New Sci., 218: 29.

Schwarze, F.W., Engels, J. and Mattheck, C. (2013). Fungal Strategies of Wood Decay in Trees, Springer Science.

Sekrecka-Belniak, A. and Toczyłowska-Mamińska, R. (2018). Fungi-based microbial fuel cells. Energies, 11(10): 2827.10.3390/en11102827.

Shukla, A.K., Suresh, P., Berchmans, S. and Rajendran, A. (2004). Biological fuel cells and their applications. Curr. Sci., 87: 455–468.

Silverman, J., Cao, H. and Cobb, K. (2020). Development of mushroom mycelium composites for footwear products. Clothing and Textiles Research Journal, 38(2): 119–133.

Staikos, T., Heath, R., Haworth, B. and Rahimifard, S. (2006). End-of-life management of shoes and the role of biodegradable materials. Proceedings of Life Cycle Engineering, 497–502.

Thurston, C.F. (1994). The structure and function of fungal laccases. Microbiol., 140: 19–26.

Toczyłowska-Mamińska, R., Szymona, K. and Kloch, M. (2018). Bioelectricity production from wood hydrothermal-treatment wastewater: Enhanced power generation in MFC-fed mixed wastewaters. Sci. Tot. Environ., 634: 586–594.

Tudryn, G.J., Smith, L.C., Freitag, J., Bucinell, R. and Schadler, L.S. (2018). Processing and morphology impacts on mechanical properties of fungal based biopolymer composites. J. Polym. Environ., 26: 1473–1483.

Väisänen, T., Das, O. and Tomppo, L. (2017). A review on new bio-based constituents for natural fiber-polymer composites. J. Cleaner Prod., 149: 582–596.

Velíšek, J. and Cejpek, K. 2011. Pigments of higher fungi—A review. Czech J. Food Sci., 29(2): 87–102.

Wang, S., Huang, L., Gan, L., Quan, X., Li, N. et al. (2012). Combined effects of enrichment procedure and non-fermentable or fermentable co-substrate on performance and bacterial community for pentachlorophenol degradation in microbial fuel cells. Bioresource Technol., 120: 120–126.

Weber, C., Haugaard, V., Festersen, R. and Bertelsen, G. (2002). Production and applications of biobased packaging materials for the food industry. Food Add. Cont., 19: 172–177.

Wesenberg, D., Kyriakides, I. and Agathos, S.N. (2003). White-rot fungi and their enzymes for the treatment of industrial dye effluents. Biotechnol. Adv., 22: 161–187.

Wilkinson, S., Klar, J. and Applegarth, S. (2006). Optimising biofuel cell performance using a targeted mixed mediator combination. Electroanalysis, 18: 2001–2007.

World Footwear. (2016). Worldwide footwear production reached 23.0 billion pairs. https://www.worldfootwear.com/news.asp?id¼1817&Worldwide_footwear_production_ reached_230_billion_pairs_in_2015.

Yang, Z., Zhang, F., Still, B., White, M. and Amstislavski, P. (2017). Physical and mechanical properties of fungal mycelium-based biofoam. J. Mat. Civil Eng., 29(7): 04017030.

Zeller, P. and Zocher, D. (2012). Ecovative's breakthrough biomaterials. Fungi Mag., 5: 51–56.

Zhao, L., Schaefer, D., Xu, H., Modi, S.J., LaCourse, W.R. and Marten, M.R. (2005). Elastic properties of the cell wall of *Aspergillus nidulans* studied with atomic force microscopy. Biotechnol. Prog., 21: 292–299.

6

Industrial Applications of Truffles and Truffle-like Fungi

Paul Thomas,[1,2] *Waill Elkhateeb*[3,*] *and Ghoson Daba*[3]

1. INTRODUCTION

Although the term 'truffle' is used to define large fruit bodies of fungi from the genus *Tuber*, its use is broader in terms often used to describe the hypogeal fruit bodies of many other fungal species. These fruit bodies often have a dependence on mycophagy for spore dispersal and the group is dominated by the Ascomycetes (Trappe and Claridge, 2010). Although the exact life cycle of some of these species has not yet been fully elucidated, we understand that the ability to form structures with a host plant's root system, the ectomycorrhiza is widespread (Daba et al., 2019). Many truffle species are highly valued for their culinary properties and may play an important role in the local cultural identity. Some of the European species are amongst the world's most expensive foodstuffs and less aromatic desert truffle species may be used as a source of local foods and medicines (Martin et al., 2010; Thomas et al., 2019). The potential and current industrial applications of truffles are discussed here with topics ranging from food products to cosmetics and medicines.

2. Application in Foods

Truffles can be quite nutritious, for example, *Terfezia boudieri* contains all the essential amino acids, 14 per cent protein, 8 per cent fat and 54 per cent carbohydrates

[1] Mycorrhizal Systems Ltd, Lancashire, PR25 2SD, UK.
[2] University of Stirling, Stirling, FK9 4LA, UK.
[3] Chemistry of Natural and Microbial Products Department, Pharmaceutical Industries Researches Division, National Research Centre, El Buhouth St., Dokki, 12311, Giza, Egypt.
* Corresponding author: waillahmed@yahoo.com

(Dundar et al., 2012). Slama et al. (2009) also reported that the fruit bodies are rich in Ca^{2+} (1,423), K^+ (1,346), P (346), Mg^{2+} (154) and Na^+ (77) mg/100 g dry mass. Antioxidant levels can also be quite high as found in *Tirmania nivea, Tirmania pinoyi* and *Picoa* spp. (Al-Laith, 2010; Stojković et al., 2013), though care must be taken as losses in antioxidant activity of truffles has been shown owing to methods employed in industrial food processing (Murcia et al., 2002). *Tuber melanosporum* also contains vitamin D2 precursor ergosterol (ergosta-5,7,22-trienol) (Harki et al., 1996).

However, it is not only for the nutritional qualities that truffles are so highly regarded. Their culinary qualities are what make truffles so desirable, so much so that prices for the most aromatic species often exceed 1000 EURO/kg (Martin et al., 2010) and it's estimated that the distribution of European truffle species may be worth 6 billion US$ in the next 10–20 years (Rupp, 2016). Truffle cultivation is broadly practiced with the majority of all the truffles coming from France being cultivated and cultivation has also spread to areas in which these species do not naturally occur (Thomas and Büntgen, 2017). However, the economic impact of fresh truffle distribution is likely to be eclipsed by the commercial exploitation of truffle-based food products. The ubiquity of truffle-flavoured foodstuffs is evident globally, where most of the major supermarkets will stock a range of truffle-flavoured foods but the majority of them rely on artificial flavouring.

The synthesis of 2,4-dithiapentane, also known as bis(methylthio)methane, the odour principle in white truffles (*Tuber magantum*) was first reported in 1941. The ease of production and low cost of manufactured 2,4-dithiapentane compared to the expense of real white truffles led to a boom in its use as a truffle flavouring in a wide range of food products. Truffle production in some countries is declining and it is forecasted to decline further with accelerated climate change (Thomas and Büntgen, 2019). However, truffle products from truffle oils, to truffle sauces and truffle flavoured crisps (Figs. 1) have never been more popular. Truffles contain a broad number of aromatic compounds (Torregiani et al., 2017). Although many truffle products may contain traces of real truffles, they often bear little resemblance to their claimed source as 2,4-dithiapentane is a potent aromatic often used in levels far higher than that found in naturally occurring in truffles. The manufactured (petroleum-based) 2,4-dithiapentane has also been found to differ to the naturally occurring form, in the $^{12}C/^{13}C$ isotope ratio (Sciarrone et al., 2018).

Within the European Union, the following laws state that the name and list of ingredients of truffle-flavoured products must not mislead a consumer and must be safe for human consumption: EU laws EC 2000/13 (Directive 2000/13/EC of 20 March 2000), EC 1334/08 (Regulation # 1335/2008 of 16 December 2008) and EC 1169/2011 (Regulation # 1169/2001 of 25 October 2011). However, these laws appear to be widely flouted. For instance, in a recent study of truffle oils, five out of eight commercial truffle oils were found to contain dimethyl sulphoxide, a compound not found to occur naturally in truffles but that has a garlicky aroma and may be used in part to convey 'truffle-like' qualities (Wernig et al., 2018). The authors also concluded that commercial oils are distinguishable from home-made oils by concentrations of 2,4-dithiapentane that are markedly higher as compared to those obtained in home-made truffle oil using real fruit bodies as a source of aroma. A recent study focusing on truffle sauces and truffle oils has confirmed this disregarded for EU laws (Wernig et al., 2018). On comparing the volatile organic compounds of raw truffles and their respective oils or sauces, Wernig et al. (2018) concluded that there were no similarities between the two for a given species. Further, 2,4-dithiapentane was not detected or was detected in low amounts in black truffles but was very high in sauces (59+%) (Torregiani et al., 2017). Additionally, during analysis of commercial truffle oil claiming to contain the species *Tuber aestivum*, 2,4-dithiapentane was found even though *T. aestivum* does not produce this volatile substance at all (Wernig et al., 2018).

3. Medicinal and Antimicrobial Features

Extracts from a number of truffle species were shown to have significant antimicrobial properties. Extracts of *Terfezia boudieri* displayed antimicrobial properties against all nine tested bacteria, although the most pronounced reported impact was on the yeast, *Candida albicans* (Doðan and Aydin, 2013). *Terfezia claveryi*-derived extracts have displayed particularly noteworthy impacts on the pathogen *Chlamydia trachomatis* (Al-Marzooky, 1981) and extracts of *Tirmania pinoyi* exhibited potent antimicrobial activities against Gram-positive bacteria (Stojković et al., 2013). However, no extract has yet been used commercially.

Truffles may also be a source of novel medicinal products, for example, extracts of *Tirmania pinoyi* aid the healing of open wounds and treat stomach ulcers (Stojković et al., 2013), whilst antimutagenic and anticarcinogenic properties have been discovered in the genus *Picoa* (Murcia et al., 2002). A study on the *Tuber* genus, isolated 52 polysaccharides with the significant potential to promote antitumor activity (Zhao et al., 2014). Further, medicinal compounds do not need to be extracted from fruit bodies, but in some cases they can be isolated from cultured mycelium, for example, *Tuber melanosporum* under submerged fermentation produces exo-polysaccharides (EPS). These EPS have biomedical activity, such as anti-inflammatory, antitumor and blood sugar-alleviating capabilities. The level of EPS produced in submerged culture can also be increased significantly by using various additives in the culture medium (Zhang et al., 2019).

In terms of commercial exploitation, although the authors are not aware of any regulated truffle-derived medicines, there are a wide number of patents submitted for the preparation of various products containing truffle, with purported health benefits. In 2014, 29 of these applications were summarised by Gajos et al. (2014), who concluded that the largest group of patents involves immunological resistance and the second largest group focuses on increased vigour and vitality. Patents also included those with a focus on cancer and cardiovascular diseases. The degree to which these applications were based on scientific evidence of the purported benefits is unknown, but a large number of patent applications is evidential of a desire to commercially exploit truffles and truffle-containing products through processing industrially.

4. Production of Silver Nanoparticles

The silver nanoparticles (AgNPs) play an important role in nanoscience and nanotechnology, with a particularly strong focus within the literature for use in nanomedicine. Their unique size and surface-to-volume ratio has led to many applications, including medical device coatings, optical sensors, cosmetics, textile production, diagnostics and in the pharmaceutical and food industries (Zhang et al., 2016). Conventional physical and chemical methods to produce AgNPs may be expensive and hazardous and thus demand alternative approaches. One such method is biosynthesis and many desert truffles have been used for production of AgNPs, such as *Tirmania nivea* and *Terfezia claveryi* (Muhsin and Hachim, 2016; Khadri et al., 2017). Moreover, a desert truffle of the *Tirmania* genus has been successfully used to produce AgNPs quickly and easily at room temperature (Owaid et al., 2018). Desert truffles are widely appreciated in areas where they occur naturally and incorporate a broad number of species with different levels of appreciation (Thomas et al., 2019). However, desert truffles may now also have a novel industrial application for use in quick and inexpensive production of AgNPs.

5. Cosmetics and Perfumes

The expense, prestige and reverence for the more fragrant species of truffles seem to have been a driver in the use of truffles and truffle-derived compounds in the cosmetics and perfume industries. As an example, on 14th of April 2020, the online perfumes magazine, *Frangatia* (https://www.fragrantica.com/notes/Truffle-335) listed 61 perfumes currently available to purchase and which claim to contain the scent of truffles, although the degree to which truffles are actually included in these products is unknown. The fragrance 'Black Orchid' by Tom Ford claims to contain 'Black Truffle' as does 'Red Truffle 21' by Jo Malone (Fig. 1A–D). However, the available ingredients listed for both products do not seem to contain any mention of truffles. The situation is similar in the cosmetics industry where a wide variety of products, from face masks to eye cream, claim to be based on or to contain truffles (Fig. 1E–H). One of the products, a 'Périgord truffle magnetic mask' by Truffoire, retails for US$ 6,900 (50 g) with the ingredient list naming '*Tuber melanosporum* (Black Truffle) Extract' and as such, this is likely to be a synthetically-derived compound as used in food products, such as 2,4-dithiapentane. In available truffle cosmetics and perfumes, the researchers were unable to find any examples that seemed to contain real truffles. The truffle component was often defined as 'truffle extract' or 'truffle aroma'. This leads us to conclude that the majority of products in the market do not contain real truffle. Although no figures are available for the size but scale of this market and the number and value of available products suggest that the scale is second only to truffle food products.

6. Production of Enzymes

Enzymes are one of the key industrial products to be obtained from truffles, arising from the complex interaction which takes place between the truffle and its host plant

Fig. 1. Examples of products claiming to contain truffle or truffle extract: (A) Perfume 'Black Orchid' by Tom Ford (www.tomford.co.uk); (B) perfume 'Red Truffle 21' by Jo Malone (www.jomalone. co.uk); (C) 'Re-Neutriv' skin cream by Estee Lauder (www.esteelauder.co.uk); (D) face mask 'Periogrd truffle magnetic mask' by Truffoire (https://truffoire.com); (E) 'Truffle Noir' skin cream by Temple Spa (www.templespa.com); (F) 'Eye Truffle' eye cream by Temple Spa (www.templespa.com); (G) Truffle crisps by Patatas Fritas Torres (http://patatastorres.com); (H) Truffle sauce 'Tartufo Nero' by Tartuflanghe (www.tartuflanghe.com).

species. In general, the enzymes of fungal origin have advantages over those from other sources due to their higher yield, efficient production using inexpensive media, and the fact that fungal cells act as small factories with a wide variety of catalytic activities (Nadim et al., 2015; Daba et al., 2019; Thomas et al., 2019). Enzymes have wide applications in a broad range of industries, including food, dairy, confectionary, textiles and leather production. Several truffle species have been investigated for enzyme production, such as *Tuber maculatum* (produces amylase, xylanase, laccase, lipase, peroxidase, cellulase and catalase) and *Tuber aestivum* (produces amylase, peroxidase and catalase) (Nadim et al., 2015). Further studies are needed to elucidate the benefits of truffles as enzymes producers.

7. Conclusion and Outlook

Truffles are amongst the most highly valued of all fungi within the phylum Ascomycota. Incorporating a number of genera, there is a broad range of current and potential industrial applications, which include cosmetics, food, perfumes, medicinal products as a source of enzymes and in efficient production of silver nanoparticles. However, the supply of truffles for these industries may be a growing problem. Although some species may be cultivated, the majority are only harvested from the wild and some of these may be declining as a consequence of a number of anthropogenic interferences or due to climate change. Growing mycelium in culture may be a solution to these problems, but the methodology is not suitable for all uses. A further complexity is that for some industrial applications, notably in cosmetics, food products and perfumes, 'truffle products' may not actually contain any truffle at all and instead incorporate a limited range of synthetically-derived analogues of aromatic compounds. The limited range of aromatic analogues currently utilised in truffle products provides only a narrow window into the volatile-complexity found within these fungi and therefore there is the possibility to isolate and exploit many other compounds. Clearly truffles can be a rich source of industrially important compounds but much work is needed to further unlock their potential. However, at the same time there is much about truffle biology that we do not yet know—ranging from culturing methods for the mycelium of some species to the identification of as yet undescribed species. These unknowns also represent limitations in the ability to effectively research and elucidate the full potential of these hypogeous fungi. For example, some of these, as yet undescribed species may, in the future represent sources of novel metabolites with possible medicinal or industrial interest but without identification and the tools to culture the mycelium, the full potential may remain undiscovered. Two other issues that require attention are the protection of species from human interference and regulation as well as marketing of 'truffle products', many of which may not contain any truffle product at all.

References

Al-Laith, A.A. (2010). Antioxidant components and antioxidant/antiradical activities of desert truffle (*Tirmania nivea*) from various Middle Eastern origins. J. Food Comp. Anal., 23(1): 15–22.

Al-Marzooky, M.A. (1981). Truffles in eye disease. Proc. Int. Islamic Med., 1: 353–357.

Daba, G.M., Elkhateeb, W.A., Wen, T.C. and Thomas, P.W. (2019). The continuous story of truffle-plant interaction. pp. 375–383. *In*: Kumar, V., Prasad, R., Kumar, M. and Choudhary, D.K. (eds.). Microbiome in Plant Health and Disease, Springer, Singapore.

Doðan, H.H. and Aydin, S. (2013). Determination of antimicrobial effect, antioxidant activity and phenolic contents of desert truffle in Turkey. Afr. J. Trad. Complem. Alt. Med., 10(4): 52–58.

Dundar, A., Yesil, O.F., Acay, H., Okumus, V., Ozdemir, S. and Yildiz, A. (2012). Antioxidant properties, chemical composition and nutritional value of *Terfezia boudieri* (Chatin) from Turkey. Food Sci. Technol. Int., 18(4): 317–328.

Gajos, M., Ryszka, F. and Geistlinger, J. (2014). The therapeutic potential of truffle fungi: A patent survey. Rev. Acta Mucol., 49(2): 305–318.

Harki, E., Klaebe, A., Talou, T. and Dargent, R. (1996). Identification and quantification of *Tuber melanosporum* Vitt. Sterols. Steroids, 61(10): 609–612.

Khadri, H., Aldebasi, Y.H. and Riazunnisa, K. (2017). Truffle mediated (*Terfezia claveryi*) synthesis of silver nanoparticles and its potential cytotoxicity in human breast cancer cells (MCF-7). Afr. J. Biotechnol., 16(22): 1278–1284.

Martin, F., Kohler, A., Murat, C., Balestrini, R., Coutinho, P.M., Jaillon, O. and Porcel, B. (2010). Périgord black truffle genome uncovers evolutionary origins and mechanisms of symbiosis. Nature, 464(7291): 1033–1038.

Muhsin, T.M. and Hachim, A.K. (2016). Characterisation and antibacterial efficacy of mycosynthesized silver nanoparticles from the desert truffle *Tirmania nivea*. Proc. Ann. Res. Conf., Qatar Foundation, HBPP1149. org/10.5339/qfarc.2016.HBPP1149.

Murcia, M.A., Martinez-Tome, M., Jimenez, A.M., Vera, A.M., Honrubia, M. and Parras, P. (2002). Antioxidant activity of edible fungi (truffles and mushrooms): losses during industrial processing. J. Food Prot., 65(10): 1614–1622.

Nadim, M., Deshaware, S., Saidi, N., Abd-Elhakeem, M.A., Ojamo, H. and Shamekh, S. (2015). Extracellular enzymatic activity of *Tuber maculatum* and *Tuber aestivum* mycelia. Adv. Microbiol., 5(07): Article ID 57817. 10.4236/aim.2015.57054.

Owaid, M.N., Muslim, R.F. and Hamad, H.A. (2018). Mycosynthesis of silver nanoparticles using *Terminia* sp. desert truffle, pezizaceae, and their antibacterial activity. Jordan J. Biol. Sci., 11: 401–405.

Rupp, R. (2016). The trouble with truffles. National Geographic. www.NationalGeographic.com (accessed on 20 September 2017).

Sciarrone, D., Schepis, A., Zoccali, M., Donato, P., Vita, F., Creti, D., Alpi A. and Mondello, L. (2018). Multidimensional gas chromatography coupled to combustion-isotope ratio mass spectrometry/quadrupole MS with a low-bleed ionic liquid secondary column for the authentication of truffles and products containing truffle. Anal. Chem., 90(11): 6610–6617.

Slama, A., Neffati, M. and Boudabous, A. (2009). Biochemical composition of desert truffle *Terfezia boudieri* Chatin. Acta Hortic., 853: 285–290.

Stojković, D., Reis, F.S., Ferreira, I.C., Barros, L., Glamočlija, J. et al. (2013). *Tirmania pinoyi*: Chemical composition, *in vitro* antioxidant and antibacterial activities and *in situ* control of *Staphylococcus aureus* in chicken soup. Food Res. Int., 53(1): 56–62.

Thomas, P.W., Elkhateeb, W.A. and Daba, G. (2019). Truffle and truffle-like fungi from continental Africa. Acta Mycol., 54(2): 1–15.

Thomas, P. and Büntgen, U. (2017). First harvest of Périgord black truffle in the UK as a result of climate change. Clim. Res., 74(1): 67–70.

Thomas, P. and Büntgen, U. (2019). A risk assessment of Europe's black truffle sector under predicted climate change. Sci. Tot. Environ., 655: 27–34.

Torregiani, E., Lorier, S., Sagratini, G., Maggi, F., Vittori, S. and Caprioli, G. (2017). Comparative analysis of the volatile profile of 20 commercial samples of truffles, truffle sauces, and truffle-flavored oils by using HS-SPME-GC-MS. Food Anal. Meth., 10(6): 1857–1869.

Trappe, J.M. and Claridge, A.W. (2010). The hidden life of truffles. Sci. Am., 302(4): 78–85.

Wernig, F., Buegger, F., Pritsch, K. and Splivallo, R. (2018). Composition and authentication of commercial and home-made white truffle-flavoured oils. Food Cont., 87: 9–16.

Zhang, X.F., Liu, Z.G., Shen, W. and Gurunathan, S. (2016). Silver nanoparticles: Synthesis, characterisation, properties, applications, and therapeutic approaches. Int. J. Mol. Sci., 17(9): 1534.

Zhang, Z., Wang, J., Liu, L., Shun, Q., Shi, W., Liu, X. and Wang, F. (2019). The optimum conditions and mechanism for increasing exo-polysaccharide production of *Truffles melanosporum* by *Dioscorea saponins*. LWT, 107(5): 331–339.

Zhao, W., Wang, X.H., Li, H.M., Wang, S.H., Chen, T., Yuan, Z.P. and Tang, Y.J. (2014). Isolation and characterisation of polysaccharides with the antitumor activity from tuber fruiting bodies and fermentation system. Appl. Microbiol. Biotechnol., 98(5): 1991–2002.

7

Industrial and Biotechnological Applications of *Pleurotus*

Jegadeesh Raman,[1,2,*] *Kab-yel Jang,*[2] *Shin Hyun-Jae,*[1]
Wong-sik Kong[1] *and Hariprasath Lakshmanan*[3]

1. INTRODUCTION

Pleurotus (Jacq.: Fr.) Kumm. (oyster mushrooms) are commercially important, edible species and cultivated on a large scale. They are the second most distributed and most consumed mushrooms globally due to their superior flavour, nutraceutical and medicinal properties (Pawlik et al., 2012; Jegadeesh et al., 2018). *Pleurotus* species are applicable in food industries, bio-restoration of the forest ecosystem, pharmaceutical, cosmetic and nanobiotechnological industries (Fig. 1).

Mushroom cultivation is a natural process for recycling of lignocellulosic organic waste. *Pleurotus* species can degrade a wide range of agro-wasted substrates (Ritota and Manzi, 2019). Additionally, these species are highly adaptive and selective degraders, requiring no additional growth requirements and can be easily grown in a wide range of temperature ranges (Raman et al., 2020). At the same time, *Pleurotus* require a shorter growth time in comparison to other commercial mushrooms. They convert a high percentage of the agro-substrate to basidiocarps, increasing production yield. Most of the *Pleurotus* species show a high biological efficacy compared to outer cultivated species (Girmay et al., 2016). While fresh harvested

[1] Department of Biochemical and Polymer Engineering, Chosun University, Gwangju, Republic of Korea.
[2] Mushroom Research Division, National Institute of Horticultural and Herbal Science, Rural Development Administration, Eumseong, Republic of Korea.
[3] Department of Biochemistry, Karpagam Academy of Higher Education, Coimbatore 641021, India.
* Corresponding author: jegadeesh_ooty@rediffmail.com

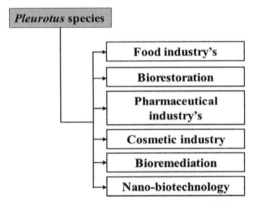

Fig. 1. Industrial and biotechnological application of *Pleurotus.*

Pleurotus are highly perishable, they require an additional preservation process. The high-degree moisture content and polyphenol oxidase may reduce the self-life and alter the physicochemical constituents. The best postharvest techniques may enhance shelf-life and improve *Pleurotus* mushroom's physicochemical attributes (Jafri et al., 2013). In recent years, the mushroom industry is looking forward to value-added products. They are commercially valuable and profitable than fresh mushrooms. The dried *Pleurotus* mushroom powder is used in bakery products, like healthy muffins and cookies (Kim and Joo, 2012; Salehi, 2019). On the other hand, bread, pastries, noodles, tortillas, etc., are made additive with cereals to increase their functional properties.

Pleurotus mushrooms are considered functional foods, mainly due to the high percentage of protein, fat, carbohydrate, minerals and vitamins (Raman et al., 2020; Ho et al., 2020). Mushroom-derived compounds are considered be low in toxicity and valuable natural drug resources with the intercellular and extracellular bioactive compounds having a wide range of biological activity. Also, many *Pleurotus* species have been documented and show tremendous medicinal properties. Polysaccharides, proteins and polysaccharide-bounded proteins are considered as high-molecular-weight bioactive compounds and terpenoids, fatty acid esters and polyphenols are low-molecular weight compounds that have been reported by many authors (Wu et al., 2011; Chen et al., 2012; Erjavec et al., 2012; Hassan et al., 2015; Sari et al., 2017; Jagadeesh et al., 2020). Most of the bioactive compounds are isolated from basidiocarps, while mycelium shows equal importance (Zhang et al., 2012; Facchini et al., 2014). Various enzymes isolated from *Pleurotus* mushrooms show promising health benefits. They are used as probiotics, preservatives, or ingredients in functional foods. Of late, nanotechnology has become the most promising area for generating new applications in biotechnology and nanomedicine (Chen et al., 2008). The most preferred bioreduction method is to use bacteria and fungi due to their fast production and eco-friendliness. Fungi have a high capacity to bioreduce the metal ions because their mycelia offer a large surface area for interaction.

Additionally, fungi/mushrooms secrete many enzymes; thus, metal nanoparticles' bioreduction is very fast. Simultaneously, mushroom-derived nanoparticles (NPs) show strong stability due to amide linkages and protein capping (Gurunathan et al., 2013; Jegadeesh et al., 2014). There are several reports available on metal NPs synthesis using *Pleurotus* species (Owaid, 2019). Various metal ions can be bio-reduced by the *Pleurotus* species (Al-Bahrani et al., 2017; Kapahi and Sachdeva, 2017). Mycoremediation is a new technology in the bioremediation process and can be practiced with fungus to manage polluted soil and water. Fungi-based approach is considered eco-friendly, cost-effective and one of the promising methods. The *Pleurotus* mushroom has unique mycelia characteristics (extensive biomass) and high enzyme capacity (laccase and Mn-peroxidase). The saprophytic fungi are involved in the bioremediation process and they obtain nutrients by degradation/decomposing. Bioremediation is done through elemental transformation, precipitation and active uptake by enzymes and organic acid secretion (Kapahi and Sachdeva, 2017). They can uptake and detoxify heavy metals through natural processes (Raj et al., 2011). Gayosso-Canales et al. (2012) reported that ligninolytic enzymes from *Pleurotus* species could degrade the polycyclic aromatic hydrocarbons. *Pleurotus* species hold a promise for use as a bioremediation tool and the biosorption potential of the species has not yet been commercialised.

2. Sustainable Production of *Pleurotus* Mushroom

Pleurotus species are common edible mushrooms with many properties that can fulfil the nutritional gap prevalent among a large population of Asia and Africa. The agro-industry generates vast volumes of waste, especially crop residues and plant waste left in the field after harvest. Direct incorporation of agro-waste increases methane emissions from the irrigated fields and impacts global warming (Yusuf et al., 2012). The modern biotechnological approach and mushroom production include reuse and valorise to reduce their environmental impact. Agro-waste substances are rich in lignocelluloses, resulting in a promising substrate for solid-state fermentation of mushrooms. *Pleurotus* species can rapidly decompose the waste wood under forestry conditions. Most of them are white-rot fungi on wood and agro-waste food wastes. For instance, white-rot fungi can degrade lignocellulosic materials.

2.1 *Pleurotus Cultivation on Agricultural Industrial Wastes*

Globally, about 998 million tons of agricultural wastes are generated every year and are disposed of in the environment without pretreatment. The major composition of agro-wastes includes cellulose, hemicellulose and lignin and this depends upon the plant species and their types and maturity (Ravindran and Jaiswal, 2016; Sadh et al., 2018) (Fig. 2).

Most commonly, agro-wastes are accumulations burned in the field and may negatively impact the environment. They exhibit a toxic effect on humans and wildlife (Anwar et al., 2014). So unused agro-wastes are considered significant sources of environmental pollution and waste generation. To overcome this problem, mushroom cultivation on these agricultural wastes is the most ecofriendly approach for solid waste management and also beneficial to the agro-industries (Kamthan and Tiwari, 2017). At the same time, the bioconversion process may promote a sustainable agricultural economy. Mushroom industries utilise these agro-wastes as basal substrates for various mushroom production and thus, this technique may convert agro-wastes into human food. The chemical composition and moisture in these agro-wastes make them a suitable substrate for mushroom production. May agricultural wastes, such as paddy straw, wheat straw, bagasse, cotton stalk, and rice/wheat bran comprise lignin, cellulose and hemicellulose materials (Mussatto et al., 2012) (Fig. 3). These substrates are inexpensive, renewable biomass and are

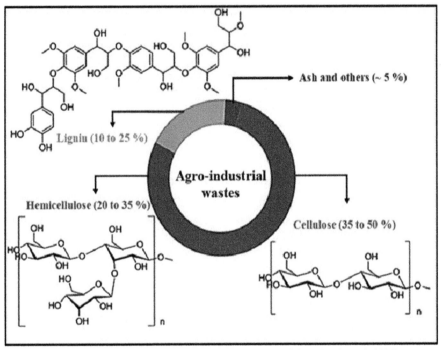

Fig. 2. Main composition of agro-industrial wastes (adapted from Kumla et al., 2020).

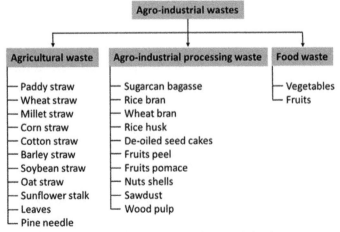

Fig. 3. Major agro-wastes produced by agro-industries.

abundantly generated from agro-industries. White-rot fungi have a high capacity to degrade and modify the chemical composition of the lignocellulosic biomass. They have high-enzyme capacity to degrade lignocellulosic substrates, while the degradation process requires a series of hydrolytic and oxidative enzymes (Tengerdy and Szakacs, 2003; Andlar et al., 2018). Nevertheless, *Pleurotus* is characterised by its fast colonisation on various agro-waste and its yield and efficacy are appreciable. Yadav et al. (2012) reported that fruit waste composts support better *Pleurotus* mushroom production. Many scientific works report use of different substrates for *Pleurotus* mushroom production (Yadav et al., 2012; Girmay et al., 2016; Jegadeesh et al., 2018; Ritota and Manzi, 2019).

2.2 Optimum Culture Condition for Industrial Production

Optimum *Pleurotus* mushroom production is related to several environmental factors, which may have individual or combined effect. Optimisation of synergistic effect of the factors, such as temperature, moisture, substrates nutrient conditions would yield mushrooms without consequent loss and cost reduction. *Pleurotus* cultivation and commercial production account for 27 per cent of its global production (Royse, 2014). Environmental factors, including temperature and humidity, control *Pleurotus* mushroom production (Fig. 4). The optimum temperature for primordial initiation and basidiocarps development vary according to their substrate. *Pleurotus* adopt a wide range of temperatures, so they are cultivated worldwide throughout the year

| Mycelia growth | Spawn production | Linear growth and spawn run |
| 22 – 24 °C | 22 – 25 °C | 18 – 35 °C |

Temperature range for mycelial growth and fruiting of *Pleurotus* spp.

| Plastic bags | Tray | Polypropylene bottles |

[20 – 35 °C, RH 85 - 95%, photoperiod 12 h/day and light density 15 - 350 lux]

Fig. 4. The optimum culture conditions for *Pleurotus* mushroom production.

(Qu et al., 2016). The optimum temperature is 20–35°C and the humidity ranges between 85–95 per cent (Raman et al., 2020).

On the other hand, high temperatures can reduce mycelial growth and yield of the mushrooms (Urben, 2004). The optimum pH for mycelial growth and subsequent basidiocarps formation range between 4.0–7.0, 3.5–5, respectively (Urben, 2004). Balanced moisture content and air circulation allow them to colonise fast with high basidiocarp formation. Low moisture content will result in the mycelium's death and fail to initiate the primordia, while a high moisture content limits the growth on the substrate (Patel et al., 2009). According to Chang and Miles (2004), the appropriate moisture in the substrate should range between 50–75 per cent for elevated yield and biological efficacy of *Pleurotus*. Ryu et al. (2015) reported that the incubation room's relative humidity should be 85–95 per cent. High humidity is also favourable for pinning and a high number of fruiting development. A photoperiod of mycelia stimulation promotes the primordial pinning and fruiting body formation (Nakano et al., 2010). While the light intensity and colour range of light may alter the fruiting body structure and yield. Koyyalamudi et al. (2009) reported that UV lamp with a broad spectrum (100–800 nm) of light intensity converts ergosterol into vitamin D_2 content in mushrooms for a short time. A gaseous environment is essential for aerobic fermentation. They require oxygen for their survival and development. The optimum CO_2 concentration for spawn-running and fruiting body development are at 2000–2500 mg/l and 1500–2000 mg/l, respectively (Li et al., 2015).

2.3 Post-harvest Management

Pleurotus mushroom consumption has rapidly increased due to their aroma, organoleptic, nutritive and medicinal properties. The self-life of this mushroom is one to three days at ambient temperature after harvest. Even after harvesting, mushrooms have a high respiration rate and physiological and biochemical changes are observed at a fast rate (Yang et al., 2020). The post-harvest qualities that cause concern on *Pleurotus* are browning, texture, opening veil, weight loss, ash, protein and crude fibre contents. The oxidation of phenolic compounds by the enzyme oxidase is the reason behind the post-harvest browning and discolouration (Xiao et al., 2011). Simultaneously, their high-level moisture and delicate texture may invade the microbes and reduce the mushrooms' profit and quality. Ambient temperature and acceptable post-harvest practices can overcome these economic losses. Drying is the most traditional preservation technique. The dried powder isused in many dishes due to nutritive content and flavour of the *Pleurotus* mushroom (Muyanja et al., 2012). The drying process can increase the self-life and quality of *Pleurotus* mushrooms. Refrigeration is an alternative method that can slow down the metabolic rate and deterioration process (Wakchaure, 2011). *Pleurotus* mushroom's shelf-life can be extended to 8–11 days at 0°C or one to two days at 20°C (Choi and Kim, 2003). Modified atmosphere packing (MAP) combined with chemical treatments have been shown to delay senescence and help to maintain *Pleurotus* mushrooms' quality. Controlled packing, chemical treatment, blanching, radiation, coating (minimal processing) and canning and drying (conventional processing) are other popular mushroom preservation techniques (Diamantopoulou and Philippoussis, 2015; Choi et al., 2018). Mushrooms produced for human consumption

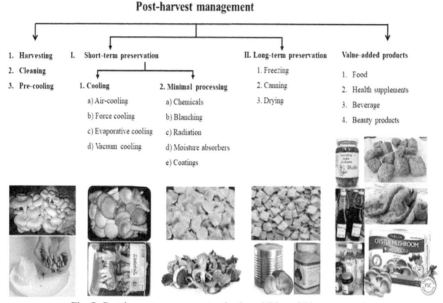

Post-harvest management

1. Harvesting
2. Cleaning
3. Pre-cooling

I. Short-term preservation

1. Cooling
   a) Air-cooling
   b) Force cooling
   c) Evaporative cooling
   d) Vacuum cooling

2. Minimal processing
   a) Chemicals
   b) Blanching
   c) Radiation
   d) Moisture absorbers
   e) Coatings

II. Long-term preservation
1. Freezing
2. Canning
3. Drying

Value-added products
1. Food
2. Health supplements
3. Beverage
4. Beauty products

Fig. 5. Post-harvest management and value addition of *Pleurotus* species.

are processed and packed mostly in canned form. Many previous studies have attempted to increase the quality of processed foods containing mushrooms. The modern post-harvest practices may enhance the shelf-life and marketability of commercial mushrooms, while additional profits will provide adequate returns to the growers and processors (Choi et al., 2018). In recent years, ready-made or ready-to-make value-added mushroom products have attracted the mushrooms industry's attention (Fig. 5). Dry mushroom powder forms of *Pleurotus* species are used in bakery products (Salehi, 2019). In the peak production seasons, surplus fresh mushrooms can be processed/converted into value-added products, resulting in distress sales (Wakchaure, 2011).

3. Conventional and Novel Application of *Pleurotus* Mushrooms in Food Industry

For centuries, *Pleurotus* mushrooms have been considered as a high nutritive diet and of high economic significance. More than 40 species of *Pleurotus* were cultivated worldwide because they can be grown on a wide range of temperatures and are economically profitable. Shukla and Biswas (2000) and Jegadeesh et al. (2018) cultivated *Pleurotus* mushrooms and gained popularity in India due to the low-cost technology and readily available substrates. *Pleurotus* species can include wood decay due to enzymes, such as laccase and versatile peroxidases, that can modify the lignin structure. They are considered to be high delicacy mushrooms due to their umami taste and fibre. *Pleurotus* contain polysaccharides, peptides proteins, terpenoids, ergosterol, fatty acid esters and polyphenols. Many authors have indicated that *Pleurotus* mushrooms could promote human health and act as a functional food

(Golak-Siwulska et al., 2018; Raman et al., 2020). Recently, *Pleurotus* mushrooms have been directly applied in food processing and biotechnological industries and indirectly in environmental applications.

3.1 Nutritional Composition

Among the lignocellulosic mushrooms, oyster mushrooms are easier to grow in a short span of time and require less cost and labour. *Pleurotus* species recycle the agro-wastes as food and reduce environmental pollution. They are considered a good source of nutrition and are called the 'poor man's protein' because of their high protein content, fibre, minerals and low fat. The percentage of protein, carbohydrate and dietary fibre range between 11.95–35.5 per cent, 34–63.03 per cent and 6.2–28.29 per cent, respectively (FAO, 1973; Atri et al., 2012, Jegadeesh et al., 2018, Raman et al., 2020). *Pleurotus* species are a good source of essential dietary fiber, the non-digestible cell walls carbohydrates constituent of dietary fiber. Mushroom dietary fiber and including cell wall polysaccharide and protein complexes show a wide range of health benefits to humans (Okolo et al., 2016). The proximate, vitamin and mineral contents in *Pleurotus* species cultivated on different agro-wastes are described in Table 1. *Pleurotus* mushrooms are a good source of B complex vitamins, which involve human metabolites and maintain neural health (Caglarirmak, 2007; Calderon-Ospina and Nava-Mesa, 2019). Folic acid helps and protects against diabetes and hypertension and reduces neural tube defect (WHO-FAO, 2014). The micro and macro mineral contents are considerable in *Pleurotus* species. They are essential sources of nutrients and work as a co-factor in enzymatic reaction and antioxidant protection. Also Selenium (Se) is an essential trace element and essential for human health at the trace level. Se functions as a co-factor in reducing antioxidant stress and preventing various immunological disorders (Kora, 2020). Therefore, the *Pleurotus* mushroom is suitable for people suffering from malnutrition, hypertension and hypoglycemia, because of its unique medicinal attributes; also, folic acid and potassium in the mushroom is considered a good source of food recommended for anaemic people.

3.2 Enzyme Production and their Application

In recent years, *Pleurotus* enzymes have received great attention in food and biotechnological industries. The versatile enzymes from *Pleurotus* species are applied in several biotechnological processes. The lignicolous mushrooms are primarily cultivated on lignocellulosic agro-wastes, tree logs, wood chips, etc. Agro-wastes are generally lignocellulosic materials that include cellulose, hemicelluloses and lignin. Among these, cellulose is the most abundant component, followed by hemicellulose and lignin. Simultaneously, lignin is a complex organic polymer, structurally crucial in forming plant and algal cell walls (Rico-Garcia et al., 2020). Lignin is the second most abundant plant biopolymer in the biosphere after cellulose, accounting for up to 10–25 per cent of the agro-wastes (Fig. 2). The heterogeneous polymer structural features support the vascular plants and protect cellulose from hydrolytic attack by microbes. While the lignin degradation process is complicated, a few ligninolytic fungi can be degraded selectively (Martínez et al., 1994). The

Table 1. Chemical composition (g/100 g dw) of *Pleurotus* species grown on different agro-waste substrates.

Agro-industrial wastes	Biological efficacy	Crude protein	Carbohydrate	Fat	Fibre	Ash	References
Paddy straw	120.07	34.50	44.75	1.72	14.60	5.90	Jegadeesh et al., 2018
Wheat straw	74.86	22.90	56.0	2.55	7.10	6.65	Patil, 2013
Corn straw	31.6	26.3	31.3	0.5	19.6	5.2	Salami et al., 2017
Bajra straw	46.66	15.49					Telang et al., 2010
Jowar straw	50.33	22.80	50.60	2.00	6.75	5.85	Telang et al., 2010
Soybean straw	72.86	21.37	50.50	2.60	9.0	6.50	Telang et al., 2010
Corn cob	55.0	1.50	25.30	3.06	12.10	23.30	Itelima, 2012
Sunflower stalk	45.86	19.19	2.20	7.92	52.00	5.25	Telang et al., 2010
Sugarcane bagasse	ND	15.54	45.25	2.30	22.79	7.48	Hoa et al., 2015
Sawdust	ND	10.99	55.92	2.05	20.05	6.30	Hoa et al., 2015
Extracted olive-press Cake	18.8	21.41	70.21	1.64	13.68	6.98	Koutrotsios et al., 2014
Almond and walnut shells (1:1)	46.9	31.36	56.64	2.49	13.00	9.86	Koutrotsios et al., 2014
Date palm tree leaves	ND	16.13	72.77	3.41	19.89	7.83	Koutrotsios et al., 2014
Pine needles	24.7	22.74	2.44	13.68	75.88	7.50	Koutrotsios et al., 2014

white-rot *Pleurotus* has a high capacity to degrade lignocellulosic biomass and convert it into utilisable carbohydrates. Therefore, this genus is an essential player in lignocellulose degradation by producing laccases, manganese peroxidases, versatile alcohol oxidases, peroxidases and many other biocatalysts (Kumla et al., 2020). The Mn^{2+} amendment enhance lignin degradation during *Pleurotus* species grown on agro-wastes, even though Mn^{2+} ions are naturally present in lignocellulosic substrates (Camarero et al., 1999). Morais et al. (2005) demonstrated that different agro-industrial wastes influence enzyme production (Fig. 6).

Laccases are the most explored microbial enzymes and are utilised in various food industries and in the bioremediation process. Fungal laccases can modify and enhance food and beverages' colour appearance for the food industry (Mayolo-Deloisa et al., 2020). Minussi et al. (2002) have highlighted the laccase applications in the beverage, baking and biosensor industries. Simultaneously, lignin peroxidase and manganese peroxidase (oxidoreductase enzyme) enhance the natural aromatic flavour. Lignin peroxidase is applied in biotechnological industries and the importance of LiP in the degradation of lignin has been demonstrated in several studies. Also,

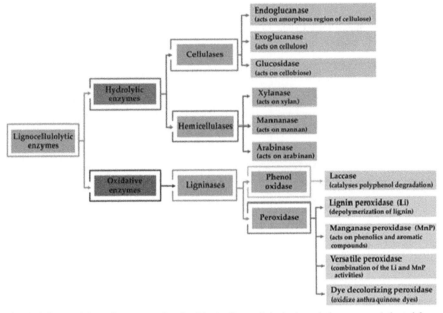

Fig. 6. Scheme of the main enzymes involved in the lignocellulosic degradation process (adapted from Kumla et al., 2020).

Pleurotus species produce extracellular enzymes, xylanases and cellulases, during solid-state fermentation (Masutti et al., 2015). On the other hand, other studies demonstrated that thermostable proteases extracted from *Pleurotus* species play an important role in industrial application (Omrane et al., 2019). Proteases are vital in detergent, food, pharmacy, leather and silk industries (Razzaq et al., 2019). Apart from industrial applications, proteases are applied in many physiological methods, such as protein turnover, sporulation and conidial discharge, germination, enzyme modification, nutrition and gene expression (Rao et al., 1998). These enzymes have a close relationship with saprophytic fungi's lifestyle, as observed in *Pleurotus citrinopileatus* (Cui et al., 2007). However, the production of these enzymes is regulated by different genes and the enzyme yield is highly dependent on the type of agro-wastes utilised during cultivation (El Enshasy et al., 2019). Furthermore, these enzymes have a wide range of potential applications in different industries. The expression of ligninolytic peroxidases, including lignin peroxidase (LiP), manganese peroxidase (MnP), versatile peroxidase (VP) and addition with multicopper oxidases (laccases) of *Pleurotus* genomes confirms its central role in lignin biodegradation. Regarded Safe (GRAS) according to the Food and Drug Administration (FDA), the produced enzyme has no limitations in terms of safety issue for application in food and feed industries (El Enshasy et al., 2019).

3.3 Bioactive Compounds from Pleurotus

Pleurotus is viewed as a promising medicinal mushroom. It has various biotechnological applications in medicine, food and drug industry (Golak-Siwulska et al., 2018). *Pleurotus* is rich in carbohydrates, proteins, amino acids, water-soluble vitamins and minerals and has key nutritional enhancement factors. Also, the *Pleurotus* species has low calories and is advisable in the diet for daily consumption and is also cultivated on a large scale globally. While the mushroom-derived polysaccharides, terpenoids, fatty acids, amino acids, fatty acid esters, steroids and polyphenols act as pharmacological agents, the mushrooms have been consumed for many years in folk medicine. Most of the bioactive metabolites were isolated from mycelium and fruiting bodies (Morris et al., 2017). Among these bioactive compounds, polysaccharides are the main active compounds and display antitumor, antioxidative, immunomodulatory, anti-inflammatory, antiviral and hepatoprotective effects (Karaman et al., 2018; Pandya et al., 2018) (Table 2). However, the composition and combination are essential characteristics of polysaccharides and they clearly show bioactivity. Daba (2006) isolated the anticancer compound from *Pleurotus* mushroom and obtained the pattern (WO 2006119783 A1). The commercially important pleuran (beta-glucan) was extracted from *Pleurotus* mushroom. The compound exhibited anti-neoplastic properties against various cells, including colon and prostate cancer cells (Lavi et al., 2006; Gu and Sivan, 2006). The high molecular compounds, like proteins, peptides and lectins isolated from the *Pleurotus* genus exhibit multidirectional health-promoting effects (Hassan et al., 2015). More interestingly, a novel HIV-1 reverse transcriptase inhibitory protein was isolated from *P. ostreatus* (Wang and Ng, 2004).

Pleurotus mushrooms are significant sources of natural bioactive peptides and proteins. According to Zhou et al. (2019), edible and medicinal mushrooms contain many proteins, especially enzymes. They have less toxic and exhibit antihypertensive, antioxidant and antimicrobial activities. Recently mushroom-derived peptides have been highly accepted by the pharmaceutical industry (Zhou et al., 2020). Also, antifungal peptides and proteins have been isolated from *Pleurotus* species. Antifungal peptides, like eryngin and pleurostrin, inhibit the pathogenic fungal growth (Wang and Ng, 2004; Erjavec et al., 2012). Also, oligopeptides, lectins, amino acids exhibit antihypertensive and antioxidant properties (Hassan et al., 2015; Lin et al., 2016). More interestingly, ribonuclease and laccase enzymes derived from

Table 2. Bioactive metabolites and pharmacological effects of *Pleurotus* mushrooms (adapted from Golak-Siwulska et al., 2018).

Activity	Bioactive compounds
Anticancer	α and β-glucans and proteins
Antitumor	Polysaccharides, proteoglycans and lectin
Immunomodulatory	Polysaccharides and heteroglycan
Antihypertensive	D-mannitol and peptides
Antiviral	Proteins, polysaccharides and lectins
Antimicrobial	Protein, polysaccharides, lectins and fatty acid esters
Anti-oxidative	Phenol, lectins and polysaccharides
Anti-inflammatory	Polysaccharide
Hypoglycemic	β-glucans and polysaccharide-peptide
Cholesterol lowering	Lovastatin and ergosterol

Pleurotus species exhibit anticancer and antiviral activity (Li et al., 2008). They have been considered a potential dietary fibre source since fungal cell walls, rich in non-starch polysaccharides (beta-glucan), exhibit favourable pharmacological properties. Besides, *Pleurotus* mushrooms contain various minerals and trace elements, such as potassium and copper and vitamins like riboflavin, niacin and folate (Patil et al., 2010). Folate is an essential supplement during ovulation and controls nutritional anemia (WHO-FAO, 2014). Simultaneously, a significant amount of sodium and potassium may reduce hypertension, diabetes and obesity. Notably, the essential fatty acids and statins display cholesterol-lowering activity (Golak-Siwulska et al., 2018). Thus, they might be used directly in the diet to promote health, taking advantage of the additive and synergistic effects of all the bioactive compounds present. The phenolic compounds are used as additives in the food industry or as components of pharmaceutical and cosmetic formulations. The multi-copper enzyme, tyrosinase, has been reported in *Pleurotus* species (Alam et al., 2010). Tyrosinase plays an essential role in melanogenesis and enzymatic browning. Therefore, tyrosinase inhibitors and phenolic compounds are attractive in cosmetics and in medices as they act as depigmentation agents and are used also in the food and agriculture industries as antibrowning compounds (Zolghadri et al., 2019). Several studies confirm that the *Pleurotus* genus exhibits multidirectional health-promoting effects and can be classified as a functional food and as a source of new innovative drugs (Patel and Goyal, 2012). Recent studies confirm that the *Pleurotus* species can serve as a raw material in the cosmetics industry (Taofiq et al., 2016; Morris et al., 2017). Overall, the *Pleurotus* genus is a high nutraceutical and medicinal food for daily consumption and is useful in various biotechnological industries.

4. Bioremediation

The bioremediation process whereby the fungi/mushrooms degrade into organic/ inorganic wastes is termed as mycoremediation. The fungal/mushroom-based approach has become increasingly popular owing to its eco-friendly approach and cost-effectiveness. *Pleurotus* species tend to be promising candidates in the biosorption/bioremediation of heavy metals from the environment (Kapahi and Sachdeva, 2017). They have a high capacity to tolerate heavy metals. The wood-degrading *Pleurotus* species utilises polyurethane (PU) instead of carbon. White rot fungi can degrade and accumulate the aromatic hydrocarbons (PAHs), dioxins, synthetic dyes and pesticides from the polluted ecosystem (Johannes and Majcherczyk, 2000; Kanaly and Hur, 2006). During degradation, mushrooms can metabolise the initial PAH by cleaving the aromatic ring to form ring fission compounds with subsequent mineralisation (Nikiforova et al., 2009). Adenipekun et al. (2015) reported that *Pleurotus ostreatus* degrades PHSs in spent and contaminated soil fluid.

On the other hand, other studies demonstrated that the *Pleurotus pulmonarius* strain isolated from tropical rainforest can be used to degrade pyrene and PHS (Hadibarata and Teh, 2014). They have an unusual combination of enzymes that catalyse the metabolism and degradation of PAHs. Also, *Pleurotus* species can uptake and assimilate the soils' heavy metals, such as Cu, Zn, Mn, and Fe

(Boamponsem et al., 2013). *Pleurotus ostreatus* cultivated on the waste newspaper can utilise the substrate, while heavy metals accumulate in the fruiting body (Kopinski and Kwiatkowska-Marks, 2012). However, despite extensive research on its biotechnological applications, a large portion of the biochemical pathways, that make the genus *Pleurotus* an attractive biological target for future research purposes, remain undescribed.

5. Nanotechnology

Microorganisms are the best model of action for nano-biotechnological applications. Fungi/mushrooms are common sources of industrial enzymes applied in various fields, such as biomedicine, environment, drug designing, drug delivery, cosmetics, textiles, food and pharmaceutical industry (Khandel et al., 2018). Successful studies and novel approaches in this area have provided a better understanding of fungi in nanobiotechnology (Saglam et al., 2016). At the same time, the mushroom-based biosynthesis-approach for metal nanoparticle synthesis has attracted considerable interest over other conventional physicochemical methods because this approach excludes the use of toxic chemicals. Mushrooms, containing a wide variety of polysaccharides, peptides and proteins, have been utilised in metal nanoparticles' biosynthesis (NPs) (Raman et al., 2015). The protein and polysaccharide-coated metal NPs have found prominent biomedical applications due to their oxidation resistance, multiple surface functionality and stability. Mushroom-based functional metallic NPs are used in different industrial and medical applications because of the presence of enormous amounts of mycelium or fruiting bodies (Bhat et al., 2011). The biosynthesised metallic NPs are the preferred choice as they have excellent biocompatibility and are economical, eco-friendly and have easy downstream processing. Also, the extracellular metabolites and redox proteins can reduce/ stabilise agents in the formation of nanocrystals. Besides, many researchers have reported on the synthesis of metal NPs by *Pleurotus* species (Owaid, 2019). They are widely used NPs with many therapeutic applications, like drug delivery, cosmetics, biosensors and allied areas. *Pleurotus ostreatus* silver nanoparticles (AgNPs) are applicable in various industrial fields, such as azo dye degradation (Karthikeyan et al., 2019). Laccase from *Pleurotus* species is capable of decolourising several industrial dyes to about 80 per cent. The other studies demonstrated gold NPs synthesis from *P. ostreatus* laccase enzyme, proving that laccase is potentially applicable in various industries (El-Batal et al., 2015).

Fungal mushrooms-based metallic NPs (viz., Ag-NP, Au-NP, Se-NP, CdS-NP, Fe-NP, Pa-NP, and ZnS-NP) have been tested in biomedical- (79 per cent) and industrial- (21 per cent) based applications. Mushroom metal NPs investigated industrial applications, such as for inorganic NPs, carbon nanotubes and waste treatment as nano-biosorbents for adsorption of toxic metals to clean our environment by using natural ecological materials and reduce environmental pollution in the future (Prabhu et al., 2019). In recent decades, nanoparticles have met a significant expansion and serious investigations due to their potential in a broad spectrum of industrial applications.

6. Conclusion and Future Perspective

The *Pleurotus* genus is a biologically versatile fungus that could serve as the basis for various biotechnological and industrial applications. The *Pleurotus* species can be cultivated on a large-scale globally and they are directly/indirectly converted into various food products. At the same time, mushroom production is considered as successful implementation of agro-waste biodegradation. *Pleurotus* species are good candidates for bioremediation applications, such as soil decontamination, degradation of industrial dyes, phenols, PAHs and wastewater treatment. Ligninolytic enzymes are involved in the degradation process and are applied in a wide variety of industries. Also, laccase is widely used in various food industry processes, such as beverage processing, baking, stabilisation of wine and beer, and sugar beet pectin gelation. The bioactive secondary metabolites and enzymes can be used in the food and drug industries. *Pleurotus* enzyme has a high potential for future application in the treatment and utilisation of agricultural waste. They have a high capacity to degrade and convert the agro-waste into valuable products for animal feed and other food products. Besides, biosynthesised NPs can treat various diseases and open a new avenue in the future biomedical field. A significant bottleneck in the development of biotechnological applications of *Pleurotus* species is the lack of information on classic and molecular genetics. The chapter covered here could be of enormous benefit in the use of *Pleurotus* species in the nutraceutical, medicinal, environmental, bio-nano technological industries.

References

Adenipekun, C.O., Ipeaiyeda, A.R., Olayonwa, A.J. and Egbewale S.O. (2015). Biodegradation of polycyclic aromatic hydrocarbons (PAHs) in spent and fresh cutting fluids contaminated soils by *Pleurotus pulmonarius* (Fries). Quelet and *Pleurotus ostreatus* (Jacq.) Fr. P. Kumm. Afr. J. Biotechnol., 14(8): 661–666. 10.5897/AJB2014.14187.

Alam, N., Yoon, K.N., Lee, K.R., Shin, P.G., Cheong, J.C. et al. (2010). Antioxidant activities and tyrosinase inhibitory effects of different extracts from *Pleurotus ostreatus* fruiting bodies. Mycobiol., 38(4): 295–301. 10.4489/MYCO.2010.38.4.295.

Al-Bahrani, R., Raman, J., Lakshmanan, H., Hassan, A.A. and Sabaratnam, V. (2017). Green synthesis of silver nanoparticles using tree oyster mushroom *Pleurotus ostreatus* and its inhibitory activity against pathogenic bacteria. Mater. Lett., 186: 21–25. 10.1016/j.matlet.2016.09.069.

Andlar, M., Rezic, T., Mardetko, N., Kracher, D., Ludwig, R. and Santek B. (2018). Lignocellulose degradation: An overview of fungi and fungal enzymes involved in lignocellulose degradation. Engineering, 18(11): 768–778. 10.1002/elsc.201800039.

Anwar, Z., Gulfraz, M. and Irshad, M. (2014). Agro-industrial lignocellulosic biomass a key to unlock the future bio-energy: A brief review. J. Rad. Res. Appl. Sc., 7(2): 163–173. 10.1016/j.jrras.2014.02.003.

Atri, N.S., Sharma, S.K., Joshi, R., Gulati, A. and Gulati, A. (2012). Amino acid composition of five wild *Pleurotus* species chosen from north-west India. Eur. J. Biol. Sci., 4(1): 31–34. 10.5829/idosi.ejbs.2012.4.1.1102.

Bhat, R., Deshpande, R., Ganachari, S.V., Huh, D.S. and Venkataraman, A. (2011). Photo-irradiated biosynthesis of silver nanoparticles using edible mushroom *Pleurotus florida* and their antibacterial activity studies. Bioinorg. Chem. Appl., 650979. 10.1155/2011/650979.

Boamponsem, G.A., Obeng, A.K., Osei-Kwateng, M. and Badu, A.O. (2013). Accumulation of heavy metals by *Pleurotus ostreatus* from soils of metal scrap sites. Int. J. Curr. Res. Rev., 5: 1–9.

Caglarirmak, N. (2007). The nutrients of exotic mushrooms (*Lentinula edodes* and *Pleurotus* species) and an estimated approach to the volatile compounds. Food Chem., 105(3): 1188–1194. 10.1016/j.foodchem.2007.02.021.

Calderon-Ospina, C.A. and Nava-Mesam, M.O. (2019). B vitamins in the nervous system: Current knowledge of the biochemical modes of action and synergies of thiamine, pyridoxine, and cobalamin. CNS Neurosci. Ther., 26(1): 5–13. 10.1111/cns.13207.

Camarero, S., Sarkar, S., Ruiz-Duenas, F.J., Martinez, M.J. and Martinez, A.T. (1999). Description of a versatile peroxidase involved in the natural degradation of lignin that has both manganese peroxidase and lignin peroxidase substrate interaction sites. J. Biol. Chem., 274(15): 10324–10330. 10.1074/jbc.274.15.10324.

Chang, S.T. and Miles, P.G. (2004). Mushrooms: Cultivation, Nutritional Value Medicinal Effect and Environmental Impact, CRC Press, Boca Raton, FL, USA, p. 451.

Chen, S.Y., Ho, K.J., Hsieh, Y.J., Wang, L.T. and Mau, J.L. (2012). Contents of lovastatin, γ-aminobutyric acid and ergothioneine in mushroom fruiting bodies and mycelia. LWT Food Sci. Technol., 47(2): 274–278. 10.1016/j.lwt.2012.01.019.

Chen, X. and Schluesener, H.J. (2008). Nanosilver: A nanoproduct in medical application. Toxicol. Lett., 176(1): 1–12. 10.1016/j.toxlet.2007.10.004.

Choi, J.W., Yoon, Y.J., Lee, J.H., Kim, C.K., Hong, Y.P. and Shin, S. (2018). Recent research trends of post-harvest technology for king oyster mushroom (*Pleurotus eryngii*). J. Mushr., 16(3): 131–139.

Choi, M.H. and Kim, G.H. (2003). Quality changes in *Pleurotus ostreatus* during modified atmosphere storage as affected by temperatures and packaging material. Acta Hort., 628: 357–362. 10.17660/ActaHortic.2003.628.43.

Cui, L., Liu, Q.H., Wang, H.X. and Ng, T.B. (2007). An alkaline protease from fresh fruiting bodies of the edible mushroom *Pleurotus citrinopileatus*. Appl. Microbiol. Biotechnol., 75(1): 81–85. 10.1007/s00253-006-0801-z.

Daba, A.S. (2006). Production of Anticancer Compound from Oyster Mushroom Fruit Bodies. Patent # WO2006119783A1.

Diamantopoulou, P. and Philippoussis, P. (2015). Cultivated mushrooms: Preservation and processing. pp. 495–525. *In*: Hui, Y.H. and Evranuz, E.Ö. (eds.). Handbook of Vegetable Preservation and Processing, CRC press, Florida.

El-Batal, A.I., Elkenawy, N.M., Yassin, A.S. and Amin, M.A. (2015). Laccase production by *Pleurotus ostreatus* and its application in synthesis of gold nanoparticles. Biotechnol. Rep., 5: 31–39. 10.1016/j.btre.2014.11.001.

El Enshasy, H., Agouillal, F., Mat, Z., Malek, R.A., Hanapi, S.Z. et al. (2019). *Pleurotus ostreatus*: A biofactory for lignin-degrading enzymes of diverse industrial applications. *In*: Yadav, A., Singh, S., Mishra, S. and Gupta, A. (eds.). Recent Advancement in White Biotechnology through Fungi, Fungal Biology, Springer, Cham. 10.1007/978-3-030-25506-0_5.

Erjavec, J., Kos, J., Ravnikar, M., Dreo, T. and Sabotic, J. (2012). Proteins of higher fungi—From forest to application. Tr. Biotechnol., 30(5): 259–273. 10.1016/j.tibtech.2012.01.004.

Facchini, J.M., Alves, E.P., Aguilera, C., Gern, R.M.M., Silveira, M.L.L. et al. (2014). Antitumor activity of *Pleurotus ostreatus* polysaccharide fractions on Ehrlich tumor and Sarcoma 180. Int. J. Biol. Macromol., 68: 72–77. 10.1016/j.ijbiomac.2014.04.033.

FAO. (1973). Energy and protein requirements, report of a joint FAO/WHO ad-hoc Committee. Food Nutr. Rep. Ser. # 52. Food Agric Organ, UN, Rome.

Gayosso-Canales, M., Rodriguez-Vazquez, R., Esparza-Garcia, F.J. and Bermudez-Cruz, R.M. (2012). PCBs stimulate laccase production and activity in *Pleurotus ostreatus* thus promoting their removal. Folia Microbiol., 57(2): 149–158. 10.1007/s12223-012-0106-9.

Girmay, Z., Gorems, W., Birhanu, G. and Zewdie, S. (2016). Growth and yield performance of *Pleurotus ostreatus* (Jacq. Fr.) Kumm (oyster mushroom) on different substrates. AMB Expr., 6(1): 87. 10.1186/s13568-016-0265-1.

Golak-Siwulska, I., Kałuzewicz, A., Spizewski, T., Siwulski, M. and Sobieralski, K. (2018). Bioactive compounds and medicinal properties of Oyster mushrooms (*Pleurotus* sp.). Folia Hortic., 30(2): 191–201. 10.2478/fhort-2018-0012.

Gu, Y.H. and Sivam, G. (2006). Cytotoxic effect of oyster mushroom *Pleurotus ostreatus* on human androgen independent prostate cancer PC-3 cells. J. Med. Food, 9(2): 196–204. 10.1089/jmf.2006.9.196.

Gurunathan, S., Raman, J., Abd Malek, S.N., John, P.A. and Vikineswary, S. (2013). Green synthesis of silver nanoparticles using *Ganoderma neo-japonicum* Imazeki: A potential cytotoxic agent against breast cancer cells. Int. J. Nanomed., 8: 4399–413. 10.2147/IJN.S51881.

Hadibarata, T. and Teh, Z.C. (2014). Optimisation of pyrene degradation by white-rot fungus *Pleurotus pulmonarius* F043 and characterisation of its metabolites. Bioproc. Biosyst. Eng., 37: 1679–1684. 10.1007/s00449-014-1140-6.

Hassan, M., Rouf, R., Tiralongo, E., May, T. and Tiralongo, J. (2015). Mushroom lectins: Specificity, structure and bioactivity relevant to human disease. Int. J. Mol. Sci., 16(4): 7802–7838. 10.3390/ijms16047802.

Ho, L.H., Zulkifli, N.A. and Tan, T.C. (2020). Edible Mushroom: Nutritional Properties, Potential Nutraceutical Values, and its Utilisation in Food Product Development [Online First], IntechOpen.

Hoa, H.T., Wang, C.L. and Wang, C.H. (2015). The effects of different substrates on the growth, yield, and nutritional composition of two oyster mushrooms (*Pleurotus ostreatus* and *Pleurotus cystidiosus*). Mycobiol., 43(4): 423–434. 10.5941/ MYCO.2015.43.4.423.

Itelima, J.U. (2012). Cultivation of mushroom (*Pleurotus ostreatus*) using corn cobs and saw dust as the major substrate. Global J. Agric. Sci., 11(1): 51–56. 10.4314/gjass.v11i1.9.

Jafri, M., Jha, A., Bunkar, D.S. and Ram, R.C. (2013). Quality retention of oyster mushrooms (*Pleurotus florida*) by a combination of chemical treatments and modified atmosphere packaging. Posthar. Biol. Technol., 76: 112–118. 10.1016/ j.postharvbio.2012.10.002.

Jagadeesh, R., Babu, G., Lakshmanan, H., Oh, O.M., Jang, J.K. et al. (2020). Bioactive sterol derivatives isolated from the *Pleurotus djamor* var. *roseus* induced apoptosis in cancer cell lines. Cardiovasc. Hematol. Ag. Med. Chem., 18(2): 124–134. 10.2174/1871525718666200303123557.

Jegadeesh, R., Lakshmanan, H., Kab-yeul, J., Sabaratnam, V. and Raaman, N. (2018). Cultivation of pink oyster mushroom *Pleurotus djamor* var. *roseus* on various agro-residues by low cost technique. J. Mycopathol. Res., 56(3): 213–220.

Johannes, C. and Majcherczyk, A. (2000). Laccase activity tests and laccase inhibitors. J. Biotechnol., 78(2): 193–199. 10.1016/S0168-1656(00)00208-X.

Kamthan, R. and Tiwari, I. (2017). Agricultural wastes—Potential substrates for mushroom cultivation. Eur. J. Exp. Biol., 7(5): 1–4. 10.21767/2248-9215.100031.

Kanaly, R.A. and Hur, H.G. (2006). Growth of *Phanerochaete chrysosporium* on diesel fuel hydrocarbons at neutral pH. Chemospherem., 63(2): 202–211. 10.1016/j.chemosphere.2005.08.022.

Kapahi, M. and Sachdeva, S. (2017). Mycoremediation potential of *Pleurotus* species for heavy metals: A review. Bioresour. Bioprecess., 4(1): 32–41. 10.1186/s40643-017-0162-8.

Karaman, M., Janjusevic, L., Jakovljević, D., Sibul, F. and Pejin, B. (2018). Anti-hydroxyl radical activity, redox potential and anti-AChE activity of *Amanita strobiliformis* polysaccharide extract. Nat. Prod. Res., 33(10): 1522–1526. 10.1080/14786419.2017.1422183.

Karthikeyan, V., Ragunathan, R., Jesteena, J. and Kabesh, K. (2019). Green synthesis of silver nanoparticles and application in dye decolourisation by *Pleurotus ostreatus* (MH591763). Glob. J. Bio-Sci. Biotechnol., 8: 80–86.

Khandel, P. and Shahi, S.K. (2018). Mycogenic nanoparticles and their bio-prospective applications: Current status and future challenges. J. Nanostruct. Chem., 8: 369–391. 10.1007/s40097-018-0285-2.

Kim, B.R. and Joo, N.M. (2012). Optimisation of sweet rice muffin processing prepared with oak mushroom (*Lentinus edodes*) powder. J. Korean Soc. Food Cult., 27(2): 202–210.

Kopinski, L. and Kwiatkowska-Marks, S. (2012). Utilisation of waste newspaper using oyster mushroom mycelium. Ind. Eng. Chem. Res., 51(11): 4440–4444. 10.1021/ie202765b.

Kora, A.J. (2020). Nutritional and antioxidant significance of selenium-enriched mushrooms. Bull. Natl. Res. Cent., 44: 34. 10.1186/s42269-020-00289-w.

Koutrotsios, G., Mountzouris, K.C., Chatzipavlidis, I. and Zervakis, G.I. (2014). Bioconversion of lignocellulosic residues by *Agrocybe cylindracea* and *Pleurotus ostreatus* mushroom fungi-assessment of their effect on the final product and spent substrate properties. Food Chem., 161: 127–35. 10.1016/j.foodchem.2014.03.121.

Koyyalamudi, S.R., Jeong, S.C., Song, C.H., Cho, K.Y. and Pan, G. (2009). Vitamin D_2 formation and bioavailability from *Agaricus bisporus* button mushrooms treated with ultraviolet irradiation. J. Agric. Food Chem., 57(8): 3351–3355. 10.1021/jf803908q.

Kumla, J., Suwannarach, N., Sujarit, K., Penkhrue, W., Kakumyan, P. et al. (2020). Cultivation of mushrooms and their lignocellulolytic enzyme production through the utilisation of agro-industrial waste. Molecules, 25(12): 2811. 10.3390/molecules25122811.

Lavi, I., Friesemm, D., Gereshm, S., Hadarm, Y. and Schwartzm, B. (2006). An aqueous polysaccharide extract from the edible mushroom *Pleurotus ostreatus* induces anti-proliferative and pro-apoptotic effects on HT-29 colon cancer cells. Canc. Lett., 244(1): 61–70. 10.1016/j.canlet.2005.12.007.

Li, W., Li, X., Yang, Y., Zhou, F., Liu, L. et al. (2015). Effects of different carbon sources and C/N values on nonvolatile taste components of *Pleurotus eryngii*. Int. J. Food Sci. Technol., 50: 2360–2366. 10.1111/ijfs.12901.

Li, Y.R., Liu, Q.H., Wang, H.X. and Ng, T.B. (2008). A novel lectin with potent antitumor, mitogenic and HIV1 reverse transcriptase inhibitor activities from the edible mushroom *Pleurotus citrinopileatus*. Biochim. Biophys. Acta Gen. Sub., 1780(1): 51–57. 10.1016/j.bbagen.2007.09.004.

Lin, S.Y., Chien, S.C., Wang, S.Y. and Mau, J.L. (2016). Non-volatile taste components and antioxidant properties of fruiting body and mycelium with high ergothioneine content from the culinary-medicinal golden oyster mushroom *Pleurotus citrinopileatus* (Agaricomycetes). Int. J. Med. Mushr., 18(8): 689–698.

Martinez, A.T., Camarero, S., Guillen, F., Gutierrez, A., Munoz, C. and Varela, E. (1994). Progress in biopulping of non-woody materials: chemical, enzymatic and ultrastructural aspects of wheat-straw delignification with ligninolytic fungi from the genus *Pleurotus*. FEMS Microbiol. Rev., 13: 265–274. 10.1111/j.1574-6976.1994.tb00047.x.

Masutti, D., Borgognone, A., Scardovi, F., Vaccari, C. and Setti, L. (2015). Effects on the enzymes production from different mixes of agro-food wastes. Chem. Eng. Trans., 43: 487–498. 10.3303/CET1543082.

Mayolo-Deloisa, K., González-González, M. and Rito-Palomares, M. (2020). Laccases in food industry: Bioprocessing, potential industrial and biotechnological applications. Front. Bioeng. Biotechnol., 8: 222. 10.3389/fbioe.2020.00222.

Minussi, R.C., Pastore, G.M. and Duran, N. (2002). Potential applications of laccase in the food industry. Tr. Food Sci. Technol., 13(6-7): 205–216. 10.1016/S0924-2244(02)00155-3.

Morais, H., Forgacs, E. and Cserhati, T. (2005). Enzyme production of the edible mushroom *Pleurotus ostreatus* in shaken cultures completed with agro-industrial wastes. Eng. Life Sci., 5(2): 152–157. 10.1002/elsc.200420065.

Morris, H.J., Beltran, Y., Llaurado, G., Batista, P.L., Perraud-Gaime, I. et al. (2017). Mycelia from *Pleurotus* sp. (oyster mushroom): A new wave of antimicrobials, anticancer and antioxidant bioingredients. Int. J. Phytocosm. Nat. Ingred., 4(1): 3. 10.15171/ijpni.2017.14.

Mussatto, S.I., Ballesteros, L.F., Martins, S. and Teixeira, J.A. (2012). Use of agro-industrial wastes in solid-state fermentation processes. pp. 121–140. *In*: Show, K.-Y. and Guo, X. (eds.). Industrial Waste. InTech. http://www.intechopen.com/books/industrial-waste/use-of-agro-industrialwastes-in-solid-state-fermentation-processes.10.5772/36310.

Muyanja, C., Kyambadde, D. and Namugumya, B. (2012). Effect of pretreatments and drying methods on chemical composition and sensory evaluation of oyster mushroom (*Pleurotus ostratus*) powder and soup. J. Food Proc. Preser., 38(1): 457–465. 10: 1111/j.1745-4549.2102.00794.x.

Nakano, Y., Fujii, H. and Kojima, M. (2010). Identification of blue-light photoresponse genes in oyster mushroom mycelia. Biosci. Biotechnol. Biochem., 74(10): 2160–2165. 10.1271/bbb.100565.

Nikiforova, S.V., Pozdnyakova, N.N. and Turkovskaya, O.V. (2009). Emulsifying agent production during PAHs degradation by the white rot fungus *Pleurotus ostreatus* D1. Curr. Microbiol., 58(6): 554–558. 10.1007/s00284-009-9367-1.

Okolo, K.O., Siminialayi, I.M. and Orisakwe, O.E. (2016). Protective effects of *Pleurotus tuber-regium* on carbon-tetrachloride induced testicular injury in Sprague Dawley Rats. Front. Pharmacol., 7: 480. 10.3389/fphar.2016.00480.

Omrane, M., Mechri, S., Jaouadi, N.Z., Elhoul, M.B., Rekik H. et al. (2019). Purification and biochemical characterization of a novel thermostable protease from the oyster mushroom *Pleurotus sajor-caju* strain CTM10057 with industrial interest. BMC Biotechnol., 19: 43. 10.1186/s12896-019-0536-4.

Owaid, M.N. (2019). Green synthesis of silver nanoparticles by *Pleurotus* (oyster mushroom) and their bioactivity: Review. Environ. Nanotechnol. Monit. Manag., 12: 100256. 10.1016/j.enmm.2019.100256.

Pandya, U., Dhuldhaj, U. and Sahay, N.S. (2018). Bioactive mushroom polysaccharides as antitumor: An overview. Nat. Prod. Res., 33(18): 2668–2680. 10.1080/14786419.2018.1466129.

Patel, H., Gupte, A. and Gupte, S. (2009). Effect of different culture conditions and inducers on production of laccase by a Basidiomycete fungal isolate *Pleurotus ostreatus* HP-1 under solid-state fermentation. BioResources, 4(1): 268–284.

Patel, S. and Goyal, A. (2012). Recent developments in mushroom as anti-cancer therapeutics: A review. 3 Biotech, 2(1): 1–15. 10.1007/s13205-011-0036-2.

Patil, S.S., Ahmed, S.A., Telang, S.M. and Baig, M.M.V. (2010). The nutritional value of *Pleurotus ostreatus* (Jacq.:Fr.) Kumm cultivated on different lignocellulosic agro-wastes. Innov. Rom. Food Biotechnol., 7: 66–76.

Patil, S.S. (2013). Productivity and proximate content of *Pleurotus sajor-caju*. Biosci. Discov., 4(2): 169–172.

Pawlik, A., Janusz, G., Koszerny, J., Malek, W. and Rogalski, J. (2012). Genetic diversity of the edible mushroom *Pleurotus* sp. by amplified fragment length polymorphism. Curr. Microbiol., 65(4): 438–445. 10.1007/s00284-012-0175-7.

Prabhu, N., Karunakaran, S., Devi, S.K., Ravya, R., Subashree, P. and Vadivu, M.S. (2019). Biogenic synthesis of myconanoparticles from mushroom extracts and its medical applications: A review. Int. J. Pharm. Sci. Res., 10(5): 2108–2018. 10.13040/IJPSR.0975-8232.10(5).2108-18.

Qu, J., Huang, C. and Zhang, J. (2016). Genome-wide functional analysis of SSR for an edible mushroom *Pleurotus ostreatus*. Gene, 575: 524–530. 10.1016/j.gene.2015.09.027.

Raj, D.D., Mohan, B. and Shetty, V. (2011). Mushrooms in the remediation of heavy metals from soil. Int. J. Environ. Pollut. Cont. Manag., 3(1): 89–101.

Raman, J., Lakshmanan, H., John, P.A., Zhijian, C., Periasamy, V. et al. (2015). Neurite outgrowth stimulatory effects of mycosynthesized AuNPs from *Hericium erinaceus* (Bull.: Fr.) Pers. on pheochromocytoma (PC-12) cells. Int. J. Nanomed., 10(1): 5853–5863. 10.2147/IJN.S88371.

Raman, J., Jang, J.-Y., Oh, Y.-L., Oh, M., Im, J.-H. et al. (2020). Cultivation and nutritional value of prominent *Pleurotus* spp.: An overview. Mycobiol., 49(1): 1–14. 10.1080/12298093.2020.1835142.

Rao, M.B., Tanksale, A.M., Ghatge, M.S. and Deshpande, V.V. (1998). Molecular and biotechnological aspects of microbial proteases. Microbiol. Mol. Biol. Rev., 62(3): 597–635.

Ravindran, R. and Jaiswal, A.K. (2016). Exploitation of food industry waste for high-value products. Tr. Biotechnol., 34(1): 58–69. 10.1016/j.tibtech.2015.10.008.

Razzaq, A., Shamsi, S., Ali, A., Ali, Q., Sajjad, M. et al. (2019). Microbial proteases applications. Front. Bioeng. Biotechnol., 7: 110. 10.3389/fbioe.2019.00110.

Rico-Garcia, D., Ruiz-Rubio, L., Perez-Alvarez, L., Hernandez-Olmos, S.L., Guerrero-Ramirez, G.L. and Vilas-Vilela, J.L. (2020). Lignin-based hydrogels: Synthesis and applications. Polymers, 12: 81. 10.3390/polym12010081.

Ritota, M. and Manzi, P. (2019). *Pleurotus* spp. cultivation on different agri-food by-products: Example of biotechnological application. Sustainability, 11(18): 5049. 10.3390/su11185049.

Royse, D.J. (2014). A global perspective on the high five: *Agaricus, Pleurotus, Lentinula, Auricularia* and *Flammulina*. Proceedings of the 8th International Conference on Mushroom Biology and Mushroom Products (ICMBMP8), 1: 1–6.

Ryu, J., Kim, M.K., Im, C.H. and Shin, P. (2015). Development of cultivation media for extending the shelf-life and improving yield of king oyster mushrooms (*Pleurotus eryngii*). Sci. Hortic., 193: 121–126. 10.1016/j.scienta.2015.07.005.

Sadh, P.K., Duhan, S. and Duhan, J.S. (2018). Agro-industrial wastes and their utilisation using solid state fermentation: A review. Bioresour. Bioprocess., 5: 1. 10.1186/s40643-017-0187-z.

Saglam, N., Yesilada, O., Cabuk, A., Sam, M., Saglam, S. et al. (2016). Innovation of strategies and challenges for fungal nanobiotechnology. *In*: Prasad, R. (ed.). Advances and Applications through Fungal Nanobiotechnology, Fungal Biology. Springer, Cham. 10.1007/978-3-319-42990-8_2.

Salami, A.O., Bankole, F.A. and Salako, Y.A. (2017). Nutrient and mineral content of oyster mushroom (*Pleurotus florida*) grown on selected lignocellulosic agro-waste substrates. J. Adv. Biol. Biotechnol., 15(1): 1–7. 10.9734/JABB/2017/35876.

Salehi, F. (2019). Characterisation of different mushrooms powder and its application in bakery products: A review. Int. J. Food Prop., 22(1): 1375–1385. 10.1080/10942912.2019.1650765.

Sari, M., Prange, A., Lelley, J.I. and Hambitzer, R. (2017). Screening of beta-glucan contents in commercially cultivated and wild growing mushrooms. Food Chem., 216: 45–51. 10.1016/j. foodchem.2016.08.010.

Shukla, C.S. and Biswas, M.K. (2000). Evaluation of different techniques for oyster mushroom cultivation. J. Mycol. Plant Pathol., 30(3): 431–435.

Taofiq, O., Gonzalez-Paramas, A.M., Martins, A., Barreiro, M.F. and Ferreira, I.C.F.R. (2016). Mushroom extracts and compounds in cosmetics, cosmeceuticals and nutricosmetics—A review. Ind. Crops Prod., 90: 38–48. 10.1016/j.indcrop.2016.06.012.

Telang, S.M., Patil, S.S. and Baig, M.M.V. (2010). Biological efficiency and nutritional value of *Pleurotus sapidus* cultivated on different substrates. Food Sci. Res. J., 1: 127–129.

Tengerdy, R.P. and Szakacs, G. (2003). Bioconversion of lignocellulose in solid substrate fermentation. Biochem. Eng. J., 13: 169–179. 10.1016/S1369-703X(02)00129-8.

Urben, A.F. (2004). Mushroom production using modified Chinese technology Embrapa. Genetic Resources and Biotechnology, Brasília.

Wakchaure, G.C. (2011). Mushrooms-value added products. pp. 235–238. *In*: Singh., M., Vijay., B., Kamal, S. and Wakchaure, G.C. (eds.). Mushrooms Cultivation, Marketing and Consumption, Directorate of Mushroom Research, Chambaghat, Solan, India.

Wang, H. and Ng, T.B. (2004). Eryngin, a novel antifungal peptide from fruiting bodies of the edible mushroom *Pleurotus eryngii*. Peptides, 25(1): 1–5. 10.1016/j.peptides.2003.11.014.

WHO-FAO. (2004). Vitamin and Mineral Requirements in Human Nutrition. Geneva, 2nd ed. http://www. who.int/nutrition/publications/micronutrients/9241546123/en/.

Wu, J.Y., Chen, C.H., Chang, W.H., Chung, K.T., Liu, Y.W. et al. (2011). Anticancer effects of protein extracts from Calvatia lilacina, *Pleurotus ostreatus* and *Volvariella volvacea*. Evid. Based Compl. Alt. Med., 10.1093/ecam/ neq057.

Xiao, G., Zhang, M., Shan, L., You, Y. and Salokhe, V.M. (2011). Extension of the shelf-life of fresh oyster mushrooms (*Pleurotus ostreatus*) by modified atmosphere packaging with chemical treatments. Afr. J. Bitechnol., 10(46): 9509–9517. 10.5897/AJB08.974.

Yadav, K.K., Garg, N., Shukla, P.K., Kumar, S., Yadav, P. and Kumar, D. (2012). Preliminary studies on utilization of fruit-waste compost for cultivation of oyster mushroom (*Pleurotus florida* (Mont.) Singer). Mush. Res., 21(1): 75–78.

Yang, R.L., Li, Q. and Hu, Q.P. (2020). Physicochemical properties, microstructures, nutritional components and free amino acids of *Pleurotus eryngii* as affected by different drying methods. Sci. Rep., 10: 121. 10.1038/s41598-019-56901-1.

Yusuf, R.O., Noor, Z.Z., Abba, A.H., Hassan, M.A.A. and Din, M.F.M. (2012). Methane emission by sectors: A comprehensive review of emission sources and mitigation methods. Ren. Sust. Energy Rev., 16(7): 5059–5070. 10.1016/j.rser.2012.04.008.

Zhang, Y., Dai, L., Kong, X. and Chen, L. (2012). Characterisation and *in vitro* antioxidant activities of polysaccharides from *Pleurotus ostreatus*. Int. J. Biol. Macromol., 51(3): 259–265. 10.1016/j. ijbiomac.2012.05.003.

Zhou, J., Chen, M., Wu, S., Liao, X., Wang, J. et al. (2020). A review on mushroom-derived bioactive peptides: Preparation and biological activities. Food Res. Int., 134: 109230. 10.1016/j. foodres.2020.109230.

Zhou, R., Liu, Z.K., Zhang, Y.N., Wong, J.H., Ng, T.B. and Liu, F. (2019). Research progress of bioactive proteins from the edible and medicinal mushrooms. Curr. Protein Pept. Sci., 20(3): 196–219. 10.21 74/1389203719666180613090710.

Zolghadri, S., Bahrami, A., Khan, M.T.H., Munoz-Munoz, J., Garcia-Molina, F. et al. (2019). A comprehensive review of tyrosinase inhibitors. J. Enz. Inhib. Med. Chem., 34(1): 279–309. 10.1080/14756366.2018.1545767.

8

Macrofungi in the Production of Alcoholic Beverages
Beer, Wine, and Spirits

Jovana Vunduk[1,]* and *Sonja Veljović*[2,]*

1. INTRODUCTION

Humankind has a long history of alcoholic beverage production and inebriation, probably as soon as the last Ice Age terminated (Guerra-Doce, 2014; Biwer and Van Derwarker, 2015; Liu et al., 2019). Development of archeobotanical and archeochemical analyses (starch grains, phytolith remains, acids, presence of moulds and yeasts) enabled scientific evidence of sources and types of alcoholic beverages, as well as the social context of their consumption (McGovern et al., 2004; Liu et al., 2019). There are two opinions concerning the oldest type of alcoholic beverage: beer or wine. Beer seems a logical consequence of grain domestication and evidence for it is scattered all over the world in the form of grinding stones, ceramics and different forms of starch granules found incorporated in the walls of vats and jars, the oldest one going back to Natufian (also known as Levant culture) (Hayden et al., 2012). A recent discovery from China also points to beer (made from rice, honey and fruit) as the first intentionally produced alcoholic beverage dating from around 9000–7500 BC (Liu et al., 2019). Other scientists doubt this evidence having in mind the complexity of the beer production process and propose grape wine production as 'non-inventive', thus more logical as the first massively produced alcoholic beverage (Kjellgren, 2004). The oldest evidence-supported proof of wine production comes from the village of

[1] Research Associate, Institute for Food Technology and Biochemistry, Faculty of Agriculture, Nemanjina 6, 11080, University of Belgrade, Belgrade, Serbia.
[2] Institute of General and Physical Chemistry, P.O. Box 551, 11001 Belgrade, Serbia.
* Corresponding authors: vunduk.jovana@hotmail.com; pecic84@hotmail.com

Jiahu, Henan province in China (McGovern et al., 2004). It is estimated to originate from around the Early Neolithic Period (7000–6600 BC) and was based on the wild grape variety mixed with other raw materials, like rice, honey and hawthorn fruit. Scientists suggest the political and cultural importance of alcohol, but the written testimony points to its healing purpose, like the texts from ancient Mesopotamia (around 2012–2004 BC) (Scurlock and Andersen, 2005). Liu et al. (2019) also pointed out that the earliest alcoholic beverages were used as health supports. However, fungi as the kingdom, particularly macrofungi (mushrooms), predate both civilisation development and grain domestication; thus the alcohol production as well. Naturally present in different environments, convenient and nutritious, mushrooms probably were a part of the human diet long before the occurrence of the intended alcohol fermentation. Artifacts found in the bag of Öetzi (the 'Iceman') point to mushroom use other than food (Peintner et al., 1998). Contemporary bioactivity, screening of macrofungal species, including *Fomes fomentarius* and *Piptoporus betulinus* found in Öetzi's bag support the speculation about its medicinal use, dating as far as 3500 BCE (Vunduk et al., 2015; Gründemann et al., 2020). Moreover, ancient cultures like Chinese had extensive knowledge and appreciation of macrofungi. Chinese pharmacopoeia, called the *Compendium of Materia Medica* from 1578, lists diverse mushroom species, physical conditions that can be addressed and the method to prepare the medicines.

Apart from the ancient scripts, the science of macrofungi went through some intensive breakthroughs: pharmacy, food industry, agriculture waste management, mycoremediation, construction materials, medicine and even fashion, just to name a few (Wasser, 2014; Kamthan et al., 2017; Marchand and Stewart, 2018; Barh et al., 2019; Novaković et al., 2019; Silverman et al., 2020; Jones et al., 2020). Finally, hybrids between the disciplines, like in the case of medicine and food, resulted as a functional food appeared (Okhuoya, 2017). Mushrooms are extensively exploited as nutraceuticals due to their numerous bioactive molecules, like polysaccharides, peptides, triterpenes and complexes like glycoproteins (Rathore et al., 2017; Veljović and Krstić, 2020). All phases of the mushroom life cycle are used: the mycelium, fruit body and the spores. Species are not just collected in nature; significant improvement and efforts are put into the development of mushroom cultivation techniques, selection of strains, stimulation of bioactive molecules synthesis to provide a secure supply and high quality according to the market demand and safety regulations (Isikhuemhen et al., 2010; Jang and Lee, 2014; Golian et al., 2015). However, the possibility of using macrofungi as a raw material in the production of alcoholic beverages is a relatively new concept. Although popular knowledge advises not to mix mushrooms with alcohol, the latest scientific findings suggest that some mushrooms contain alcohol-dehydrogenase, the key enzyme in alcohol synthesis, which enables them to perform fermentation (Okamura-Matsui et al., 2003). Furthermore, medicinal mushrooms can be added as additives that not only improve the medicinal properties of beverages but contribute to the flavour, aroma and colour (Leskošek-Čukalović et al., 2010a; Leskošek-Čukalović et al., 2010b). These sensory changes help make the beverage more distinctive and appealing, especially in the markets with a long tradition of health drinks (Kim et al., 2004; Nguyen et al., 2019a).

The current chapter provides an overview of the state-of-the-art of the use of macrofungi in the production of three main alcoholic beverages: beer, wine and spirits. The history of medicinal use and the incorporation of mushrooms from the aspect of different addition procedures—technologies that the existing macrofungal species used, the effect of their addition on the beverage's medicinal and sensory properties, and the situation in the market are addressed. In addition, spirits based on *Ganoderma lucidum*, the most apprised and studied medicinal mushroom, are also discussed.

2. History of Macrofungi for Applying to Alcoholic Beverages

The only written historical evidence of mushroom use as a part of folk medicine, which includes alcoholic beverages, originates from China. In ancient times, mushrooms were collected in nature and many of them, like *G. lucidum*, were scarce, thus expensive and reserved for royalty (Wachtel-Galor et al., 2011). Other more abundant species found their way into folk medicine. Although some were used in the form of powder, Chinese physicians applied alcohol extract as well. Concerning the use of macrofungi as a part of alcoholic remedies, there are two essential works: (1) *Compendium of Materia Medica*; (2) *Thousand Golden Prescriptions*. Although the first one is more famous, the latter is older, dating from 652 AD, written by Sun Simiao. Three mushrooms are appearing in the recipes mentioned in this book: *Poria cocos*, *Polyporus umbellatus* and the silkworm mushroom *Cordyceps militaris*. The majority of prescriptions included *P. cocos*, and there were eleven of them. The mushroom was mixed with numerous medicinal plants and soaked or cooked in wine. This means that Chinese physicians knew two ways of extraction of mushroom's bioactive compounds. They performed cold and hot alcohol extraction and in some cases, the ingredients were cooked in water and wine, enabling double extraction. The *P. umbellatus* was used in two recipes, which combine it with wine, again applying cold or hot extraction. This one was also used in a combination with *P. cocos*. Finally, only one prescription explains the method for the preparation of a wine-based medicine, in which silkworm mushroom was soaked in wine and used for menstrual problems. The listed recipes were always based on yellow rice wine, in which alcohol content varied from 14–20 per cent. The distilled wine, or *Baidju*, as a part of medicine only appears much later in the compendium of *Materia Medica* (1596 AD), that is at the beginning of the Yuan dynasty. Interestingly, wine-based prescriptions with *C. militaris* and *P. umbellatus* evanesce from the latter folk medicine, while two new species were introduced, *Ganoderma sinensis* and *Auricularia auricula*. Cold and hot alcohol extraction of mushroom powder remained as the methods of choice in preparation.

3. The Chemistry of Macrofungi in Alcoholic Remedies

As evident from the historical sources, mushrooms in ancient China were used in the form of powder mixed with other medicinal herbs or combined with wine. Although scientific evidence for the impact of other-than-water solvent extraction of bioactive molecules is of a more recent date, a long tradition and evidence provided ancient physicians with enough support to apply cold or hot alcohol extraction in the form of soaking or boiling in rice wine (Ngo et al., 2017). A significant number of research conducted in the last forty years identified the main bioactive compounds in mushrooms (Öztürk et al., 2015). However, polysaccharides, proteins, amino acids, fatty acids, phenolic compounds, vitamins and pigments cannot be universally extracted with the same solvent or procedure, especially not with the same efficacy (Sasidharan et al., 2010). Moreover, the type of solvent is not the only parameter affecting the yield of the extract. Temperature, time of extraction, raw material particle size, all exhibit the effect on the procedure's outcome (Petrović et al., 2014; Cör et al., 2017; Ngo et al., 2017). For the extraction of secondary metabolites, from plants and mushrooms equally, use of acetone, methanol and ethanol are a common choice. They are suitable for the extraction of both polar and some non-polar molecules, like polyphenols, alkaloids, terpenes and glycoside. At the same time, these solvents can be evaporated under lower temperatures (less than 100°C), which is important for preservation of bioactivity (Rosselló-Soto et al., 2019). Mushrooms, particularly *G. lucidum* is rich in phenolic compounds, which are easily extracted with polar solvents (Kim et al., 2008; Oludemi et al., 2018). Polyphenols exhibit health-promoting effects, like blood sugar regulation, anti-tumour, anti-oxidative, cardioprotective, anti-inflammatory, antimicrobial and antiallergenic effects (Butkhup et al., 2017). Moreover, phenolics from mushrooms promote detoxification of the body (Shomali et al., 2019) and even alcoholic beverages (wine and beer), when consumed moderately, can exhibit health-promoting effects (Arranz et al., 2012). Based on these novel findings, it is no wonder that mushroom—supplemented rice wine, used almost 2,000 years ago, found its place for treating conditions like kidney disorder, typhus, edema, short breath and weak joints or liver. Having in mind the effect of the preparation procedure of health wine on phenolic compounds, there is a justified reason to believe that the ancient recipes were efficient at least as the radical scavengers, when consumed as prescribed. The same logic was applied in the last twenty years when the health-wine production procedures, which include mushrooms, exploded in China (Zhou et al., 2011). On the other hand, novel procedures and alcoholic beverages with macrofungi were developed in laboratories in Europe and Australia plus in the market (USA) (Nguyen et al., 2020; https://www.foodrepublic.com/2015/01/21/mushrooms-have-entered-the-craft-beer-game/). Since many of these new beverages are already patented, the available scientific literature is scarce.

4. Macrofungi in Production of Alcoholic Beverages—The Challenges

Medicinal mushrooms could be a source of intoxication due to their ability to accumulate toxic elements from their environment (Kalač, 2000; Lalotra et al., 2016). Heavy metals and radionuclides are often found in the fruit bodies collected from polluted areas, smelters and urban regions (Kalač et al., 2004; Falandysz et al., 2017). Some mushrooms, like *Amanita muscaria*, *Xerocomus badius*, *Laccaria amethistina* and *Suillus granulatus* are even able for hyperaccumulation of specific elements (Falandysz, 2010). Moreover, different microorganisms, like spore-forming or pathogenic bacteria, can end up on the mushroom fruit body since it grows near the soil (Ma et al., 2019). Other microorganisms, like moulds and non-pathogenic bacteria can also be prevail in the raw material, which will later be a source of beverage spoiling, especially if it contains lower alcohol amounts, like beer. Unfiltered and unpasteurised beers, that tend to preserve biologically active compounds, thus can become particularly dangerous. A tangible solution is to use mycelium obtained by submerged fermentation where all the production parameters are controlled.

Another problem with the use of the fruit body is standardisation (Chang and Miles, 2004). The size, structure and amount of biomolecules cannot be controlled if the mushroom is used directly as a powdered fruit body. Extraction serves as an easy solution since it enables precise content of biomolecules as well as the dose. Macrofungi being annual organisms are sensitive to temperature, moisture, nutrient levels and light (Sher et al., 2010). If the production of beverage relies on the collected mushrooms from nature, it can be hard to predict the season and delivery of the same quality every year. Moreover, mushrooms are perishable goods with their tendency to spoil in one to three days, when stored under room temperature (Vunduk et al., 2018). Even when refrigerated, they need to be used in no more than two weeks (Djekic et al., 2016). Every day of cold storage affects the change in their nutritive and functional value. One option is to use dried and powdered fruit bodies. Extracts are even better since they can be standardised and stored for one to two years. Finally, with the development of technology for submerged fermentation (air-lift fermentation), the production can be faster, cleaner and standardised (Vunduk et al., 2019). Both mycelium and fermentation broth can be used as raw materials in alcoholic beverage production.

4.1 Beer

Beer is considered as the oldest and the most consumed alcoholic beverage worldwide (Lorencová et al., 2019). In 2012, Hayden et al. published a paper in which they speculated about 13,000 year-long history of beer brewing in Natufian although more solid evidence dates it back to Early Neolithic China (about 9000–7500 BC) (Liu et al., 2019). Mesopotamian and Egyptian cultures were the first to report about beer in the form of text (el-Guebaly and el-Guebaly, 1981; Damerow, 2012). Although a part of political and social culture, beer was a sort of medicine as well, starting from the period of the Middle Kingdom (2160–1580 BC) (el-Guebaly and el-Guebaly, 1981). The raw materials for beer brewing differentiated according to the local conditions and available sources. In southern Peru, the Chicha beer was made from maize with a lower percentage of alcohol, while in ancient China, rice beer has been produced from 1700–1100 BC with much higher alcohol content (10–20 per cent) and barley in Euro-Asia (Jennings et al., 2005; Biwer and Van Derwarker, 2015; Guerra-Doce, 2014). These early beers have all sorts of problems, ranging from unsatisfactory taste to being prone to spoilage. Every culture had its approach toward these problems but in the majority of cases, additives like fruit, honey and herbs were used (el-Guebaly and el-Guebaly, 1981; Jennings et al., 2005; Leskošek-Čukalović et al., 2010b; Guerra-Doce, 2014; Liu et al., 2019). Diverse additives were always present in the beer production out of necessity as well as taste. Only recently, when the German pure beer law ('*reinheitsgebot*', 1516) was introduced, beer is considered to be made only from four ingredients: barley malt, water, yeast and hops. However, with the revival of craft beer (so-called 'craft beer renaissance'), novel scientific findings of its health benefits, expanding of the market over India and Asia, and an enormous economic potential (593.024 million US$ in 2017 to 685–354 million US$ in 2025) beer differentiation throughout new ingredients seems inevitable (Arranz et al., 2012; Gómez-Corona et al., 2016; Donadini and Porretta, 2017; Gatrell et al., 2018; Bertuzzi et al., 2020; https://www.alliedmarketresearch.com/beer-market). As stated by Capece et al. (2018), a new beer category emerged as functional beers. The distinguishing characteristic of this type of beer is that it contains health-promoting ingredients and if consumed moderately, health beneficial effects should be expected (Sánchez-Muniz et al., 2019). Functional food represents a relatively new food category, but in the case of beer, the situation is different (Hasler, 2002). It appeared in the 18th century, but under a different name 'medicinal beer' as noted by Wouter van Lis, who was a Dutch medical doctor. In his apothecary handbook, van Lis proposed two methods for making medicinal beer (Peeters, 1988). Herbs were to be soaked in a drink or added during the fermentation process. New research attempts followed to apply the same principles (Đorđević et al., 2015). However, contemporary producers and researchers are exploring a wide range of biologically active compounds, different sources, extraction techniques and macrofungi represent one of them (Leskosek-Cukalovic et al., 2010a; Đorđević et al., 2015; Veljović et al., 2015). New attempts in making functional beer are aiming to go beyond health benefits. Sensory properties, like colour, taste, flavour and alcohol content are considered since the new consumer category has evolved (Jaeger et al., 2020).

4.1.1 Macrofungi in Beer Production

To date, research on the potential of mushroom use in any phase of beer production is scarce and mainly limited to Europe. On the other side, there are several patents for mushroom beer and the majority of them were submitted in China. The reason for this should be looked at in the Chinese culture, where medicinal mushrooms do not represent an exotic ingredient, but as well-known traditional medicine and food. Besides, consumers are already accustomed to their specific taste. There are several points of view when it comes to the existing evidence of mushroom use.

Macrofungi consist of the vegetative part (mycelium) and a reproductive part (fruit body), which have different constituents according to the role they serve in an organism's life. The dominating constituents in the mushroom body express specific biological activity based on its chemical nature. This topic is extensively researched and reported elsewhere, so it is not covered in this chapter (*see* Wasser, 2002; Zjawiony, 2004; Moradali et al., 2007; Kozarski et al., 2015; Thu et al., 2020). For example, fruit body and especially spores are rich in phenolic compounds and unsaturated fatty acids, which are known as good antioxidants and cardio-protective substances (Carroll and Roth, 2002; Kozarski et al., 2015; Sánchez, 2017; Sande et al., 2019). Thus, which part of the mushroom is to be used in beer production is of great significance if the main purpose is to have a drink with pharmacodynamical properties. This is not the only matter of concern, but physicochemical interactions between phenolic compounds and proteins could cause turbidity in the resulting beer, while polysaccharides cause the turbidity only by its presence (Belščak-Cvitanović et al., 2017). This visual parameter can be of great importance depending on the market. Donadini and Porretta (2017) proved that Italians do not like turbid beer (especially women). The second important factor is the exact form of mushroom which is to be exploited. There are different options, like dried mushroom powder, extract and microencapsulated extract. In the case of submerged-fermentation-produced mycelium, even the fermentation broth can be used (Barkov and Vinokurov, 2017). However, every material has an effect on the final product and needs to be extensively examined before introduction in commercial practice. When it comes to the phase of production in which to add fungal material, there are two options—before the primary fermentation or after it. The following discussion reflects on a combination of all the listed options exiting in the published research.

4.1.2 *Beer Fermentation using Macrofungi*

To convert pyruvate to ethanol and CO_2, yeasts employ alcohol dehydrogenase. This enzyme finishes the two-step process of alcohol fermentation, catalysing the transformation of pyruvic acid into carbon dioxide and ethanol. The discovery of alcohol dehydrogenase in macrofungi at the beginning of the 21st century opens a new area of possibilities when it comes to ethanol production (Okamura et al., 2001a). The first attempt in examining the real effectiveness of this novel enzyme source in beer production was performed by a group of Japanese scientists (Okamura et al., 2001a). They used two popular edible mushrooms, *Tricholoma matsutake* and *Flammulina velutipes* in a small-scale experiment. A cell-free extract of these two species was added to a hopped malt extract and fermented under the same conditions as for regular beer brewing. After two weeks of fermentation, a beer-like beverage was produced. It was reported that *T. matsutake* is more suitable for fetching a higher percentage of alcohol (4.6 per cent v/v) compared with *F. velutipes* (3 per cent v/v), with a dominant taste of mushroom's fruit body. A similar experiment was repeated now with more macromycetes species (Okamura-Matsui et al., 2003). However, *Agaricus blazei* and *Pleurotus ostreatus* produced no alcohol when added to hopped malt extract, implying that not all mushroom species can be used for this purpose. The mushroom-based beer had higher β-glucan content and exhibited thrombosis-preventing activity. Interestingly, the mentioned research is the only existing so far; all others are patents and they are diverse according to the macrofungi used and its preparation before being added to the wort. The majority of patents use the mushroom fruit body prepared in different ways. The existing options are:

- to powder dry fruit body and add it in the saccharification phase
- to prepare hot water extract and add it into the non-hopped wort
- to sterilise powdered mushroom with non-hopped wort without filtration
- to sterilise powdered mushroom with hopped wort
- to prepare water extract heated for a short time and hydrolyse it with enzymes

The most recent patent, from 2017, included mycelium produced by submerged fermentation (Barkov and Vinokurov, 2017). Again, several options were developed:

(1) the use of mycelium (solid part); (2) mycelium and fermentation broth; (3) only fermentation broth, in its natural form or as an extract (in water and ethanol). Species used in other patents were *Lentinus edodes* and *Ganoderma lucidum*. The downfall of these methods varies from the lowering of biological activity due to the aggressive preparation procedure, non-standard material with possible contamination of chemical or microbiological nature to the unpleasant or strange taste (Lan, 2001; Jiang, 2002; Yin et al., 2013; Su, 2014).

4.1.3 Macrofungi Added as a Beer-Flavouring Agent

The second, easier and more often examined, means of using macrofungi in beer production is to add them when the fermentation is done, in a form of flavoring agent. This method was well known and often used since ancient times. Medical herbs, spices and fruit were added to improve beer flavour or to prevent spoilage (Jennings et al., 2005; Guerra-Doce, 2014). Leskosek-Cukalovic et al. (2010a) explored this option using *G. lucidum* ethanol extract aseptically adding into the commercial pilsner-type beer. The question was, how would the consumers accept this novel beer? The *G. lucidum* is known for its high content of triterpenes, particularly ganoderic and lucidenic acid. These phenolic compounds have a bitter taste and a long aftertaste, so the type of extraction and the amount of extract added to beer is of highest importance. Surprisingly, a beer produced in this way was well accepted and perceived as better compared with commercial pilsner beer based on all tested parameters, especially bitterness and body (Leskosek-Cukalovic et al., 2010a). In another study conducted by the same group, beer with *G. lucidum* extract was characterised as refreshing with a pleasant bitterness. However, these results were preliminary and included a small size sample. Another limitation of these studies is the lack of colour analysis from a consumer's point. Visual appearance, especially colour, strongly affects other sensory properties of beer and sets expectations toward its hedonic characteristics (Reinoso-Carvalho et al., 2019). Lelièvre et al. (2009) reported both trained and untrained assessors to judge beer mainly based on its colour. Dark beer is expected to be sweeter, which might explain the better average grades for *G. lucidum* flavoured beer in the research by Leskosek-Cukalovic et al. (2010a) and Reinoso-Carvalho et al. (2019). Having in mind the colouring properties of *G. lucidum*, the colour aspect of its addition to beer and its effect on consumers' acceptance should be studied as well.

More in-depth research on the same topic was carried out by Despotović (2017), who examined not only the perceived sensorial properties by beer consumers, but also the different methods of *G. lucidum* fruit body preparation and the effect of the several doses and time of its adding into the beer production. Research showed that the addition of mushroom extract in the phase of fermentation does not influence the fermentation speed. The activity of yeast was neither affected, and the consumers did not notice any sensory difference, depending on the parameters varied in this experiment. Additionally, physicochemical parameters of the beer were evaluated and showed as being strongly influenced by the mushroom extract. It has been speculated that beer's improved viscosity was due to *G. lucidum* addition. Moreover, new beer had almost four times more β- and α-glucans, a bigger foam surface

(~ 15 per cent more) and higher and more stable foam, which was also creamy and fine-granulated.

Following the trends in the food industry, Belščak-Cvitanović (2017) tested the effect of addition of encapsulated *G. lucidum* extract to the beer. The aim was to exploit all the benefits from the encapsulation method, especially the preservation and masking of the unpleasant taste of phenolic compounds. The outcome was semi-successful: encapsulation was not an effective method for the phenolic compound content increase in beer, but it did mask the off-taste as well as haziness. The resulting beer was visually attractive due to the more pronounced and darker colour of hydrogel microbeads.

Existing patents used only *G. lucidum* as a beer-flavouring agent. In all of them, the powdered fruit body was boiled (Gao and Qian, 1996; Fu, 2008). They differed according to the temperature and time of cooking. Downsides were the decomposition of thermo-sensitive compounds (polyphenols), turbidity and unconventional medicine-like taste. Finally, the mycelium produced by submerged fermentation (alone or with fermentation broth) enabled the production of beer with physiological function and with a prolonged shelf-life. The latter is especially important since the beer was not filtered or pasteurised. Beers produced with macrofungi added as spice (Leskosek-Cukalovic et al., 2010a; Despotović, 2017) were also tested for biological activity and proved to have the following properties:

- antimicrobial activity against *Listeria monocytogenes*, *Kocuria rhizophila* and several bacteria that cause beer spoilage
- antioxidative activity
- hypotensive
- cardioprotective activity (in animal and human studies)

4.1.4 *The Beer Market: Challenges and Perspectives*

Unlike the science, the market already accepted and launched mushroom-based beer labels. The fact that the craft beers include premium and unconventional ingredients posed a solid ground for the microbreweries to start experimenting with mushrooms (Donadini and Porretta, 2017). Another feature typical for these small productions is their willingness to innovate. Furthermore, craft beer consumers expect a strong and more complex flavour (Jaeger et al., 2020). Gómez-Corona et al. (2016) stating that 'the quest for authenticity' appears as the strongest motivating force in craft beer drinking. There are two following main characteristics of this market:

Ingredient variety: Theoretically, any edible and non-toxic species can be used. The existing beers with *G. lucidum*, *L. edodes*, *Grifola frondosa*, *Craterellus cornucopioides*, *P. ostreatus*, *C. cibarius*, *Inonotus obliquus*, species of *Lactarius*, *Morchella* and black truffles offer a great variety in aroma, taste, texture and flavour. This is just a small fragment of species and a potential of about 11,000 macrofungal species await exploitation (Mueller et al., 2007).

Easy-to apply technology: When it comes to the brewing technology, mushrooms are added as whole fresh or dried fruit bodies, water decoction before the fermentation or powdered and subjected to a sort of hot water extraction (similar to tea making) and combined with beer when the fermentation is done; however, limitations also exist. Fungi are usually from a local area and have seasonal character. Often a high price is inevitable when the resources are time-dependent. Quality varies from season to season as well as the place of mushroom collection. Fresh mushrooms are bulky, voluminous and high in water content, which makes them hard to preserve and store (Singh et al., 2016). The taste of the resulting beer is not always satisfying because of an unpleasant odour. The result is a short market-life (one season at the most), implying that more research is necessary to provide a better understanding of the effects of macrofungal volatiles and aromatic compounds on beer's sensory profile. The science has to establish which species are best to be used, in which form and when to be added, how to combine fungi with other ingredients in beer brewing, raw materials and seasoning as well. Health-promoting effects are poorly understood and often not properly accessed or represent a small sample result. At the moment the pharmacokinetic effects of mushroom beer are not promoted at all. Moreover, the chemical composition of craft beer is under-researched (Bertuzzi et al., 2020). Having in mind that craft beer has a steady two-decade-long exponential growth, especially in the USA and China, the room for novel findings and recommendations will be welcomed (Chan et al., 2019). For some commercialised versions of macrofungi flavoured-beer consult the following links:

- https://portsmouthbrewery.com/whats_on_tap/smutlabs-satchmo/
- https://jesterkingbrewery.com/our-beer
- http://www.5rabbitbrewery.com/las-chingonas
- https://www.rogue.com/stories/santa-s-back-again
- https://www.scratchbeer.com/single-post/2019/04/12/Mushroom-Beer-Extravaganza-April-27th
- https://www.laugarbrewery.com/beers/
- https://brasserieharricana.com/bieres/n75-saison-forestire

4.2 Wine

This several millennia-old commodity and source of enjoyment and religious practice is facing a new challenge. The Old World wine needs to expend to the new markets and even more to compete with the New World wines. In the meantime, Asia, and especially China, appeared as the most promising new market (Ma, 2017). In just a decade, wine consumption in China doubled, mainly due to the health connotation (Somogyi et al., 2011). On the other side, knowledge about this market is limited (Cohen and Lockshin, 2017). Based on existing fragmentary research, Asian wine consumers have different sensorial preferences and pay close attention to the health benefits of drink (Somogyi et al., 2011; Higgins and Llanos, 2015; Chang et al., 2016). This opens a space for innovation in wine production and consideration of unusual mixtures of wine with other ingredients. The idea of mixing wine with additives, like mushrooms, might appear as blasphemy for a Western wine producer and consumer with established wine culture. However, change is inevitable in the conquest of the new markets.

Since it was discovered about 9,000 years ago, the wine went through many changes. The oldest wine found in today's China was not even wine. It was a combination of unidentified wild vine, hawthorn fruit, rice and honey (Guerra-Doce, 2014). Latest evidence for grape-wine production originates from Gadachrili Gora, in Georgia, from around 6000–5800 BC (McGovern et al., 2017). It served many purposes and medicinal one was probably one of them (el-Guebaly and el-Guebaly, 1981; McGovern et al., 2004; Liu et al., 2019). This hypothesis found its written confirmation much later in the oldest pharmacopeia from Sumer 2200–2100 BC (Kramer, 1959). Ancient Egyptians even had wine treatment (oenotherapy) (Chang et al., 2016).

For sure there were numerous problems with taste and shelf-life; thus additives like herbs, spices, honey, fruit and resin were used (Guerra-Doce, 2014). Some cultures like Asian, maintained the strong wine-health connection, evident through the practice of drinking (defined amounts) and wine preferences (Somogyi et al., 2011). On the other hand, Western nations have deep roots in wine production and consumption and just recently its health-promoting aspect became important in the purchase and consumption decisions (Higgins and Llanos, 2015). There is a well-known French paradox that connects moderate wine drinking with benefits for the cardiovascular system. It is proven that this alcoholic beverage, especially red wine, contains significant amounts of polyphenols. Resveratrol was identified as the most effective antioxidant. Snopek et al. (2018) associated resveratrol with brain and nerve cells protection, reduction of platelet aggregation and prevention of blood clots. However, epidemiological, intervention and clinical studies gave contradictory results. As Arranz et al. (2012) stated, more randomised studies with long-term interventions are needed for the definitive conclusions.

4.2.1 *Macrofungi in Wine Production*

How traditional are the contemporary grape-wine production and establishment of wine styles reveal the fact that only a few research groups are working on alternative grape-wine produced with macrofungi. Pioneers in this filed were Okamura-Matsui et al. (2003) and Matsui et al. (2009), who produced wine using macrofungi, instead of *S. cerevisiae*. This extensive research included 200 species of mushroom and *Schizophyllum commune* showed the highest alcohol-dehydrogenase activity. Researchers reported the mushroom's ability to ferment grape juice under aerobic and anaerobic conditions. However, *S. commune* was more efficient in ethanol production when oxygen was available. The screening of amino acids present in wine showed that sweet-taste alanine represents the most abundant amino acid in wine produced with mushroom, instead of yeast. This wine was also characterised as a functional food due to the fibrinolytic activity expressed in *in vitro* assay.

Nguyen et al. (2019a) first tested consumers' attitudes toward new wine products among Chinese, Vietnamese and Australian nationality. They needed a positive input from the real market and followed a consumer-centric approach in product creation (Yeretzian et al., 2004). The researchers identified a mainly positive attitude toward wine with *G. lucidum*, especially among Vietnamese respondents. All nationalities included in this study considered health benefits from wine as very important. Next, Nguyen et al. (2019b) explored the use of *G. lucidum* in wine production. Mushroom in the form of extract was added prior to primary fermentation as well as after it. This extensive research covered different aspects of the small-, medium- and large-scale production and the wine-base was Australian Shiraz grape.

In the first and second phases (100 ml volume batches), different concentrations of *G. lucidum* extract (4.5–36 g/l) were added to grape must. The fermentation kinetic was not influenced by mushroom addition in any of the concentrations tested. The same fermentation kinetics was observed in the case of a medium-scale (5 l) experiment. Finally, the addition of mushroom extract in a 28 L fermentation vessel went without any interference; neither yeast nor lactic acid bacteria were affected by addition of *G. lucidum*. On the other side, sensory profile and chemical composition of the resulting wines were strongly affected by the extract addition. Notably, 39 among 54 sensory attributes evaluated differed significantly for the control and *G. lucidum* wine. Both the amount of extract and the time of its addition influenced the wine sensory profile, thus the acceptability. Wine with less *G. lucidum* was perceived as sweeter and more acceptable, while higher extract concentration contributed to more bitter and sour taste. Since the amount of residual sugar was in the range of dry wine, it was speculated that the sweet taste could only originate from the mushroom extract. The same was noticed on adding *G. lucidum* pre- and post-fermentation stages. The colour was affected as well. The colour intensity of wines with mushroom extract added before fermentation was lower as well as the extent of green-red and blue-yellow in comparison with wine, where *G. lucidum* was added after the fermentation. Furthermore, chemical parameters and their correlation with sensory attributes were measured. Among 29 volatile compounds isolated from *G. lucidum* wine, ethyl- and acetate-esters were the most dominant. When mushroom was added, the amount of ethyl acetate doubled, while the concentration of 3-methyl butyl acetate decreased as compared with the control and *G. lucidum* pre-fermentation addition. All wines with mushroom extract had more ethyl decanoate than the control as well as diethyl succinate. Volatile compounds of wine were in correlation with sensory attributes, limonene and 1-octanol appeared positively correlated with bitterness and dry fruit aroma, the mushroom taste was due to the presence of 1-octanol. As Nguyen et al. (2019b) reported, *G. lucidum*-supplemented wines had a different aromatic profile, were more complex, darker, with tones of dry fruit, toasted, earth and wood.

In their third study, Nguyen et al. (2020) examined consumers liking wine with *G. lucidum* extract. As they reported, the majority of consumers were very likely to be interested in this novel wine. The most acceptable amount of mushroom extract added before or post-fermentation was 1 g/l. However, they identified two groups of consumers with a more individualised taste and who preferred a higher *G. lucidum* concentration in Shiraz wine. Concerning the existing differences, more research, especially on technology development and the use of different grape varieties, is necessary to develop innovative products for the emerging wine markets.

An older study conducted by Okamura et al. (2001b) examined the wine-producing potential of different mushroom species since the same group of authors discovered the presence of alcohol dehydrogenase in macromycetes. Sterilised grape must was inoculated with a cell-free extract of three mushroom species, *A. blazei*, *F. velutipes* and *P. ostreatus* instead of the yeast *S. cerevisiae*. The experiment was extended to one more species, *T. matsutake* in 2003, by the same research group. The original three species produced alcohol and the highest concentration was obtained when *P. ostreatus* was used (12.2 per cent v/v). Interestingly, *A. blazei* was able to synthesise ethanol under aerobic as well as anaerobic conditions.

Japanese group (2001b) was also interested in medicinal aspects of the resulting wine; they measured its anti-coagulative and fibrinolytic activity. Furthermore, the wine produced with *A. blazei* instead of yeast contained 0.68 per cent of β-glucan, which is known for its anticarcinogenic activity (Lemieszek and Rzeski, 2012; Friedman, 2016). Macrofungal-wine showed the potential for thrombosis prevention in *in vitro* conditions as well.

The use of macrofungi in wine-like-beverages' fermentation has another advantage as demonstrated by Lin et al. (2010). Namely, they explored the possibility of using stipes of *L. edodes*, which is the edible and medicinal mushroom. Stipe is often regarded as waste material although it contains different nutrients, especially N-compounds, which are a limiting factor for yeast activity in alcohol fermentation. In this study, water extract from *L. edodes* stipes served as N-source for several *S. cerevisiae* strains. As demonstrated by actual fermentation, all examined strains were able to utilise mushroom extract as a nitrogen source. Moreover, the amount of N assimilated was double above the minimum required concentration. Even the acceptability of the resulting wine was a bit higher than that of the commercial white wine originating from France.

4.3 Rice Wine

Rice wine is a traditional alcoholic beverage produced all over Asia. Although its name points to a drink similar to Western style, wine from a technological point of view resembles beer more. Moreover, the Chinese National Standard for Wine and Beverages (GB/T 10789-2015) declare it as a special group of beverages, known as Chinese rice wine (Section 4.1.4) (National Standards—Wine & Beverage, Appendix # A7. 2015). It can be further sub-classified according to alcohol and sugar contents, raw ingredients and product style. Wine flavoured with supplementary ingredients, like medicinal substances, is classified as Special Type Chinese Rice Wine (Section 4.1.4.3.3). In comparison, the US and EU Regulation (No. 251/2014) classify wine with herbs added (with a minimum of 75 per cent wine base) as aromatised wine (No. 1601/91), while Australian and New Zealand Food Standards Code (Standard 2.7.4) categorise it as a wine product (The European Parliament and the Council, 2019). According to the South African Regulations, wine flavoured with herbs is designated as grape-based liquors (Liquor Products Act 60 of 1989, amended by GG42260 1/3/2019).

In most of the Asian countries, rice wine is considered as a national beverage and although its name points to rice-base, it can be produced from other grains as well. Usually, wheat, sticky or red rice, millet or corn are subjected to fermentation with mixed microbial culture consisting of moulds, yeasts and bacteria (known as *Koji* or *Jiuqu*). This combination enables two-in-one action. Saccharification and fermentation are done simultaneously, while in traditional brewing indigenous microorganisms perform key processes, in industrial setting single pure cultures are preferred. The quality of rice wine depends mainly on raw materials and microbial starter, but the production technology consisting of many steps which are sensitive to a specific factor (e.g., time, specific surface, O_2 and temperature) also influence the final quality (Liu et al., 2014; Hong et al., 2016).

Starch containing raw materials form the basis for wine's aroma and flavour due to the volatile compounds formed during fermentation. Very often additional flavouring ingredients are added. He et al. (2019) identified 103 plant species used in rice wine production by indigenous people in southeast China. In Borneo, cinnamon, pepper, dried chili and herbs are added to the starter to provide a unique aroma of rice wine (Palaniveloo and Vairappan, 2013). In Korea, flavouring ingredients are even promoted on the label, especially if they have a medicinal effect (Buglass and Caven-Quantrill, 2012).

Due to the rising interest in health-promoting effects of beverages, with unusual and premium ingredients, there are several attempts at creating rice wine with macrofungi. Although the founding block for this type of wine was set a long time ago as a part of Chinese *Materia Medica*, these new attempts elaborate on: (a) macrofungi species; (b) type of material; (c) time of addition. Additionally, researchers examined the *in vitro* biological activity of mushroom wine.

Not many mushrooms are present in available research. As in other disciplines, the ingredients/materials with the most abundant research and folk history are the first to test. Macromycetes are in scientific focus for the last 30 years and a significant amount of evidence of its biological activity has been gathered. Activities like antioxidant, antiviral, antibacterial, immunomodulatory, antifungal, hepatoprotective, cholesterol-lowering and others were confirmed for about 126 mushrooms *in vitro* as well as *in vivo* tests (Wasser, 2010). Furthermore, they are rich in aromatic compounds, proteins, minerals, vitamins and carbohydrates, which altogether make them an interesting natural flavouring ingredient. At the same time, some mushrooms possess alcohol dehydrogenase, enabling them to be more than flavour-adding supplements (Okamura-Matsui, 2003). With around 53000–110000 estimated species, macrofungi are an inexhaustible niche for the new product development (Mueller et al., 2007). However, only a few species are reported to be used in experimental rice-wine production with *G. lucidum* being used the most often. This is probably due to its long use in traditional Chinese mdicine. In the case of patents, species are more diverse. Table 1 gives an overview of macrofungi used in rice-wine production in connection with the referent research or patent.

Fungal material used in rice-wine production has been prepared in different ways, from alcohol maceration of the fresh or dry fruit body, over-powdered fruit body (PFB), spore powder, water, alcohol or combined extraction of PFB, cold or hot PFB extraction, ultrasound extraction, to mycelium obtained by submerged fermentation and its fermentation broth (Table 2). The material of choice was added at one of the two distinct moments of wine production: during fermentation or after it. Okamura-Matsui et al. (2003) even used several mushrooms for fermentation during sake production. The *A. blazei*, *T. matsutake*, and *F. velutipes* were all able to synthesise ethanol, although in different concentrations, with *A. blazei* being the most productive mushroom (8 per cent v/v). The resulting wine even has anticoagulative

Table 1. Macrofungi used in rice wine production (*Health Care Beer).

Macrofungi	Reference/Patent number
Ganoderma lucidum	Gao, 2004; Kim et al., 2004; Wang, 2005; Yang et al., 2016; Guo et al., 2019; CN1629280A (Fu et al., 2005)
Ganoerma lucidum and *Cordyceps*	Zeng et al., 2007
Cordyceps sinensis	CN103756820A (Zhang et al., 2014)*
Cordyceps militaris	Tian et al., 2001; Wang and Ning, 2013; Zhang et al., 2015; Yang et al., 2016; Patent No. CN103540503A (Cheng, 2014); CN103421655A (Zhou et al., 2013); CN103351972A (Song, 2013); CN103103091B (Wei, 2014)
Lentinula edodes	CN103756855B (Hu, 2014)
Poria cocos	CN103497870A (Chen et al., 2015)
Pleurotus ostreatus	CN103232923A (Meng, 2013)
Tricholoma matsutake	CN1150176A (Yang, 2000)
Morchella sp.	CN1017348B (Song, 1992)
Phellinus linteus	CN105062798A (Chen, 2017)
Hericium erinaceus	CN105505673A (Yu, 2016)
Phellinus igniarius	CN104403898A (Meng, 2015)
Phallus duplicates	CN1059126C (Yang, 2000)
Seven mushroom formula (*Ganoderma lucidum, Coriolus versicolor, Poria cocos, Phellinus linteus, Antrodia camphorate, Phallus rugulosus* and *Cordyceps sinensis*)	CN103525634A (Chen, 2014)

activity, especially when *T. matsutake* was applied in the form of cell-free extract and this sake has the taste of mushroom's fruit body.

From the technology point, the least demanding way to use macrofungi in rice-wine production is to submerge the whole fruit body in alcohol (Wang, 2005). Although fast and easy, this procedure does not enable standardisation. Fruit bodies differentiate in active substance concentration and the active surface is relatively small, so the transfer rate of the compounds of interest from fruit body to wine is low. In the case of expensive ingredients, like *Cordyceps* species, the presented method is often a waste of material. Zhang et al. (2015) proposed the use of *C. militaris* fermentation broth since it contains active compounds secreted during submerged fermentation of mycelium and starch, which can serve as a sugar source during wine fermentation. Moreover, this method enables the use of material regarded as waste. Similarly, Yang et al. (2016) reported the development of rice wine obtained by submerged fermentation of *C. militaris* and the subsequent use of fermentation broth (rich in cordycepin) in rice-wine fermentation. In this way, the final product can be standardised for the active compound amount. The mycelium fermentation broth was also exploited by Tian et al. (2001), but in this case, the wine was produced with *G. lucidum*. As the authors reported, the final product was rich not just in expected active compounds, but contained a significantly higher amount of amino acids (50 per cent more), essential amino acids (18 per cent more), proteins (the amount

Table 2. The type of preparation method of mushrooms used in patents.

Mushroom species	Method of preparation	Time of addition	Additional ingredients
Tricholoma matsutake	Combined cold alcohol (in basic wine) and hot water extraction	Wine	/
Morchella sp.	Fermentation broth	Fermentation	/
Phellinus linteus	Mycelium	Fermentation	/
Seven mushroom formula	Cold alcohol extraction (in wine)	Wine	/
Cordyceps militaris	Cold alcohol extraction (in wine) Hot water extraction Fermentation broth Pupa of silkworm	Wine Wine Fermentation Fermentation	/ fruit juice / / wolfberry fruit, red ginseng, bamboo water and sha shen root
Corcyceps sinensis	Fruit body	Fermentation	pear juice
Hericium erinaceus	Powder mix with auxiliary materials	Fermentation	mulberry leaf
Phellinus igniarius	Cold ethanol extraction	Wine	brown rice
Phallus duplicatus	Cold ethanol extraction (in wine)	Wine	medicinal herbs
Pleurotus ostreatus	Double hot water extraction	Fermentation	coffee powder, mulberry leaf, pear flower, hawthorn powder, ginger, Du Zhong bark, Mao Li leaf, sunflower, flos cucurbitae and orange peel
Poria cocos	Steamed and cooked with rice	Fermentation	yam root, hawthorn, red dates and goji
Lentinula edodes	Cold ethanol extraction	Wine	/
Ganoderma lucidum	Fermentation broth	Wine	wolfberry fruit

doubled) and minerals (like Fe and Zn). Besides, the resulting wine had a pleasant taste with a rich taste of mushroom, rice and honey. They also reported a unique bitter taste of *G. lucidum* and a long aftertaste. Overall, the fermentation method is characterised as fast, efficient, easy to apply and does not have seasonal restrictions.

Other researchers experimented with mushroom powder extract and added it after the fermentation. The use of powdered fruit bodies is a usual method, when working with *G. lucidum* and the most common in the case of other species as well (Zhou et al., 2011; Zhang et al., 2015). Fruit bodies of this macrofungus are large and have a hard structure. Milling enables the physical destruction of cell walls, so the active compounds are more available. It also enhances the active compounds extraction rate. Kim et al. (2004) used plain *G. lucidum* fruit body powder during the second fermentation in the production of Korean traditional rice wine. They found that the increase of mushroom-powder concentration inhibits alcohol fermentation and has a negative effect on the sensory properties (bitter, grassy and chlorogenic aroma). They proposed the addition of 0.1 per cent of mushroom powder for obtaining the optimal characteristics of the wine. In other studies, the PFB was extracted with ethanol. Gao (2004) reported rice wine produced with subsequent extraction of *G. lucidum* powder in 95, 85 and 70 per cent ethanol. In a study conducted by Zeng et al. (2007) on *G. lucidum*, the PFB was extracted with 70 per cent ethanol only and after 35 days, the optimal amount of saponins and polysaccharide was obtained, while another option was with 55 per cent ethanol (Zhang et al., 2015). Extraction was also the method of choice by Guo et al. (2019). However, the concentration of solvent has not been reported. As summarised by Zhang et al. (2015), macrofungal rice wine has been prepared by hot as well as cold alcohol extract and each method has its advantages or disadvantages. Cold extract should be the method of choice if the aim is to preserve mushroom's original taste as well as to maintain temperature-sensitive bioactive compounds. This method is very simple too. In the case of industrial production, hot extract might be more preferred due to the higher amounts of extracted compounds. The downside is the change, often lowering of bioactivity.

As a combination of complex main and auxiliary ingredients, rice wine is prone to precipitation and haziness (Zhang et al., 2015). Turbidity is considered as unappealing for consumers and needs to be addressed. High protein concentration in some mushroom species can lead to precipitation. Another problem is polysaccharides, especially in water extracts (Belščak-Cvitanović et al., 2017). Thus, clarification and filtration are often used as additional steps in production to ensure an acceptable final product.

4.3.1 Medicinal Benefits of Macrofungi Rice Wine

Rice wine has always been connected with health promotion and even traditional Chinese medicine supports a moderate amount of alcoholic beverage consumption for maintaining good health (Somogyi et al., 2011; Chang et al., 2016). Old medicinal texts, like *Compendium of Materia Medica* and *Thousand Golden Prescriptions* list many prescriptions based on wine and even macrofungi are mentioned. Moreover, modern technologies are used to prove and scientifically support folk and traditional claims. Kang et al. (2012) isolated angiotensin I-converting enzyme inhibitory peptide from Korean traditional rice wine. It was a proof of antihypertension activity of this traditional alcoholic beverage. On the other hand, Que et al. (2006) demonstrated high antioxidant activity of several rice wines and employing HPLC, identified 10 phenolic compounds responsible for oxidative protection. Syringic acid and catechin were the compounds with the highest correlation with demonstrated antioxidant activity. Another group of biologically active compounds from rice wine is polysaccharides. Bae et al. (2010) reported the immune-modulatory activity of this polysaccharide by measuring its anticomplementary activity *in vitro* in comparison with potent immunomodulatory polysaccharide from the mushroom *Coriolus versicolor*.

Based on existing findings for the biological properties of rice wine and macrofungi, new rice wine supplemented with mushrooms was also expected to show bioactivity. Thus, rice wine with ethanol extract of *G. lucidum* showed antioxidative activity and the peak was achieved after 35 days of cold extraction (Yang et al., 2016). In another study, Guo et al. (2019) demonstrated immune-function-promoting activity of wine with *G. lucidum* on the mice model. Several immune responses were examined—thymus index, ear swelling, phagocytic index, lymphocyte proliferation and natural killer cell activity. The biological activity of mushroom materials was even preserved in an experiment, where rice wine was fermented with cell-free mushroom extract (Okamura et al., 2001b). In their study, wine fermented with *T. matsutake* cell-free extract demonstrated the highest anticoagulative activity.

4.3.2 *Wine Market: Challenges and Perspectives*

In the last two decades, the wine market experienced constant growth, which was very intensive before 2008. World economic crisis slowed this trend in the next seven years and finally, significant growth has been achieved (Castellini and Samoggia, 2018). The most recent forecasts for the period 2020–2025 are promising with CAGR of 5.8–9.8 per cent (depending on the company). Interestingly, the EU leader status in the wine trade is changing in favour of Chile and New Zealand, as commented by Pomarici (2016). At the same time, as the fifth biggest wine consumer, China is becoming an important yet still poorly understood market (Ma, 2017). From the consumer perspective, several changes occurred: the change of taste and preference; search for new and exotic flavours; the demand for health-promoting properties (Higgins and Llanos, 2015). Furthermore, new generations of consumers are the millenials who look for something different, adventurous, experiential, innovative and affordable (Chang et al., 2016; Castellini and Samoggia, 2018).

The reviewed articles demonstrated that in the competitive atmosphere, like the Australian wine market, product differentiation and innovation, like Shiraz wine with *G. lucidum* might find its place (Nguyen et al., 2020). Another successful attempt to deliver a new beverage was wine obtained from *L. edodes* stipe extract and sugar from sugarcane. The authors produced a beverage with a higher acceptability rate than the commercial French wine from the store (Lin et al., 2010). On the other side, rice wine has been known as a national wine in Asia (Jiao et al., 2017). The contemporary technology is quite similar to that of traditional brewing method of 7000-year-old and has many downfalls. From the market point of view, sensory characteristics of rice wine produced in traditional ways are hard to manage with decreasing sales and grim forecasts. Consumers look for healthy, nutritious, tasty and safe products. Wine became a daily use commodity due to the increase in health conscience (Zeng et al., 2007). Therefore, a strong interest in developing new rice wine potential exists in Asia (Jiao et al., 2017).

Based on discussed evidence, macrofungi have a high potential to fulfil all requirements in both segments of grape and rice-based wine. However, more research is necessary to develop production techniques that are efficient and cost-effective. Questions, such as what kind of mushroom material to use, in what quantity, when to add and how to combine with other ingredients—need to be answered, based on scientific data. In-depth chemical analysis of all compounds developed in many production steps combined with novel extraction techniques needs to be coupled with traditional analysis of consumer preferences. Finally, as suggested by Pomarici

(2016), the experimental economy has to be involved to better understand market tendencies and subsequent responses to them.

5. Distilled Alcoholic Beverages—Spirits

The breakthrough in the history of alcoholic beverage production was the invention of the distillation process, which appeared in Mesopotamia around 4000 BCE (Standage, 2006). Even though this process is primarily used in the perfume industry, distillation enables the concentration of alcohol produced by fermentation (Wolf et al., 2008). According to written and archeological evidence, it is assumed that Arab alchemists had the most important role in perfecting the distillation process (Hudson and Buglass, 2011). Knowledge of distillation in practice from the Middle East was spread to China and Europe between 800–1300 AD (Comer, 2000). In the European literature, distilled spirits were named *aqua ardens* (the water that burns) or *aqua vitae* (the water of life), which indicate its huge importance in various segments (Hudson and Buglass, 2011). Originally, alcohol was mainly used for medicinal or research purpose, but until the latter part of 19th century to date, alcohol has been commonly consumed as a means of enjoyment.

Nowadays, the largest production of spirits is concentrated in Europe (Pielech-Przybylska and Balcerek, 2019) and under Articles 43(2) and 114(1) of Regulation (EU) No. 2019/787, spirit drink is defined as an alcoholic beverage with a minimum alcoholic strength by volume of 15 per cent. Although the distillation process of spirits has a slight difference in various countries, the world alcoholic beverage map is very complex with numerous high-quality alcoholic beverages (*see* https://www.tasteatlas.com/beverages). As the same for beer production, the raw material used as a source of sugar in the fermentation process is very broad and has a crucial influence on the profile of primary aromatic compounds (Wiśniewska et al., 2016). Even more, distinguishing among alcoholic spirits is based on the origin of the aromatic compounds. Besides the local tradition, the selection of raw material also depends on the price and availability in a specific region. Thus, Eastern countries mainly use grain for spirit production, such as Chinese liquor (ethanol content 38–65 per cent v/v), while in Western counties, different fruits are used for brandy as well as grain for whisky production (ethanol content higher than 37.5 per cent).

In Balkan countries, fruit brandy is a traditional spirit which is important for cultural and religious rituals, but also as an important part of a daily diet. The plum is one of the most valuable raw materials for spirit production in many countries. It is characterised by high sugar content and abundant aromatic compounds. In the Slavic countries of eastern and central Europe, local blue plum variety, *Prunus domestica*, is commonly used as the raw material for brandy production. A recent study estimated that the difference in alcoholic-beverage preference is diminishing with globalisation, but the spirits are still the predominant beverage in Nordic and some East European countries (Sierksma et al., 2003).

5.1 Macrofungi in Spirit Production

Differently from beer and wine, distillates have a negligible content of bioactive compounds (Veljović et al., 2019a). The distillation process partially separates the volatile compounds from the fermented mixture to increase the content of desirable aromatic compounds (Pecić et al., 2016). To improve healthy properties, mushrooms and herbs were added to spirit as a source of bioactive compounds throughout the history. According to this heritage, these products are treated as forms of folk medicine used in various disease treatments. In traditional Chinese medicine, spirit-based beverages are mostly enriched with plants (> 80 per cent). Usually, the single-herb formula is rare in practice; more common are multiple formulated mixtures with four to 12 individual herbs (Lee et al., 2003). Macrofungi were also appreciated as a component of the herbal formula. Ever more, macrofungi belong to upper-class herbs in *Shen Nung Pen Tsao Ching* (The Book of Herbs by Shen Nung) (including 365 herbs), which are non-toxic for human consumption (Lee et al., 2003).

Utilisation of herbs in the production of medicinal spirit was also a part of traditional European medicine for more than five centuries (Egea et al., 2015). These products are traditionally consumed as a cure in medicinal purposes as well as a tonic for enhancing general health condition (Petrović et al., 2019). A large number of spirits produced with the addition of medicinal herbs have a bitter taste, which is refreshing and desirable for many consumers and this kind of spirit is widely known as a 'bitter'. More than 50 extracts (alcohol or/and water) obtained from different parts of herbs can be used for bitter preparation (Vukosavljević et al., 2009; Pecić et al., 2012b). Under Articles 43(2) and 114(1) of Regulation (EU) No. 2019/787, a bitter-tasting spirit drink or bitter is defined as 'a spirit drink with a predominantly bitter taste produced from ethyl alcohol of agricultural origin, with the minimum alcoholic strength of 15 per cent v/v, with the addition of flavouring agent.' On the other side, utilisation of macrofungi in spirit production has a long history in Asian counties. The macrofungi used are mainly inedible medicinal species with the woody texture, such as *G. lucidum, P. linteus*, and *T. versicolor* (Smith et al., 2002). Generally, the alcohol-water mixture (such as spirits) is a suitable solution for the extraction of numerous bioactive compounds. Therefore, these spirit-based beverages have been used as remarkable Orient elixir during many centuries (Veljović et al., 2019b).

According to the traditional recipes, medicinal fungus, *P. linteus*, was soaked in whisky to improve the functional characteristics of Korean whisky (Halpern, 2009). Additionally, *G. lucidum* (Reishi) is used to improve the overall quality of spirits. According to Mizuno et al. (1995), 'Reishi liquor' was produced by addition of *G. lucidum* into the low-quality spirit with an ethanol content higher than 35 per cent, which has been sweetened with sugar and/or molasses. The addition of macrofungi in spirit has multiple positive effects, including enrichment of the sensory characteristics (taste and colour) and improvement of the health properties of final spirits.

Contrary to Asian culture, the traditional utilisation of macrofungi in alcoholic-beverage production is rare as in European culture. The exception comes from the Bulgarian folk mycological oral tradition that is testimony to the utilisation of 'Morus fungus' in the production of traditional brandy 'Rakiya' in mountain regions of Bulgaria, near the two towns, Russe and Troyan (Stoyneva-Gärtner et al., 2018). After extensive research of electronic bases, Stoyneva-Gärtner et al. (2018) found that under the name 'Morus fungus' is hidden fungus *Fomitiporia robusta*. This mushroom is appreciated as a source of colourant and can be used for hastening the aging of brandy. Even more, utilisation of fungus improves the sensory properties, since it prevents the woody taste and smoke smell, typical for the aged spirit in a wooden cask.

5.2 *Ganoderma lucidum (Lingzhi/Reishi) in Spirit Production*

Lingzhi has gained wide popularity as an ingredient in spirits not only in Eastern countries (China, Japan, Korea and Taiwan), but its popularity constantly rises in Western countries as well (Europe and the Northern United States). In the past 40 years, the improvement of artificial production, the increase of yield and quality has created the possibility for commercial utilisation of this macrofungus. Therefore, the numerous spirits produced with the addition of *G. lucidum* can be found as a part of the portfolio of small distilleries with a competitive price all over the world. Additionally, spirit-based beverages enriched with medicinal fungus *G. lucidum* are marketed as a symbol of healthy product (Pecic et al., 2016).

The particular popularity of *G. lucidum* is due to the wide variety of its bioactive compounds. To date, more than 400 different bioactive compounds were detected in fruit bodies, cultured mycelia and cultured broth of *G. lucidum*, including triterpenoids, polysaccharides, proteins, amino acids, nucleosides, alkaloids, steroids, lactones, fatty acids and enzymes (Wasser, 2005). Besides improving the healthy characteristic of spirits, recent studies confirmed that Reishi also has a notable influence on the sensory properties of spirit-based beverages, such as in taste and colour (Niksic et al., 2001; Pecic et al., 2016; Veljović et al., 2019a; Veljović et al., 2019b).

Nowadays, *G. lucidum* is used as a natural flavouring agent in spirit production. Of particular interest is the fact that Reishi is a rich source of bitter triterpenes, with ganoderic acids being the most abundant (Kim and Kim, 1999; Veljović et al., 2019a). Depending on the intensity of bitterness, triterpenes are divided into three groups: (1) intensively bitter triterpenes (lucidinic acid A and D1, ganoderic acids A, J, and C1, lucidones C and A); (2) slightly bitter (lucidinic acid I, ganoderic acids B, C2, and K); (3) no bitter (Nishitoba et al., 1988).

After distillation, a colourless new spirit (distillate) has an unharmonious taste and sharp odour. To improve the quality of distillate, the old practice is aging in wooden casks for many years (Pecić et al., 2012a; Coldea et al., 2020). Besides the taste and odour, the extracted wood compounds contribute to the evolution of the different intensity of the yellow colour of aged spirits. To speed up the aging process, the possibility of using *G. lucidum* in production of spirit-based beverages and comparing their quality (mainly colour) with aged spirits was studied in-depth by Pecić (2015). The pioneer in this subject were Niksic et al. (2001), who first proposed that the addition of this medicinal fungus has the potential to be an innovative method for replacing expensive and durable aging method. According to these results, it was pointed out that 60 per cent of alcohol has the best colour (similar to cognac) in comparison with 40, 50 and 70 per cent of alcohol-water extracts. These data were used as a base point in the study of Pecić (2015). *Ganoderma* extracts were produced with the addition of a fragmented mushroom fruit body (40 g/l), of different particle sizes (milled, 0.130 mm; chopped, ~ 1 cm) and extracted with 60 per cent ethanol during one, seven and 40 days. After extraction, aliquots were filtered through the filter paper and concentrated by using a vacuum evaporator. Generally the more effective colourant of grain spirit was *Ganoderma* extracts made from milled mushroom, as the extraction from smaller particles was more intensive. Although the utilisation of *Ganoderma* extracts as an innovative method for accelerating spirit aging has indisputable potential, the disadvantage of this method was that it was not as effective as the maceration of *G. lucidum* in spirit. The effective dosage of produced extract is very high, which is not practical for the spirit industry. During concentration on vacuum evaporator, the colourant compounds extracted from *G. lucidum* degrade. Consequently, the use of unprocessed *G. lucidum* is recommended as more effective and less expensive.

Therefore, the innovative aging process of spirit production using fragmented *G. lucidum* was created by Pecić et al. (2016). In this research, wine, grape, plum and grain distillates were used for spirit-based beverage preparation. The influence of different extraction parameters, time (seven, 21 and 60 days) and concentration (10, 25 and 40 g/l) on colour intensity was examined. Besides colour determination, physicochemical characteristics of produced spirit-based beverages enriched with *G. lucidum* (*Ganoderma* spirit) were also tested. Generally, the higher amount of *G. lucidum* increased the colour intensity of all the used distillates, while the influence of extraction time increased with a higher amount of added mushrooms. Comparing the colour intensity of the spirit-based beverages with wooden aged brandies, it can be concluded that the spirit-based beverage enriched with 40 g/l of copped *G. lucidum* had the same intensity as the plum brandy aged in sessile oak during 11 years and in mulberry cask during 18 years (Pecić et al., 2012a).

The colour of *Ganoderma* spirit was also examined by a more sophisticated CIE L*a*b* method. Generally, the brightness of *Ganoderma* spirit decreased with the addition of a higher quantity of mushroom in all used distillates, while the intensity of red and yellow increased (Pecić, 2015; Veljović et al., 2019b). Previous studies showed that ethanol extract of *G. lucidum* contained flavanones (hesperetin

Fig. 1. The effect of *G. lucidum* addition on grape distillate colour. *From left to right:* pure grape distillate, grape distillate enriched with 10, 25 and 40 g/L of *G. lucidum* fruit body extracted for seven days.

and naringin) and flavonoids (quercetin, morin and myricetin), which cumulatively contribute to the yellow colour of *Ganoderma* spirit (Saltarelli et al., 2015; Veljović et al., 2019a). Therefore, the results of both colour analyses confirm that the addition of *G. lucidum* can be regarded as an innovative process and can replace the long period of aging in wooden casks.

The colour of the *Ganoderma* spirit is an important feature, primarily the impression, which influences consumer choice. However, the taste of spirit is the essential parameter for consumer acceptance of innovative products. Thus, the sensory properties (colour, clarity, typicality, smell and taste) of Reishi spirit were evaluated. The sensory quality score of the *Ganoderma* spirit was in the range from very good to excellent quality (score, 16.20 and 18.26) (maximum score, 20). According to the sensory evaluators, the addition of *G. lucidum* to all distillates (grain, grape, plum and wine distillate) enriched the sensory quality. The amount of added mushroom had a slight effect on sensory quality scores, while the aromatic compounds of used distillate had a stronger contribution. In addition, wine distillate was evaluated as the most adequate for producing high-quality *Ganoderma* spirit.

As we previously mentioned, the bitter triterpenes influence the sensory properties of Reishi spirits, mainly the taste. The content of bitter triterpenes acids in *Ganoderma* spirits has a slight effect on sensory assessments among these samples. Therefore, the chemical composition of distillate has a significant effect on the sensory quality score of *Ganoderma* spirits. Therefore, the qualitative triterpene profile and content in grain spirit enriched with *G. lucidum* (seven, 21 and 40 g/l) and extracted (10, 21 and 40 days) were determined using HPLC-DAD/ESI-ToF-MS. In all nine samples of grain *Ganoderma* spirit, the same 15 triterpene compounds were identified, including ganoderic acid A, B, C2, C6, G, D, F and J; lucidenic acid A, E, D2 and LM1; 12-hydroxy-ganoderic acid D, ganoderenic acid D and elfvingic acid A. The triterpene profile of *Ganoderma* spirit did not defer, regardless of the different extraction condition, but their amounts were significantly different and were in the range of 3.02–4.40 mg/100 mg on a dry weight basis. The most abundant triterpene was ganoderic acid A, which has an intensively bitter taste and reputable medicinal effect. In addition, the summarised amount of identified bitter triterpenoid acids (lucidenic acid A, ganoderic acid A and B) in *Ganoderma* spirits ranged between 0.71–1.11 mg/100 mg, while the percentage of bitter triterpenes in the total amount of detected triterpenes was very similar (22.20–27.48 per cent). The highest amount of bitter triterpene was measured in the *Ganoderma* spirit sample produced with 25 g/l and extracted during 60 days.

Although the study by Taskın et al. (2013) reported that *G. lucidum* mycelia and fruit bodies are a negligible source of volatile compounds, the volatile composition of distillate significantly changed after addition of the mushroom. Almost all detected volatile compounds of *Ganoderma* spirit already existed in the initial distillate, but their amounts were different. Moreover, aromatic terpene eugenol (characteristic quaternary compounds of aged spirit) was present in all *Ganoderma*-enriched spirits (Veljović et al., 2019b).

As we pointed out, the herbs are an excellent source of aromatic as well as bioactive compounds. The compatibility of extractible herbal compounds with *Ganoderma* spirits was also examined. The sensory quality score of the spirits with

G. lucidum and herbal extracts ranged from very good to excellent quality (score 16.85 and 18.55). The addition of *G. lucidum* and herbal extracts enriched the distillate with a wide range of fatty ethyl esters, which contribute to pleasant fruity and floral odour (Veljović et al., 2019b).

The excellent connection of traditional knowledge, scientific data and industrial practice in the production of bitter-enriched beverage with *Ganoderma* (Bitter 55) are presented by Vukosavljević et al. (2009). Small-scale product, Bitter 55 is a multi-component mixture consisting of 46 selected aromatic herbs, eight fruits and medicinal mushroom *G. lucidum* (35 per cent v/v of alcohol). The industrial production is based on the traditional process used from the ancient times in order to maximise the preservation of bioactive compounds of Bitter 55. The sensory quality of Bitter 55 is greatly appreciated, especially for its bitter, sophisticated, appealing flavour and for the unique aroma. Like almost all herbal brandies, Bitter 55 has a positive impact on digestion and it is used both as an appetiser as well as digestive. The antioxidant capacity of Bitter 55 was determined by three methods: linoleic acid oxidation, DPPH and the TLC-fluorescence. The results showed that its antioxidant capacity is significantly higher in comparison with the commercial bitter from the supermarket (Vukosavljević et al., 2009; Gorjanović et al., 2010). The storage of Bitter 55 in a green bottle in the dark during 150 days did not have any significant influence on the antioxidant potential.

Spirit-based beverage enriched with *G. lucidum* and herbal extract (Pecić, 2015) was also tested for its biological activity and proved to have:

- antimicrobial activity towards *Pseudomonas aeruginosa* (ATCC 35032), *Enterococcus faecalis* (ATCC 29212) and *Staphylococcus aureus* (ATCC 6538)
- antioxidant activities
- antiproliferative activity towards EA.hy 926, A549, FemX and HeLa cell lines

5.3 *Modern Trends*

Taking into account the worldwide demand for craft healthy products, innovative spirit beverages with macrofungi can be a potential solution. Over the last three decades, globalisation has influenced the market structure, with more involving untraditional ingredients from the distinct world regions. Therefore, the global market has recognised the traditional healthy Asian products, mostly produced with macrofungi as a very prominent and potential innovation. Even more, spirit-based beverages enriched with macrofungi build a strong bridge between Eastern and Western cultures.

Although in online trade can be found many different spirits produced in many countries, mainly China, the scientific data about their chemical composition and effects are limited. Moreover, in recent years, the testimony is that novel commercial spirit-based beverages enriched with macrofungi, such as Chaga (*Inonotus obliquus*), *G. lucidum* and *Tuber aestivum* can be found in the Serbian market. On the other hand, the North Korean whisky—Paektusan *Ganoderma* liquor gained significant attention on the online portals. For some of the commercialised versions of spirit-based beverages enriched with different macrofungi we suggest a visit to the following sites:

- http://www.alternativnamedicina.rs/proizvodi/dunavska-fantazija-rakija-od-tartufa-i-kordicepsa/
- http://vinarijaaleks.co.rs/index.php/moja-prodavnica/rakije/rakije-travarice/%C4%8Dagi-rakija-detail
- https://sg.carousell.com/p/rare-north-korea-liquor-dprk-201786714/

6. Conclusion and Outlook

The alcoholic beverages enriched with macrofungi are the results of a successful connection between ancient, mostly Chinese, and modern practice. Although historical evidence showed that the first attempt to utilise macrofungi as a raw material in the production of alcoholic beverages is very old, using macrofungi in the alcoholic beverage industry is a relatively new idea. One process that positively influenced development of this innovation is globalisation since the exotic ingredients, including macrofungi, are now available on the world markets at an acceptable price. Another is the modernisation of the artificial cultivation of macrofungi, which produced a sufficient amount of macrofungi. Finally, the most important influence is the increase in the worldwide demand for craft healthy products; thus innovative alcoholic beverages with macrofungi can be a potential solution.

Nowadays, only a few commercial alcoholic beverages enriched with macrofungi are to be found in the local or global market. Although the majority of them are produced in China, Western countries are more involved in the production and development of these beverages. Historical evidence reports that macrofungi, mostly medicinal mushrooms, are primarily used to improve the medical properties of alcoholic beverages. Besides bioactive compounds, macrofungi are also a rich source of aromatic compounds which contribute to sensory characteristics (colour, aroma and flavour) of alcoholic beverages. During the last decade, the purpose of macrofungi has expanded as they are also used as a flavouring agent in alcoholic-beverage production. Keeping in mind that only a few mushrooms are commercially used, the possibility for novel recipes has unlimited scope. It must be noted that scientific studies of alcoholic beverages enriched with macrofungi are rare; thus further exploration is necessary. Questions, such as what kind of mushroom material to use, in what quantity, when to add and how to combine with other ingredients need to be answered on the basis of scientific data.

Acknowledgments

This work was supported by the Ministry of Education, Science and Technological Development, Republic of Serbia (Contact No. 200051 and 200116). The authors would like to thank Ganoherb International Inc. from Fuzhou, China, for providing the research articles from Chinese research database as well as the prescriptions which include macrofungi and alcoholic beverages.

References

Arranz, S., Chiva-Blanch, G., Valderas-Martínez, P., Medina-Remón, A., Lamuela-Raventós, R.M. and Estruch, R. (2012). Wine, beer, alcohol and polyphenols on cardiovascular disease and cancer. Nutrients, 4(7): 759–781. https://www.ncbi.nlm.nih.gov/pmc/articles/PMC3407993/ (accessed on January 15, 2020).

Australian and New Zealand Food Standards Code–Standard 2.7.4—Wine and wine product (n.d.). https://www.legislation.gov.au/Series/F2015L00391 (accessed on May 22, 2020).

Bae, S.H., Jung, E.Y., Kim, S.Y., Shin, K.S. and Suh, K.J. (2010). Antioxidant and immune-modulating activities of Korean traditional rice wine, Takju. J. Food Biochem., 34: 233–248.

Barh, A., Kumari, B., Sharma, S., Annepu, S.K., Kumar, A., Kamal, S. and Sharma, V.P. (2019). Mushroom mycoremediation: Kinetics and mechanism. pp. 1–22. *In*: Bhatt, P. (ed.). Smart Bioremediation Technologies: Microbial Enzymes, Academic Press, Elsevier Inc.

Barkov, A.A. and Vinokurov, V.A. (2017). Mushroom beer and preparation thereof. https://patents.google.com/patent/RU2608497C1/en.

Belščak-Cvitanović, A., Nedović, V., Salević, A., Despotović, S., Komes, D., Nikšić, M., Bugarski, B. and Leskošek-Čukalović, I. (2017). Modification of functional quality of beer by using microencapsulated green tea (*Camellia sinensis* L.) and *Ganoderma* mushroom (*Ganoderma lucidum* L.) bioactive compounds. Chem. Ind. Chem. Eng. Quart., 23(4): 457–471.

Bertuzzi, T., Mulazzi, A., Rastelli, S., Donadini, G., Rossi, F. and Spigno, G. (2020). Targeted healthy compounds in small and large-scale brewed beers. Food Chem., 25: 310. 125935.10.1016/j.foodchem.2019.125935.

Biwer, M.E. and Van Derwarker, A.M. (2015). Paleoethnobotany and ancient alcohol production: A mini-review. Ethnobiol. Lett., 6(1): 28–31.

Buglass, A.J. and Caven-Quantrill, D.J. (2012). Applications of natural ingredients in alcoholic drinks. pp. 358–416. *In*: Baines, D. and Seal, R. (eds.). Natural Food Additives, Ingredients and Flavourings, Woodhead Publishing Series in Food Science, Technology and Nutrition, Cambridge, UK.

Butkhup, L., Samappito, W. and Jorjong, S. (2017). Evaluation of bioactivities and phenolic contents of wild edible mushrooms from northeastern Thailand. Food Sci. Biotechnol., 27(1): 193–202.

Capece, A., Romaniello, R., Pietrafesa, A., Siesto, G., Pietrafesa, R., Zambuto, M. and Romano, P. (2018). Use of *Saccharomyces cerevisiae* var. *boulardii* in co-fermentations with *S. cerevisiae* for the production of craft beers with potential healthy value-added. Int. J. Food Microbiol., 284: 22–30.

Carroll, D.N. and Roth, M.T. (2002). Evidence for the cardioprotective effects of omega-3 fatty acids. Ann. Pharmacother., 36(12): 1950–1956.

Castellini, A. and Samoggia, A. (2018). Millenial consumers' wine consumption and purchasing habits and attitude towards wine innovation. Wine Econ. Policy, 7: 128–139.

Chan, M.Z.A., Chua, J.Y., Toh, M. and Liu, S.-Q. (2019). Survival of probiotic strain *Lactobacillus paracasei* L26 during co-fermentation with *S. cerevisiae* for the development of a novel beer beverage. Food Microbiol., 82: 541–550.

Chang, K.J., Liz Thach, M. and Olsen, J. (2016). Wine and health perceptions: Exploring the impact of gender, age and ethnicity on consumer perceptions of wine and health. Wine Econ. Pol., 5(2): 105–113.

Chang, S.T. and Miles, P.G. (2004). *Ganoderma lucidum*—A leader of medicinal mushrooms. pp. 357–372. *In*: Mushrooms: Cultivation, Nutritional Value, Medicinal Effect and Environmental Impact, 2nd ed., CRC Press, Boca Raton, USA.

Chen, F. (2017). Preparation method of *Phellinus linteus* sweet wine. https://patents.google.com/patent/CN105062798A/en?oq=CN105062798A (accessed on May 17, 2020).

Chen, M., Fang, S., Cao, J., Shilu, R., Zhang, J. and Yan, J. (2015). Healthcare yellow wine with *Poria cocos* and preparation method of health-care yellow wine with *Poria cocos*. https://patents.google.com/patent/CN103497870A/en?oq=CN103497870A (accessed on May 18, 2020).

Chen, Y. (2014). Seven-mushroom wine formula. https://patents.google.com/patent/CN103525634A/en?oq=CN103525634A+ (accessed on May 18, 2020).

Cheng, F. (2014). Production method for *Cordyceps militaris* wine. https://patents.google.com/patent/CN103540503A/en?oq=CN103540503A (accessed on May 17, 2020).

Cohen, J. and Lockshin, L. (2017). Conducting wine marketing research with impact in China: Guidelines for design, execution and dissemination. Wine Econ. Pol., 6(2): 77–79.

Coldea, T.E., Socaciu, C., Mudura, E., Socaci, S.A., Ranga, F., Pop, C.R., Vriesekoop, F. and Pasqualone, A. (2020). Volatile and phenolic profiles of traditional Romanian apple brandy after rapid ageing with different wood chips. Food Chem., 320: 126643. https://doi.org/10.1016/j.foodchem.2020.126643.

Comer, J. (2000). Distilled beverages. pp. 653–664. *In*: Kiple, K.F. and Ornelas, K.C. (eds.). The Cambridge World History of Food. Cambridge University Press, Cambridge, UK.

Cör, D., Botić, T., Gregori, A., Pohleven, F. and Knez, Ž. (2017). The effects of different solvents on bioactive metabolites and *in vitro* antioxidant and anti-acetylcholinesterase activity of *Ganoderma lucidum* fruiting body and primordia extracts. Maced. J. Chem. Chem. Eng., 36(1). 10.20450/mjcce.2017.1054.

Damerow, P. (2012). Sumerian beer: The origins of brewing technology in ancient Mesopotamia. Cuneiform Digital Library Journal, 2: 1–20. http://www.cdli.ucla.edu/pubs/cdlj/2012/cdlj2012_002.html.

Despotović, S.M. (2017). Biochemical and functional properties of beer with the addition of *Ganoderma lucidum* mushroom. Ph.D dissertation, Ph.D Thesis, Faculty of Agriculture, University of Belgrade, Serbia. http://nardus.mpn.gov.rs/handle/123456789/8572?locale-attribute=sr_RS.

Djekic, I., Vunduk, J., Tomašević, I., Kozarski, M., Petrovic, P., Niksic, M., Pudja, P. and Klaus, A. (2016). Total quality index of *Agaricus bisporus* mushrooms packed in modified atmosphere. J. Sci. Food Agric., 97(9): 3013–3021.

Đorđević, S., Popovic, D., Despotovic, S., Veljovic, M., Atanackovic, M., Cvejic, J., Nedovic, V. and Leskosek-Cukalovic, I. (2016). Extracts of medicinal plants as functional beer additives. Chem. Ind. Chem. Eng. Quart., 22(3): 301–308.

Donadini, G. and Porretta, S. (2017). Uncovering patterns of consumers' interest for beer: A case study with craft beers. Food Res. Int., 91: 183–198.

Egea, T., Signorini, M.A., Bruschi, P., Rivera, D., Obón, C., Alcaraz, F. and Palazón, J.A. (2015). Spirits and liqueurs in European traditional medicine: Their history and ethnobotany in Tuscany and Bologna (Italy). J. Ethnopharmacol., 175: 241–255.

El-Guebaly, N. and el-Guebaly, A. (1981). Alcohol abuse in ancient Egypt: The recorded evidence. Int. J. Addic., 16(7): 1207–1221.

Falandysz, J. (2010). Mercury in certain mushroom species in Poland. pp. 349–383. *In*: Progress in Mycology. Springer, Dordrecht.

Falandysz, J., Chudzińska, M., Barałkiewicz, D., Drewnowska, M. and Hanć, A. (2017). Toxic elements and bio-metals in *Cantharellus* mushrooms from Poland and China. Environ. Sci. Poll. Res., 24(12): 11472–11482.

Friedman, M. (2016). Mushroom polysaccharides: Chemistry and antiobesity, antidiabetes, anticancer, and antibiotic properties in cells, rodents, and humans. Foods, 5(4): 80.10.3390/foods5040080.

Fu, J., Zou, H., Hu, Z., Xie, G., Meng, Z., Dong, H. and Wu, J. (2005). Low alcohol content health yellow rice wine and its preparation. https://patents.google.com/patent/CN1629280A/en?oq=CN1629280A (accessed on May 17, 2020).

Fu, M. (2008). *Ganoderma* beer and its preparation method. https://patents.google.com/patent/CN100422303C/en?oq=CN+100422303 (accessed on May 17, 2020).

Gao, C. and Qian, R. (1996). Glossy *Ganoderma* extract and glossy *Ganoderma* beer. https://patents.google.com/patent/CN1115612A/en?oq=CN+1115612A (accessed on May 17, 2020).

Gao, F. (2004). Development of *Ganoderma lucidum* health wine. Liquor-Making Sci. Technol., 123(6): 101–102.

Gatrell, J., Reid, N. and Steiger, T.L. (2018). Branding spaces: Place, region, sustainability and the American craft beer industry. Appl. Geogr., 90: 360–370. https://www.sciencedirect.com/science/article/pii/S0143622816301977(accessed on March 16, 2019).

Golian, M., Andrejiová, A., Mezeyová, I. and Hegedűsová, A. (2015). Design of oyster (*Pleurotus ostreatus*) production unit taking into account its agrotechnic of growing and quality and quantity of its production. J. Microbiol. Biotechnol. Food Sci., 4(special issue 3): 48–51.

Gómez-Corona, C., Escalona-Buendía, H.B., García, M., Chollet, S. and Valentin, D. (2016). Craft vs. industrial: Habits, attitudes and motivations towards beer consumption in Mexico. Appetite, 96: 358–367. http://europepmc.org/abstract/MED/26455311 (accessed on March 16, 2019).

Gorjanović, S.Z., Novaković, M.M., Vukosavljević, P.V., Pastor, F.T., Tešević, V.V. and Sužnjević, D.Z. (2010). Polarographic assay based on hydrogen peroxide scavenging in determination of antioxidant activity of strong alcohol beverages. J. Agric. Food Chem., 58(14): 8400–8406.

Gründemann, C., Reinhardt, J.K. and Lindequist, U. (2020). European medicinal mushrooms: Do they have potential for modern medicine? An update. Phytomed., 66: 153131.

Guerra-Doce, E. (2014). The Origins of inebriation: Archaeological evidence of the consumption of fermented beverages and drugs in prehistoric Eurasia. J. Archaeol. Meth. Theor., 22(3): 751–782. https://link.springer.com/article/10.1007%2Fs10816-014-9205-z (accessed on August 10, 2019).

Guo, W., Rong, Z.Z., Beili, W., Ping, Z., Bao, L.S. and Hua, L. (2019). Development of *Dendrobium candidum Ganoderma lucidum* wine with enhanced immune function. Food Res. Develop., 21: 115–118.

Halpern, G.M. (2009). Healing Mushrooms, Square One Publishers, New York.

Hasler, C.M. (2002). Functional foods: Benefits, concerns and challenges—A position paper from the American Council on Science and Health. J. Nutr., 132(12): 3772–3781. https://academic.oup.com/jn/article/132/12/3772/4712139.

Hayden, B., Canuel, N. and Shanse, J. (2012). What was brewing in the Natufian? An archaeological assessment of brewing technology in the epipaleolithic. J. Archaeol. Meth. Theor., 20(1): 102–150.

He, J., Zhang, R., Lei, Q., Chen, G., Li, K., Ahmed, S. and Long, C. (2019). Diversity, knowledge, and valuation of plants used as fermentation starters for traditional glutinous rice wine by Dong communities in Southeast Guizhou, China. J. Ethnobiol. Ethnomed., 15: 20. https://doi.org/10.1186/s13002-019-0299-y.

Higgins, L.M. and Llanos, E. (2015). A healthy indulgence? Wine consumers and the health benefits of wine. Wine Econ. Pol., 4(1): 3–11.

Hong, X., Chen, J., Liu, L., Wu, H., Tan, H., Xie, G., Xu, Q., Zou, H., Yu, W., Wang, L. and Qin, N. (2016). Metagenomic sequencing reveals the relationship between microbiota composition and quality of Chinese rice wine. Sci. Rep., 6: 26621. https://doi.org/10.1038/srep26621.

Hu, B. (2014). Preparation method of health *Lentinula edodes* wine. https://patents.google.com/patent/CN103756855B/en?oq=CN103756855B (accessed on May17, 2020).

Hudson, J.A. and Buglass, A.J. (2011). Introduction, background and history. pp. 630–932. *In*: Buglass, A.J. (ed.). Handbook of Alcoholic Spirits, Technical, Analytical and Nutritional Aspects, John Wiley & Sons, London.

Isikhuemhen, O.S., Adenipekun, C.O. and Ohimain, E.I. (2010). Preliminary studies on mating and improved strain selection in the tropical culinary-medicinal mushroom *Lentinus squarrosulus* Mont. (Agaricomycetideae). Int. J. Med. Mush., 12(2): 177–183.

Jaeger, S.R., Worch, T., Phelps, T., Jin, D. and Cardello, A.V. (2020). Preference segments among declared craft beer drinkers: Perceptual, attitudinal and behavioural responses underlying craft-style vs. traditional-style flavor preferences. Food Qual. Prefer. 10.1016/j.foodqual.2020.103884.

Jang, M.-J. and Lee, Y.-H. (2014). The suitable mixed LED and light intensity for cultivation of oyster mushroom. J. Mush., 12(4): 258–262.

Jennings, J., Antrobus, K.L., Atencio, S.J., Glavich, E., Johnson, R., Loffler, G. and Luu, C. (2005). Drinking beer in a blissful mood. Curr. Anthropol., 46(2): 275–303.

Jiang, X. (2002). *Ganoderma* beer and its production method. https://patents.google.com/patent/CN1084382C/en?oq=CN+1084382C (accessed on May 16, 2020).

Jiao, A., Xu, X. and Jin, Z. (2017). Research progress on the brewing techniques of new type rice wine. Food Chem., 215: 508–515.

Jones, M., Mautner, A., Luenco, S., Bismarck, A. and John, S. (2020). Engineered mycelium composite construction materials from fungal biorefineries: A critical review. Mat. Des., 187: 108397.

Kalač, P. (2000). A review of trace element concentrations in edible mushrooms. Food Chem., 69(3): 273–281.

Kalač, P., Svoboda, L. and Havlíčková, B. (2004). Contents of cadmium and mercury in edible mushrooms. J. Appl. Biomed., 2(1): 15–20.

Kamthan, R. and Tiwari, I. (2017). Agricultural wastes—Potential substrates for mushroom cultivation. Eur. J. Exp. Biol., 7(5): 31. 10.21767/2248-9215.100031.

Kang, M.G., Kim, J.H., Ahn, B.H. and Lee, J.S. (2012). Characterisation of new antihypertensive angiotensin I-converting enzyme inhibitory peptides from Korean traditional rice wine. J. Microbiol. Biotechnol., 22(3): 339–342.

Kim, H.W. and Kim, B.K. (1999). Biomedicinal triterpenoids of *Ganoderma lucidum* (Curt.: Fr.) P. Karst. (Aphyllophoromycetideae). Int. J. Med. Mush., 1(2): 121–138.

Kim, J.H., Lee, D.H., Lee, S.H., Choi, S.Y. and Lee, J.S. (2004). Effect of *Ganoderma lucidum* on the quality and functionality of Korean traditional rice wine, Yakju. J. Biosc. Bioeng., 97(1): 24–28.

Kim, M.-Y., Seguin, P., Ahn, J.-K., Kim, J.-J., Chun, S.-C., Kim, E.-H., Seo, S.-H., Kang, E.-Y., Kim, S.-L., Park, Y.-J., Ro, H.-M. and Chung, I.-M. (2008). Phenolic compound concentration and antioxidant activities of edible and medicinal mushrooms from Korea. J. Agric. Food Chem., 56(16): 7265–7270.

Kjellgren, B. (2004). Drunken modernity: Wine in China. Anthropol. Food, December 03, 2004.

Kozarski, M., Klaus, A., Jakovljevic, D., Todorovic, N., Vunduk, J., Petrović, P., Niksic, M., Vrvic, M. and van Griensven, L. (2015). Antioxidants of Edible mushrooms. Molecules, 20(10): 19489–19525.

Kramer, S.N. (1959). History Begins at Sumer. Doubleday Anchor Books, Doubleday Co. Inc., New York.

Lalotra, P., Gupta, D., Yangdol, R., Sharma, Y. and Gupta, S. (2016). Bioaccumulation of heavy metals in the sporocarps of some wild mushrooms. Curr. Res. Environ. Appl. Mycol., 6(3): 159–165.

Lan, T.W. (2001). Formula for the production of beer from *Ganoderma*. https://patents.google.com/patent/ EP1067178A1/en?oq=EP+1067178+A1 (accessed on May 16, 2020).

Lee, K.H., Itokawa, H. and Kozuka, M. (2003). Oriental herbal products: The basis for development of dietary supplements and new medicines in the 21st century. pp. 2–31. *In*: Ho, C.T., Lin J.K. and Zheng, Q.Y. (eds.). Oriental Foods and Herbs: Chemistry and Health Benefits, ACS Symposium Series # 859), American Chemical Society, Washington DC.

Lelièvre, M., Chollet, S., Abdi, H. and Valentin, D. (2009). Beer-trained and untrained assessors rely more on vision than on taste when they categorise beers. Chemosen. Percep., 2(3): 143–153.

Lemieszek, M. and Rzeski, W. (2012). Anticancer properties of polysaccharides isolated from fungi of the Basidiomycetes class. Współczesna Onkologia, 4: 285–289.

Leskosek-Cukalovic, I., Despotovic, S., Lakic, N., Niksic, M., Nedovic, V. and Tesevic, V. (2010). *Ganoderma lucidum*—Medical mushroom as a raw material for beer with enhanced functional properties. Food Res. Int., 43(9): 2262–2269.

Leskošek-Čukalović, I., Despotović, S., Nedović, V., Lakić, N. and Nikšić, M. (2010). New type of beer—Beer with improved functionality and defined pharmacodynamic properties. Food Technol. Biotechnol., 48(3): 384–391.

Lin, P.-H., Huang, S.-Y., Mau, J.-L., Liou, B.-K. and Fang, T.J. (2010). A novel alcoholic beverage developed from shiitake stipe extract and cane sugar with various Saccharomyces strains. LWT Food Sci. Technol., 43(6): 971–976.

Liquor Products Act 60 of 1989 amended by GG42260. 2019. http://www.sawis.co.za/winelaw/download/ Regulations,_annotated_05_2019.pdf (accessed on May 22, 2020).

Liu, D., Zhang, H.-T., Xiong, W., Hu, J., Xu, B., Lin, C.-C., Xu, L. and Jiang, L. (2014). Effect of temperature on chinese rice wine brewing with high concentration presteamed whole sticky rice. BioMed Res. Int., 2014: 426929. 10.1155/2014/426929.

Liu, L., Wang, J., Levin, M.J., Sinnott-Armstrong, N., Zhao, H., Zhao, Y., Shao, J., Di, N. and Zhang, T. (2019). The origins of specialised pottery and diverse alcohol fermentation techniques in early neolithic China. Proc. Nat. Acad. Sci., 116(26): 12767–12774. https://www.pnas.org/ content/116/26/12767 (accessed on November 23, 2019).

Lorencová, E., Salek, R.N., Černošková, I. and Buňka, F. (2019). Evaluation of force-carbonated Czech-type lager beer quality during storage in relation to the applied type of packaging. Food Cont., 106: 106706. https://doi.org/10.1016/j.foodcont.2019.106706.

Ma, H. (2017). A letter by the regional editor for Asia: China trends. Wine Econ. Pol., 6(1): 1–2.

Ma, Y.J., Zheng, L.P. and Wang, J.W. (2019). Bacteria associated with Shiraia fruiting bodies influence fungal production of hypocrellin. A. Front. Microbiol., 10: 2023. https://doi.org/10.3389/ fmicb.2019.02023.

Marchand, L.R. and Stewart, A. (2018). Breast cancer. Integr. Med., 772–784. e7.https://doi.org/10.1016/ B978-0-323-35868-2.00078-5.

Matsui, T., Kagemori, T., Fukuda, S., Ohsugi, M. and Tabata, M. (2009). Characteristics of wine produced by mushroom fermentation using *Schizophyllum commune* NBRC 4929. Mush. Sci. Biotechnol., 17(3): 107–111.

McGovern, P.E., Zhang, J., Tang, J., Zhang, Z., Hall, G.R., Moreau, R.A., Nunez, A., Butrym, E.D., Richards, M.P., Wang, C.-S., Cheng, G., Zhao, Z. and Wang, C. (2004). Fermented beverages of pre- and proto-historic China. Proc. Nat. Acad. Sci., 101(51): 17593–17598.

Meng, J. (2015). Brown rice and *Phellinus igniarius* wine. https://patents.google.com/patent/CN104403898A/en?oq=CN104403898A (accessed on May 17, 2020).

Meng, L. (2013). Flavoured rice wine containing oyster mushroom and coffee, and preparation method thereof. https://patents.google.com/patent/CN103232923A/en?oq=CN103232923A+ (accessed on May 17, 2020).

Mizuno, T., Sakai, T. and Chihara, G. (1995). Health foods and medicinal usages of mushrooms. Food Rev. Int., 11(1): 69–81.

Moradali, M.-F., Mostafavi, H., Ghods, S. and Hedjaroude, G.-A. (2007). Immunomodulating and anticancer agents in the realm of macromycetes fungi (macrofungi). Int. Immunopharmacol., 7(6): 701–724.

Mueller, G.M., Schmit, J.P., Leacock, P.R., Buyck, B., Cifuentes, J., Desjardin, D.E., Halling, R.E., Hjortstam, K., Iturriaga, T., Larsson, K.-H., Lodge, D.J., May, T.W., Minter, D., Rajchenberg, M., Redhead, S.A., Ryvarden, L., Trappe, J.M., Watling, R. and Wu, Q. (2006). Global diversity and distribution of macrofungi. Biodiver. Conser., 16(1): 37–48.

National Standards—Wine & Beverage, Appendix # A7. 2015. https://www.mpi.govt.nz/dmsdocument/14395/direct.

Ngo, T.V., Scarlett, C.J., Bowyer, M.C., Ngo, P.D. and Vuong, Q.V. (2017). Impact of different extraction solvents on bioactive compounds and antioxidant capacity from the root of *Salacia chinensis* L. J. Food Qual., 2017: 1–8.

Nguyen, A.N.H., Johnson, T.E., Jeffery, D.W., Danner, L. and Bastian, S.E.P. (2019a). A cross-cultural examination of Australian, Chinese and Vietnamese consumers' attitudes towards a new Australian wine product containing *Ganoderma lucidum* extract. Food Res. Int., 115: 393–399.

Nguyen, A.N.H., Capone, D.L., Johnson, T.E., Jeffery, D.W., Danner. L. and Bastian S.E.P. (2019b). Volatile composition and sensory profiles of a Shiraz wine product made with pre- and post-fermentation additions of *Ganoderma lucidum* extract. Foods, 8(11): 538. 10.3390/foods8110538.

Nguyen, A.N.H., Johnson, T.E., Jeffery, D.W., Capone, D.L., Danner, L. and Bastian, S.E.P. (2020). Sensory and chemical drivers of wine consumers' preference for a new Shiraz Wine product containing *Ganoderma lucidum* extract as a novel ingredient. Foods, 9(2): 10.3390/foods9020224.

Niksic, M., Nikicevic, N., Tesevic, V. and Klaus, A. (2001). Evaluation of alcoholic beverages based on *Ganoderma lucidum* (Curt.: Fr.) P. Karst. Extract. Int. J. Med. Mush., 3: 192.

Nishitoba, T., Sato, H. and Sakamura, S. (1988). Bitterness and structure relationship of the triterpenoids from *Ganoderma lucidum* (reishi). Agric. Biol. Chem., 52(7): 1791–1795.

Novakovic, S., Djekic, I., Klaus, A., Vunduk, J., Djordjevic, V., Tomović, V., Šojić, B., Kocić-Tanackov, S., Lorenzo, J.M., Barba, F.J. and Tomasevic, I. (2019). The effect of *Cantharellus cibarius* addition on quality characteristics of Frankfurter during refrigerated storage. Foods, 8(12): 635. 10.3390/foods8120635.

Okamura, T., Ogata, T., Minamimoto, N., Takeno, T., Noda, H., Fukuda, S. and Ohsugi, M. (2001a). Characteristics of beer-like drink produced by mushroom fermentation. Food Sci. Technol. Res., 7(1): 88–90.

Okamura, T., Ogata, T., Minamoto, N., Takeno, T., Noda, H., Fukuda, S. and Ohsugi, M. (2001b). Characteristics of wine produced by mushroom fermentation. Biosci. Biotechnol. Biochem., 65(7): 1596–1600.

Okamura-Matsui, T., Tomoda, T., Fukuda, S. and Ohsugi, M. (2003). Discovery of alcohol dehydrogenase from mushrooms and application to alcoholic beverages. J. Mol. Catal. B: Enzy., 23(2-6): 133–144.

Okhuoya, J. (2017). Edible mushrooms: As functional foods and nutriceuticals. Trop. J. Nat. Prod. Res., 1(5): 186–187.

Oludemi, T., Barros, L., Prieto, M.A., Heleno, S.A., Barreiro, M.F. and Ferreira, I.C.F.R. (2018). Extraction of triterpenoids and phenolic compounds from *Ganoderma lucidum*: Optimisation study using the response surface methodology. Food Func., 9(1): 209–226.

Öztürk, M., Tel-Çayan, G., Muhammad, A., Terzioğlu, P. and Duru, M.E. (2015). Mushrooms. Stud. Nat. Prod. Chem., 45: 363–456.

Palaniveloo, K. and Vairappan, C.S. (2013). Biochemical properties of rice wine produced from three different starter cultures. J. Trop. Biol. Conser., 10: 31–41.

Pecić, S., Veljović, M., Despotović, S., Leskošek-Čukalović, I., Jadranin, M., Tešević, V., Nikšić, M. and Nikićević, N. (2012a). Effect of maturation conditions on sensory and antioxidant properties of old Serbian plum brandies. Eur. Food Res. Technol., 235(3): 479–487.

Pecić, S., Veljović, M., Despotović, S., Leskošek-Čukalović, I., Nikšić, M., Vukosavljević, P. and Nikićević, N. (2012b). Antioxidant capacity and sensory characteristics of special herb brandy. *In*: Proceedings of 6th Central European Congress on Food, CE Food Congress, Institute of Food Technology, Novi Sad, Serbia.

Pecić, S. (2015). The effect of fruit body of *Ganoderma lucidum* on chemical composition and sensory properties of special brandy, PhD Dissertation, Faculty of Agriculture, University of Belgrade, Serbia. http://nardus.mpn.gov.rs/handle/123456789/4542 (accessed on June 24, 2020).

Pecic, S., Nikicevic, N., Veljovic, M., Jardanin, M., Tesevic, V., Belovic, M. and Niksic, M. (2016). The influence of extraction parameters on physicochemical properties of special grain brandies with *Ganoderma lucidum*. Chem. Ind. Chem. Eng. Quart., 22(2): 181–189.

Peeters, F.A.H. (1988). Wouter van Lis: Apotheker, Bierbrouwer En Stadsmedicus, Kring Voor de Geschiedenis van de Pharmacie in de Benelux. Bulletin, 73: 1–21.

Peintner, U., Pöder, R. and Pümpel, T. (1998). The iceman's fungi. Mycol. Res., 102(10): 1153–1162.

Petrović, J., Papandreou, M., Glamočlija, J., Ćirić, A., Baskakis, C., Proestos, C., Lamari, F., Zoumpoulakis, P. and Soković, M. (2014). Different extraction methodologies and their influence on the bioactivity of the wild edible mushroom *Laetiporus sulphureus* (Bull.) Murrill. Food Funct., 5(11): 2948–2960.

Petrović, M., Vukosavljević, P., Đurović, S., Antić, M. and Gorjanović, S. (2019). New herbal bitter liqueur with high antioxidant activity and lower sugar content: innovative approach to liqueurs formulations. J. Food Sci. Technol., 56(10): 4465–4473.

Pielech-Przybylska, K. and Balcerek, M. (2019). New trends in spirit beverages production. pp. 65–111. *In*: Grumezescu, A. and Holban, A.M. (eds.). Alcoholic Beverages, Woodhead Publishing, Cambridge, UK.

Pomarici, E. (2016). Recent trends in the international wine market and arising research questions. Wine Econ. Policy, 5: 1–3.

Que, F., Mao, L. and Pan, X. (2006). Antioxidant activities of five Chinese rice wines and the involvement of phenolic compounds. Food Res. Int., 39(5): 581–587.

Rathore, H., Prasad, S. and Sharma, S. (2017). Mushroom nutraceuticals for improved nutrition and better human health: A review. Pharma Nutrition, 5(2): 35–46.

Regulation (EU) # 251/2014 of the European Parliament and of the Council (2014). https://eur-lex.europa.eu/legal-content/EN/TXT/?uri=CELEX%3A32014R0251 (accessed on May 24, 2020).

Reinoso-Carvalho, F., Dakduk, S., Wagemans, J. and Spence, C. (2019). Dark vs. light drinks: The influence of visual appearance on the consumer's experience of beer. Food Qual. Pref., 74: 21–29.

Roselló-Soto, E., Martí-Quijal, F., Cilla, A., Munekata, P., Lorenzo, J., Remize, F. and Barba, F. (2019). Influence of temperature, solvent and ph on the selective extraction of phenolic compounds from tiger nuts by-products: Triple-TOF-LC-MS-MS characterisation. Molecules, 24(4): 797.10.3390/molecules24040797.

Saltarelli, R., Ceccaroli, P., Buffalini, M., Vallorani, L., Casadei, L., Zambonelli, A., Iotti, M., Badalyan, S. and Stocchi, V. (2015). Biochemical characterisation and antioxidant and antiproliferative activities of different *Ganoderma* collections. J. Mol. Microbiol. Biotechnol., 25(1): 16–25.

Sánchez, C. (2017). Reactive oxygen species and antioxidant properties from mushrooms. Synt. Syst. Biotechnol., 2(1): 13–22.

Sanchez-Muniz, F.J., Macho-González, A., Garcimartín, A., Santos-López, J.A., Benedí, J., Bastida, S. and González-Muñoz, M.J. (2019). The nutritional components of beer and its relationship with neurodegeneration and Alzheimer's disease. Nutrients, 11(7): 1558. 10.3390/nu11071558.

Sande, D., Oliveira, G.P. de, Moura, M.A.F.E, Martins, B. de A., Lima, M.T.N.S. and Takahashi, J.A. (2019). Edible mushrooms as a ubiquitous source of essential fatty acids. Food Res. Int., 125: 108524. https://doi.org/10.1016/j.foodres.2019.108524.

Sasidharan, S., Chen, Y., Saravanan, D., Sundram, K. and Latha, L. (2010). Extraction, isolation and characterisation of bioactive compounds from plants' extracts. Afr. J. Trad. Compl. Alt. Med., 8(1): 1–10.

Scurlock, J. and Andersen, B. (2005). Diagnoses in Assyrian and Babylonian Medicine: Ancient Sources, Translations, and Modern Medical Analyses, University of Illinois Press, Urbana, USA.

Sher, H., Al-Yemeni, M., Bahkali, A.H.A. and Sher, H. (2010). Effect of environmental factors on the yield of selected mushroom species growing in two different agro ecological zones of Pakistan. Saudi J. Biol. Sci., 17(4): 321–326.

Shomali, N., Onar, O., Alkan, T., Demirtaş, N., Akata, I. and Yıldırım, Ö. (2018). Investigation of polyphenol composition, biological activities and detoxification properties of some medicinal mushrooms from Turkey. Turk. J. Pharmaceut. Sci., 16(2): 155–160.

Sierksma, A., Hulshof, K.F.A.M., Grobbee, D.E. and Hendriks, H.F.J. (2003). Alcohol consumption. pp. 119–126. In: Caballero, B. (ed.). Enc. Food Sci. Nutr., Academic Press.

Silverman, J., Cao, H. and Cobb, K. (2020). Development of mushroom mycelium composites for footwear products. Cloth. Tex. Res. J., 38(2): 119–133.

Singh, N., Vaidya, D., Mishra, V. and Thakur, K. (2016). Shelf-life and storage quality of white button mushrooms (*Agaricus bisporus*) as affected by packaging material. Int. J. Adv. Res., 4(11): 1790–1799.

Smith, J.E., Rowan, N.J. and Sullivan, R. (2002). Medicinal mushrooms: A rapidly developing area of biotechnology for cancer therapy and other bioactivities. Biotechnol. Lett., 24: 1839–1845.

Snopek, L., Mlcek, J., Sochorova, L., Baron, M., Hlavacova, I., Jurikova, T., Kizek, R., Sedlackova, E. and Sochor, J. (2018). Contribution of red wine consumption to human health protection. Molecules, 23(7): 1684. 10.3390/molecules23071684.

Somogyi, S., Li, E., Johnson, T., Bruwer, J. and Bastian, S. (2011). The underlying motivations of Chinese wine consumer behaviour. Asia Pac. J. Market. Logist., 23(4): 473–485.

Song, J. (2013). Method for preparing *Cordyceps militaris* wine. https://patents.google.com/patent/CN103351972A/en?oq=CN103351972A (accessed on May 17, 2020).

Song, S. (1992). Process for obtaining edible fungus fermented wine. https://patents.google.com/patent/CN1017348B/en?oq=CN1017348B (accessed on May 17, 2020).

Standage, T. (2006). A History of the World in 6 Glasses. Atlantic Books, London.

Stoyneva-Gärtner, M.P., Uzunov, B.A., Dimitrova, P.H. and Chipev, N. (2018). Ethno mycological notes on the use of polypore fungi in domestic production of alcoholic beverages in Bulgaria. Res. J. Food Nutr., 2(3): 13–17.

Su, L. (2014). *Flammulina velutipes* beer and preparation method thereof. https://patents.google.com/patent/CN104073385A/en?oq=CN+104073385+A (accessed on May 16, 2020).

Taskin, H., Kafkas, E., Çakiroglu, Ö. and Büyükalaca, S. (2013). Determination of volatile aroma compounds of *Ganoderma lucidum* by gas chromatography mass spectrometry (HS-GC/MS). Afr. J. Trad. Compl. Alt. Med., 10(2): 353–355.

The European Parliament and the Council. (2019). On the definition, description, presentation and labelling of spirit beverages, the use of the names of spirit beverages in the presentation and labelling of other foodstuffs, the protection of geographical indications for spirit beverages, the use of ethyl alcohol and distillates of agricultural origin in alcoholic beverages, 2019/787 and Repealing Regulation (EC) # 110/2008.

Thu, Z.M., Myo, K.K., Aung, H.T., Clericuzio, M., Armijos, C. and Vidari, G. (2020). Bioactive phytochemical constituents of wild edible mushrooms from Southeast Asia. Molecules, 25(8): 10.3390/molecules25081972.

Tian, J., He, J., Zhao, Q. and Zhao, S. (2001). Study of the processing technique of *Ganoderma lucidum* rice wine and its fermentation conditions. J. Henan Agric. Uni., 35(1): 43–46.

Veljovic, M., Despotovic, S., Stojanovic, M., Pecic, S., Vukosavljevic, P., Belovic, M. and Leskosek-Cukalovic, I. (2015). The fermentation kinetics and physicochemical properties of special beer with addition of Prokupac grape variety. Chem. Ind. Chem. Eng. Quart., 21(3): 391–397.

Veljović, S. and Krstić, J. (2020). Elaborating on the potential for mushroom-based product market expansion: Consumers' attitudes and purchasing intentions. pp. 643–663. In: Singh, J., Meshram, V. and Gupta, M. (eds.). Bioactive Natural Products in Drug Discovery, Springer Nature, Singapore Pte Ltd.

Veljović, S., Nikićević, N. and Nikšić, M. (2019a). Medicinal fungus *Ganoderma lucidum* as raw material for alcohol beverage production. pp. 161–197. *In*: Grumezescu, A.H. and Holban, A.M. (eds.). Alcoholic Beverages, Woodhead Publishing, Cambridge, USA.

Veljović, S.P., Belović, M.M., Tešević, V.V., Nikšić, M.P., Vukosavljević, P.V., Nikićević, N.J. and Tomić, N.S. (2019b). Volatile composition, colour, and sensory quality of spirit-based beverages enriched with medicinal fungus *Ganoderma lucidum* and herbal extract. Food Technol. Biotechnol., 57(3): 408–417.

Vukosavljevic, P., Novakovic, M., Bukvic, B., Niksic, M., Stanisavljevic, I. and Klaus, A. (2009). Antioxidant activities of herbs, fruit and medicinal mushroom *Ganoderma lucidum* extracts produced by microfiltration process. Journal of Agricultural Sciences, Belgrade, 54(1): 45–62.

Vunduk, J., Klaus, A., Kozarski, M., Petrovic, P., Zizak, Z., Niksic, M. and Van Griensven, L.J.L.D. (2015). Did the iceman know better? Screening of the medicinal properties of the birch polypore medicinal mushroom, *Piptoporus betulinus* (higher Basidiomycetes). Int. J. Med. Mush., 17(12): 1113–1125.

Vunduk, J., Djekic, I., Petrović, P., Tomašević, I., Kozarski, M., Despotović, S., Nikšić, M. and Klaus, A. (2018). Challenging the difference between white and brown *Agaricus bisporus* mushrooms. Br. Food J., 120(6): 1381–1394.

Vunduk, J., Wan-Mohtar, W.A.A.Q.I., Mohamad, S.A., Abd Halim, N.H., Mohd Dzomir, A.Z., Žižak, Ž. and Klaus, A. (2019). Polysaccharides of *Pleurotus flabellatus* strain Mynuk produced by submerged fermentation as a promising novel tool against adhesion and biofilm formation of food-borne pathogens. LWT Food Sci. Technol., 112: 108221. https://doi.org/10.1016/j.lwt.2019.05.119.

Wachtel-Galor, S., Yuen, J., Buswell, J. and Benzie, I.F. (2011). *Ganoderma lucidum* (Lingzhi or Reishi): A medicinal mushroom. *In*: Benzie, I.F. and Wachtel-Galor, S. (eds.). Herbal Medicine: Biomolecular and Clinical Aspects, CRC Press/Taylor & Francis, Boca Raton, Florida, USA. https://www.ncbi. nlm.nih.gov/books/NBK92757/?report=reader (accessed on April 15, 2020).

Wang, M. (2005). Research on the preparation technique of healthy wine made from Se-enriched *Ganoderma lucidum* herb. China Brewing, 24: 58–59.

Wang, Y. and Ning, Z. (2013). Development of *Ganoderma lucidum* and fleece-flower root wine. Liquor-making Sci. Technol. http://en.cnki.com.cn/Article_en/CJFDTotal-NJKJ201305024.htm.

Wasser, S.P. (2002). Medicinal mushrooms as a source of antitumor and immunomodulating polysaccharides. Appl. Microbiol. Biotechnol., 60(3): 258–274.

Wasser, S.P. (2005). Reishi or Ling Zhi (*Ganoderma lucidum*). pp. 603–622. *In*: Coates, P.M., Betz, J.M., Blackman, M.R., Grass, G.M., Levine, M., Moss J. and White, J. (eds.). Encyclopedia of Dietary Supplements, Marcel Dekker, New York.

Wasser, S.P. (2010). Current findings, future trends, and unsolved problems in studies of medicinal mushrooms. Appl. Microbiol. Biotechnol., 89(5): 1323–1332.

Wasser, S.P. (2014). Mushroom pharmacy. Access Science. https://doi.org/10.1036/1097-8542.900145.

Wei, Y. (2014). Silkworm chrysalis *Cordyceps* bamboo wine and preparation method thereof. https:// patents.google.com/patent/CN103103091B/en?oq=CN103103091B (accessed on May 18, 2020).

Wiśniewska, P., Śliwińska, M., Dymerski, T., Wardencki, W. and Namieśnik, J. (2016). The analysis of raw spirits—A review of methodology. J. Inst. Brew., 122(1): 5–10.

Wolf, A., Bray, G.A. and Popkin, B.M. (2008). A short history of beverages and how our body treats them. Obes. Rev., 9(2): 151–164.

Yang, D., Lu, S., Xia, Q., Zheng, M. and Zhou, J. (2016). Changes of the main functional components and antioxidant capacity of *Ganoderma lucidum* wine during the extraction with ethanol solution. J. Food Sci. Biotechnol., 35(2): 205–210.

Yang, G. (2000a). Pine mushroom rice wine and compounding method thereof. https://patents.google. com/patent/CN1150176A/en?oq=CN1150176A (accessed on May 17, 2020).

Yang, L. (2000b). 'Shuangzhuyuye' liquid-health beverage. https://patents.google.com/patent/ CN1059126C/en?oq=CN1059126C (accessed on May 18, 2020).

Yeretzian, C., Pollien, P., Lindinger, C. and Ali, S. (2004). Individualisation of flavour preferences: Toward a consumer-centric and individualized aroma science. Compre. Rev. Food Sci. Food Saf., 3(4): 152–159.

Yin, H., Lu, W., Liu, H., Wang, Y., Ji, Z. and Li, X. (2013). Mushroom beer and production method. thereof. https://patents.google.com/patent/CN103305359A/en?oq=CN+103305359+A (accessed on May 16, 2020).

Yu, S. (2016). *Hericium erinaceus* and mulberry leaf rice wine and preparation method thereof. https://patents.google.com/patent/CN105505673A/en?oq=CN105505673A (accessed on May 18, 2020).

Zeng, L., Zhao, Z., Chu, S. and Zeng, F. (2007). Research and development of *Ganoderma lucidum Cordyceps* health wine. Liquor-making Sci. Technol., 162(12): 88–90.

Zhang, H., Ma, W., Feng, J. and Zhang, Y. (2014). Preparation method and standard of Chinese caterpillar fungus health care beer. https://patents.google.com/patent/CN103756820A/en?oq=CN103756820A (accessed on May 20, 2020).

Zhang, Y., Kuang, X., Shen, C., Ao, L., Ao, Z. and Luo, L. (2015). Research progress in *Cordyceps militaris* wine. Liquor-making Sci. Technol., 248(2): 90–93.

Zhou, X., Hu, S., Zhou, W., Lang, H., Lu, Y., Liu, Y., Liu, Y. and Liu, W. (2013).*Cordyceps militaris* juice wine and production method thereof. https://patents.google.com/patent/CN103421655A/en?oq=CN103421655A (accessed on May 17, 2020).

Zhou, X.-W., Su, K.-Q. and Zhang, Y.-M. (2011). Applied modern biotechnology for cultivation of *Ganoderma* and development of their products. Appl. Microbiol. Biotechnol., 93(3): 941–963.

Zjawiony, J.K. (2004). Biologically active compounds from Aphyllophorales (Polypore) Fungi. J. Nat. Prod., 67(2): 300–310.

9

Biochemical Profile of Six Wild Edible Mushrooms of the Western Ghats

Venugopalan Ravikrishnan,[1] Kandikere R Sridhar[1,2,]* and Madaiah Rajashekhar[1]

1. INTRODUCTION

Cultivated and wild mushrooms have been integral part of human diet. They have long been appreciated for their pleasant taste, flavour, texture, nutritional and medicinal properties (Strapáč et al., 2016). Among the 7,000 familiar mushrooms occurring around the world, about 42 per cent are major edible mushrooms (Badalyan et al., 2019). Up to 3 per cent of identified mushrooms are poisonous, whereas nearly 700 species among 2,000 edible mushrooms have potential medicinal attributes (Boa, 2004; Barceloux, 2008; Chang and Wasser, 2017). Health-promoting bioactive components are of special interest as they drastically differ from animal and plant sources (Pavithra et al., 2016).

Wild mushrooms possess a broad range of medicinal potential to protect human health and combat many human diseases (Pavithra et al., 2016). Antioxidant potential is one of the key beneficial characteristics of mushrooms to prevent many aggressive human ailments (Zhang et al., 2007). Bioactive molecules have been isolated, not only from members of edible Agaricaceae and Ganodermataceae, but also from non-edible species belonging to Paxillaceae, Polyporaceae, Scutigeraceae, Thelephoraceae and Xylariaceae (Quang et al., 2006). The role of bioactive compounds in promoting human health is well-known from many studies. For example, carotenoids (lycopene

[1] Department of Biosciences, Mangalore University, Mangalagangotri, Mangalore, Karnataka, India.

[2] Centre for Environmental Studies, Yenepoya (deemed to be) University, Mangalore, Karnataka, India.

* Corresponding author: kandikere@gmail.com

and astaxanthin) contribute significantly to the reduction of risks caused by cardiovascular diseases (CVD) (Riccioni, 2009; Mordente et al., 2011). Carotenoids are also known for their therapeutic potential against eye defects (e.g., xerophthalmia, cataract, macular degeneration and light-induced erythema) (Namitha and Negi, 2010; Adams et al., 2008). Polysaccharides of *Pleurotus pulmonarius* delay the progression of hepatocellular carcinoma; polysaccharides of *Pholiota nameko* have anti-inflammatory potential in rodents; polysaccharides of *Agaricus bisporus* is are known to inhibit prostate tumor growth in mice and some mushrooms prevent the development of atherosclerosis (e.g., *Pleurotus eryngii*, *Grifola frondosa*, and *Hypsizygus marmoreus*) (Li et al., 2008; Mori et al., 2008; Wasonga et al., 2008).

Since mushrooms are natural sources of nutritional and pharmaceutical components, it is pertinent to screen bioactive nutraceuticals and pharmaceuticals by appropriate methods (Robertson, 2005). Over 130 components of therapeutic significance have been recognised in mushrooms (e.g., hypoglycemic, antimicrobial, antiviral, analgesic, immune modulation, antioxidant, hypocholesterolemic, cytotoxic, hypotensive and so on) (Badalyan et al., 2019). Presence of a wide array of high- as well as low-molecular weight bioactive components in wild mushrooms is responsible for nutritional and health benefits (e.g., vitamins, lipids, terpenoids, polysaccharides, polysaccharide-peptides, polysaccharide-protein, lanostane-type triterpenoids, polychetides, lectins, phenolics, steroids, flavonoids and other secondary metabolites).

Identification of different components in the biological system is an essential basic step of industries pertaining to pharmaceutical, cosmetic, flavour and essence products. Gas chromatography and Mass spectroscopy (GC-MS) is an established method of assessment to elucidate the bioactive constituents of biological sources (Robertson, 2005). It is the combination of an ideal separation (by GC) with spectra (by MS) for qualitative and quantitative analysis of compounds (volatile and semi-volatile) (Balamurugan et al., 2017). Identification of compounds will be on the basis of peak area, retention time, molecular weight and molecular formula. It is obvious to assess bioactive components of unprocessed and processed mushrooms to follow nutritionally and pharmaceutically feasible procedures without loss of active principles.

Being a hotspot of biodiversity, the Western Ghats with diverse geographical setup, ecological versatility and climatic conditions result in harbouring a huge wealth of wild mushrooms. Evaluation of wild mushrooms of the Western Ghats for their bioactive components is still in its infancy. Thus, the present chapter concentrates on the biochemical composition of nutritionally-valued six wild mushrooms that occur in the forests at the foothills of the Western Ghats. Biochemical profile provides evidence of resourcefulness of a wild mushroom and its biological functions beyond nutritional value. Such insights stimulate further research towards conservation, exploitation and domestication for beneficial purposes.

2. Mushrooms and Processing

2.1 Sampling Locations

Two potential locations were chosen from the foothills of the Agumbe Ghat in the Western Ghats of Karnataka to discover wild edible mushrooms. Agumbe

Ghat possesses mountainous terrain with rainforest and waterfalls comparable to Cherrapungi of northeast India. The first location, Hebri in Udupi district (13°28'N, 74°59'E), is known for virgin forests with rainfall almost throughout the year. It is in the vicinity of the River Seeta. The average temperature ranges from 23–24°C with 97 per cent humidity. The second location, Someshwara, is a wild life sanctuary in Udupi district (13°29'N, 74°50'E) and forest reserve situated along the banks of River Seeta having an approximate area of 315 km². The climatic conditions are similar to Hebri forests. Someshwara sanctuary is known for many endangered animals as well as plants. It comprises semi-evergreen as well as moist mixed deciduous forests. The climatic conditions and secondary products of forests are the main requirements for growth of mushrooms. Although these locations are well known for tourism and wildlife, little attention has been paid to understand the occurrence and diversity of wild mushrooms. However, during the wet season, the local people and tribals take advantage of wild mushrooms by collecting them for their food as well as livelihood. For instance, the tribals of Karkala (foothills of Western Ghats) are involved in collection of the wild mushroom, *Astraeus hygrometricus* for their nutrition as well as trade (Pavithra et al., 2015). A broad survey was carried out in about 3 km² of forests of Hebri and Someshwara to trace the potential region for availability of select edible wild mushrooms. Trial collections were performed to ascertain the identity followed by edibility based on literature and discussion with local people, tribes and forest personnel. Subsequently, mushrooms were sampled in batches to process for assessment of nutritional and biochemical potential.

2.2 *Harvest and Processing*

The whole fruit bodies of mushrooms were harvested from three locations of the forests as replicates, stored in cool packs and transferred to laboratory within two to three hours for processing (Fig. 1). Care was taken not to exhaust all the mushrooms in a specific collection site and about 10–20 per cent of the fruit bodies were left

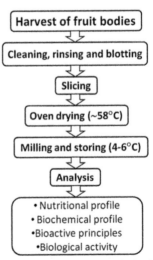

Fig. 1. General scheme of processing and analysis of wild mushrooms.

without disturbance. The transported fruit bodies were spread over a blotting paper and the visible debris was removed. Later, part of the fruit bodies (about three to five) from triplicate samples were processed for moisture content gravimetrically. The rest of the fruit bodies were immersed in distilled water for 5 min to remove the fine debris followed by a rinse. After rinsing, they were blot-dried followed by drying at $58 \pm 2°C$ in an hot-air oven by spreading on aluminium foil until attaining a constant weight. Triplicate dried fruit bodies were pulverised from coarse to fine powder. The powder of fruit bodies was transferred into air-tight containers and preserved in the refrigerator for further analysis. The overall gross analysis of fruit bodies included nutritional profile, biochemical profile, bioactive principles and biological activity.

3. Assessment of Biochemical Qualities

3.1 Total Phenolics

Total phenolic content was determined according to the method outlined by Rosset et al. (1982). Fifty mg of mushroom powder was extracted in 5 ml of 50 per cent methanol on water bath ($95 \pm 1°C$) for 10 min, centrifuged (1500 rpm) and the supernatant was collected. The extraction was repeated for the remaining residue, supernatant were pooled and made up to 10 ml. An aliquot of 0.5 ml extract was mixed with 0.5 ml distilled water and treated with 5 ml Na_2CO_3 (in 0.1 N NaOH). After 10 min of incubation at room temperature, 0.5 ml Folin-Ciocalteu's reagent (diluted, 1:2 v/v) was added and the absorbance was read at 725 nm. Gallic acid was used to prepare the standard curve and the results were expressed as mg of gallic acid equivalents (GAEs) per gram of the mushroom sample (mg GAEs/g).

3.2 Tannins

Vanillin-HCl method was employed to determine the tannins of mushroom samples (Burns, 1971). One gram of mushroom powder was extracted with 50 ml methanol (28°C; 20–28 hours), centrifuged and supernatant was collected. One ml of extract was treated with 5 ml of vanillin hydrochloride reagent mixture (4 per cent vanillin in methanol and 8 per cent concentrated HCl in methanol; 1:1). After 20 min, the developed colour was read at 500 nm with 50–250 µg catechin as standard. The tannin content was expressed as catechin equivalents (CEs) in milligram per gram of the sample (mg CEs/g).

3.3 Phytic Acid

Extraction and estimation of phytic acid in the mushroom sample was performed by standard procedures (Deshpande et al., 1982; Sathe et al., 1983). In brief, a known amount of the mushroom sample (2 g) was extracted (2 hours) with 1.2 per cent HCl (10 ml) containing sodium sulphite (10 per cent) at room temperature ($25 \pm 1°C$) and centrifuged. The volume was made up to 10 ml with the same extract. Phytic phosphorus was estimated before and after precipitation of phytic acid by $FeCl_3$. Five ml of the above extract was taken and 3 ml of $FeCl_3$ solution ($FeCl_3$, 2 g + concentrated HCl, 16.3 diluted to 1 litre) was added, stirred, boiled for 75 min on a boiling water bath, cooled and left at room temperature (one hour) prior to centrifuge

(2000 g, 10 min) and filtered (Whatmann # 1). The supernatant (made up to 10 ml with distilled water) was used for assay. Analysis of soluble phosphorus was carried out by the method by Bartlett (1959) using ammonium molybdate reagent. The absorbance was read at 430 nm after 30 min with KH_2PO_4 as standard to determine phytate:

$$\text{Phytate phosphorous} = (A \times 28.18)/\ 100\ (\text{where, A} = \text{phytic acid})$$

3.4 Flavonoids

The flavonoid content in the mushroom powder was determined by the method outlined by Jia et al. (1999) with a slight modification. The mushroom extract (250 µl) was mixed with 1.25 ml of distilled water and 75 µl of 5 per cent $NaNO_2$ solution, incubated at room temperature for 5 min and then 150 µl of 10 per cent $AlCl_3$ and H_2O solution was added. After 6 min, 500 µl of 1 M NaOH and 275 µl of distilled water were added to the mixture. The solution was mixed well and the colour developed was measured at 510 nm. Quercetin dehydrate was used to prepare the standard curve and the flavonoid content was expressed as quercetin equivalents (QEs) in milligram per gram of the sample (mg QEs/g).

3.5 Vitamin C

The vitamin C content of mushrooms was estimated according to Roe (1954) with a minor modification. One gram of mushroom powder was extracted in 10 ml of 5 per cent trichloroacetic acid (TCA). An aliquot (0.2 ml) was made up to 1 ml by adding 5 per cent TCA followed by addition of 1 ml of 2, 4-dinitrophenylhydrazine (DNPH). The reaction mixture was boiled for 10 min, cooled to laboratory temperature; 4 ml of 65 per cent sulphuric acid was added, incubated up to 30 min at laboratory temperature and the absorbance was measured at 540 nm. Ascorbic acid was used to prepare the standard curve and vitamin C content was expressed as ascorbic acid equivalents (AAEs) in milligram per gram of the sample (mg AAEs/g).

3.6 Pigments

The β-carotene and lycopene contents in mushroom were assessed by the method of Barros et al. (2007). The dried methanolic extract of mushrooms (100 mg) was vigorously shaken with acetone-hexane mixture (4:6; 10 ml) for 1 min and filtered through Whatman # 4 filter paper. The absorbance of the filtrate was read at three wavelengths (453 nm, 505 nm and 663 nm). Contents of β-carotene and lycopene were calculated thus:

$$\text{Lycopene (mg/100 ml)} = (-0.0458\ A_{663} + 0.372\ A_{505} - 0.0806\ A_{453})$$
$$\text{β-carotene (mg/100 ml)} = (0.216\ A_{663} - 0.304\ A_{505} + 0.452\ A_{453})$$

3.7 Trypsin Inhibition Activity

Trypsin-inhibition activity was measured according to Kakade et al. (1974). The mushroom powder (1 g) was constantly stirred with NaOH (0.01 N; 50 ml; 10 min). The extract (1 ml) was diluted with distilled water (1:1) followed by addition of enzyme standard 2 ml (2 mg trypsin/100 ml 0.001 M NaOH; 2 ml) and incubated in water bath (37°C, 10 min), BAPNA (40 mg Nα-benzoy-L-arginine 4-nitroanilide hydrochloride) dissolved in dimethyl sulphoxide and made up to 100 ml with tris-buffer (at 37°C; 5 ml), mixed and the absorbance was measured (410 nm). The control was prepared as per protocol without the addition of the mushroom extract. Trypsin inhibition unit (TIu) per mg of mushroom powder was calculated.

TIu/mg = $[(A_{C410} - A_{S410}) \times 100]$ per ml extract/mg sample per ml of extract

3.8 Haemagglutinion Activity

Haemagglutination activity was assessed based on Occenã et al. (2007) method. Mushroom powder (2 g) was suspended in NaCl (0.9 per cent, 20 ml) shaken vigorously (1 min) and allowed to stand (1 hour) followed by centrifugation (2000 g; 10 min) to obtain a clear solution, which was filtered and used as crude agglutinin extract. Heparinised human blood samples (5 ml) of different groups (A[+], B[+], AB[+] and O[+]) were collected and RBCs were separated by centrifugation (2000 g; 10 min). One volume of RBCs was diluted with four volumes of cold saline (0.9 per cent), centrifuged (2000 g; 10 min) and the supernatant was discarded. The pellet was washed with saline (0.9 per cent) thrice until the supernatant became colourless. Washed erythrocytes (4 ml) were suspended in phosphate buffer (0.0006 M; pH 7.4) (100 ml). Trypsin solution (2 per cent, 1 ml) was added to washed erythrocytes (10 ml), mixed and incubated (37°C; 1 hour). The trypsinised erythrocytes were washed (four to five times in 0.9 per cent saline) to remove traces of trypsin. The packed cells (1.2–1.5 ml) obtained were suspended in saline (0.9 per cent; 100 ml). The round-bottomed 96 well micro titre-plate was used for the assay. Initially, phosphate buffer (50 µl) was added in the well # 1 to 12, followed by the addition of crude agglutinin extract (50 µl) to the well # 1, mixed and twofold serial dilution till well # 11. The suspension of erythrocytes (50 µl) was added to well # 1 to 12. The well # 12 served as control where crude agglutinin was absent. The contents in the wells were gently mixed and incubated at room temperature for 4 hours and then observed for haemagglutination in each well. Haemagglutination unit per gram (Hu/g) was calculated.

Hu/g = (Da × Db × S)/V (Where Da, dilution factor of extract in well # 1; Db, dilution factor of well containing 1 Hu is the well in which the haemagglutination was observed; S, initial extract/g mushroom powder; V, volume of the extract in well # 1).

4. Chemical Constituents

Chemical composition of mushrooms was investigated through GC-MS/MS and the analysis was carried out on a Scion 436-GC Bruker model coupled with a triple quadrupole mass spectrophotometer. The GC-MS was operated by using the following conditions: fused silica capillary column BR-5MS [30 m (length) × 0.25 mm (internal diameter) × 0.25 µm (thickness) composed of 5 per cent Diphenyl/95 per cent Dimethyl poly siloxane]. Helium gas (99.999 per cent) was used as carrier gas at a constant flow of 1 ml/min and an injection volume of 2 µl was employed (split ratio of 10:1). The column oven temperature programme was as follows: 110°C hold for 3.5 min, up to 200°C at the rate of 10°C/min-no hold, up to 280°C at the rate of 5°C/min–9 min hold. The injector temperature was 280°C; inlet and source temperature was 290°C and 250°C, respectively. The mass spectrometer was operated in the positive electron ionisation (EI) mode with ionisation energy of 70 eV. The solvent delay was 0–3.5 min. The fragments from mass scan (m/z) 50–500 amu were programmed. The total GC-MS running time was 37.5 min. The relative percentage amount of each component was calculated by comparing its average peak area to the total areas. Software adopted to handle mass spectra and chromatograms was MS Workstation 8. The NIST version 2.0 library database of National Institute of Standard and Technology (NIST), having more than 62,000 patterns, was used for identifying the chemical components. The spectrum of the unknown component was compared with the spectrum of the known components stored in the NIST library. The name, molecular weight and structure of the components of the test materials were ascertained (Srinivasan and Kumaravel, 2016). The therapeutic activities of the identified compounds (NIST Mass Spectral Search Programme for the NIST/EPA/NIH Mass Spectral Library Version # 2.0; Gaithersburg, MD, USA) were found by using Dr Duke's phytochemical and ethnobotanical databases (www.ars-grin.gov/cgi-bin/duke).

5. Discussion

There is tremendous interest to study the bioactive components of cultivated and wild mushrooms. However, thus far the potential of bioactive components of mushrooms is not fully understood or exploited. As many wild mushrooms are ethnically used, studies are more from the east Asian region (Kalač, 2009). In view of utilisation of synthetic antioxidants in food industries (those are carcinogenic) (e.g., butylated hydroxytoleune, butylated hydroxyanisole and tertiary butyl hydroquinone), studies on natural antioxidant agents present in mushrooms assume importance (Botterweck et al., 2000; Ren et al., 2014). Various other components in mushrooms, like phytic acid, flavonoids, vitamins, pigments and L-DOPA are valuable for health-promoting potential.

5.1 Phenolics

The total phenolics was found to be high in *T. schimperi* followed by *H. alwisii*, *B. edulis* and *L. squarrosulus* (8.3–48.3 mg/g), while its content in the rest of mushrooms ranged between 1.9 and 3.7 mg/g (Table 1). The total phenolics serve as major antioxidant components in mushrooms due to their hydroxyl groups. They have been linked to major antioxidant function in three Portuguese wild edible mushrooms (*Agaricus arvensis, Leucopaxillus giganteus* and *Sarcodon imbricatus*) (Barros et al., 2007). Phenolic content in wild mushrooms of Asia has been connected to antioxidant significance by many researchers (Cheung and Cheung, 2005; Lo and Cheung, 2005). Phenolic compounds also play an important role in stabilising lipid peroxidation (Yen and Wu, 1993) and its consumption in diets controls progression of atherosclerosis as well as reduces the risk of cardiovascular diseases (Visioli et al., 2000; Meng et al., 2002).

Table 1. Bioactive components, trypsin inhibition and haemagglutinin activity of six wild edible mushrooms of the Western Ghats (n = 3 ± SD).

	Amanita hemibapha	*Boletus edulis*	*Hygrocybe alwisii*	*Lentinus polychrous*	*Lentinus squarrosulus*	*Termitomyces schimperi*
Total phenolics (mg/g)	3.70 ± 0.46	18.10 ± 1.22	23.50 ± 0.92	1.90 ± 0.46	8.30 ± 0.46	48.3 ± 0.46
Tannins (mg/g)	3.45 ± 0.09	13.07 ± 0.06	11.96 ± 0.06	1.04 ± 0.10	1.32 ± 0.10	24.98 ± 0.08
Phytic acid (%)	0.163 ± 0.006	0.180 ± 0.0	0.063 ± 0.002	0.003 ± 0.0	0.021 ± 0.001	0.133 ± 0.0
Flavonoids (mg/g)	3.36 ± 0.06	33.47 ± 0.09	3.62 ± 0.26	2.30 ± 0.00	1.86 ± 0.00	28.04 ± 0.03
Vitamin C (mg/g)	3.73 ± 0.06	5.15 ± 0.09	11.33 ± 0.46	3.40 ± 0.00	9.50 ± 0.00	3.60 ± 0.46
β-carotene (μg/g)	4.80 ± 0.01	5.04 ± 0.02	1.22 ± 0.00	1.60 ± 0.00	1.56 ± 0.00	5.06 ± 0.00
Lycopene (μg/g)	3.32 ± 0.00	3.84 ± 0.03	0.58 ± 0.00	1.28 ± 0.00	1.20 ± 0.00	4.08 ± 0.00
Trypsin inhibition activity (TIu/mg)	0.04 ± 0.007	0.02 ± 0.005	0.02 ± 0.023	0.10 ± 0.000	0.01 ± 0.001	0.01 ± 0.001
Haemagglutinin activity against human blood groups (Hu/g)						
A$^+$	100	100	50	0	0	0
B$^+$	50	25	0	0	100	0
AB$^+$	100	0	100	0	100	0
O$^+$	50	100	50	0	100	0

Quantity of total phenolics in wild and cultivated mushrooms showed drastic variations. Moreover the same mushroom in different geographic locations possesses variable quantities of total phenolics. For example, its content was lower than the present study in *L. polychrous* and *L. squarrosulus* collected from Thailand (Attarat and Phermthai, 2015), while it was opposite for the same species in Nigeria (Okoro, 2012). Hussein et al. (2015) also reported lower quantities of phenolics in *L. squarrosulus* than the present study. The phenolic content of *T. schimperi* in this study is higher as compared to *Cantharellus friessi*, *C. subcibarius* and *C. cinerius* collected from north-western Himalayan region (Kumari et al., 2011). Methanolic and petroleum ethers extract of *Hygrocybe conica* showed higher quantity of total phenolics than *H. alwisii* in the present study (Wong and Chye, 2009). Total phenolic content in *B. edulis, H. alwisii, L. squarrosulus* and *T. schimperi* is higher as compared to another edible wild mushroom, *Astraeus hygrometricus* on the foothills of the Western Ghats, while rest of the mushrooms studied are comparable (Pavithra et al., 2016). The quantity of total phenolics in *A. hemibapha* possesses substantially lower quantities than wild edible mushroom *Amanita* sp. of southwest India (Greeshma et al., 2018). Total phenolic in *L. squarrosulus* is very low as compared to the same species from the southwest coast of India (Ghate and Sridhar, 2017).

5.2 Tannins

As seen in total phenolics, the tannin content was high in *T. schimperi* followed by *B. edulis* and in *H. alwisii* (12–25 mg/g), while its content in rest of the mushrooms ranged between 1 and 3.5 mg/g (Table 1). Tannins are common in a variety of foodstuffs (e.g., banana, sorghum, grapes, raisins, spinach, red wine, persimmons and so on). Beverages, especially tea, has been recognised to be a rich source of tannins (http://www.livestrong.com/article/527907-tannin-levels-in-teas/).

Tannins are also an important component of wild and cultivated mushrooms. Tannin content of mushrooms is comparable with other mushrooms, like *Pleurotus florida* and *Pleurotus tuber-regium* (Dandapat et al., 2015; Pandimeena et al., 2015). However, Woldegirogis et al. (2015) reported less tannin in a few edible mushrooms of Ethiopia. Tannin content is low in *Rusulla* sp. As compared to mushrooms evaluated in this study, so also in *Pleurotus* sp. and *Cheimonophyllum candidissimus* when compared to *A. hemibapha, B. edulis* and *H. alwisii*, while it is opposite when compared to *L. polychrous, L. squarrosulus* and *T. schimperi* (Okwulehie and Ogoke, 2013). All the mushrooms studied possess a higher quantity of tannins as compared with *A. hygrometricus* occurring on the foothills of the Western Ghats (Pavithra et al., 2016). It is interesting to note that a difference in quantity of tannins was found in different parts of wild mushroom *A. hygrometricus* (Singh, 2011). Tannin content of *A. hemibapha* was higher than in wild edible mushroom *Amanita* sp. of southwest India (Greeshma et al., 2018). Tannin content of *L. squarrosulus* is lower as compared to the same species from southwest coast of India (Ghate and Sridhar, 2017).

5.3 Phytic Acid

The phytic acid content was high in *B. edulis* followed by *A. hemibapha* and *T. schimperi* (0.13–0.18 per cent), while its content in the rest of the mushrooms ranged between 0.003 and 0.063 per cent (Table 1). Besides antioxidant potential, phytic acid serves in checking the kidney stones and deposition of calcium in the arteries (Knekt et al., 2004; Ye and Song, 2008).

The phytic acid content is low as compared to the quantity in other edible mushrooms, like *Agaricus bisporus, Calocybe indica, Lentinula edodes, Macrocybe gigantea, Lentinus sajor-caju* and *Pleurotus sajor-caju* (Gaur et al., 2016). Phytic acid content in the present study is insufficient as the safe limit is 22.1 mg/100 g (WHO, 2003). *Amanita hemibapha* and *B. edulis* possess a higher quantity of phytic acid as compared to *A. hygrometricus* occurring on the foothills of the Western Ghats (Pavithra et al., 2016). The quantity of phytic acid in *A. hemibapha* is comparable with wild edible mushroom *Amanita* sp. of southwest India (Greeshma et al., 2018).

5.4 Flavonoids

Boletus edulis possesses a high quantity of flavonoids followed by *T. schimperi* (28–33.5 mg/g), while its content in rest of the mushrooms ranged between 1.9 and 3.6 mg/g (Table 1). Studies on flavonoids of mushrooms are meagre (Barros et al., 2008). Flavonoids have several health-protective qualities, especially cardioprotective, hepatoprotective, anti-inflammatory and antidiabetic properties (Champ, 2002; Tapas et al., 2008).

Flavonoid content of *L. squarrosulus* of Nigeria was higher than the present study (Okoro, 2012). However, flavonoid content in the present study is higher than edible *Cantharellus mushrooms* (Kumari et al., 2011). Hussein et al. (2015) reported higher quantity of flavonoids (25.62 ± 1.78 mg/100 g) for in *L. squarrosulus* than the present study. All the mushrooms studied possess a higher quantity of flavonoids as compared to *A. hygrometricus* occurring on the foothills of the Western Ghats (Pavithra et al., 2016). Flavonoid content of *A. hemibapha* is higher than wild edible mushroom *Amanita* sp. of southwest India (Greeshma et al., 2018). Flavonoid in *L. squarrosulus* is substantially lower as compared to same species from the southwest coast of India (Ghate and Sridhar, 2017).

5.5 Vitamin C

Hygrocybe alwisii consists of a high quantity of vitamin C followed by *L. squarrosulus* and *B. edulis* (5.2–9.5 mg/g), while its content in rest of the mushrooms ranges between 3.4 and 3.7 mg/g (Table 1). The high level of vitamins in mushrooms, particularly vitamin C and D, has been reported as being responsible for its antioxidant activity. Although vitamin C is supreme in antioxidant and radical-scavenging power, it will succumb to degradation by increased heat (Gregory, 1996; Podmore et al., 1998).

All the mushrooms studied possess a higher quantity of vitamin C as compared to *A. hygrometricus* occurring on the foothills of the Western Ghats (Pavithra et al., 2016). Vitamin C content of *A. hemibapha* is lower than in wild edible mushroom *Amanita* sp. of southwest India (Greeshma et al., 2018). Vitamin C content of *L. squarrosulus* is substantially higher when compared to the same species occurring on the southwest coast of India (Ghate and Sridhar, 2017).

5.6 Pigments

All mushrooms evaluated possess β-carotene as well as lycopene (Table 1). The β-carotene was high in *T. schimperi* as well as *B. edulis* followed by *A. hemibapha* (4.8–5.1 μg/g), while its content in the rest of the mushrooms range between 1.22 and 1.6 μg/g. The lycopene content was highest in *T. schimperi* followed by *B. edulis* and *A. hemibapha* (3.8–4.1 μg/g), while its content in the rest of the mushrooms ranges between 0.58 and 1.28 μg/g. Similar to vegetables, carotenoids present in mushrooms also serve as antioxidants. Quantity of carotenoids in edible wild mushrooms of Portugal has been related to antioxidant activity (Barros et al., 2007).

All the mushrooms studied possess comparable quantities of β-carotene as well as lycopene than *A. hygrometricus* occurring on the foothills of the Western Ghats (Pavithra et al., 2016). The β-carotene and lycopene contents of *A. hemibapha* are lower than wild edible mushroom *Amanita* sp. of southwest India (Greeshma et al., 2018).

5.7 Trypsin Inhibition Activity

Trypsin inhibition activity was the highest in *L. polychrous* (0.1 Tiu/mg), while its content in rest of the mushrooms ranges between 0.01 and 0.04 TIu/mg (Table 1).

Foodstuffs consist of trypsin inhibitors which are known to utilise sulphur amino acid synthesis of trypsin and chymotrypsin, which leads to deficiency of sulphur amino acids (Liener and Kakade, 1980).

Trypsin inhibition activity of all the mushrooms studied is comparable with *A. hygrometricus* occurring on the foothills of the Western Ghats (Pavithra et al., 2016). Trypsin inhibition activity of *A. hemibapha* is higher than wild edible mushroom *Amanita* sp. of southwest India (Greeshma et al., 2018). Interestingly, unlike *L. squarrosulus* in the present study, same species from the west coast of India did not show trypsin inhibition activity (Ghate and Sridhar, 2017).

5.8 Haemagglutinin Activity

Among six mushrooms, *L. polychrous* and *T. schimperi* did not show haemagglutinin activity against all the blood groups assessed (Table 1). *Amanita hemibapha* possess agglutination activity in all the blood groups (range between 50–100 Hu/g). Rest of the species showed agglutination for a maximum of three blood groups, ranging from 25–100 Hu/g. Absence or below the threshold level in mushrooms denotes nutritional benefit, especially by *L. polychrous* and *T. schimperi*. Hemagglutinins or lectins are widely distributed in the plant kingdom and pose a high degree of specificity towards sugar component and thus have diagnostic importance. On ingestion, hemagglutinins exhibit a unique property to bind carbohydrate-containing molecules and resist digestion (WHO, 2003).

Uncooked wild mushroom *A. hygrometricus* on the foothills of the Western Ghats are also devoid of haemagglutinin activity against blood groups B^+, AB^+ and O^+ (Pavithra et al., 2016). Haemagglutinin activity of *A. hemibapha* is higher than of cooked wild edible mushroom *Amanita* sp. of southwest India (Greeshma et al., 2018). It is likely the cooking of *A. hemibapha* leads to decrease in the haemagglutinin activity similar to *Amanita* sp. (Greeshma et al., 2018). Unlike *L. squarrosulus* in the present study, the same species from the west coast of India (uncooked as well as cooked) did not show haemagglutinin activity (Ghate and Sridhar, 2017).

6. Chemical Composition

The GC-MS analysis of the bioactive components in the ethanol extract of six wild edible mushrooms reveals that these mushrooms contain a considerable amount of bioactive compounds (Tables 2–7). Most of the isolated and identified compounds by GC-MS in the crude extracts of mushrooms are biologically important. The identified compounds were different types of terpenes, alcohols, fatty acids, amines, sterols and others. *Termitomyces schimperi* composed of the highest number of compounds (33) followed by *A. hemibapha* (31), *Boletus edulis* presented (30), *Hygrocybe alwisii* (25), *L. polychrous* (20) and *L. squarrosulus* (17).

Common name and specific active principle in mushrooms along with biological activity and applications are given for each of the mushrooms evaluated by GC-MS (Tables 2-7). Based on the established biological value, compounds present in the mushrooms studied broadly possess compounds useful in cancer prevention, health-promotion, pharmacological value (e.g., analgesic, sedatives, hallucinogen, anti-angiogenic and antioxidant), food industry (emulsifiers, flavours,

Table 2. Bioactive compounds, medicinal properties and applications of *Amanita hemibapha* (?, Not defined) (www.ars-grin.gov/cgi-bin/duke).

Compound	Active principle	Biological activity and application
Maltose (disaccharide)	D-Glucose, 4-O-α-D-glucopyranosyl-	Detoxicant, diuretic, anticancer, sweetening agent in food and beverages
Indole	Indole	Anti-acne, antibacterial (anti-streptococci and anti-*Salmonella*), anticariogenic and cancer preventive
Galactonic acid derivative	2-Acetamide-2-deoxygalactono-1,4-lactone	N-acetylglucosaminidase inhibitor
Acetic ester	Acetic acid, 2-phenylethyl ester	Aroma component (cosmetic uses)
Fatty acid methyl ester	Hexadecanoic acid, methyl ester	Nutrient, energy storage and membrane stabilizer
Saturated fatty acid	n-hexadecanoic acid	Antioxidant, hypocholestrolemic, antiandrogenic and flavoring agent
Fatty acid methyl ester	9, 12-Octadecadienoic acid, methyl ester	Personal care products
Fatty acid methyl ester	9-Octadecenoic acid, methyl ester, (E)-	Industrial use
Fatty acid methyl ester	Stearic acid, methyl ester	Flavouring agents and industrial use
Fatty acid ethyl ester	9, 12-Octadecadienoic acid, ethyl ester	Anti-inflammatory agent
Fatty acid	9, 12-Octadecadienoic acid (Z, Z)-	Anti-inflammatory, hypocholesterolemic, cancer preventive, hepatoprotective, antihistamine, antieczema, antiandrogenic, antiarthritic, anticoronary, insect repellent andnematicide
Amide	9-Octadecenamide, (Z)-	Food additives, sedatives, anti-inflammatory, antibacterial and antioxidant
Fatty amide	Octadecanamide	?
Triterpene	Squalene	Cancer preventive, immunostimulant, chemopreventive and diuretic
Flavonoid fraction	2H-Pyran, 2-(7-heptadecynyloxy) tetrahydro-	Antimicrobial, hallucinogen, haemagglutinator and anti-inflammatory and antioxidant
Terpenoids	Urs-12-en-28-al	Antimicrobial
Triterpene	Urs-12-en-28-oic acid, 3-hydroxy-, methyl ester, (3β)-	Anti-inflammatory effects
Phytosterol	Ergosterol	Antiangiogenic, antiflu, antitumor and antiviral
Sterols	7, 22-Ergostadienol	Anti-proliferative
Fungisterol	γ-Ergostenol	?

Table 3. Bioactive compounds, medicinal properties and applications of *Boletus edulis* (?, Not defined) (www.ars-grin.gov/cgi-bin/duke).

Compound	Active principle	Biological activity and application
Maltose (disaccharide)	D-Glucose, 4-O-α-D-glucopyranosyl-	Detoxicant, diuretic, anticancer, sweetening agent in food and beverages
1-benzofurans	Benzofuran, 2, 3-dihydro-	In treatment of diabetic retinopathy and arthritis
Indole	Indole	Anti-acne, antibacterial, anticariogenic, anti-*Salmonella*, anti-streptococci and cancer preventive
Phenol	Eugenol	Analgesic, antibacterial, anticancer, antimelanomic, antiseptic, antiedemic, antigenotoxic, anti-inflammatory, antiviral, antiulcer, antithromboxane, hepatoprotective, trypsin enhancer and perfumery
Long chain fatty alcohol	Falcarinol	Allergenic, analgesic, antibacterial, anticancer, antimelanomic, antiseptic, antithromboxane, anti-tuberculosis and antitumor
Phenolic compound	Phenol, 2-methoxy-4-(1-propenyl)-	Sensitizer, allergen, flavour and fragrance
Saturated fatty acid	Tetradecanoic acid	Antifungal, antioxidant, cancer preventive, nematicide, hypercholesterolemic and lubricant
Cyclic purine nucleotide	Adenosine 3,5-cyclic monophosphate	Second messenger
Diester	Dibutyl phthalate	Antimicrobial and antifouling
Saturated fatty acid	n-hexadecanoic acid	Antioxidant, hypocholestrolemic, antiandrogenic and flavor
Fatty acid ethyl ester	Hexadecanoic acid, ethyl ester	Antioxidant, hypocholesterolemic, hemolytic, alpha-reductase inhibitor and antiandrogenic
Triterpene	Squalene	Cancer preventive, immunostimulant, chemopreventive and diuretic
Flavonoid fraction	2H-Pyran, 2-(7-heptadecynyloxy)tetrahydro	Antimicrobial agent, hallucinogen, haemagglutination, anti-inflammatory and antioxidant
Triterpene	Urs-12-en28-oic acid, 3-hydroxy-, methyl ester, (3β)-	Anti-inflammatory
Terpenoids	Urs-12-en-28-al	Antimicrobial effects
Phytosterol	Ergosterol	Antiangiogenic, anti-flu, antitumor and antiviral
Sterols	7, 22-Ergostadienol	Anti-proliferative
Fungisterol	γ-Ergostenol	?

Table 4. Bioactive compounds, medicinal properties and applications of *Hygrocybe alwisii* (?, Not defined) (www.ars-grin.gov/cgi-bin/duke).

Compound	Active principle	Biological activity and application
Purine nucleobase	6H-Purin-6-one, 1, 7-dihydro-	Anti-inflammatory and cytoprotective
Indole	Indole	Anti-acne, antibacterial, Anticariogenic, anti-*Salmonella*, anti-treptococci and cancer preventive
Phenol	Eugenol	Analgesic, antibacterial, anticancer, antimelanomic, antiseptic, antiedemic, antigenotoxic, anti-inflammatory, antiviral, antiulcer, antithromboxane, hepatoprotective, trypsin enhancer and perfumery
Fatty acid methyl ester	Hexadecanoic acid, methyl ester	Nutrient, energy storage and membrane stabilizer
Fatty acid ethyl ester	Hexadecanoic acid, ethyl ester	Antioxidant, hypocholesterolemic, hemolytic, alpha-reductase inhibitor andantiandrogenic
Fatty acid methyl ester	9, 12-Octadecadienoic acid, methyl ester	Personal care products
Fatty acid methyl ester	9-Octadecenoic acid, methyl ester, (E)-	Industrial use
Fatty acid methyl ester	Stearic acid, methyl ester	Flavouring agents and industrial use
Fatty acid ethyl ester	9, 12-Octadecadienoic acid, ethyl ester	Anti-inflammatory
Triterpenic acid	Z-(13, 14-Epoxy) tetradec-11-en-1-ol acetate	Antioxidant and haemolytic
Acetate ester/Acetate compound	7-Methyl-Z-tetradecen-1-ol acetate	Anticancer, anti-inflammatory and hepatoprotective
Fatty acid methyl ester	Tetracosanoic acid, methyl ester	?
Triterpene	Squalene	Cancer preventive, immunostimulant, chemopreventive and diuretic
Triterpene	Urs-12-en-28-oic acid, 3-hydroxy-,methyl ester, (3β)-	Anti-inflammatory
Phytosterol	Ergosterol	Antiangiogenic, anti-flu, antitumor and antiviral

Table 5. Bioactive compounds, medicinal properties and applications of *Lentinus polychrous* (www.ars-grin.gov/cgi-bin/duke).

Compound	Active principle	Biological activity and application
Cinnamaldehyde	Octanal, 2-(phenylmethylene)-	Flavouring agent
Diester	Dibutyl phthalate	Antimicrobial and antifouling
Fatty acid ethyl ester	Hexadecanoic acid, ethyl ester	Antioxidant, hypocholesterolemic, hemolytic, alpha-reductase inhibitor and antiandrogenic
Fatty acid ethyl ester	9,12-Octadecadienoic acid, ethyl ester	Anti-inflammatory
Oleic acid ester	Ethyl Oleate	Flavoring agent and cancer preventive
Fatty acid ethyl ester	Linoleic acid ethyl ester	Anti-inflammatory agent
Pentacyclic triterpenoid	β-Amyrin	Analgesic, antiedemic, anti-inflammatory, antinociceptive, antiulcer, gastroprotective and hepatoprotective
Triterpene	Betulin	Analgesic, antiviral agent, anti-inflammatory, anti-neoplastic
Unsaturated fatty acid ester	9-Octadecenoic acid (Z)-, phenylmethyl ester	Anti-inflammatory and cancer preventive
Triterpene	Urs-12-en-28-oic acid, 3-hydroxy-, methyl ester, (3β)-	Anti-inflammatory
Steroid	1-Monolinoleoylglycerol trimethylsilyl ether	Antimicrobial, antioxidant, anti-inflammatory, antiarthritic, antiasthma and diuretic
Oleanane triterpenoids	Olean-12-en-28-oic acid, 3-oxo-, methyl ester	Surfactant, emulsifier and membrane stabilizer
Terpenoids	Urs-12-en-28-al	Antimicrobial effects
Phytosterol	Ergosterol	Antiangiogenic, anti-flu, antitumor and antiviral
Phytosterol	Stigmasterol	Cholesterol lowering, cancer prevention, antioxidant, hypogycemic and thyroid-inhibitions, precursor of progesterone, antimicrobial, antiarthritic, anti-asthma, anti-inflammatory and diuretic

Table 6. Bioactive compounds, medicinal properties and applications of *Lentinus squarrosulus* (www.ars-grin.gov/cgi-bin/duke).

Compound	Active principle	Biological activity and application
Maltose (disaccharide)	D-Glucose, 4-O-α-D-glucopyranosyl-	Detoxicant, diuretic, anticancer, sweetening agent in food and beverages
Cinnamaldehyde	Octanal, 2-(phenylmethylene)-	Flavour
Diester	Dibutyl phthalate	Antimicrobial and antifouling
Saturated fatty acid	n-Hexadecanoic acid	Antioxidant, hypocholestrolemic, antiandrogenic and flavor
Fatty acid ethyl ester	Hexadecanoic acid, ethyl ester	Antioxidant, hypocholesterolemic, hemolytic, alpha-reductase inhibitor and antiandrogenic
Fatty acid methyl ester	9,12-Octadecadienoic acid, methyl ester	Personal care products
Fatty acid methyl ester	5, 8, 11, 14-Eicosatetraenoic acid, methyl ester, (all-Z)-	Nutrient and membrane stabilizer
Flavonoid fraction	2H-Pyran, 2-(7-heptadecynyloxy) tetrahydro-	Antimicrobial agent, hallucinogen, haemagglutinator, anti-inflammatory and antioxidant
Phytosterol	Ergosterol	Antiangiogenic, anti-flu, antitumor and antiviral
Sterols	7,22-Ergostadienol	Anti-proliferative

Table 7. Bioactive compounds, medicinal properties and applications of *Termitomyces schimperi* (?, Not defined) (www.ars-grin.gov/cgi-bin/duke).

Compound	Active principle	Biological activity and application
Purine nucleobase	6H-Purin-6-one, 1,7-dihydro	Anti-inflammatory and cytoprotective
Piperidinones	2-Piperidinone	?
1-benzofurans	Benzofuran, 2,3-dihydro-	Used in treatment of diabetic retinopathy and arthritis
Indole	Indole	Antiacne, antibacterial, anticariogenic, anti-*Salmonella*, anti-streptococci and cancer preventive
Phenolic compound	Phenol, 2-methoxy-4-(1-propenyl)	Sensitizer, allergen, flavour and fragrance
Saturated fatty acid	Tetradecanoic acid	Antifungal, antioxidant, cancer preventive, nematicide, lubricant and hypercholesterolemic
Cyclic purine nucleotide	Adenosine 3,5-cyclic monophosphate	Second messenger
Fatty acid methyl ester	Hexadecanoic acid methyl ester	Nutrient, energy storage and membrane stabilizer
Saturated fatty acid	n-Hexadecanoic acid	Antioxidant, hypocholestrolemic, antiandrogenic and flavor
Fatty acid methyl ester	9,12-Octadecadienoic acid, methyl ester	Personal care products
Fatty acid ethyl ester	9, 12-Octadecadienoic acid, ethyl ester	Anti-inflammatory
Acetate ester	7-Methyl-Z-tetradecen-1-ol acetate	Anticancer, anti-inflammatory and hepatoprotective
Triterpene	Squalene	Cancer preventive, immunostimulant, chemopreventive and diuretic
Flavonoid fraction	2H-Pyran, 2-(7-heptadecynyloxy)tetrahydro-	Antimicrobial agent, hallucinogen, haemagglutinator, anti-inflammatory and antioxidant
Ursolic acid derivative	Urs-12-en-28-al	Antimicrobial
Phytosterol	Ergosterol	Antiangiogenic, anti-flu, antitumor and antiviral
Sterol	γ-Ergostenol	?

enzyme inhibition, fragrance, precursors and detoxicants), cosmetics (antiseptic, antimicrobial and antiviral) and other environmental and industrial applications (anti-fouling, nematicide, lubricant and perfumes). The details given below are the major chemical compounds with unique biological role, especially pharmacological properties, like antimicrobial, anticarcinogenic, anti-inflammatory agent, anti-viral agent, analgesic, anti-neoplastic agent, urinary acidulant, endocrine protective, endocrine-tonic, diuretic, detoxicant, antidote, increase NK cell activity, antioxidant and antihepatotoxic and so on. In future, further qualitative and quantitative studies should be preformed for harnessing the wild edible mushrooms of ethnic value.

Details of bioactive components identified in six mushrooms by GC-MS were Indole; Eugenol; Falcarinol; Adenosine 3,5-cyclic monophosphate; Squalene; Urs-12-en-28-oic acid, 3 hydroxy-,methyl ester (3β); Ergosterol; γ-Ergostenol; Hexadecanoic acid, methyl ester; 6H-Purin-6-one, 1,7-dihydro; Z-(13,14-Epoxy) tetradec-11-en-1-ol acetate; 7-methyl-Z-tetradecen-1-ol-acetate; 9,12-octadecadienoic acid, ethyl ester; linoleic acid ethyl ester; β-amyrin; Betulin; 9-Octadecenoic acid (Z)-phenyl methyl ester; 1-monolinoleoylglycerol trimethylsilyl ether; Stigmasterol; 5,8,11,14-Eicosatetraenoic acid, methyl ester, (all Z); Tetradecanoic acid; n-Hexadecanoic acid; 7,22-Ergostadienol; Acetic acid, 2-phenylethyl ester; 9,12-Octadecadienoic acid, methyl ester; 9-Octadecanoic acid, methyl ester (E); Stearic acid methyl ester; 9-Octadecenamide, (Z)-; Octadecanimde; Tetracosanoic acid, methyl ester; Dodecanoic acid, methyl ester; Benzofuran, 2, 3-dihydro; 9,12-Octadecadienoic acid (Z-Z)-; Ethyl oleate; D-Glucose, 4-O-α-D-glucopyranosyl-; Hexadecanoicacid, ethyl ester; 2H-Pyran, 2-(7-heptadecynyloxy) tetrahydro-; Octanal, 2-(phenylmethylene)-; Olean-12-en-28-oic acid,3-oxo-, methyl ester; Urs-12-en-28-al; Phenol, 2-methoxy-4-(1-propenyl)-; Dibutyl phthalate; 2-Acetamide-2-deoxygalactono-1,4-lactone.

7. Conclusion

Wild mushrooms are the nutritional and medicinal choice throughout geographically versatile India by different ethnic groups. Western Ghats possess several untapped wild mushrooms of nutritional and medicinal versatility. Besides nutritional security, bioactive components in wild mushrooms provide health security. Six wild mushrooms evaluated from the foothills of Western Ghats consist of phenolics, tannins, phytic acid, flavonoids, vitamins and pigments. In addition, they also possess meagre trypsin-inhibition activity, while two mushrooms are devoid of haemagglutinin activity. In spite of recognising some bioactive components as anti-nutritional, many of them serve as antioxidants in appropriate quantities to combat disease. Wild mushrooms are also known for possess in specific chemical constituents of medicinal, pharmaceutical and industrial significance. Many of them are known to prevent cancer, possess pharmacological potential, are useful in food industries, valuable in cosmetics manufacture and have potential in environmental and industrial applications. Wild mushrooms are an untapped source of bioactive components; therefore attention needs to be focused on developing the blueprint of their geographic distribution, extent of availability and extent for harvesting. Attentions need to be focused towards the dual benefits of wild mushrooms (nutrition and health) by diet-management for the benefit of ethnic population or the tribals. Occurrence of wild mushrooms is dependent on the characteristics of eco-regions (e.g., forest type, soil condition, substrate availability and wild fauna); thus conservation measures should not be overlooked.

References

Adams, L.S., Phung, S., Wu, X. et al. (2008). White button mushroom (*Agaricus bisporus*) exhibits antiproliferative and proapoptotic properties and inhibits prostate tumor growth in athymic mice. Nutr. Canc., 60: 744–756.

Attarat, J. and Phermthai, T. (2015). Bioactive compounds in three edible lentinus mushrooms. Walailak J. Sci. Technol., 12: 491–504.

Badalyan, S.M., Barkhudaryan, A. and Rapior, S. (2019). Recent progress in research on the pharmacological potential of mushrooms and prospects for their clinical application. pp. 1–70. *In*: Agrawal, D.C. and Dhanasekaran, M. (eds.). Medicinal Mushrooms, Springer Nature, Singapore.

Balamurugan, A., Evanjaline, M.R., Arthipan, B. and Mohan, V.R. (2017). GC-MS analysis of bioactive compounds from the ethanol extract of leaves of *Niebuhria apetala* Dunn. Int. Res. J. Pharm., 8: 72–78.

Barceloux, D.G. (2008). Medical Toxicology of Natural Substances: Foods, Fungi, Medicinal Herbs, Plants and Venomous Animals, Wiley, Hoboken, New Jersey. 10.1002/9780470330319.

Barros, L., Ferreira, M.J., Queirós, B. et al. (2007). Total phenols, ascorbic acid, β-carotene and lycopene in Portuguese wild edible mushrooms and their antioxidant activities. Food Chem., 103: 413–419.

Barros, L., Correia, D.M., Ferreira, I.C.F.R. et al. (2008). Optimisation of the determination of tocopherols in *Agaricus* sp. edible mushrooms by a normal phase liquid chromotographic method. Food Chem., 110: 1046–1050.

Bartlett, G.R. (1959). Phosphorus assay in column chromatography. J. Biol. Chem., 234: 466–468.

Boa, E.R. (2004). Wild Edible Fungi a Global Overview of their Use and Importance to People, Food and Agriculture Organisation of the United Nations, Rome.

Botterweck, A.A.M., Verhagen, H., Goldbohm, R.A. et al. (2000). Intake of butylated hydroxyanisole and butylated hydroxytoluene and stomach cancer risk: Results from analyses in The Netherlands cohort study. Food Chem. Toxicol., 38: 599–605.

Burns, R. (1971). Methods for estimation of tannins in grain sorghum. Agron. J., 63: 511–512.

Champ, M.M. (2002). Non-nutrient bioactive substances of pulses. Br. J. Nutr., 88: 307–319.

Chang, S.T. and Wasser, S.P. (2017). The cultivation and environmental impact of mushrooms. *In*: Oxford Research Encyclopedia of Environmental Science. https://doi.org/10.1093/acrefore/9780199389414.013.231.

Cheung, L.M. and Cheung, P.C.K. (2005). Mushroom extracts with antioxidant activity against lipid peroxidation. Food Chem., 89: 403–409.

Dandapat, S., Sinha, M.P., Kumar, M. and Jaggi, Y. (2015). Hepatoprotective efficacy of medicinal mushroom Pleurotus tuber-regium. Environ. Exp. Biol., 13: 103–108.

Deshpande, S.S., Sathe, S.K., Salunkhe, D.K. and Cornforth, D.P. (1982). Effects of dehulling on phytic acid, polyphenols, and enzyme inhibitors of dry beans (*Phaseolus vulgaris* L.). J. Food Sci., 47: 1846–1850.

Gaur, T. and Rao, P.B. (2016). Antioxidant potential of giant mushroom, *Macrocybe gigantiea* (Agaricomycetes) from India in different drying methods. Int. J. Med. Mush., 18: 133–140.

Ghate, S.D. and Sridhar, K.R. (2017). Bioactive potential of *Lentinus squarrosulus* and *Termitomyces clypeatus* from the southwestern region of India. Ind. J. Nat. Prod. Res., 8: 120–131.

Greeshma, A.A., Sridhar, K.R., Pavithra, M. and Tomita-Yokotani, K. (2018). Bioactive potential of non-conventional edible wild mushroom *Amanita*. pp. 719–738. *In*: Gehlot, P. and Singh, J. (eds.). Fungi and their Role in Sustainable Development: Current Perspectives, Springer Nature, Singapore.

Gregory III, J.F. (1996). Vitamins. pp. 531–616. *In*: Fennema, O.R. (ed.). Food Chemistry, 3rd ed., Dekker, New York.

Hussein, J.M., Tibuhwa, D.D., Mshandete, A.M. and Kivaisi, A.K. (2015). Antioxidant properties of seven wild edible mushrooms from Tanzania. Afr. J. Food Sci., 9: 471–479.

Jia, Z., Tang, M. and Wu, J. (1999). The determination of flavonoid contents in mulberry and their scavenging effects on superoxide radicals. Food Chem., 6: 555–559.

Kakade, M.L., Rackis, J.J., McGhee, J.E. and Puski, G. (1974). Determination of trypsin inhibitor activity of soy products: A collaborative analysis of an improved procedure. Cereal Chem., 51: 376–382.

Kalač, P. (2009). Chemical composition and nutritional value of European species of wild growing mushrooms: A review. Food Chem., 113: 9–16.

Knekt, P., Ritz, J., Pereira, M.A. et al. (2004). Antioxidant vitamins and coronary heart disease risk: A pooled analysis of 9 cohorts. Am. J. Clin. Nutr., 80: 1508–1520.

Kumari, D., Reddy, M.S. and Upadhyay, R.C. (2011). Antioxidant activity of three species of wild mushroom genus *Cantharellus* collected from north-western Himalaya, India. Int. J. Agric. Biol., 13: 415–418.

Li, H., Lu, S., Zhang, S. et al. (2008). Anti-inflammatory activity of polysaccharide from *Pholiota nameko*. Biochem., 73: 669–675.

Liener, I.E. and Kakade, M.L. (1980). Protease inhibitors. pp. 7–71. *In*: Liener, I.E. (ed.). Toxic Constituents of Plant Foodstuffs, Academic Press, New York.

Lo, K.M. and Cheung, P.C.K. (2005). Antioxidant activity of extracts from the fruiting bodies of *Agrocybe aegerita* var. *alba*. Food Chem., 89: 533–539.

Meng, C.Q., Somers, P.K., Rachita, C.L. et al. (2002). Novel phenolics antioxidants as multifunctional inhibitors of inducible VCAM-1 expression for use in atherosclerosis. Bioorg. Med. Chem. Lett., 12: 2545–2548.

Mordente, A., Guantario, B., Meucci, E. et al. (2011). Lycopene and cardiovascular diseases: An update. Curr. Med. Chem., 18: 1146–1163.

Mori, K., Kobayashi, C., Tomita, T. et al. (2008). Antiatherosclerotic effect of the edible mushrooms *Pleurotus eryngii* (Eringi), *Grifola frondosa* (Maitake), and *Hypsizygus marmoreus* (Bunashimeji) in apolipoprotein E deficient mice. Nutr. Res., 28: 335–342.

Namitha, K.K. and Negi, P.S. (2010). Chemistry and biotechnology of carotenoids. Crit. Rev. Food Sci. Nutr., 50: 728–760.

Occenã, I.V., Mojica, E.-R. and Merca, F. (2007). Isolation and partial characterisation of a Lectin from the seeds of *Artocarpus camansi* Blanco. Asian J. Pl. Sci., 6: 757–764.

Okoro, I.O. (2012). Antioxidant activities and phenolic contents of three mushroom species, *Lentinus squarrosulus* Mont., *Volvariella esculenta* (Massee) Singer and *Pleurocybella porrigens* (Pers.) Singer. Int. J. Nutr. Met., 4: 72–76.

Okwulehie, I.C. and Ogoke, J.A. (2013). Bioactive, nutritional and heavy metal constituents of some edible mushrooms found in Abia state of Nigeria. Int. J. Appl. Micorbiol. Biotechnol. Res., 1: 7–15.

Pandimeena, M., Prabu, M., Sumathy, R. and Kumuthakalavalli, R. (2015). Evaluation of phytochemicals and *in vitro* anti-inflammatory, antidiabetic activity of the white oyster mushroom, *Pleurotus florida*. Int. Res. J. Pharma. Appl. Sci., 5: 16–21.

Pavithra, M., Sridhar, K.R., Greeshma, A.A. and Tomita-Yokotani, K. (2016). Bioactive potential of the wild mushroom *Astraeus hygrometricus* in the southwest India. Mycology, 7: 191–202.

Podmore, I.D., Griffiths, H.R., Herbert, K.E. et al. (1998). Vitamin C exhibits pro-oxidant properties. Nature, 392: 559.

Quang, D.N., Hashimoto, T. and Asakawa, Y. (2006). Inedible mushrooms: A good source of biologically active substances. Chem. Rec., 6: 79–99.

Ren, L., Hemar, Y. and Perera, C.O. et al. (2014). Antibacterial and antioxidant activities of aqueous extracts of eight edible mushrooms. Bioact. Carbohyd. Diet Fib., 3: 41–51.

Riccioni, G. (2009). Carotenoids and cardiovascular disease. Curr. Atherosc. Rep., 11: 434–439.

Robertson, D.G. (2005). Metabonomics in toxicology: A review. Toxicol. Sci., 85: 809–822.

Roe, J.H. (1954). Chemical determination of ascorbic, dehydroascorbic and diketogluconic acids. pp. 115–139. *In*: Glick, D. (ed.). Methods of Biochemical Analysis, vol. 1, InterScience, New York.

Rosset, J., Bärlocher, F. and Oertli, J.J. (1982). Decomposition of conifer needles and deciduous leaves in two Black Forest and two Swiss Jura streams. Int. Rev. Ges. Hydrobiol., 67: 695–711.

Sathe, S.K., Deshpande, S.S., Reddy, N.R. et al. (1983). Effect of germination on proteins, raffinose oligosaccharides and antinutritional factors in Great Northern beans (*Phaseolus vulgaris* L). J. Food Sci., 48: 1796–1800.

Singh, N. (2011). Wild edible plants: A potential source of nutraceuticals. Int. J. Phar. Sci. Res., 2: 216–225.

Srinivasan, K. and Kumaravel, S. (2016). Unravelling the potential phytochemical compounds of *Gymnema sylvestre* through GC-MS study. Int. J. Pharm. Sci., 8: 1–4.

Strapáč, I., Baranová, M., Smrčová, M. and Bedlovičová, Z. (2016). Antioxidant activity of honey mushrooms (*Armillaria mellea*). Folia Veter., 60: 37–41.

Tannin levels in tea. http://www.livestrong.com/article/527907-tannin-levels-in-teas/.

Tapas, A.R., Sakarkar, D.M. and Kakde, R.B. (2008). Flavonoids as nutraceuticals: A review. Trop. J. Pharma. Res., 7: 1089–1099.

Visioli, F., Borsani, L. and Galli, C. (2000). Diet and prevention of coronary heart disease: The potential role of phytochemicals. Cardiovas. Res., 47: 419–425.

Wasonga, C.G.O., Okoth, S.A., Mukuria, J.C. and Omwandho, C.O.A. (2008). Mushroom polysaccharide extracts delay progression of carcinogenesis in mice. J. Exp. Ther. Oncol., 7: 147–152.

WHO. (2003). Post harvest and pressing technology of staple food. Technical Compendium of WHO Agricultural Science Bulletin 88: 171–172.

Woldegiorgis, A.Z., Abate, D., Haki, G.D. and Ziegler, G.R. (2015). Major, minor and toxic minerals and anti-nutrients composition in edible mushrooms collected from Ethiopia. Food Process. Technol., 6: 1–8.

Wong, J.Y. and Chye, F.Y. (2009). Antioxidant properties of selected tropical wild edible mushrooms. J. Food Comp. Anal., 22: 269–277.

Ye, Z. and Song, H. (2008). Antioxidant vitamins intake and the risk of coronary heart disease: Meta-analysis of cohort studies. Eur. J. Card. Prev. Rehab., 15: 26–34.

Yen, G.C. and Wu, J.Y. (1993). Antioxidant and radical scavenging properties of extracts from Ganodermatsugae. Food Chem., 65: 375–379.

Zhang, M., Cui, S.W., Chueng, P.C. and Wang, K.Q. (2007). Antitumor polysaccharides from mushrooms: A review on their isolation process, structural characteristics and antitumor activity. Tr. Food Sci. Technol., 18: 4–19.

Truffles
A Potential Source of
Bioactive Compounds

*Elsayed Ahmed Elsayed,[1] Ramzi Abdel Alsaheb,[2]
Ting Ho,[3] Siti Zulaiha Hanapi,[4] Dalia Sukmawati,[5]
Racha Wehbe[6] and Hesham Ali El-Enshasy[4,7,8,*]*

1. INTRODUCTION

Truffles are hypogeous macrofungi that grow below the earth surface. They are highly appreciated culinary foods, which have long been used in many cultures as food and medicine (El-Enshasy et al., 2013). The history of truffle usage goes back to ancient times, i.e., they have been reported to be used by indigenous Africans, Aborigines in Australia and Bedouins in the Middle East (Shavit, 2014). Furthermore, records showing their medical applications were traced back to the Bronze Age Amorites (Lönnqvist, 2008). Nowadays, special truffle types are highly priced foods and

[1] Chemistry of Natural and Microbial Products Department, National Research Centre, Cairo, Egypt.
[2] Al-khwarizmi College of Engineering, University of Baghdad, Baghdad, Iraq.
[3] Global Agro-innovation (HK) Limited, Hong Kong.
[4] Institute of Bioproduct Development (IBD), Universiti Teknologi Malaysia (UTM), Johor Bahru, Johor, Malaysia.
[5] Biology Department, Laboratory of microbiology, 9th Floor Hasyim Ashari Building, Faculty of Mathematics and Natural Sciences, Universitas Negeri Jakarta, Jakarta, Indonesia.
[6] Laboratory of Cancer Biology and Molecular Immunology, Faculty of Science I, Lebanese, University, Hadath, Lebanon.
[7] School of Chemical and Energy Engineering, Faculty of Engineering, Universiti Teknologi Malaysia (UTM), Johor Bahru, Johor, Malaysia.
[8] City of Scientific Research and Technology Application, New Burg Al Arab, Alexandria, Egypt.
* Corresponding author: henshasy@ibd.utm.my

can reach up to an original price of 90 and 1800 €/kg for black and white truffles, respectively (Vidale, 2019). However, these prices can be doubled or tripled upon reaching the final consumer.

Unlike mushrooms, truffles grow beneath the earth surface (5–10 cm) and therefore, they require special skills in collection (Elsayed et al., 2014). In some countries, trained farm animals are used to search for truffles, depending on their characteristic odour due to their volatile compounds (El-Enshasy et al., 2013). Truffles are ectomycorrhizal symbiotic fungi that grow in close relation to a wide range of plants. Plants benefit from fungal metabolites necessary for their growth, such as auxins and other growth hormones (Patel et al., 2017). Truffles have been found in moderate, tropical as well as harsh desert environment. In certain areas they require specific amounts of rainfall to grow (Bradai et al., 2015a), even though the time and distribution for rainfall plays a crucial role in the development and growth of truffles (El-Enshasy et al., 2013). Furthermore, their growth also requires specific conditions of soil structure and composition. Generally, their distribution depends mainly on climatic and soil conditions as well as the abundance of host plants (Le Tacon et al., 2014; Kagan-Zur et al., 2014; Bradai et al., 2015b). For a long time, truffles were considered as highly priced food commodities as well as important medicinal items for treatment in folk medicinal cultures (Federico et al., 2015). This is mainly due to the fact that they are rich in all the necessary constituents, i.e., proteins, amino acids, fibres, minerals, vitamins, important terpenoids, steroids and many other bioactive phytochemicals. Furthermore, recent research focusing on bioactive potentials of compounds isolated from truffles reveal that truffles exhibit a wide range of bioactive properties, including antimicrobial, anti-inflammatory, antioxidant and anticancer (Patel et al., 2017; Elsayed et al., 2019a; Lee et al., 2020). This chapter focuses on the background and ongoing research on ecological, ethnopharmacological and medicinal aspects of truffles.

2. Classification and Distribution

The fact that truffles have a widespread growth in the world has resulted in the presence of an overwhelming number of species depending on their location. The three major truffle families (Tuberaceae, Pezizaceae and Terfeziaceae) comprise most of the edible truffles found. Generally, European truffles belong to the genus *Tuber*, while those found in desert countries belong to the genus *Terfezia* and *Tirmania* (Patel et al., 2017). Figure 1 represents some macro-morphological features of native desert truffle samples collected from Riyadh province of Saudi Arabia. The black truffle (*T. melanosporum*) is a common culinary-grade truffle consumed in European countries, while desert truffles are most commonly used in semi-arid and arid geographical regions (Khalifa et al., 2019). The general distribution of most widely-consumed truffles is presented in Table 1. However, the quality of truffles is highly dependent on the source and environmental conditions. For example, it has been reported that dehydration and low temperature induce the production of dehydrins in *Tuber borchii* as a protection mechanism to survive under stress conditions (Abba et al., 2006). On the other hand, high temperature regulates the heat chock proteins to reduce the thermal stress. It has also been reported that change in temperature from

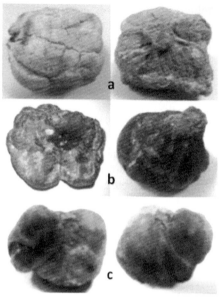

Fig. 1. Photographic images of some truffle species—(a) *Tirmana nivea*, (b) *Terfezia claveryi*, (c) *Phaeangium lefebvrei*—collected from Riyadh, Saudi Arabia.

Table 1. Distribution of key genera of truffles in Nature (Modified from Khalifa et al., 2019).

Key genera of truffles		Distribution
Tuber (Comprise of more than 180 species)		Widely distributed and found in many places in: North America, Asia, Europe, with limited parts in South America and Africa.
Choiroyces (Pig truffle)		Widely found in the Northern atmosphere.
Desert truffles	*Delastria* sp. *Balsamia* sp. *Mattirolomyces* sp. *Picoa* sp. *Phaeangium* sp. *Leucangium* sp. *Terfezia* sp. *Delastreopis* sp. *Tirmania* sp.	Widely distributed in countries around the Mediterranean area (Spain, Portugal, Italy, France, Hungary, Turkey), North Africa (from Morocco to Egypt), Middle East, South Africa, and North America.

optimal (22°C) up to 34°C results in an increase in hyphal septation concomitant with significant reduction in growth (Leonardi et al., 2017). Other factors, such as soil moisture content, pH, salt composition and mineral contents play significant role in truffle growth and development (Ge et al., 2017; Lee et al., 2020).

3. Truffles, Plants, and Microbial Relationships

Truffles grow symbiotically with plant species and develop a mycorrhizal relationship. Tuber species grow near the roots of oak trees, while desert truffles grow in the vicinity of plants belonging to the *Helianthemum* and *Cistus* genus (Bradai et al.,

2015a). Truffle spores colonise the plant roots and grow, benefiting from such mutual relationship. Truffles absorb sugars produced by photosynthetic plants for their ascocarp development, while plants get their water and mineral uptake facilitated through the fungal cells (De la Varga and Murat, 2015). Moreover, it has been found that some plants can be adversely affected by their absorption of truffle's volatile compounds. In such a case, plant growth gets inhibited, resulting in bleaching of plant leaves (Splivallo et al., 2007).

Furthermore, volatile organic compounds have been extracted and identified from both truffle fruiting bodies as well as plant roots (Splivallo et al., 2011; Kanchiswamy et al., 2015). These compounds were found to be the main reason for different aroma distinguishing various truffle species (Daba et al., 2019). Fu et al. (2015) reported that truffle species secrete certain auxin hormones, which are indole-3-acetic acid and ethylene. They explained that such hormones induce the formation of ectomycorrhiza through the induction of certain morphological changes in plant roots. After secretion, roots are shortened with formation of side root branches. Afterwards, these side branches enlarge, leading to the formation of elongated root hairs. Truffle auxins are believed to be the principal signal for controlling root development prior to mycelial contact (Daba et al., 2019).

However, it is believed that the plant-truffle relationship is also highly dependent on plant location and soil's chemical and physical features. Studies show that some broad host spectrum truffles, such as *Tuber indicum* are capable of completing its fruiting body cycle in non-native soil (Bonito et al., 2011). However, beside the known relationship between truffles and plants, it is also interesting to observe that truffles include a wide range of bacteria and yeast in their tissues. Therefore, truffles' microbiome is a matter of interest for many researchers in the recent years (Splivallo et al., 2014; Vahdatzadeh et al., 2015; Benucci and Bonito, 2016). Microbes heavily colonise all parts of truffle and have been isolated at all stages of truffle development: fruiting body and free living mycelial stage (Antony-Babu et al., 2014). The real functional role of truffle microbiota in terms of development, differentiation and metabolism are still not yet fully understood. However, most of studies show a strong link between truffle aroma, volatile organic compounds (VOC) production and the presence/diversity of truffle microbiota (Splivallo et al., 2014; Buzzini et al., 2005). The presence of some specific VOC and sulphur-volatile compounds is strongly associated with the presence of specific microbial groups within the truffle tissues (Becher et al., 2012; Splivallo and Ebeler, 2015). Interestingly, it was reported that different types of endoviruses have been isolated from different parts of truffles (Stielow and Menzel, 2010; Stielow et al., 2011, 2012).

3.1 Nutritional Value

Truffles are considered as one of the most valuable food sources appreciated by several human cultures. They are used typically as foods in northern African, Asia and the Arabian peninsular countries (Khalifa et al., 2019). They can be used as a main dish, salad, or for their oil content (Wernig et al., 2018). Generally, it has been found that different truffle species are varied in terms of their chemical and nutritional contents. Desert truffles can have up to two-thirds of their dry matter of carbohydrates, while protein can range between 20–27 per cent, fats range between 3–7.5 per cent, fibres between 7–13 per cent and ascorbic acid, 2–5 per cent (El-Enshasy et al., 2013). Truffles are also considered as a source of essential amino acids (valine, isoleucine, methionine, threonine and serine) in addition to essential metals, such as K, Fe, Cu, Mn and Zn. In studies on the nutritional value of desert truffles, Iraqi truffles were found to have lower carbohydrate contents (16–25 per cent), while having increased percentages of phosphorus ranging between 9–25.5 per cent (Hussain and Al-Ruqaie, 1999). Furthermore, Sawaya et al. (1985) reported that desert truffles varied widely in their chemical composition. They found that the Zubaidi white truffle (*Tirmania nivea*) is richer than the other two types of black truffles (Gibeah and Kholeissi) in protein, fat and crude fibre contents. Furthermore, truffles contain all the essential amino acids and therefore can be considered an important food supplement (Wang and Macrone, 2011). Additionally, ergosteroids are reported to be present in truffles. These ergosteroids are potential source of vitamins as they can be easily converted into vitamin D inside the human body (Dogan and Aydin, 2013; Patel et al., 2017). Nevertheless, truffles contain volatile aromatic compounds which are responsible for different odours, a characteristic of truffles (Xiao et al., 2015). Terpenoids (e.g., carveol, cymene and limonene) have been detected in various truffle species (Kanchiswamy et al., 2015).

4. Bioactives of Potential Medical Applications

Since centuries, truffles have been used in many ancient cultures for the treatment of different microbial and non-microbial diseases. They have been mainly used to treat ophthalmic diseases, and as aphrodisiac agents (El-Enshasy et al., 2013). Moreover, sub-Saharan African and Middle Eastern countries have used their extracts to treat skin and eye diseases, such as trachoma. Further, nomadic cultures in Africa used indigenous truffles to treat some gastrointestinal problems (Al-Marzooky, 1981). However, with the exception of few studies on their medical use, truffles have not been intensively evaluated for their different biological activities during the last century. Recently, focus has been on the bioactive potentials of truffles and this can be mainly attributed to the advanced developments achieved in their extraction methods, identification, metabolic profiling and consequently *in vitro* evaluation (Li et al., 2019; Lee et al., 2020). The following sections will shed some light on different therapeutic potentials of truffles.

4.1 Antimicrobial and Antiviral Potential

Antimicrobial activities of various truffle species have been intensively investigated and researchers have reported on their antimicrobial potentials. The methanolic extract of *T. claveryi* is reported to have inhibitory effect on gram positive *Bacillus subtilis* and *Staphylococcus aureus* (Hussain and Al-Ruqaie, 1999). Furthermore, *Pseudomonas aeruginosa* cells were found to be inhibited by the same truffle extract (Janakat et al., 2005). Additionally, it has been proven that the effect of truffle extract on pathogenic bacteria is solvent-dependent. Janakat et al. (2005) reported that methanolic extract of *Tirmania* truffles possesses the most potential antimicrobial effects when compared to other solvent extraction processes, i.e., ethanol, water and ethyl acetate. However, it should be noted that such effects also depend on the species itself as well as the climatic region from where it was collected. Janakat et al. (2004) found that water extract of *T. claveryi* inhibited the growth of *S. aureus* by about 66 per cent, while the same extract inhibited *P. aeruginosa* by about 50 per cent. On the other hand, they reported the ineffectiveness of the methanolic extract against the studied microbial cells. Recently, antimicrobial activities of *T. boudieri* against both Gram-negative and Gram-positive bacteria as well as *Candida albicans* has been evaluated (Dogan and Aydin, 2013). Malik et al. (2018) revealed the antibacterial inhibition of different extracts of desert truffle, *Terfezia* spp. against methicillin-resistant *Staphylococcus aureus* (MRSA) strain. In another study, Hussain and Al-Ruqaie (1999) found that *Terfezia* truffles have antiviral potential and suggested the application of their extracts for treating eye and skin diseases resulting from viral infection.

It has been postulated that the antimicrobial effects of truffles may be due to the presence of different fungal lectins, polysaccharides and laccases, flavonoids, VCOs, dimethyl-sulphide and different types of phenolic compounds which are responsible for their inhibitory action (Elsayed et al., 2014). Lectins can interact with bacterial cell wall exopolysaccharides, resulting in their distortion and elimination (Elsayed et al., 2014). Fungal laccases may result in phenol oxidation, which produces superoxide anion radicals and hydrogen peroxide. The latter products are possibly responsible for the inhibition of bacterial cells (Nadim et al., 2015). It was also reported that many bioactive compounds of different classes have been isolated from truffles and exhibited antiviral properties (Elkhateeb et al., 2019).

4.2 *Antioxidants' Potential*

Antioxidant properties of truffles have been widely investigated in order to evaluate their potential in neutralising the harmful effects of reactive oxygen species (ROS) that produce an oxidative stress status, leading finally to serious human disorders, ranging from mild inflammation to cancer and heart diseases (Wang and Marcone, 2011). The antioxidant effects of truffles are related to their higher percentages of antioxidants and phenolics (Patel et al., 2017). Such compounds react with harmful ROS, free radicals and peroxyl radicals. Reports demonstrate the antioxidant effects of polysaccharides isolated from *T. indicum* against PC12 cells subjected to H_2O_2 stress.

The antioxidant potentials of *T. nivea* obtained from different sub-Sahara as well as Middle Eastern countries were evaluated, using different assays, ferric reducing ability (FRAP), DPPH free radical scavenging activity, deoxyribose and nitric oxides (Al-Laith, 2010). Results showed that the antioxidant capacities were country-dependent, although the same species was collected. *T. nivea* from Iran revealed the highest DPPH values, strongest nitric oxide radical scavenging activity and highest inhibitory percentage of deoxyribose breakage. On the other hand, Saudi and Bahraini species showed the highest FRAP values. Results also showed variable contents of ascorbic acid, total carotenoids, esterified phenolics, free- and non-flavonoid phenolics in all the collected samples. Other studies showed also that the antioxidant activities of *Tuber indicum* extracts were directly correlated to the total phenolics and total flavonoids content (Guo et al., 2011).

Furthermore, *T. boudieri* extracts were found to have a scavenging activity against DPPH radicals (Dogan and Aydin, 2013). Authors attributed their antioxidant potential to the presence of catechin, ferulic acid and cinnamic acid. Additionally, methanolic extracts of *T. nivea* were found to exert DPPH-scavenging activity as well as lipid peroxidation inhibitory effects. This was attributed to the presence of ascorbic acid, carotenoinds and anthocyanins in high concentrations (Hamza et al., 2016a).

Additionally, β-carotene and linoleic acid were found in various extracts obtained from *Chondrogaster pachysporus* and *Setchelliogaster tenuipes* truffles. These two bioactive molecules exhibited potential antioxidant activities, which were much higher than the tested positive control of control α-tocopherol (Khalifa et al., 2019). Furthermore, different antioxidant compounds were isolated from *T. indicum* and *Elaphomyces granulate* truffles. These compounds included polysaccharides, syringic acid and syringaldehyde (Hamza et al., 2016b).

4.3 Anti-inflammatory Potential

Truffles have been considered as a source of anti-inflammatory compounds, as reported by many authors. Beara et al. (2014) evaluated ethanolic extracts of *T. aestivum* and *T. magnatum* for their anti-inflammatory effects, using cyclooxygenase-1 and cyclooxygenase-2 (COX-1 and COX-2), which are considered as key indicators for inflammatory pathways. They found that only *T. magnatum* extracts possess anti-inflammatory effects in comparison to positive control (quercetin). Additionally, COX-2 was inhibited by ethanolic extract of *Elaphomyces granulatus* truffles and the effect was mainly attributed to the presence of syringaldehyde and syringic acids (Stanikunaite et al., 2009). A recent study showed also that *Tuber aestivum* Vittad extracts using petroleum ether and ethanol exhibited anti-inflammatory properties, using the lipoxygenase-inhibition assay. Further studies on these fractions using GC-MS showed that the bioactivity is higher in ethanolic fraction and is related to the presence of a mixture of fatty acids, fatty acid methyl esters, trimethylsily-derivatives, hydroxyoctadecadienoic acid, benzotheophine compounds and many other fractions (Marathe et al., 2020).

4.4 Anti-depressant and Aphrodisiac Potential

The fact that the modern lifestyle and human diet have largely affected the emotional and psychological status of humans, motivated the application of natural sources as anti-depressants. Unhealthy foods and current lifestyle result in the development of unusual psychological disorders, which in turn may lead to the misuse of certain drugs that can finally lead to serious neural disorders. Truffles contain adequate amounts of L-tyrosine that acts as a catecholamine's precursor (neurotransmitters). Additionally, L-tyrosine serves as a crucial component of thyroid hormones (Patel et al., 2017). It has been recently found that when L-tyrosine is supplemented in a mouse's diet, the mouse's neurotransmitters improved. Also, this supplementation leads to an increase in serum levels of total thyrotropin and total triiodothyronine, T3 (Wang et al., 2012).

Traditionally, truffles have been used in certain cultures for their aphrodisiac properties (El-Enshasy et al., 2013). It has also been reported that local Bahrainis eat truffles to promote their sexual vitality. Dopamine and serotonin neurotransmitters improve genital reflexes and copulation instincts by activating human sexual reward learning (Hull et al., 2004), induce rostenol production, a steroid produced by males, and activates the female hypothalamus (Ngun et al., 2011). Truffles contain myco-steroids, which can lead to sexual vitality. This has been proved experimentally as this pheromone can induce excitement in female pigs.

4.5 Anticancer Potentials

Cancer is considered as one of the most threatening diseases affecting human life, with 9.6 million deaths in 2018 (Elsayed et al., 2019a). Cancer treatment usually starts with chemotherapy and radiotherapy, which are conventional treatments that pose a higher risk of side effects on patients. Therefore, research has focused recently on discovering new anticancer agents derived from natural resources to minimise the risk of conventional cancer therapy. In this regard, truffles have been recently investigated as possible anticancer agents. In 2010, Janakat and Nassar firstly reported on the hepatoprotective effects of different extracts obtained from desert truffle *T. claveryi* against liver CCl_4-induced hepatotoxicity in Wister albino rats. They concluded that aqueous extracts of *T. claveryi* show a potential hepatoprotective effect (Janakat and Nassar, 2010).

Furthermore, Patel and Goyal (2012) reported on the isolation of 52 polysaccharides from the fermentation broth of different truffles. The polysaccharides showed *in vitro* antitumor activities against different cancer cell lines that were tested, e.g., human hepatocellular carcinoma (HepG2), human lung adenocarcinoma (A549), human colon carcinoma (HCT-116), human breast cancer (SK-BR-3) and human promyelocytic leukemia (HL-60). Additionally, Beara et al. (2014) evaluated the anticancer properties of aqueous and methanolic extracts of *T. magnatum* and *T. aestivum* methanol against HeLa, MCF7 and HT-29 cancer cell lines. Results showed that aqueous extracts exhibit the most promising anticancer effects against breast adenocarcinoma (MCF7). The effect of truffle polysaccharides has been further investigated by Attia et al. (2018). Authors evaluated the *in vitro* anticancer activities of polysaccharides extracted from *T. claveryi* Chatin against Ehrlich's ascites carcinoma. The extract exerted potential dose- and time-dependent anticancer effects with IC_{50} of 77.6 and 47.6 µg/ml after 48 hours of treatment. They also concluded that the anticancer effect was mainly due to the induction of G0/G1 phase cell cycle arrest and apoptosis. Moreover, Dahham et al. (2018) investigated the possible anticancer effects of different solvent extracts of *T. claveryi* against different cancer cell lines. They showed that the hexane extract produced the most inhibiting effect against cancer cell growth in case of brain cell line (U-87 MG). They concluded that this effect was related to mitochondrial-based cell apoptosis and associated with a decline in mitochondrial membrane potential and DNA fragmentation.

Recently, we have investigated the anticancer properties of different solvent extracts of desert truffles collected from Saudi Arabia against various cancer cell lines. Results revealed that hexane and ethyl acetate extracts of *Tirmania nivea* and *Terfezia claveryi* showed a dose-dependent effect against MCF-7 human breast cancer cell line (Elsayed et al., 2019a). Furthermore, hexane extracts of *T. nivea* and *T. claveryi* exhibited the highest and comparable anticancer effects against HepG2 (human hepatocellular carcinoma) and L929 (human fibroblasts). On the other hand, ethyl acetate extracts of *T. claveryi* were more potent by about 6 per cent than *T. nivea* (Elsayed et al., 2019b). In another report, we found that different solvent extracts of *Phaeangium lefebvrei* exhibited potential anticancer effects against MCF-7, HepG2 and L929 cancer cell lines. Results showed that ethyl acetate extract produced maximal growth inhibitory effect against L929 cells (95.7 ± 0.02 per cent), while both ethyl acetate and hexane extracts inhibited MCF-7 cell growth by about 92.5 ± 0.03 and 91.7 ± 0.04 per cent, respectively (Elsayed et al., 2019c). Besides the direct anticancer properties of the truffles bioactive extracts, truffle extracts have been also successfully used in nano-anticancer drug production. Khadri et al. (2017) have reported the potential use of *Terfezia claveryi* extract in the synthesis of silver nanoparticles which exhibited potent cytotoxicity against the human breast cancer cell line MCF-7.

5. Conclusion

Different reasons explain the growing demand for truffles. In fact, they are rich in several essential nutrients, such as polysaccharide, proteins, amino acids, vitamins, fats, fibres and many other compounds. In addition to their valuable nutrient content, truffles are identified as an important source of bioactive compounds of high therapeutic value. Traditionally, different types of truffles have been used in folk medicine and show potent activity in the treatment of cancer, inflammation, depression, diabetes and other diseases. However, the biosynthesis of these bioactive compounds and the role of host plant and truffle microbiome on truffle development still need further investigation. In addition, most of the research undertaken is based on the study of the activities of bioactive metabolite mixture in truffle extracts. Therefore, more studies are still needed for isolation, identification and full molecular structure characterisation of truffle bioactive compounds. These studies are important to understand in depth the relationship between the molecular structure and bioactivity. In addition, for upgrading the application of truffle bioactives up to the drug level, more intensive clinical trials need to be undertaken to determine the effective dose and any potential risks or side effects. These all together will help to develop a new series of novel drugs in the pharmaceutical market based on truffles as a safe and natural medicine.

References

Abba, S., Ghighone, S. and Bonfante, P. (2006). A dehydration-inducible gene in the truffle *Tuber borchii* identifies a novel group of dehydrins. BMC Genomics, 7: 39. 10.1186/1471-2164-7-39.

Al-Laith, A.A.A. (2010). Antioxidant components and antioxidant/antiradical activities of desert truffle (*Tirmania nivea*) from various Middle Eastern origins. J. Food Comp. Anal., 23: 15–22.

Al-Marzooky, M.A. (1981). Truffles in eye diseases. pp. 353–357. *In*: Bulletin of Islamic Medicine, vol. 1, Kuwait Ministry of Public Health and National Council for Culture, Arts and Letters, Kuwait.

Antony-Babu, S., Deveau, A., Nostrand, J.D., Zhou, J., Le Tacon, F. et al. (2014). Black truffle-associated bacterial communities during the development and maturation of *Tuber melanosporum* ascocarps and putative functional roles. Environ. Microbiol., 16: 2831–2847.

Attia, W.Y., El-Naggar, R.E., Bawadekji, A. and Al Ali, M. (2018). Evaluation of some *in vitro* anti-carcinogenic activities of polysaccharides extracted from ascomata of the desert truffle *Terfezia claveryi* Chatin. J. Appl. Biol. Sci., 8: 152–159.

Beara, I.N., Lesjak, M.M, Četojević-Simin, D.D., Marjanović, Ž.S., Ristić, J.D. et al. (2014). Phenolic profile, antioxidant, anti-inflammatory andcytotoxic activities of black (*Tuber aestivum* Vittad.) and white (*Tuber magnatum* Pico) truffles. Food Chem., 165: 460–466.

Becher, P.G., Flick, G., Rozpedowska, E., Schmidt, A., Hagman, A. et al. (2012). Yeast, not fruit volatiles mediates *Drosophila melanogaster* attraction, ovi-position and development. Funct. Ecol., 26: 822–828.

Benucci, G.M.N. and Bonito, G.M. (2016). The truffle microbiome: Species and geography effects on bacteria associated with fruiting bodies of hypogenous pezizales. Microb. Ecol., 72: 4–8.

Bonito, G., Trappe, J.M., Donovan, S. and Vilgalys, R. (2011). The Asian black truffle *Tuber indicum* can form ectomycorrhizas with North American host plants and complete its life cycle in non-native soils. Fungal Ecol., 4: 83–93.

Bradai, L., Neffar, S., Amrani, K., Bissati, S. and Chenchouni, H. (2015a). Ethnomycological survey of traditional usage and indigenous knowledge on desert truffles among the native Sahara Desert people of Algeria. J. Ethnopharmacol., 162: 31–38.

Bradai, L., Bissati, S., Chenchouni, H. and Amrani, K. (2015b). Effects of climate on the productivity of desert truffles beneath hyper-arid conditions. Int. J. Biometeorol., 59: 907–915.

Buzzini, P., Gasparetti, C., Turchetti, B., Cramarossa, M.R., Vaughan-Martini, A. et al. (2005). Production of volatile organic compounds (VOCs) by yeasts isolated from the ascocarps of black (*Tuber melanosporum* Vitt.) and white (*Tuber magnatum* Pico) truffles. Arch. Microbiol., 184: 187–193.

Daba, G.M., Elkhateeb, W.A., Wen, T.-C. and Thomas, P.W. (2019). The continuous story of truffle-plant interaction. pp. 375–383. *In*: Kumar, V., Prasad, R., Kumar, M. and Choudhary, D. (eds.). Microbiome in Plant Health and Disease, Springer Nature, Singapore.

Dahham, S.S., Al-Rawi, S.S., Ibrahim, A.H., Majid, A.S.A. and Majid, A.M.S.A. (2018). Antioxidant, anticancer, apoptosis properties and chemical composition of black truffle *Terfezia claveryi*. Saudi J. Biol. Sci., 25: 1524–1534.

De la Varga, H. and Murat, C. (2015). Identification and *in situ* distribution of a fungal gene marker: The mating type genes of the black truffle. Meth. Mol. Biol., 1399: 141–149.

Dogan, H.H. and Aydin, S. (2013). Determination of antimicrobial effect, antioxidantactivity and phenolic contents of desert truffle in Turkey. Afr. J. Trad. Complemen. Alter. Med., 10: 52–58. 10.4314/ajtcam.v10i4.9.

El-Enshasy, H.A., Elsayed, E.A., Aziz, R. and Wadaan, M.A. (2013). Mushrooms and truffles: Historical biofactories for complementary medicine in Africa and in the Middle East. Evid. Based Complement. Alter. Med., 2013: 620451. 10.1155/2013/620451.

Elkhateeb, W.A., Daba, G.M., Elmahdy, E.M., Thomas, P.W., Wen, T.-C. and Shaheen, M.N.F. (2019). Antiviral potential of mushrooms in the light of their biological active compounds. ARC J. Pharmaceut. Sci., 5: 8–12.

Elsayed, E.A., El-Enshasy, H.A., Wadaan, M.A. and Aziz, R. (2014). Mushrooms: A potential natural source of anti-inflammatory compounds for medical applications. Mediators Inflamm., 2014: 805841. 10.1155/2014/805841.

Elsayed, E.A., Alsahli, F.D., Barakat, I.A., El-Enshasy, H.A. and Wadaan, M.A. (2019a). Assessment of *in vitro* antimicrobial and anti-breast cancer activities of extracts isolated from desert truffles in Saudi Arabia. J. Sci. Ind. Res., 78: 419–425.

Elsayed, E.A., Alsahli, F.D., El-Enshasy, H.A. and Wadaan, M.A. (2019b). Cytotoxic activities of different solvent extracts of *Tirmania nivea* and *Terfezia claveryi* against HepG2 and L929 cells. J. Sci. Ind. Res., 78: 454–457.

Elsayed, E.A., Alsahli, F.D., Amr, A.E., El-Enshasy, H.A. and Wadaan, M.A. (2019c). *In vitro* anti-proliferative potentials of *Phaeangium lefebvrei* desert truffle towards MCF-7, HepG2 and L929 cell lines. J. Sci. Ind. Res., 78: 858–862.

Federico, V., Cosimo, T., Antonio, P., Nadia, B., Valentina, L. et al. (2015). Volatile organic compounds in truffle (*Tuber magnatum* Pico): Comparison of samples from different regions of Italy and from different seasons. Sci. Rep., 5: 12629. 10.1038/srep12629.

Fu, S.F., Wei, J.Y., Chen, H.W., Liu, Y.Y., Lu, H.Y. and Chou, J.Y. (2015). Indole-3-acetic acid: A widespread physiological code in interactions of fungi with other organisms. Pl. Signal Behaviour, 10(8): e1048052. 10.1080/15592324.2015.1048052.

Ge, Z.-W., Brenneman, T., Bonito, G. and Smith, M.E. (2017). Soil pH and mineral nutrients strongly influence truffles and other ectomycorrhizal fungi associated with commercial pecans (*Caryaillino inensis*). Pl. Soil., 418: 493–505.

Guo, T., Wei, L., Sun, J., Hou, C.-I. and Fan, L. (2011). Antioxidant activities of extract and fractions from *Tuber indicum* Cooke & Massee. Food Chem., 127: 1634–1640.

Hamza, A., Jdir, H. and Zouari, N. (2016a). Nutritional, antioxidant and antibacterial properties of *Tirmania nivea*, a wild edible desert truffle from Tunisia arid zone. Med. Aromatic Pl., 5(4): 1000258. 10.4172/2167-0412.1000258.

Hamza, A., Zouari, N., Zouari, S., Jdir, H., Zaidi, S., Gtari, M. and Neffati, M. (2016b). Nutraceutical potential, antioxidant and antibacterial activities of *Terfezia boudieri* Chatin, a wild edible desert truffle from Tunisia arid zone. Arab. J. Chem., 9: 383–389.

Hull, E.M., Muschamp, J.W. and Sato, S. (2004). Dopamine and serotonin: Influences on male sexual behaviour. Physiol. Behaviour, 83(2): 291–307.

Hussain, G. and Al-Ruqaie, I.M. (1999). Occurrence, chemical composition and nutritional value of truffles: An overview. Pak. J. Biol. Sci., 2: 510–514.

Janakat, S., Al-Fakhiri, S. and Sallal, A.K. (2004). A promising peptide antibiotic from *Terfezia claveryi* aqueous extract against *Staphylococcus aureus in vitro*. Physiother. Res., 18: 810–813.

Janakat, S., Al-Fakhiri, S. and Sallal, A.K. (2005). Evaluation of antibacterial activity of aqueous and methanolic extracts of the truffle *Terfezia claveryi* against *Pseudomonas aeruginosa*. Saudi Med. J., 26: 952–955.

Janakat, S. and Nassar, M. (2010). Hepatoprotective activity of desert truffle (*Terfezia claveryi*) in comparison with the effect of *Nigella sativa* in the rat. Pak. J. Nutr., 9: 52–56.

Kagan-Zur, V., Roth-Bejerano, N., Sitrit, Y. and Morte, A. (2014). Desert Truffles: Phylogeny, Physiology, Distribution and Domestication, vol. 38, Springer, Heidelberg, Berlin.

Kanchiswamy, C.N., Malnoy, M. and Maffei, M.E. (2015). Chemical diversity of microbial volatiles and their potential for plant growth and productivity. Front. Pl. Sci., 10.3389/fpls.2015.00151.

Khadri, H., Aldebasi, Y.H. and Riazunnisa, K. (2017). Truffle mediated (*Terfezia claveryi*) synthesis of silver nanoparticles and its potential cytotoxicity in human breast cancer cells (MCF-7). Afr. J. Biotechnol., 16: 1278–1284.

Khalifa, S.A.M., Farag, M.A., Yosri, M., Sabir, I.S.M., Saeed, A. et al. (2019). Truffles: From Islamic culture to chemistry, pharmacology and food trends in recent times. Tr. Food Sci. Technol., 91: 193–218.

Le Tacon, F., Marçais, B., Courvoisier, M., Murat, C., Montpied, P. and Becker, M. (2014). Climate variations explain annual fluctuations in French Périgord black truffle wholesale markets but do not explain the decrease in black truffle production over the last 48 years. Mycorrhiza, 24: 115–125.

Lee, H., Nam, K., Zahra, Z. and Farooqi, M.Q.U. (2020). Potentials of truffles in nutritional and medicinal applications: A review. Fungal Biol. Biotechnol., 7(1): 10.1186/s40694-020-00097-x.

Leonardi, P., Iotti, M., Donati Zeppa, S., Lancellotti, E., Amicucci, A. and Zambonelli, A. (2017). Morphological and functional changes in mycelium and mycorrhiza of *Tuber borchii* due to heat stress. Fungal Ecol., 29: 20–29.

Li, X., Zhang, X., Kang, L.Y.Z., Jia, D., Yang, L. and Zhang, B. (2019). LC-MS-based metabolomics approach revealed the significantly different metabolic profiles of five commercial truffle species. Front. Microbiol., 10: 2227. 10.3389/fmicb.2019.02227.

Lönnqvist, M.A. (2008). Were nomadic Amorites on the move? Migration, invasion and gradual infiltration as mechanisms for cultural transitions. pp. 1–215. *In*: Kuhne, H., Czichon, M. and Kreppner, F.J. (eds.). Proceedings of the 4th International Congress of the Archaeology of the Ancient Near East, March 3, 2004, Harrassowitz Verlag, Weisbaden.

Malik, H.M., Munazza, G., Omar, U., Kumosani, T.A., Al-Hejin, A.M. et al. (2018). Evaluation of the antibacterial potential of desert truffles (*Terfezia* spp.) extracts against methicillin resistant *Staphylococus aureus* (MRSA). J. Exp. Biol. Agric. Sci., 6: 652–660.

Marathe, S.J., Hamzi, W., Bashein, A.M., Deska, J., Seppänen-Laakso, T. et al. (2020). Anti-angiogenic and anti-inflammatory activity of the summer truffle (*Tuber aestivum* Vittad.) extracts and a correlation with the chemical constituents identified therein. Food Res. Int., 137: 109699. 10.1016/j. foodres.2020.109699.

Nadim, M., Deshaware, S., Saidi, N., Abd-Elhakeem, M.A., Ojamo, H. and Shamekh, S. (2015). Extracellular enzymatic activity of *Tuber maculatum* and *Tuber aestivum* mycelia. Adv. Microbiol., 5: 523e530. 10.1016/j.foodres.2020.109699.

Ngun, T.C., Ghahramani, N., Sanchez, F.J., Bocklandt, S. and Vilain, E. (2011). The genetics of sex differences in brain and behaviour. Front. Neuroendocrinol., 32(2): 227–246.

Patel, S. and Goyal, A. (2012). Recent developments in mushrooms as anti-cancer therapeutics: A review. 3 Biotech, 2(1): 1–15. 10.1007/s13205-011-0036-2.

Patel, S., Rauf, A., Khan, H., Khalid, S. and Mubarak, M.S. (2017). Potential health benefits of natural products derived from truffles: A review. Tr. Food Sci. Technol., 70: 1–8.

Sawaya, W.N., Al-Shalhat, A., Al-Sogair, A. and Mohammad, M. (1985). Chemical composition and nutritive value of truffles of Saudi Arabia. J. Food Sci., 50: 450–453.

Shavit, E. (2014). The history of desert truffle use. pp. 217–241. *In*: Kagan-Zur, V., Roth-Bejerano, N., Sitrit, Y. and Morte, A. (eds.). Desert Truffles: Phylogeny, Physiology, Distribution and Domestication, Springer, Heidelberg, Berlin.

Splivallo, R., Bossi, S., Maffei, M. and Bonfante, P. (2007). Discrimination of truffle fruiting body versus mycelial aromas by stir bar sorptive extraction. Phytochemistry, 68: 2584–2598.

Splivallo, R., Ottonello, S., Mello, A. and Karlovsky, P. (2011). Truffle volatiles: From chemical ecology to aroma biosynthesis. New Phytol., 189: 688–699.

Splivallo, R., Deveau, A., Valdez, N., Kirchhoff, N., Frey-Klett, P. and Karlovsky, P. (2014). Bacteria associated with truffle-fruiting bodies contribute to truffle aroma. Environ. Microbiol., 17: 2647–2660.

Splivallo, R. and Ebeler, S.E. (2015). Sulphur-volatiles of microbial origin are key contributors to human-sensed truffle aroma. Appl. Microbiol. Biotechnol., 99: 2583–2592.

Stanikunaite, R., Khan, S.I., Trappe, J.M. and Ross, S.A. (2009). Cyclooxygenase-2 inhibitory and antioxidant compounds from the truffle *Elaphomyces granulatus.* Phytother. Res., 23: 575–578.

Stielow, B. and Menzel, W. (2010). Complete nucleotide sequence of TaV1, a novel totivirus isolated from a black truffle ascocarp (*Tuber aestivum* Vittad.). Arch. Virol., 155: 2075–2078.

Stielow, B., Klenk, H.-P. and Menzel, W. (2011). Complete genome sequence of the first endovirus from the ascocarp of the ectomycorrhizal fungus *Tuber aestivum* Vittad. Arch. Virol., 156: 343–345.

Stielow, J.B., Bratek, Z., Klenk, H-P., Winter, S. and Menzel, W. (2012). A novel mitovirus from the hypogenous ectomycorrhizal fungus *Tuber excavatum.* Arch. Virol., 157: 787–790.

Vahdatzadeh, M., Deveau, A. and Splivallo, R. (2015). The role of the microbiome of truffles in aroma formation: A meta-analysis approach. Appl. Environ. Microbiol., 81: 6946–6952.

Vidale, E. (2019). Truffles: The precious mushroom. pp. 79–81. *In*: Pullanikkatil, D. and Shackleton, C.M. (eds.). Poverty Reduction Through Non-timber Forest Products, Sustainable Development Goals Series, Springer Nature, Switzerland.

Wang, S. and Marcone, M.F. (2011). The biochemistry and biological properties of the world's most expensive underground edible mushroom: Truffles. Food Res. Int., 44: 2567–2581.

Wang, Z., Li, J., Wang, Z., Xue, L., Zhang, Y. et al. (2012). L-tyrosine improves neuroendocrine function in a mouse model of chronic stress. Neural Regen. Res., 7(18): 1413–1419.

Wernig, F., Buegger, F., Pritsch, K. and Splivallo, R. (2018). Composition and authentication of commercial and home-made white truffle-flavoured oils. Food Cont., 87: 9–16.

Xiao, D.R., Liu, R.S., He, L., Li, H.-M., Tang, Y.-L. and Liang, X.-H. (2015). Aroma improvement by repeated freeze-thaw treatment during *Tuber melanosporum* fermentation. Sci. Rep., 5: 17120. 10.1038/srep17120.

Enzymes

11

Potential Biotechnological Applications of Enzymes from Macrofungi

Naif Abdullah Al-Dhabi,[1,] Mariadhas Valan Arasu,[1,2,*] Lekshmi R,[3] Savarimuthu Ignacimuthu,[2] Rajakrishnan R,[1] Veeramuthu Duraipandiyan,[1,4] Ameer Khusro,[5] Paul Agastian,[5] Soosaimanickam Maria Packiam[3] and Ponnuswamy Vijayaraghavan[6]*

1. INTRODUCTION

Macrofungi are an important group of the fungal kingdom and play a significant role in various ecosystems. They usually produce colloidal and fleshy morphologically distinct fruit bodies representing sexual reproductive structures. Most of the reported macrofungi belong to Ascomycota or Basidiomycota while a few are Zygomycota. The fruit bodies of macrofungi located below or above the ground (Mueller and Schmit, 2007). About 100,000 macrofungi were described and 6,000 can produce fruit bodies and sclerotia (Ainsworth, 2008). Macrofungi have the tendency to grow

---

[1] Department of Botany and Microbiology, College of Science, King Saud University, P.O. Box 2455, Riyadh 11451, Saudi Arabia.
[2] Xavier Research Foundation, St. Xavier's College, Palayamkottai, Thirunelveli, India.
[3] Department of Botany and Biotechnology, MSM College, Kayamkulam, Kerala, India.
[4] Entomology Research Institute, Loyola College, Chennai, India.
[5] Research Department of Plant Biology and Biotechnology, Loyola College, Nungambakkam, Chennai, India.
[6] Bioprocess Engineering Division, Smykon Biotech Pvt LtD, Nagercoil, Kanyakumari, India.
* Corresponding authors: naldhabi@ksu.edu.sa; mvalanarasu@gmail.com

during the wet season. These macrofungi follow parasitic, saprophytic or symbiotic mode of lifestyle. Many macrofungi, especially the symbiotic groups, cannot reproduce independently for want of suitable host partners to assist them reproduce (e.g., *Termintomyces* sp.). Macrofungi play a major role of terrestrial ecosystems, with a major share of their species diversifying and involve in various ecosystem processes. They prefer decaying organic matter, like leaf litter, dead twings, wood, bark and various type of soils for their growth. Basidiomycetes and Ascomycetes are the main macrofungi have spore-bearing structures and easy to observe (Senn-Irlet et al., 2007).

Macrofungi have numberous economic and biological significances (Miles and Chang, 2004). From the commercial point of view, macrofungi are very important to produce metabolites, chemical compounds, bio-ethanol, medicine, nutraceuticals and other biotechnological products. Despite the significance of macrofungi in the process of nutrient recycling and succession in the forest environment, the present understanding of enzymes from macrofungi remains fairly limited. Enzymes from macrofungi are widely used for various industrial processes such as cellulase, laccase, xylanase, amylase and proteases. Laccases are widely used in the degradation of a wide number of textile dyes, chemical synthesis, treatment of effluents polluted with lignin, bioremediation of recalcitrant aromatic compounds, biosorbents in wastewater treatment, in beverage and food industries (Demierege et al., 2015; Schneider et al., 2018). Laccases also act as cleaning agents in various purification systems, in medical diagnostic tools, as catalysts in drug manufacturing, in bioremediation of herbicides/pesticides and as ingredients in cosmetics (Demain and Adrio, 2008). In nature various white rot fungi produce types of lignocellulolytic enzymes, thus involved in carbon recycling. White rot fungi, such as *Trametes pubescens*, *Trametes ochracea*, *Trametes versicolor*, *Trametes hirsuta*, *Lentinus edodes*, *Phanerochaete chrysosporium*, *Pycnoporus sanguineus*, *Pleurotus ostreatus*, *Ganoderma applanatum*, *Irpex lacteus* and *Trametes villosa* are involved in the hydrolysis of lignocellulosic biomass. These macrofungi have the potential to generate ethanol from lignocellulose materials more competitively with the fossil fuel industry (Cardoso et al., 2018; Giorgio et al., 2018; Coniglio et al., 2020). Proteases extracted from macrofungi have therapeutic and industrial applications. A serine protease characterised from *Cordyceps sobolifera* showed HIV-1 reverse transcriptase inhibitory activity (Wang et al., 2012), whereas a protease identified from *Hypsizigus marmoreus* has anticancer activity (Zhang et al., 2010).

2. Macrofungal Enzymes

In recent decades, researchers have conducted various studies on macrofungal enzymes and their potential in biotechnological and industrial applications. Perspective directions and practical value of the application of various enzymes from Macromycetes has been analysed by various researchers (Madhavi and Lele, 2009; Thakur, 2012). Basidiomycetes produce various lignolytic enzymes (Elisashvili et al., 2002, 2009; Kalmiş et al., 2008; Erden et al., 2009; Kumari et al., 2012; Rana and Rana, 2011) and various proteases (Goud et al., 2009; Denisova, 2010), amylases (Goud et al., 2009; Chen et al., 2012), cellulase (Elisashvili et al., 1999; Kobakhidze et al., 2012), xylanase (Elisashvili et al., 2008, 2009; Kobakhidze et al., 2012), phytase (Collopy, 2003), lipase (Thakur, 2012; Shu et al., 2006), urease (Mukchaylova and Buchalo, 2012) and nitrate reductase (Mukchaylova and Buchalo, 2012). Macrofungi have the ability to degrade the available lignocellulosic biomass from the environment. The combined effect of various hydrolytic enzymes, hemicellulases, cellulases and pectinase and oxidative enzymes, such as laccases and peroxidises act in a synergistic and co-operative mechanism. In general, cellobiohydrolases and cellulases act in a synergistic way (Dashtban et al., 2009). Enzymes from macrofungi, such as glycoside hydrolase, carbohydrate esterase and polysaccharide lyase are involved in the degradation of plant polysaccharide. In macrofungi, lignin biodegradation is an important process and various enzymes, including phenol oxidases are involved. These enzymes are commonly called ligninases and include peroxidises and laccases, mangansese peroxidase and lignin peroxidase and versatile peroxidase which show the catalytic properties of manganese peroxidase and lignin peroxidase (Leonowicsz et al., 1999). The macrofungi, including *Agrocybe aegerita*, *Ganoderma applanatum*, *Lepista luscina*, *Pycnoporus sanguineus* and *Lactarius deliciosus* synthesise various commercial and therapeutic enzymes (Table 1). These enzymes have various applications, including, degradation of plant polysaccharide,

Table 1. Commercial and therapeutic enzyme producing macrofungi.

Macrofungi	Enzymes	References
Grifola frondosa (Dicks.) S.F. Gray	Amylase, laccase, protease, lipase	Montoya et al., 2012
Hohenbuehelia myxotricha (Lev.) Singer	Amylase, lipase, urease	Montoya et al., 2012
Hohenbuehelia myxotricha (Lev.) Singer	Amylase, lipase, protease	Krupodorova et al., 2014
Lepista luscina (Fr.) Singer	Amylase, laccase, lipase, urease	Krupodorova et al., 2014
Phellinus igniarius (Fr.) Quel	Anticarcinogenic enzymes	Park et al., 2002
Pleurotus ostreatus (Jacq.) Kumm	Amylase, laccase, lipase, urease	Krupodorova et al., 2014
Inonotus obliquus (Pers.) Pilat	Cellulases	Xu et al., 2018
Xylaria poitei	Cellulase, pectinase, xylanase	Gutiérrez-Soto et al., 2015
Cordyceps militaris	Fibrinolytic enzymes	Kim et al., 2006
Schizophyllum commune Fr.	Glucanolytic enzymes	Chiu and Tzean, 1995
Fomes fomentarius (Fr.) Gill	1, 4-β-glucosidase	Vetrovsky et al., 2013
Agrocybe cylindracea	Laccase	Hu et al., 2011
Coprinus comatus (Mull.) S.F. Gray	Laccase	Zhao et al., 2014
Ganoderma applanatum (Pers.) Pat.	Laccase	Krupodorova et al., 2014
Pycnoporus sanguineus	Laccases	Lu et al., 2007
Pleurotus djamor (Rumph. ex Fr.)	Ligninolytic enzymes	Hongbo, 2009
Morchella esculenta	Lignocellulolytic enzyme	Papinutti and Lechner, 2008
Trametes versicolor	Ligninolytic enzymes	Moredo et al., 2003
Pleurotus ostreatus	Manganese peroxidase, laccase	Valášková et al., 2006a

degradation of lignin, degradation of dye, clarifying agent, thrombolytic agent and wastewater treatment. The important macrofungi enzymes and their applications are described in Table 2.

Table 2. Macrofungi enzymes and their applications.

Macrofungi enzymes	Applications	References
Glycoside hydrolase	Degradation of plant polysaccharide	Leonowicsz et al., 1999
Carbohydrate esterase	Degradation of plant polysaccharide	Leonowicsz et al., 1999
Polysaccharide lyase	Degradation of plant polysaccharide	Leonowicsz et al., 1999
Ligninases	Lignin biodegradation, degradation of hazadous compounds	Papinutti and Lechner, 2008
Laccases	Detoxification, Dye degradation, Biosensor, wastewater treatment	Lu et al., 2007
Glyoxal oxidase	Degradation of lignin	Tuomela and Hatakka, 2019
Aryl alcohol oxidase	Degradation of lignin	Tuomela and Hatakka, 2019
Xylanase	Biodegradation of wood, biofuels	Gutiérrez-Soto et al., 2015
Cellulase	Detergent additives	Krupodorova et al., 2014
Manganese peroxidase	Textile, pulp and paper industries	Valášková et al., 2006a
Amylases	Clarifying agent	Montoya et al., 2012
Proteases	Food processing, wastewater treatment	Montoya et al., 2012
Tyrosinase	Removal of phenolic substances from wastewater	Mueller et al., 1996
Fibrinolytic enzymes	Cardiovascular diseases	Kim et al., 2006

3. Ligninolytic Enzymes

Lignicolous macrofungi are important as they are specifically involved in the degradation of wood, possess various adaptations to various ecological niches. Degradation of lignin of plant material has the added advantage to support the degradation of cellulose and also increase the availability of nitrogen and other wood components (Deacon, 2006). Ligninolytic enzymes are widely used in degradation and utilisation of lignin, the highly complex polymer. They involve many oxidative enzymes, such as manganese peroxidase, lignin peroxidase and laccase. Recently, the demand for ligninolytic enzymes has increased rapidly because of their potential biotechnological application in wastewater treatment, for example, degradation of hazardous compounds, such as xenobiotic, phyenols and dyes. The enzymes, such as glyoxal oxidase and aryl alcohol oxidase, are called accessory enzymes as they are involve in the degradation of lignin. These are also used in production of hydrogen peroxide (Tuomela and Hatakka, 2019). The enzymes from the wood-rotting fungi are widely used in bio-pulping and various industrial processes (Ahuja et al., 2004).

3.1 Laccases

Laccases are copper-containing enzymes which catalyse and degrade lignin from the wastewaters. Laccases from mushrooms have potential in chemical synthesis, bioremediation, detoxification of polluted water, synthetic dye decolouration, bleaching of textile dyes and paper-pulp and serve as biosensors (Annunziatini et al., 2005; Nagai et al., 2002; Timur et al., 2004; Palmieri et al., 1994; Reid and Paice, 1994; Martirani et al., 1996; Jaouani et al., 2005; Karamyshev et al., 2003; Araújo et al., 2008; Couto and Herrera, 2006; Mainardi et al., 2018). Laccases consist of a group of multi-copper enzymes, with p-diphenol oxidase belonging to the oxidoreductase class or oxygen oxidoreductase. The enzyme reaction takes place by one oxygen molecule oxidised into water. The oxidation ability of these laccases includes various substrates, such as polyphenols, methoxy-substituted monophenols and aromatic amines (Bourbonnais et al., 1995; Bourbonnais and Paice, 1990). The oxidation process generates free radicals and converts to quinine (Gianfreda et al., 1999). These laccases play a critical role in industries with specific applications in the synthetic chemistry, textile, pulp, paper, food and cosmetic industry and as pollutant-removal agents, removal of endocrine distrupors and biodegradation of phenolic compounds (Couto and Herrera, 2006). Laccases are also capable of oxidising non-phenolic compounds (Eggert et al., 1996). Macrofungal laccases generally contribute to various processes, such as pigment production, sporulation, lignin degradation, plant pathogenicity and fruit body formation (Mayer and Staples, 2002). The molecular weight of laccases from fungal origin ranges between 60–80 kDa and the pI value ranges between 3–4. Many macrofungi belong to the white rot group which produces laccase with varying specificity (Khushal et al., 2010). Macrofungi which produce laccases include *Phanerochaete, Stereum, Trametes, Pycnoporus cinnabarinus, Trametes versicolor Pleurotus* and *Ganoderma lucidum* (Khushal et al., 2010; Osma et al., 2011; Kuhara and Papinutti, 2014). These laccases are oxidoreductases involved in oxidation of aromatic compounds, mainly phenoic substances, with the reduction of oxygen to water (Valeriano et al., 2009). Macrofungi, such as *Pleurotus eryngii, T. versicolor* and *G. Lucidum*, produce ligin-degrading enzymes (Pelaez et al., 1995; Kalmis et al., 2008; Addleman and Archibald, 1993). The woodrot fungi simultaneously produce manganese peroxidase, lignin peroxidase and laccase. These cleary indicate that most of the fungi do not require all the three enzymes at a specific time during the process of lignin degradation. Most of woodrot fungi synthesise these three groups of enzymes; however, other macrofungi produce only two or one group of these enzymes (De Jong et al., 1994).

3.2 *Lignin peroxidase*

Lignin peroxidase is generally described as ligninase as it is one of the very important enzymes to participate in the utilisation of lignin. Both non-phenolic and phenolic compounds can be readily oxidised by cleaving the propyl side chain of the substrate of lignin (Schoemaker et al., 1985). This enzyme depolymerises lignin *in vivo* (Hammel et al., 1993). Oxidation of non-phenolic substances by lignin peroxidase with high redox potential has been interpreted by various researchers as the result of a high redox potential of various oxidised enzyme intermediates, which are termed as lignin peroxidase compound I and lignin peroxidase compound II (Schoemaker and Piontek, 1996). This enzyme also has the potential to solubilise the initial product of veratryl alcohol oxidation and veratryl alcohol radical cation. Macrofungi, such as *Trametes, Phlebia, Phanerochaete* and *Bjerkandera* produce lignin peroxidases (Riley et al., 2014; Duenas et al., 2013; Floudas et al., 2012). Other macrofungi producing lignin peroxidases include, *Bjerkandera adusta, Phanerochaete chrysosporium, Podoscypha elegans* and *Ganoderma lucidum* (Shaheen et al., 2017; Bouacem et al., 2018; Kadri et al., 2017; Agrawal et al., 2017). Lignin peroxidase is heme-containing enzyme and was initially screened from the lignin-degrading *Phanerochaete chrysosporium*. These catalyse the oxidation of various nonphenolic aromatic substances and further cation radicals are decomposed chemically.

3.3 Manganese peroxidase

Manganese peroxidase are produced by the lignin-degrading macrofungi. Manganese peroxidase shows strong affinity for Mn (II) as its reducing substrate. The redox potential of manganese peroxides and Mn-system is lower than that of lignin peroxidase and genrally it does not oxidise the various non-phenolic lignins (Glenn and Gold, 1985). In general, the polymer of lignin gives maximum strength to higher plants and the important function of manganese peroxides is effective degradation of the matrix (Sundaramoorthy et al., 1997). Macrofungi, such as *Irpex lacteus, Stereum ostrea, Dichomitus squalens, Polyporus* sp., *T. versicolor, Phanerochaete chrysosporium, Phlebia radiata* and *Trichaptum abietinum* are producers of manganese peroxidase (Sukarta and Sastrawidana, 2014; Kuuskeri et al., 2015; Mali et al., 2017). Manganese peroxidise oxidises lignin to polycyclic aromatic hydrocarbons and the molecular weight of enzymes ranges between 40–50 kDa and the isoelectric point ranged between 3–4 and neutral types are also reported. Species of *Pleurotus* also produce manganese peroxidase (Kamitsuji et al., 2005). The interest in the analysis of manganese peroxidase has grown due to its potential application in various biotechnological processes. The phenol oxidases are used to treat wastewater effluent from textile, pulp and paper industries; also used in biobleaching cellulose pulp and biodelignification of wood chips, polymerisation reaction as well as to clarify juices and wines. They can also be used in the blanching of fabrics and have very promising applications in the bioremediation of xenobiotics (Gomes et al., 2009). These enzymes are also used in the pre-teatment of biomass to generate biofuels (Camassola and Dillon, 2009). The wood-colonising many soild substrate colonising and litter-decomposing fungi are known to produce manganese peroxidase (Hofrichter, 2002).

3.4 Tyrosinases

Tyrosinase is a copper-containing enzyme that involves oxidation with types of phenolic substrates. Intracellular tyrosinase produced by *Agaricus bisporus* shows a lot of biotechnological potential. It is a heterotetramer with two light and heavy polypeptide chains. The molecular weight of tyrosinase is 120 kDa (Seo et al., 2003; Soler-Rivas et al., 1999; Haq and Ali, 2006). In *Amanita muscaria*, this enzyme has been purified and its molecular weight was determined. The purified tyrosinase has heterodimer and the molecular weight is 50 kDa. In *Neurospora crassa*, the molecular weight of the enzyme was reported as 46 kDa and it is a monomer (Mueller et al., 1996; Kupper et al., 1989). Tyrosinase was extracted from macrofungi, such as *Lentinula boryana, Pycnoporus sanguineus, Amanita muscaria* and *Lentinula edodes* (Faria et al., 2007; Halaouli et al., 2005; Mueller et al., 1996; Kanda et al., 1996). Tyrosinases have various environmental and biotechnological applications. These are used in biosynthesis of L-DOPA, removal of phenolic substances from wastewaters, quantification and detection of phenolic substances from wastewater and the production of cross-linked protein networks. The L-DOPA has been widely used as a drug for the treatment of Parkinson's disease (Gelb et al., 1999). This drug is also used for controlling the myocardium after neurogenic injury. The worldwide L-DOPA market is about 250 metric tons per year. However, most of the commercially available L-DOPA was derived from hydantoin and vanillin (Reinhold et al., 1987). Tyrosine phenol-lyase converts a mixture of catechol, ammonia and pyruvate into L-DOPA after various reactions. This method has several advantages because it does not have a diphenolase activity and no enzymatic reaction with L-DOPA. Tyrosinases also have various applications in electrochemical workstations as biosensors (Halaouli et al., 2006). They are widely used to effectively activate the tyrosine residues in various polypeptides in direct protein-protein cross-linking and for protein cross-linking to chitosan films. Enzymatic bioconversion of L-tyrosine to L-DOPA is achieved by tyrosinase (Halaouli et al., 2006). Tyrosinases are also useful in treating wastewaters. In a study, tyrosinase was coated on chitosan-coated polysulphone capillary membranes for wastewater treatment. The immobilised tyrosinse oxidises various phenols from the industrial and synthetic effluents (Edwards et al., 1999). In the physical and chemical method of wastewater treatment, it is very difficult to eliminate phenol at low concentrations; besides immobilisation of tyrosinases on chitosan-alginate reduces enzyme inactivation. Tyrosinase was isolated and immobilised from *Agaricus bisporus* on a chitosan-alginate support, in pilot- and bench-scale bioreactors and air was continuously bubbled into the system. In this bioreactor, more than 60 per cent of phenol was removed in the pilot-scale reactor, while about 92 per cent was removed by bench-scale reactor.

3.5 *Amylases*

Amylases are starch-hydrolysing enzymes with significant applications in various biotechnological industries. These applications include fermentation, production of paper, animal feeds, soft drinks, textile, detergents and food (Pandey et al., 2000; James and Lee, 1997). Differnet amylases are extracted from the mycelia and fruit bodies of mushrooms were studied for their commercial applications (Diez and Alvarez, 2001). *Nothopanus hygrophanus, Pogonomyces hydnoides, Podoscypha bolleana, Pleurotus tuberregium, Termitomyces globules, Termitomyces clypeatus, Coriolus versicolor, Corilopsis occidentalis, Agaricus* sp. and *Agaricus blazei* are the macrofungi which produce amylases. Amylase production by *L. shimeji* was carried out in a liquid medium and enzyme production varied widely. The fruit-body-forming strains showed high amylase production than non-fruitbody-forming mushrooms (Wu et al., 2019). A glucoamylase was purified from the samples of *L. shimeji* extracted from liquid medium using Toyopearl-DEAE column chromatography (Terashita et al., 2000a). Earlier, glucoamylase had been charactreised from *Lyophyllum fumosum* and *T. matsutake* (Yoshida et al., 1994a, b). In a study, Lee et al. (1998)

reported the potential of amylase production of *T. matsutake*. Some fungi used barley grain starch effectively than other sources of starch (Terashita, 2000). Amylases were also purified from *T. matsutake* Z-1 and *T. matsutake* with varying substrate specificity. These amylases have potential applications in starch-processing industries, paper industries and also used as clarifying agents (Hur et al., 2001; Kusuda et al., 2003).

4. Cellulolytic Enzymes

Macrofungi produce pectinase, chitinase, hemicellulases and cellulases. These enzymes are responsible for various xylanolytic and cellulolytic activites (Mtui, 2012).

4.1 Cellulases

Cellulase catalyses the breakdown of chemical bonds and is used to reduce waste and for cleaning process. It also contributes to the production of biofuels and various chemical derivatives (Trivedi et al., 2015). These enzymes convert cellulose into a very simple form of sugars (Chinedu et al., 2005) and they hydrolyse β-1, 4 linkages in cellulose chains and are produced by macrofungi (Henrissat, 1991). Cellulases are widely used in bioconversion of lignocellulosic biomass to biofuel, production of fruit juice, beer and wine (Philippidis, 1994). The productivity of cellulases from various sources has several advantages (Rana and Kaur, 2012). Based on their crystal structure and amino acid structure, the catalytic property of cellulases has been classified into various families (Henrissat, 1991). For the complete hydrolysis of cellulose materials, the combination of β-glucosidase, exoglucanases and endoglucanases is required (Zhang and Lynd, 2006). These enzymes are applied in detergents, textiles, paper, agriculture, beverage industries and also used as an alternative to produce energy sources. The white-rot macrofungus, *Ganoderma lucidum* produces various lignocellulolytic enzymes and is involved in degradation of cellulose and lignin. Cellulose-degrading enzymes were determined from *G. lucidum* (Ko et al., 2001; You et al., 2014; Jakucs et al., 1994; Manavalan et al., 2015). Unlike bacteria and other fungi, macrofungi witness very slow growth rate and biomolecular characterisation are lacking. Recently, the whole genome sequence of *Ganoderma lucidum* was performed (Chen et al., 2012). About 13,000 protein-encoding putative genes have been predicted and about 80 per cent of them were functionally annotated. The whole genome of *Ganoderma lucidum* is enriched with various wood-degrading enzymes (Liu et al., 2012).

4.2 Endoglucanases

The endoglucanase is generally termed as carboxy methyl cellulase (CMCase) which randomly cuts the cellulose chain at the β-1,4-bonds and mainly generates new ends for other combinations of cellulase to act upon. On microcrystalline cellulose, endoglucanse works effectivity by possibly cracking cellulose chain from the crystalline structure of cellulose (Teeri, 1997). The mushrooms, such as *Shiraia bambusicola*, *Pleurotus nebrodensis*, *Grifola frondosa*, *Hericium erinaceus*, *Auricularia auricular*, *Macrolepiota procera*, *Coprinus comatus*, *Lentinus edodes*,

Pholiota nameko and *Lentinus edodes* produce CMCase at various degrees (Wu and Shin, 2016). Among these mushroomss, *Coprinus comatus* has a lot of potential for CMCase production (Fen et al., 2014). Other macrofungi, like *Pycnoporus sanguineus, Bjerkandera adusta, Irpex lacteus, Lentinus edodes, Fomitosis* sp. and *Volvariella volvacea* have been screened for production of CMCase (Chang and Steinkraus, 1982; Ding et al., 2001; Deswal et al., 2011; Yoon and Kim, 2008; Yoon et al., 2005; Kapoor et al., 2009; Dias et al., 2010; Quiroz-Casta-eda et al., 2009). The CMCase was characterised from various wood-rotting Basidomycets, whiterot fungi, brownrot fungi, basidomycetous yeast (*Rhodotorula glutinis*), plant pathogen (*Sclerotium rolfsii*) and the symbiont of termites (*Termitomyces* sp.), wood-associated yeasts (Jimenez et al., 1991), ectomycorrhizal fungi (Maijala et al., 1991; Cao and Crawford, 1993) and in cultures of litter-decomposing Basidomycetes (Steffen et al., 2007; Valášková et al., 2007). The CMCase from Basidomycetes show optimum pH between 4 and 5 (Suzuki et al., 2006), near to neutral pH in the case of *Volvariella volvacea* (Zhang, 2012). The optmimum temperature of CMCase ranged between 50–70°C (Valášková and Baldrian, 2006).

4.3 β-glucosidase

The β-D-glucoside glucohydrolase or β-glucosidase degrades polysaccharides into monosaccharide units. These can be readily absorbed and used by the organism to effectively hydrolyse soluble cellobiose and short-chain oligosaccharides into glucose units. Increase in the length of the cellulose chain readily affects the loss of enzyme activity and also involves biosynthesis of the oligosaccharide unit in glycolipids or glycoproteins and the hydrolysis of terminal β-D-glucose oligosaccharides (Melo et al., 2006). Mushrooms, like *Nothopanus hygrophanus, Pogonomyces hydnoides, Podoscypha bolleana, Pleurotus tuberregium, Termitomyces globules, Termitomyces clypeatus, Coriolus versicolor, Corilopsis occidentalis, Agaricus* sp., and *Agaricus blazei* produce β-glucosidase (Jonathan and Adeoyo, 2011). Generally, in a natural environment, cellobiose is available in large quantities, hence β-glucosidase is produced by various organisms to survive in the ambient environment (Lynd et al., 2002). Among the Basidomycetes, β-glucosidase was isolated and characereised from *Termitomyces* sp., *Sclerotium rolfsii, Tricholoma matsutake* and *Pisolithus tinctorius*. The β-glucosidase activity was detected from the litter-decomposing Basidomycetes and other ectomycorrhizal fungi screened from various environments (Mucha et al., 2006). The molecular weight of β-glucosidase varied widely, for example, the cell-wall-associated enzyme isolated from *Pisolithus tinctorius* is of homotrimer nature with 450 kDa (Cao and and Crawford, 1993) and the molecular weight of *Termitomyces clypeatus* was 110 kDa (Sengupta et al., 1991). In the case of *Trametes versicolor*, the molecular weight of β-glucosidase was 300 kDa and it varied in the case of *Volvariella volvacea* (256 kDa) and *Phanerochaete chrysosporium* (410 kDa). β-glucosidase activity varied on the basis of organisms and pH value of the medium. The isoelectric points of extracellular enzymes and cell-wall-associated enzymes are acidic in nature, typically ranging between 3.5–5.2. However, enzymes in intracellular origin ranged between 6.2–7.0 (Cai et al., 1998). These enzymes play a vital role in biofuel production and the candidates with high stability and activity over various pH ranges are preferred in industrial applications (Xia et al., 2016).

4.4 *Xylanases*

The synergistic mechanism of xylanase is specifically required for better hydrolysis of lignocellulosic residues (Ghose and Bisaria, 1979). Xylanase like endo-xylanase, β-xylosidase, α-glucuronidase, α-arabinofuranosidase and acetylxylan esterase were responsible for the hydrolysis of xylan (Uday et al., 2016). In the tropical region, biodegrading enzyme-producing *Xylaria* consists of various ligninolytic enzymes and plays a potential role in the biodegradation of leaf litter as well as in biodegradation of wood (Gutiérrez-Soto et al., 2015; Osono and Takeda, 2002; Koide et al., 2005; Bezerra et al., 2012; Chaparro et al., 2009; Moissenko et al., 2018; Negrão et al., 2014; Liers et al., 2006). Many agro-residues have been utilised as substrate for the production of enzymes by mushrooms. For example, a spent substrate was obtained after the essential oil extraction process from *Bulbophyllum graveolens*. The inexpensive substrates, such as, olives (*Olea europea*), sugar cane (*Saccharum officinarum*), wheat straw (*Triticum* spp.), banana stems (*Musa paradisiaca*) and rice cane (*Oryza sativa*) could be used for the production of wood-degrading enzymes by *Lentinula edodes* and *Xylaria guianensis* (Ntougias et al., 2015; Ullah et al., 2002; Felix et al., 2014). Xyanases are used in the manufacture of textiles, food (bread), pulp bleaching, animal feed supplement and also in the production of xylitol and ethanol (Polizeli et al., 2005). Xylanases are used in beverage, food, paper and pulp industries (Gutiérrez et al., 2009). These are widely used in the preparation of second generation biofuels. Xylanase activity is desirable to improve low-cost bioethanol production (Qing and Wyman, 2011).

5. Macrofungal Proteases

Macrofungal proteases have various applications in leather processing, detergents, preparation of medical equipments, silver recovery, chemical industry, food processing, feed industry as well as wastewater treatment (Ma et al., 2007). Apart from these industrial applications, proteases are also involved in various physiological processes, such as regulation of gene expression, nutrition, enzyme modification, germination, conidial discharge, sporulation and protein turnover (Rao et al., 1998). Basidomycetes are an important group of macrofungi in production of various proteases. These proteases play important roles in the physiology of fungi by involvement in processes, such as sporulation and germination. These enzymes are very much involved in the lifestyle of macrofungi than other hydrolytic enzymes as observed in *Pleurotus citrinopileatus* (Cui et al., 2007). The *P. pulmonarius* grows on various dead timber and secretes subtilisin type of proteases; however, it is unable to produce trypin. The *P. ostreatus* grows in living hosts and secretes trypsin-like proteases in its life. It could be noted that the production of trypsin-like proteases may be related to the presence of living tissues as hosts. Most of the nitrogen content in wood is in the form of proteins; proteases critically play an important role in the metabolism of the available proteins in the whiterot fungi in wood and has been previously reported (Inácio et al., 2018). Sudden depletion of nitrogen content in the medium generally stimulates the production of various proteases in macrofungi (Kudryavtseva et al., 2008; Watkinson et al., 2001; Wasser and Weis, 1999). Saprophytic Basidiomycetes produce various groups of proteases, including subtilases (*Coprinus* sp., *Schizophyllum commune*, *Serpula lacrymans*, *Phanerochaete chrysosporium* and *Pleurotus ostreatus*) (Faraco et al., 2005; Shaginyan et al., 1990; Watkinson et al., 2001). *Chondrostereum purpureum* and *Hypsizygus marmoreus* also produce metalloproteinases (Terashita et al., 1998). *Irpex lacteus, Amanita muscaria* and *P. chrysosporium* produce aspartate proteases (Dass et al., 1995). Proteases have been purified and characterised from the honey fungus *Armillaria mellea*. The fibrinolytic enzyme was isolated and characterised from entomopathogenic fungus *Cordyceps militaris* and the molecular weight was 52 kDa, like subtilisin types of proteases. This fibrinolytic enzyme degraded α and γ chains of fibrinogen and also partially degraded the β chains (Healy et al., 1999). This proteolytic property was different from the proteases derived from snake venom (Kim et al., 2006). Fibrinolytic enzymes have several therapeutic applications. The fibrinolytic enzymes from *Pleurotus ostreatus* were cloned, purified and characterised (Kim et al., 2006; Shen et al., 2007; Joh et al., 2004; Yin et al., 2014). Laboratory cultivation of macrofungi is mainly based on the culture of vegetative phase of macrofungi. This mode of culture is simple for medium- to small-scale and the process parameters, such as agitation, humidity and temperature can be effectively controlled (Rao et al., 1998). The application of a liquid medium allows rapid purification of polysaccharides and enzymes (Campos et al., 2011). The *P. ostreatus* utilises wheat gluten resulting in the production of proteolytic enzymes that specifically increase the solubility of the culture medium (Eisele et al., 2011). The solid state fermentation (SSF) of mycelia of mushrooms utilises enzymes and bioactive compounds by using agro-industrial wastes, municipal wastes, forestry debris for the production of proteases. Combined effect of various substrates enhances the production of proteases by fungi (Sumantha et al., 2006). Solid state cultures provide natural environment to the basidomycetous mushrooms to produce various enzymes. The *P. ostreatus* utilised tomato pulp, wheat bran fibres and soybean for enhanced protease titre (Iandolo et al., 2011). The fibrinolytic enzyme was purified and characterised from *Pleurotus* spp. and these macrofungi are non-toxic and edible (Liu et al., 2014). This enzyme was purified 29.3-fold after various steps of purification from the strain *P. eryngii* cultured in corn. This enzyme also directly degrades fibrin and could be used as a thrombolytic agent. It degrades α and β chains within 10 min of incubation in optimised conditions. This shows similarities with subtilisin-like serine protease and is the most important group of proteases (Cha et al., 2010). A fibrinolytic enzyme was purified from *P. pulmonarius* and effectively degraded α and β chains of fibrinogen substrate, followed by γ chain. Fibrinolytic enzyme also showed plasminogen-activator activity. A metalloprotease with fibrinolytic potential was purified and characterised from *P. ostreatus*. These fibrinolytic enzyme-producing mushrooms have the potential in thrombosis research and have the prospective to treat cardiovascular diseases (Shen et al., 2007). Recently, Katrolia et al. (2020) cloned and expressed fibrinolytic enzyme from the mushroom *Cordyceps militaris*. The recombinant fibrinolytic enzyme showed potent blood clot lysis activity.

6. Perspects and Future Challenges

Macrofungi are rich in high quality and quantity of industrially beneficial enzymes. In contrast to the huge number of enzymes isolated from bacteria as well as microfungi, only a few enzymes extracted from macrofungi are well known. More than 100,000 macrofungi have been reported; however, most of the studies focussed on the nutrition point than enzymatic point of view. Enzymes extracted from macrofungi are useful in paper, pulp, textile and petrochemical industries. These enzymes are used as cleaning agents in various purification systems, bioremediation of herbicides/ pesticides and applied in cosmetics. Cellullases and xylanases from macrofungi have the potential for the generation of biofuels. Degradation of ligninolytic biomass is a highly complex mechanism and multidimensional approach must be undertaken to analyse macrofungi and their evolution and mechanism. It is obvious that large-scale cellulases and xylanases can be produced through site-specific mutagenesis, random mutagenesis, look through mutagenesis, iterative saturation mutagenesis and combinatorial beneficial mutagenesis using the genetic engineering tool. It is very clear that concomitant production of xylanases and cellulases may improve various industrial processes by providing unanimous options for low-cost biofuel production from biomass. Therapeutic enzymes, such as fibrinolytic enzymes, showed lot of potential to treat cardiovascular diseases and tyrosinase effectively used to remove phenolic compounds from wastewater. Among all the properties, degradation of ligninolytic wastes, heavy metal degradation and wastewater treatment are important prospectives. In conclusion, various achievements have been made regarding macrofungal enzymes, especially ligninolytic enzymes. Except ligininolytic enzymes, commercial production of enzymes from macrofungi is very limited. Rapid screening and extraction of enzymes from macrofungi for various biotechnological applications should be the focus of future research.

References

Addleman, K. and Archibald, F. (1993). Kraft pulp bleaching and delignification by dikaryons and monokaryons of *Trametes versicolor*. Appl. Environ. Microbiol., 59(1): 266–273.

Agrawal, N., Verma, P., Singh, R.S. and Shahi, S.K. (2017). Ligninolytic enzyme production by white rot fungi Podoscypha elegans strain FTG4. Int. J. Curr. Microbiol. Appl. Sci., 6(5): 2757–2764.

Ahuja, S.K., Ferreira, G.M. and Moreira, A.R. (2004). Utilisation of enzymes for environmental applications. Critic. Rev. Biotechnol., 24: 125–154.

Ainsworth, G.C. (2008). Ainsworth and Bisby's Dictionary of the Fungi, Cabi.

Annunziatini, C., Baiocco, P., Gerini, M.F., Lanzalunga, O. and Sjögren, B. (2005). Aryl substituted N-hydroxyphthalimides as mediators in the laccase-catalysed oxidation of lignin model compounds and delignification of wood pulp. J. Mol. Catal. B Enzym., 32: 89–96.

Araujo, R., Casal, M. and Cavaco-Paulo, A. (2008). Application of enzymes for textile fibres processing. Biocatal. Biotransform., 26(5): 332–349.

Bezerra, J.D.P., Santos, M.G.S., Svedese, V.M., Lima, D.M.M., Fernandes, M.J.S., Paiva, L.M. and Souza-Motta, C.M. (2012). Richness of endophytic fungi isolated from *Opuntia ficus-indica* Mill. (Cactaceae) and preliminary screening for enzyme production. World J. Microbiol. Biotechnol., 28(5): 1989–1995.

Bouacem, K., Rekik, H., Jaouadi, N.Z., Zenati, B., Kourdali, S., El Hattab, M., Badis, A., Annane, R., Bejar, S., Hacene, H. and Bouanane-Darenfed, A. (2018). Purification and characterisation of two novel peroxidases from the dye-decolorising fungus *Bjerkandera adusta* strain CX-9. Int. J. Biol. Macromol., 106: 636–646.

Bourbonnais, R. and Paice, M.G. (1990). Oxidation of non-phenolic substrates: An expanded role for laccase in lignin biodegradation. FEBS Lett., 267(1): 99–102.

Bourbonnais, R., Paice, M.G., Reid, I.D., Lanthier, P. and Yaguchi, M. (1995). Lignin oxidation by laccase isozymes from *Trametes versicolor* and role of the mediator 2, 2'-azinobis (3-ethylbenzthiazoline-6-sulfonate) in kraft lignin depolymerisation. Appl. Environ. Microbiol., 61(5): 1876–80.

Cai, Y.J., Buswell, J.A. and Chang, S.T. (1998). β-glucosidase components of the cellulolytic system of the edible straw mushroom, *Volvariella volvacea*. Enz. Microb. Technol., 22: 122–129.

Camassola, M. and Dillon, A.J. (2009). Biological pretreatment of sugar cane bagasse for the production of cellulases and xylanases by *Penicillium echinulatum*. Ind. Crops Prod., 29: 642–647.

Campos, C., Dias, D.C., dos Santos, M.P., de Medeiros, C., do Valle, J.S., Vieira, E.S.N., Colauto, N.B. and Linde, G.A. (2011). Seleção de basidiomicetos proteolíticos. Arquivos de Ciências Veterinárias e Zoologia da UNIPAR, 14(1): 45–49.

Cao, W. and Crawford, D.L. (1993). Purification and some properties of β-glucosidase from the ectomycorrhizal fungus *Pisolithus tinctorius* strain SMF. Can. J. Microbiol., 39(1): 125–129.

Cardoso, W.S., Queiroz, P.V., Tavares, G.P., Santos, F.A., Soares, F.E., Kasuya, M.C. and Queiroz, J.H. (2018). Multi-enzyme complex of white rot fungi in saccharification of lignocellulosic material. Brazil. J. Microbiol., 49: 879–884.

Cha, W.S., Park, S.S., Kim, S.J. and Choi, D. (2010). Biochemical and enzymatic properties of a fibrinolytic enzyme from *Pleurotus eryngii* cultivated under solid-state conditions using corn cob. Bioresour. Technol., 101(16): 6475–6481.

Chan Cupul, W., Heredia Abarca, G., Martínez Carrera, D. and Rodríguez Vázquez, R. (2014). Enhancement of ligninolytic enzyme activities in a *Trametes maxima-Paecilomyces carneus* co-culture: Key factors revealed after screening using a Plackett-Burman experimental design. Electron. J. Biotechnol., 17(3): 114–121.

Chang, S.C. and Steinkraus, K.H. (1982). Lignocellulolytic enzymes produced by *Volvariella volvacea*, the edible straw mushroom. Appl. Environ. Microbiol., 43(2): 440–446.

Chaparro, D.F., Rosas, D.C. and Varela, A. (2009). Aislamiento y evaluación de la actividad enzimática de hongos descomponedores de madera (Quindío, Colombia). Rev. Iberoam. Micol., 26(4): 238–243.

Chen, S., Xu, J., Liu, C., Zhu, Y., Nelson, D.R., Zhou, S., Li, C., Wang, L., Guo, X., Sun, Y. and Luo, H. (2012). Genome sequence of the model medicinal mushroom *Ganoderma lucidum*. Nat. Com., 3(1): 1–9.

Chinedu, S.N., Okochi, V.I., Smith, H.A. and Omidiji, O. (2005). Isolation of cellulolytic microfungi involved in wood-waste decomposition: Prospects for enzymatic hydrolysis of cellulosic wastes. Int. J. Biomed. Health Sci., 1(2): 0794–4748.

Chiu, S.C. and Tzean, S.S. (1995). Glucanolytic enzyme production by *Schizophyllum commune* Fr. during mycoparasitism. Physiol. Mol. Plant Pathol., 46(2): 83–94.

Collopy, P.D. (2003). Characterisation of phytase activity from cultivated edible mushrooms and mushroom substrates, PhD Thesis, Pennsylvania State University, Pennsylvania.

Coniglio, R.O., Díaz, G.V., Fonseca, M.I., Castrillo, M.L., Piccinni, F.E., Villalba, L.L., Campos, E. and Zapata, P.D. (2020). Enzymatic hydrolysis of barley straw for biofuel industry using a novel strain of *Trametes villosa* from Paranaense rainforest. Prep. Biochem. Biotechnol., 1–10. https://doi.org/1 0.1080/10826068.2020.1734941.

Couto, S.R. and Herrera, J.L.T. (2006). Industrial and biotechnological applications of laccases: A review. Biotechnol. Adv., 24(5): 500–513.

Cui, L., Liu, Q.H., Wang, H.X. and Ng, T.B. (2007). An alkaline protease from fresh fruiting bodies of the edible mushroom *Pleurotus citrinopileatus*. Appl. Microbiol. Biotechnol., 75(1): 81–85.

Dashtban, M., Schraft, H. and Qin, W. (2009). Fungal bioconversion of lignocellulosic residues; opportunities and perspectives. Int. J. Biol. Sci., 5: 578–595.

Dass, S.B., Dosoretz, C.G., Reddy, C.A. and Grethlein, H.E. (1995). Extracellular proteases produced by the wood-degrading fungus *Phanerochaete chrysosporium* under ligninolytic and non-ligninolytic conditions. Arch. Microbiol., 163(4): 254–258.

De Jong, E.D., Field, J.A. and de Bont, J.A. (1994). Aryl alcohols in the physiology of ligninolytic fungi. FEMS Microbiol. Rev., 13: 153–188.

Deacon, J.W. (2006). Fungal Biology, 4th ed., Blackwell Publishing, Cornwell.

Demain, A.L. and Adrio, J.L. (2008). Contributions of microorganisms to industrial biology. Mol. Biotechnol., 38(1): 41–45.

Demierege, S., Toptas, A., Mavioglu Ayan, E., Yasa, I. and Yanik, J.A.L.E. (2015). Removal of textile dyes from aqueous solutions by biosorption on mushroom stump wastes. Chem. Ecol., 31(4): 365–378.

Denisova, N.P. (2010). History of the study of thrombolytic and fibrinolytic enzymes of higher Basidiomycetes mushrooms at the VL Komarov Botanical Institute in Saint Petersburg, Russia. Int. J. Med. Mush., 12(3): 317–325.

Deswal, D., Khasa, Y.P. and Kuhad, R.C. (2011). Optimisation of cellulase production by a brown rot fungus *Fomitopsis* sp. RCK2010 under solid state fermentation. Bioresour. Technol., 102(10): 6065–6072.

Dias, A.A., Freitas, G.S., Marques, G.S., Sampaio, A., Fraga, I.S., Rodrigues, M.A., Evtuguin, D.V. and Bezerra, R.M. (2010). Enzymatic saccharification of biologically pre-treated wheat straw with white-rot fungi. Bioresour. Technol., 101(15): 6045–6050.

Diez, V.A. and Alvarez, A. (2001). Compositional and nutritional studies on two wild edible mushrooms from northwest Spain. Food Chem., 75: 417–22.

Ding, S.J., Ge, W. and Buswell, J.A. (2001). Endoglucanase I from the edible straw mushroom, *Volvariella volvacea*. Purification, characterisation, cloning and expression. Eur. J. Biochem., 268(22): 5687–5695.

Duenas, F.J., Lundell, T., Floudas, D., Nagy, L.G., Barrasa, J.M., Hibbett, D.S. and Martínez, A.T. (2013). Lignin-degrading peroxidases in polyporales: An evolutionary survey based on 10 sequenced genomes. Mycologia, 105(6): 1428–1444.

Edwards, W., Leukes, W.D., Rose, P.D. and Burton, S.G. (1999). Immobilisation of polyphenol oxidase on chitosan-coated polysulphone capillary membranes for improved phenolic effluent bioremediation. Enz. Microb. Technol., 25(8-9): 769–773.

Eggert, C., Temp, U., Dean, J.F. and Eriksson, K.E.L. (1996). A fungal metabolite mediates degradation of non-phenolic lignin structures and synthetic lignin by laccase. FEBS Lett., 391: 144–148.

Eisele, N., Linke, D., Nimtz, M. and Berger, R.G. (2011). Heterologous expression, refolding and characterization of a salt activated subtilase from *Pleurotus ostreatus*. Proc. Biochem., 46(9): 1840–1846.

Elisashvili, V., Kachlishvili, E., Khardziani, T., Tsiklauri, N. and Bakradze, M. (2002). Physiological regulation of edible and medicinal higher Basidiomycetes lignocellulolytic enzyme activity. Int. J. Med. Mushroom., 4(2): 159–166.

Elisashvili, V., Kachlishvili, E. and Penninckx, M. (2008). Effect of growth substrate, method of fermentation, and nitrogen source on lignocellulose-degrading enzymes production by white-rot Basidiomycetes. J. Ind. Microbiol. Biotechnol., 35(11): 1531–1538.

Elisashvili, V., Kachlishvili, E., Tsiklauri, N., Metreveli, E., Khardziani, T. and Agathos, S.N. (2009). Lignocellulose-degrading enzyme production by white-rot Basidiomycetes isolated from the forests of Georgia. World J. Microbiol. Biotechnol., 25(2): 331–339.

Elisashvili, V.I., Khardziani, T.S., Tsiklauri, N.D. and Kachlishvili, E.T. (1999). Cellulase and xylanase activities in higher Basidiomycetes. Biochemistry New York, English Translation of Biokhimiya, 64(6): 718–722.

Erden, E., Ucar, M.C., Gezer, T. and Pazarlioglu, N.K. (2009). Screening for ligninolytic enzymes from autochthonous fungi and applications for decolourisation of Remazole Marine Blue. Braz. J. Microbiol., 40(2): 346–353.

Faraco, V., Palmieri, G., Festa, G., Monti, M., Sannia, G. and Giardina, P. (2005). A new subfamily of fungal subtilases: structural and functional analysis of a *Pleurotus ostreatus* member. Microbiology, 151(2): 457–466.

Faria, R.O., Rotuno Moure, V., Balmant, W., Lopes de Almeida Amazonas, M.A., Krieger, N. and Mitchell, D.A. (2007). The tyrosinase produced by *Lentinula boryana* (Berk. & Mont.) Pegler suffers substrate inhibition by L-DOPA. Food Technol. Biotechnol., 45(3): 334–340.

Félix, N.F.E., Vaca, M. and Izurieta, B. (2014). Obtención de extractos enzimáticos con actividad celulolítica y ligninolítica a partir del hongo *Pleurotus ostreatus* 404 y 2171 en rastrojo de maíz. Revista Politec., 33(1): 2.

Fen, L., Xuwei, Z., Nanyi, L., Puyu, Z., Shuang, Z., Xue, Z., Pengju, L., Qichao, Z. and Haiping, L. (2014). Screening of lignocellulose-degrading superior mushroom strains and determination of their CMCase and laccase activity. Sci. World J., 763108. 10.1155/2014/763108.

Floudas, D., Binder, M., Riley, R., Barry, K., Blanchette, R.A., Henrissat, B., Martínez, A.T., Otillar, R., Spatafora, J.W., Yadav, J.S. and Aerts, A. (2012). The Paleozoic origin of enzymatic lignin decomposition reconstructed from 31 fungal genomes. Science, 336(6089): 1715–1719.

Gelb, D.J., Oliver, E. and Gilman, S. (1999). Diagnostic criteria for Parkinson's disease. Arch. Neurol., 56(1): 33–39.

Ghose, T.K. and Bisaria, V.S. (1979). Studies on the mechanism of enzymatic hydrolysis of cellulosic substances. Biotechnol. Bioeng., 21(1): 131–146.

Gianfreda, L., Xu, F. and Bollag, J.M. (1999). Laccases: A useful group of oxidoreductive enzymes. Bioremed. J., 3(1): 1–26.

Giorgio, M.E., Villalba, L.L., Robledo, G.L., Zapata, P.D. and Saparrat, M.C.N. (2018). Cellulolytic ability of a promising *Irpex lacteus* (Basidiomycota: Polyporales) strain from the subtropical rainforest of Misiones Province, Argentina. Revisit. Biol. Tropic., 66: 1034–1045.

Glenn, J.K. and Gold, M.H. (1985). Purification and characterisation of an extracellular Mn (II)-dependent peroxidase from the lignin-degrading Basidiomycete, *Phanerochaete chrysosporium*. Arch. Biochem. Biophy., 242(2): 329–341.

Gomes, E., Aguiar, A.P., Carvalho, C.C., Bonfá, M.R.B., Silva, R.D. and Boscolo, M. (2009). Ligninases production by Basidiomycetes strains on lignocellulosic agricultural residues and their application in the decolourisation of synthetic dyes. Braz. J. Microbiol., 40: 31–39.

Goud, M.J.P., Suryam, A., Lakshmipathi, V. and Charya, M.S. (2009). Extracellular hydrolytic enzyme profiles of certain South Indian Basidiomycetes. Afr. J. Biotechnol., 8(3): 354–360.

Gutiérrez, A., José, C. and Martínez, A.T. (2009). Microbial and enzymatic control of pitch in the pulp and paper industry. Appl. Microbiol. Biotechnol., 82(6): 1005–1018.

Gutiérrez-Soto, G., Medina-Gonzalez, G.E., Treviño-Ramírez, J.E. and Hernández-Luna, C.E. (2015). Native macrofungi that produce lignin-modifying enzymes, cellulases, and xylanases with potential biotechnological applications. BioResources, 10(4): 6676–6689.

Halaouli, S., Asther, M., Kruus, K., Guo, L., Hamdi, M., Sigoillot, J.C., Asther, M. and Lomascolo, A. (2005). Characterisation of a new tyrosinase from *Pycnoporus* species with high potential for food technological applications. J. Appl. Microbio., 98(2): 332–343.

Halaouli, S., Asther, M., Sigoillot, J.C., Hamdi, M. and Lomascolo, A. (2006). Fungal tyrosinases: New prospects in molecular characteristics, bioengineering and biotechnological applications. J. Appl. Microbiol., 100: 219–232.

Hammel, K.E., Jensen, K.A., Mozuch, M.D., Landucci, L.L., Tien, M. and Pease, E.A. (1993). Ligninolysis by a purified lignin peroxidase. J. Biol. Chem., 268(17): 12274–12281.

Haq, I.U. and Ali, S. (2006). Mutation of *Aspergillus oryzae* for improved production of 3, 4-dihydroxy Phenyl-L-Alanine (L-Dopa) from L-Tyrosine. Braz. J. Microbiol., 37(1): 78–86.

Healy, V., O'Connell, J., McCarthy, T.V. and Doonan, S. (1999). The lysine-specific proteinase from *Armillaria mellea* is a member of a novel class of metalloendopeptidases located in Basidiomycetes. Biochem. Biophy. Res. Commun., 262(1): 60–63.

Henrissat, B. (1991). A classification of glycosyl hydrolases based on amino acid sequence similarities. Biochem. J., 280(2): 309–316.

Hofrichter, M. (2002). Lignin conversion by manganese peroxidase (MnP). Enzyme Microb. Technol., 30: 454–466.

Hongbo, C.Y.Y. (2009). Detection on laccase, manganese peroxidase and lignin peroxidase in ligninolytic enzymes of *Pleurotus djamor* [J]. Scientia Silvae Sinicae, 45(12): 154–158.

Hu, D.D., Zhang, R.Y., Zhang, G.Q., Wang, H.X. and Ng, T.B. (2011). A laccase with antiproliferative activity against tumor cells from an edible mushroom, white common *Agrocybe cylindracea*. Phytomedicine, 18(5): 374–379.

Hur, T.C., Ka, K.H., Joo, S.H. and Terashita, T. (2001). Characteristics of the amylase and its related enzymes produced by ectomycorrhizal fungus *Tricholoma matsutake*. Mycobiology, 29(4): 183–189.

Iandolo, D., Piscitelli, A., Sannia, G. and Faraco, V. (2011). Enzyme production by solid substrate fermentation of *Pleurotus ostreatus* and *Trametes versicolor* on tomato pomace. Appl. Biochem. Biotechnol., 163(1): 40–51.

Inácio, F.D., Martins, A.F., Contato, A.G., Brugnari, T., Peralta, R.M. and de Souza, C.G.M., (2018). Biodegradation of human keratin by protease from the Basidiomycete *Pleurotus pulmonarius*. Int. Biodet. Biodegrad., 127: 124–129.

Jakucs, E., Racz, I. and Lasztity, D. (1994). Some characteristics and partial purification of the *Ganoderma lucidum* cellulase system. Acta Microbiol. Immunol. Hung., 41: 23–31.

James, J.A. and Lee, B.H. (1997). Glucoamylases: Microbial sources, industrial applications and molecular biology—A review. J. Food Biochem., 21: 1–52.

Jaouani, A., Guillén, F., Penninckx, M.J., Martínez, A.T. and Martínez, M.J. (2005). Role of *Pycnoporus coccineus* laccase in the degradation of aromatic compounds in olive oil mill wastewater. Enzyme Microb. Technol., 36: 478–86.

Jimenez, M., Gonzalez, A.E., Martinez, M.J., Martinez, A.T. and Dale, B.E. (1991). Screening of yeasts isolated from decayed wood for lignocellulose-degrading enzyme activities. Mycol. Res., 95(11): 1299–1302.

Joh, J.H., Kim, B.G., Kong, W.S., Yoo, Y.B., Kim, N.K., Park, H.R., Cho, B.G. and Lee, C.S. (2004). Cloning and developmental expression of a metzincin family metalloprotease cDNA from oyster mushroom *Pleurotus ostreatus*. FEMS Microbiol. Lett., 239(1): 57–62.

Jonathan, S.G. and Adeoyo, O.R. (2011). Evaluation of ten wild Nigerian mushrooms for amylase and cellulase activities. Mycobiology, 39(2): 103–108.

Kadri, T., Rouissi, T., Brar, S.K., Cledon, M., Sarma, S. and Verma, M. (2017). Biodegradation of polycyclic aromatic hydrocarbons (PAHs) by fungal enzymes: A review. J. Environ. Sci., 51: 52–74.

Kalmis, E., Yasa, I., Kalyoncu, F., Pazarbasi, B. and Koçyigit, A. (2008). Ligninolytic enzyme activities in mycelium of some wild and commercial mushrooms. Afr. J. Biotechnol., 23(7): 4314–4320.

Kamitsuji, H., Watanabe, T., Honda, Y. and Kuwahara, M. (2005). Direct oxidation of polymeric substrates by multifunctional manganese peroxidase isoenzyme from *Pleurotus ostreatus* without redox mediators. Appl. Microbiol. Biotechnol., 65: 287–294.

Kanda, K., Sato, T., Ishii, S., Enei, H. and Ejiri, S.I. (1996). Purification and properties of tyrosinase isozymes from the gill of *Lentinus edodes* fruiting body. Biosci. Biotechnol. Biochem., 60(8): 1273–1278.

Kapoor, S., Khanna, P.K. and Katyal, P. (2009). Effect of supplementation of wheat straw on growth and lignocellulolytic enzyme potential of *Lentinus edodes*. World J. Agric. Sci., 5(3): 328–331.

Karamyshev, A.V., Shleev, S.V., Koroleva, O.V., Yaropolov, A.I. and Sakharov, I.Y. (2003). Laccase-catalysed synthesis of conducting polyaniline. Enzyme Microb. Technol., 33: 556–564.

Katrolia, P., Liu, X., Zhao, Y., Kopparapu, N.K. and Zheng, X. (2020). Gene cloning, expression and homology modeling of first fibrinolytic enzyme from mushroom (*Cordyceps militaris*). Int. J. Biol. Macromol., 146: 897–906.

Khushal, B., Anne, R. and Praveen, V.V. (2010). Fungal laccases: production, function, and applications in food processing. Enzyme Res. 2010(149748): 1–10.

Kim, J.S., Sapkota, K., Park, S.E., Choi, B.S., Kim, S., Hiep, N.T., Kim, C.S., Choi, H.S., Kim, M.K., Chun, H.S. and Park, Y. (2006). A fibrinolytic enzyme from the medicinal mushroom *Cordyceps militaris*. J. Microbiol., 44(6): 622–631.

Ko, E.M., Leem, Y.E. and Choi, H. (2001). Purification and characterisation of laccase isozymes from the white-rot Basidiomycete *Ganoderma lucidum*. Appl. Microbiol. Biotechnol., 57: 98–102.

Kobakhidze, A., Elisashvili, V., Irbe, I., Tsiklauri, N., Andersone, I., Andersons, B. and Isikhuemhen, O.S. (2012). Lignocellulolytic enzyme activity of new corticoid and poroid Basidiomycetes isolated from Latvian cultural monuments. J. Waste Con. Bioprod. Biotechnol., 1: 16–21.

Koide, K., Osono, T. and Takeda, H. (2005). Fungal succession and decomposition of *Camellia japonica* leaf litter. Ecol. Res., 20(5): 599–609.

Krupodorova, T., Ivanova, T. and Barshteyn, V. (2014). Screening of extracellular enzymatic activity of macrofungi. J. Microbiol. Biotechnol. Food Sci., 3(4): 315–318.

Kudryavtseva, O.A., Dunaevsky, Y.E., Kamzolkina, O.V. and Belozersky, M.A. (2008). Fungal proteolytic enzymes: features of the extracellular proteases of xylotrophic Basidiomycetes. Microbiology, 77(6): 643–653.

Kuhara, F. and Papinutti, L. (2014). Optimisation of laccase production by two strains of *Ganoderma lucidum* using phenolic and metallic inducers. Rev. Arg. Microbiol., 46(2): 144–149.

Kumari, B., Upadhyay, R.C. and Atri, N.S. (2012). Screening and evaluation of extra-cellular oxidases in some *Termitophilous* and lepiotoid mushrooms. World J. Agric. Sci., 8: 409–414.

Kupper, U., Niedermann, D.M., Travaglini, G. and Lerch, K. (1989). Isolation and characterization of the tyrosinase gene from *Neurospora crassa*. J. Biol. Chem., 264(29): 17250–17258.

Kusuda, M., Nagai, M., Hur, T.C., Terashita, T. and Ueda, M. (2003). Purification and some properties of α-amylase from an ectomycorrhizal fungus, *Tricholoma matsutake*. Mycoscience, 44(4): 311–317.

Kuuskeri, J., Mäkelä, M.R., Isotalo, J., Oksanen, I. and Lundell, T. (2015). Lignocellulose-converting enzyme activity profiles correlate with molecular systematics and phylogeny grouping in the incoherent genus *Phlebia* (Polyporales, Basidiomycota). BMC Microbial., 15(217): 1–18.

Lee, C.Y., Hong, O.P., Jung, M.J. and Han, Y.H. (1998). The extracellular enzyme activities in culture broth of *Tricholoma matsutake*. Kor. J. Mycol., 26: 496–501.

Leonowicz, A., Matuszewska, A., Luterek, J., Ziegenhagen, D., Wojtaś-Wasilewska, M., Cho, N.S., Hofrichter, M. and Rogalski, J. (1999). Biodegradation of lignin by white rot fungi. Fungal Genet. Biol., 27(2-3): 175–185.

Liers, C., Ullrich, R., Steffen, K.T., Hatakka, A. and Hofrichter, M. (2006). Mineralisation of 14 C-labelled synthetic lignin and extracellular enzyme activities of the wood-colonising Ascomycetes *Xylaria hypoxylon* and *Xylaria polymorpha*. Appl. Microbiol. Biotechnol., 69(5): 573–579.

Liu, D., Gong, J., Dai, W., Kang, X., Huang, Z., Zhang, H.M., Liu, W., Liu, L., Ma, J., Xia, Z. and Chen, Y. (2012). The genome of *Ganderma lucidum* provide insights into triterpense biosynthesis and wood degradation. PloS ONE, 7(5): e36146.

Liu, X.L., Zheng, X.Q., Qian, P.Z., Kopparapu, N.K., Deng, Y.P., Nonaka, M. and Harada, N. (2014). Purification and characterisation of a novel fibrinolytic enzyme from culture supernatant of *Pleurotus ostreatus*. J. Microbiol. Biotechnol., 24(2): 245–253.

Lu, L., Zhao, M., Zhang, B.B., Yu, S.Y., Bian, X.J., Wang, W. and Wang, Y. (2007). Purification and characterization of laccase from *Pycnoporus sanguineus* and decolorization of an anthraquinone dye by the enzyme. Appl. Microbiol. Biotechnol., 74(6): 1232–1239.

Lynd, L.R., Weimer, P.J., Van Zyl, W.H. and Pretorius, I.S. (2002). Microbial cellulose utilisation: Fundamentals and biotechnology. Microbiol. Mol. Biol. Rev., 66(3): 506–577.

Ma, C., Ni, X., Chi, Z., Ma, L. and Gao, L. (2007). Purification and characterisation of an alkaline protease from the marine yeast *Aureobasidium pullulans* for bioactive peptide production from different sources. Mar. Biotechnol., 9: 343–351.

Madhavi, V. and Lele, S.S. (2009). Laccase: Properties and applications. Bioresour., 4(4): 1694–1717.

Maijala, P., Fagerstedt, K.V. and Raudaskoski, M. (1991). Detection of extracellular cellulolytic and proteolytic activity in ectomycorrhizal fungi and *Heterobasidion annosum* (Fr.) Bref. New Phytol., 117(4): 643–648.

Mainardi, P.H., Feitosa, V.A., de Paiva, L.B.B., Bonugli-Santos, R.C., Squina, F.M., Pessoa Jr, A. and Sette, L.D. (2018). Laccase production in bioreactor scale under saline condition by the marine-derived Basidiomycete *Peniophora* sp. CBMAI 1063. Fungal Biol., 122(5): 302–309.

Mali, T., Kuuskeri, J., Shah, F. and Lundell, T.K. (2017). Interactions affect hyphal growth and enzyme profiles in combinations of coniferous wood-decaying fungi of Agaricomycetes. PLoS ONE, 12(9): 1–21.

Manavalan, T., Manavalan, A., Thangavelu, K.P. and Heese, K. (2015). Characterisation of a novel endoglucanase from *Ganoderma lucidum*. J. Basic Microbiol., 55: 761–771.

Martirani, L., Giardina, P., Marzullo, L. and Sannia, G. (1996). Reduction of phenol content and toxicity in olive oil mill waste waters with the ligninolytic fungus *Pleurotus ostreatus*. Water Res., 30(8): 1914–1918.

Mayer, A.M. and Staples, R.C. (2002). Laccase: New functions for an old enzyme. Phytochemistry, 60: 551–565.

Melo, E.B., da Silveira Gomes, A. and Carvalho, I. (2006). α-and β-Glucosidase inhibitors: Chemical structure and biological activity. Tetrahedron, 62(44): 10277–10302.

Miles, P.G. and Chang, S.T. (2004). Mushrooms: Cultivation, Nutritional Value, Medicinal Effect and Environmental Impact, CRC press, Boca Raton, Forida.

Moiseenko, K.V., Vasina, D.V., Farukshina, K.T., Savinova, O.S., Glazunova, O.A., Fedorova, T.V. and Tyazhelova, T.V. (2018). Orchestration of the expression of the laccase multigene family in white-rot Basidiomycete *Trametes hirsuta* 072: evidences of transcription level subfunctionalisation. Fungal Biol., 122: 353–362.

Montoya, S., Orrego, C.E. and Levin, L. (2012). Growth, fruiting and lignocellulolytic enzyme production by the edible mushroom *Grifola frondosa* (maitake). World J. Microbiol. Biotechnol., 28(4): 1533–1541.

Moredo, N., Lorenzo, M., Domínguez, A., Moldes, D., Cameselle, C. and Sanroman, A. (2003). Enhanced ligninolytic enzyme production and degrading capability of *Phanerochaete chrysosporium* and *Trametes versicolor.* World J. Microbiol. Biotechnol., 19(7): 665–669.

Mtui, G.Y. (2012). Lignocellulolytic enzymes from tropical fungi: Types, substrates and applications. Sci. Res. Essays, 7(15): 1544–1555.

Mucha, J., Dahm, H., Strzelczyk, E. and Werner, A. (2006). Synthesis of enzymes connected with mycoparasitism by ectomycorrhizal fungi. Arch. Microbiol., 185(1): 69–77.

Mueller, G.M. and Schmit, J.P. (2007). Fungal biodiversity: What do we know? What can we predict? Biodivers. Conserv., 16(1): 1–5.

Mueller, L.A., Hinz, U. and Zrÿd, J.P. (1996). Characterisation of a tyrosinase from *Amanita muscaria* involved in betalain biosynthesis. Phytochem., 42(6): 1511–1516.

Mukchaylova, O.B. and Buchalo, A.S. (2012). Ascomycetes fungi of the genus *Morchella Dill. In*: Biological Properties of Medicinal Macromycetes in Culture (in Russian), Institute of Botany, Kyiv, Russia.

Nagai, M., Sato, T., Watanabe, H., Saito, K., Kawata, M. and Enei, H. (2002). Purification and characterization of an extracellular laccase from the edible mushroom *Lentinula edodes*, and decolourisation of chemically different dyes. Appl. Microbiol. Biotechnol., 60(3): 327–335.

Negrão, D.R., da Silva Júnior, T.A.F., de Souza Passos, J.R., Sansígolo, C.A., de Almeida Minhoni, M.T. and Furtado, E.L. (2014). Biodegradation of *Eucalyptus urograndis* wood by fungi. Int. Biodeterior. Biodegrad., 89: 95–102.

Ntougias, S., Baldrian, P., Ehaliotis, C., Nerud, F., Merhautová, V. and Zervakis, G.I. (2015). Olive mill wastewater biodegradation potential of white-rot fungi—Mode of action of fungal culture extracts and effects of ligninolytic enzymes. Bioresour. Technol., 189: 121–130.

Osma, J.F., Moilanen, U., Toca-Herrera, J.L. and Rodríguez-Couto, S. (2011). Morphology and laccase production of white-rot fungi grown on wheat bran flakes under semi-solid-state fermentation conditions. FEMS Microbiol. Lett., 318(1): 27–34.

Osono, T. and Takeda, H. (2002). Comparison of litter decomposing ability among diverse fungi in a cool temperate deciduous forest in Japan. Mycologia, 94(3): 421–427.

Palmieri, G., Giardina, P., Desiderio, B., Marzullo, L., Giamberini, M. and Sannia, G. (1994). A new enzyme immobilization procedure using copper alginate gel: Application to a fungal phenol oxidase. Enz. Microb. Technol., 16: 151–158.

Pandey, A., Nigam, P., Soccol, C.R., Soccol, V.T., Singh, D. and Mohan, R. (2000). Advances in microbial amylases. Biotechnol. Appl. Biochem., 31: 135–152.

Papinutti, L. and Lechner, B. (2008). Influence of the carbon source on the growth and lignocellulolytic enzyme production by *Morchella esculenta* strains. J. Ind. Microbiol. Biotechnol., 35(12): 1715–1721.

Park, K.B., Ha, H.C., Kim, S.Y., Kim, H.J. and Lee, J.S. (2002). Induction of anticarcinogenic enzymes of waxy brown rice cultured with *Phellinus igniarius* 26005. Mycobiology, 30(4): 213–218.

Pelaez, F., Martinez, M.J. and Martinez, A.T. (1995). Screening of 68 species of Basidiomycetes for enzymes involved in lignin degradation. Mycol. Res., 99(1): 37–42.

Philippidis, G.P. (1994). Enzymatic conversion of biomass for fuel production. pp. 188–217. *In*: Himmel, M.E., Baker, J.O. and Overend, R.P. (eds.). Cellulase Production Technology, ACS Symposium Series, vol. # 566.

Polizeli, M.L.T.M., Rizzatti, A.C.S., Monti, R., Terenzi, H.F., Jorge, J.A. and Amorim, D.S. (2005). Xylanases from fungi: Properties and industrial applications. Appl. Microbiol. Biotechnol., 67(5): 577–591.

Qing, Q. and Wyman, C.E. (2011). Hydrolysis of different chain length xylooliogmers by cellulase and hemicellulase. Bioresour. Technol., 102: 1359–1366.

Quiroz-Castañeda, R.E., Balcázar-López, E., Dantán-González, E., Martinez, A., Folch-Mallol, J. and Martínez Anaya, C. (2009). Characterisation of cellulolytic activities of *Bjerkandera adusta* and *Pycnoporus sanguineus* on solid wheat straw medium. Elec. J. Biotechnol., 12(4): 5–6.

Ramírez-Anguiano, A.C., Santoyo, S., Reglero, G. and Soler-Rivas, C. (2007). Radical scavenging activities, endogenous oxidative enzymes and total phenols in edible mushrooms commonly consumed in Europe. J. Sci. Food Agric., 87(12): 2272–2278.

Rana, I.S. and Rana, A.S. (2011). Lignocellulolytic enzyme profile of *Agaricus* and *Pleurotus* species cultured on used tea leaves substrate. Adv. Biotech., 11(6): 10–14.

Rana, S. and Kaur, M. (2012). Isolation and screening of cellulase producing microorganisms from degraded wood. Int. J. Pharm. Biol. Sci. Fund., 2: 10–5.

Rao, M.B., Tanksale, A.M., Ghatge, M.S. and Deshpande, V.V. (1998). Molecular and biotechnological aspects of microbial proteases. Microbiol. Mol. Biol. Rev., 62: 597–635.

Reid, I.D. and Paice, M.G. (1994). Biological bleaching of kraft pulps by white-rot fungi and their enzymes. FEMS Microbiol. Rev., 13(2-3): 369–375.

Reinhold, D.F., Utne, T., Abramson, N.L., Merck and Co. Inc. (1987). Process for L-DOPA. US Patent # 4,716,246.

Riley, R., Salamov, A.A., Brown, D.W., Nagy, L.G., Floudas, D., Held, B.W., Levasseur, A., Lombard, V., Morin, E., Otillar, R. and Lindquist, E.A. (2014). Extensive sampling of Basidiomycete genomes demonstrates inadequacy of the white-rot/brown-rot paradigm for wood decay fungi. Proc. Nat. Acad. Sci., 111(27): 9923–9928.

Schneider, W.D.H., Fontana, R.C., Mendonça, S., de Siqueira, F.G., Dillon, A.J.P. and Camassola, M. (2018). High level production of laccases and peroxidases from the newly isolated white-rot Basidiomycete *Marasmiellus palmivorus* VE111 in a stirred-tank bioreactor in response to different carbon and nitrogen sources. Process Biochem., 69: 1–11.

Schoemaker, H.E., Harvey, P.J., Bowen, R.M. and Palmer, J.M. (1985). On the mechanism of enzymatic lignin breakdown. FEBS Lett., 183(1): 7–12.

Schoemaker, H.E. and Piontek, K. (1996). On the interaction of lignin peroxidase with lignin. Pure Appl. Chem., 68(11): 2089–2096.

Sengupta, S., Ghosh, A.K. and Sengupta, S. (1991). Purification and characterisation of a β-glucosidase (cellobiase) from a mushroom *Termitomyces clypeatus*. Biochim. Biophys. Acta, 1076(2): 215–220.

Senn-Irlet, B., Heilmann-Clausen, J., Genney, D. and Dahlberg, A. (2007). Guidance for Conservation of Macrofungi in Europe, ECCF, Strasbourg.

Seo, S.Y., Sharma, V.K. and Sharma, N. (2003). Mushroom tyrosinase: Recent prospects. J. Agric. Food Chem., 51(10): 2837–2853.

Shaginian, K.A., Alekhina, I.A. and Denisova, N.P. (1990). Serine proteinase from the higher Basidiomycetes of *Coprinus* genus, *Biokhimiia* (Moscow, Russia), 55(8): 1387–1395.

Shaheen, R., Asgher, M., Hussain, F. and Bhatti, H.N. (2017). Immobilised lignin peroxidase from *Ganoderma lucidum* IBL-05 with improved dye decolourisation and cytotoxicity reduction properties. Int. J. Biol. Macromol., 103: 57–64.

Shen, M.H., Kim, J.S., Sapkota, K., Park, S.E., Choi, B.S., Kim, S., Lee, H.H., Kim, C.S., Chun, H.S., Ryoo, C.I. and Kim, S.J. (2007). Purification, characterisation, and cloning of fibrinolytic metalloprotease from *Pleurotus ostreatus* mycelia. J. Microbiol. Biotechnol., 17(8): 1271–1283.

Shu, C.H., Xu, C.J. and Lin, G.C. (2006). Purification and partial characterisation of a lipase from *Antrodia cinnamomea*. Proc. Biochem., 41(3): 734–738.

Soler-Rivas, C., Jolivet, S., Arpin, N., Olivier, J.M. and Wichers, H.J. (1999). Biochemical and physiological aspects of brown blotch disease of *Agaricus bisporus*. FEMS Microbiol. Rev., 23(5): 591–614.

Steffen, K.T., Cajthaml, T., Šnajdr, J. and Baldrian, P. (2007). Differential degradation of oak (*Quercus petraea*) leaf litter by litter-decomposing Basidiomycetes. Res. Microbiol., 158(5): 447–455.

Sukarta, I.N. and Sastrawidana, I.D.K. (2014). The use of agricultural waste to increase the production ligninolytic enzyme by fungus *Polyporus* sp. Open Acc. Lib. J., 1(642): 1–7.

Sumantha, A., Larroche, C. and Pandey, A. (2006). Microbiology and industrial biotechnology of food-grade proteases: A perspective. Food Technol. Biotechnol., 44(2): 211–220.

Sundaramoorthy, M., Kishi, K., Gold, M.H. and Poulos, T.L. (1997). Crystal structures of substrate binding site mutants of manganese peroxidase. J. Biol. Chem., 272(28): 17574–17580.

Suzuki, M.R., Hunt, C.G., Houtman, C.J., Dalebroux, Z.D. and Hammel, K.E. (2006). Fungal hydroquinones contribute to brown rot of wood. Environ. Microbiol., 8: 2214–2223.

Teeri, T.T. (1997). Crystalline cellulose degradation: New insight into the function of cellobiohydrolases. Trend. Biotechnol., 15(5): 160–167.

Terashita, T., Nakaie, Y., Inoue, T., Yoshikawa, K. and Shishiyama, J. (1998). Role of metal proteinases in the fruit-body formation of *Hypsizygus marmoreus*. J. Wood Sci., 44(5): 379–384.

Terashita, T. (2000a). Amylase productions during the vegetative mycelial growth of *Lyophyllum shimeji*. Mush. Soc. Biotechnol., 8: 61–69.

Terashita, T. (2000b). Production of extracellular amylase from *Tricholoma matsutake* and its properties on starch hydrolysis. Mush. Soc. Biotechnol., 8: 115–120.

Thakur, S. (2012). Lipases, its sources, properties and applications: A review. Int. J. Sci. Eng. Res., 3(7): 1–29.

Thurston, C.F. (1994). The structure and function of fungal laccases. Microbiol., 140(1): 19–26.

Timur, S., Pazarlıoğlu, N., Pilloton, R. and Telefoncu, A. (2004). Thick film sensors based on laccases from different sources immobilized in polyaniline matrix. Sens Actuators B., 97: 132–136.

Trivedi, N., Reddy, C.R.K., Radulovich, R. and Jha, B. (2015). Solid state fermentation (SSF)-derived cellulase for saccharification of the green seaweed *Ulva* for bioethanol production. Algal Res., 9: 48–54.

Tuomela, M. and Hatakka, A. (2019). Oxidative fungal enzymes for bioremediation. pp. 224–239. *In*: Comprehensive Biotechnology: Environmental and Related Biotechnologies, 3rd ed., Elsevier.

Uday, U.S.P., Choudhury, P., Bandyopadhyay, T.K. and Bhunia, B. (2016). Classification, mode of action and production strategy of xylanase and its application for biofuel production from water hyacinth. Int. J. Biol. Macromol., 82: 1041–1054.

Ullah, M.A., Camacho, R., Evans, C.S. and Hedger, J.N. (2002). Production of ligninolytic enzymes by species assemblages of tropical higher fungi from Ecuador. Trop. Mycol., 1: 101–112.

Valaškova, V. and Baldrian, P. (2006). Degradation of cellulose and hemicelluloses by the brown rot fungus *Piptoporus betulinus*—Production of extracellular enzymes and characterization of the major cellulases. Microbiol., 152(12): 3613–3622.

Valášková, V. and Baldrian, P. (2006a). Estimation of bound and free fractions of lignocellulose-degrading enzymes of wood-rotting fungi *Pleurotus ostreatus*, *Trametes versicolor* and *Piptoporus betulinus*. Res. Microbiol., 157(2): 119–124.

Valášková, V., Šnajdr, J., Bittner, B., Cajthaml, T., Merhautová, V., Hofrichter, M. and Baldrian, P. (2007). Production of lignocellulose-degrading enzymes and degradation of leaf litter by saprotrophic Basidiomycetes isolated from a *Quercus petraea* forest. Soil Biol. Biochem., 39(10): 2651–2660.

Valeriano, V.S., Silva, A.M.F., Santiago, M.F., Bara, M.T. and Garcia, T.A. (2009). Production of laccase by *Pynoporus sanguineus* using 2, 5-Xylidine and ethanol. Braz. J. Microbiol., 40: 790–794.

Větrovský, T., Baldrian, P. and Gabriel, J. (2013). Extracellular enzymes of the white-rot fungus *Fomes fomentarius* and purification of 1, 4-β-glucosidase. Appl. Biochem. Biotechnol., 169(1): 100–109.

Wang, S.X., Liu, Y., Zhang, G.Q., Zhao, S., Xu, F., Geng, X.L. and Wang, H.X. (2012). Cordysobin, a novel alkaline serine protease with HIV-1 reverse transcriptase inhibitory activity from the medicinal mushroom *Cordyceps sobolifera*. J. Biosci. Bioeng., 113(1): 42–47.

Wasser, S.P. and Weis, A.L. (1999). Therapeutic effects of substances occurring in higher Basidiomycetes mushrooms: a modern perspective. Cri. Rev. Immunol., 19(1): 65–96.

Watkinson, S.C., Burton, K.S. and Wood, D.A. (2001). Characteristics of intracellular peptidase and proteinase activities from the mycelium of a cord-forming wood decay fungus, *Serpula lacrymans*. Mycol. Res., 105(6): 698–704.

Wu, F., Zhou, L.W., Yang, Z.L., Bau, T., Li, T.H. and Dai, Y.C. (2019). Resource diversity of Chinese macrofungi: Edible, medicinal and poisonous species. Fungal Div., 1–76.

Wu, Y. and Shin, H.J. (2016). Cellulase from the fruiting bodies and mycelia of edible mushrooms: A review. J. Mushroom., 14(4): 127–135.

Xia, W., Xu, X., Qian, L., Shi, P., Bai, Y., Luo, H., Ma, R. and Yao, B. (2016). Engineering a highly active thermophilic β-glucosidase to enhance its pH stability and saccharification performance. Biotechnol. Biofuel., 9(1): 1–12.

Xu, X., Lin, M., Zang, Q. and Shi, S. (2018). Solid state bioconversion of lignocellulosic residues by Inonotus obliquus for production of cellulolytic enzymes and saccharification. Bioresour. Technol., 247: 88–95.

Yin, C., Zheng, L., Chen, L., Tan, Q., Shang, X. and Ma, A. (2014). Cloning, expression, and characterisation of a milk-clotting aspartic protease gene (Po-Asp) from *Pleurotus ostreatus*. Appl. Biochem. Biotechnol., 172(4): 2119–2131.

Yoon, J.J. and Kim, Y.K. (2005). Degradation of crystalline cellulose by the brown-rot Basidiomycete *Fomitopsis palustris*. J. Microbiol., 43(6): 487–492.

Yoon, J.J., Cha, C.J., Kim, Y.S. and Kim, W. (2008). Degradation of cellulose by the major endoglucanase produced from the brown-rot fungus *Fomitopsis pinicola*. Biotechnol. Lett., 30(8): 1373–1378.

Yoshida, H., Fujimoto, S. and Hayashi, J. (1994a). Nutritional requirements for the vegetative growth of *Lyophyllum shimeji*. Trans. Mycol. Soc. Jpn., 35: 89–96.

You, L.F., Liu, Z.M., Lin, J.F., Guo, L.Q., Huang, X.L. and Yang, H.X. (2014). Molecular cloning of a laccase gene from *Ganoderma lucidum* and heterologous expression in *Pichia pastoris*. J. Basic Microbiol., 54(Suppl. 1): S134–S141.

Zhang, K. (2012). Production and characterisation of thermostable laccase from the mushroom, *Ganoderma lucidum*, using submerged fermentations. Afr. J. Microbiol. Res., 6: 1147–57.

Zhang, Y.H.P. and Lynd, L.R. (2006). A functionally based model for hydrolysis of cellulose by fungal cellulase. Biotechnol. Bioeng., 94(5): 888–898.

Zhang, X., Liu, Q., Zhang, G., Wang, H. and Ng, T. (2010). Purification and molecular cloning of a serine protease from the mushroom *Hypsizigus marmoreus*. Process Biochem., 45(5): 724–730.

Zhao, S., Rong, C.B., Kong, C., Liu, Y., Xu, F., Miao, Q.J., Wang, S.X., Wang, H.X. and Zhang, G.Q. (2014). A novel laccase with potent antiproliferative and HIV-1 reverse transcriptase inhibitory activities from mycelia of mushroom *Coprinus comatus*. BioMed Res. Int., 2014: Article ID 417461.

12

Enzymes of White Rot Fungi and their Industrial Potential

Deepak K Rahi,[1,]* *Sonu Rahi*[2] *and Ekta Chaudhary*[1]

1. INTRODUCTION

Fungi, involved in degradation of wood, have been classified into three major catagories—white rot, brown rot and the soft rot fungi, depending on the nature of wood degradation. White rot and brown rot are caused by basidiomycetous fungi, whereas soft rot by ascomycetous fungi (Hatakka, 2001). White rot fungi are capable of decomposing almost all wood fractions, including lignin and leave the wood with a white and fibrous appearance. These fungi grow on hardwoods, like ash, aspen, beech, birch, hickory, magnolia and oak. However, certain white rot fungi attack softwoods, such as cedar, Douglas fir, larch, pine, redwood and spruce (Blanchette, 1995). White rot fungi degrade lignin with two modes of action, namely selective and non-selective decays (Blanchette, 1995). Selectivity of white rot fungi regarding lignin degradation depends on the nature of lignocelluloses, cultivation time and other factors (Hatakka and Hammel, 2011). Some examples of white rot fungi responsible for selective decay of woody material are *Ceriporiopsis subvermispora, Dichomitus squalens, Phaenerochaetae chrysosporium* and *Phlebia radiata.* White rot fungi, responsible for non-selective decay, include *Fomes fomentarius* and *Trametes versicolor.* In selective decay, lignin and hemicellulose fractions get degraded, while the cellulose fraction remains essentially unaffected. In non-selective decay, approximately equal amounts of all the three fractions of lignocelluloses are degraded (Blanchette, 1995; Hatakka, 2001). The white rot fungi can be grouped

[1] Department of Microbiology, Panjab University, Chandigarh-160014, India.
[2] Department of Botany, Government Girls College, A.P.S. University, Rewa-486003, India.
* Corresponding author: deepakraahi10@gmail.com

into 14 families: Inocybaceae, Ganodermataceae, Grifolaceae, Hymenochaetaceae, Marasmiaceae, Meripilaceae, Meruliaceae, Phanerochaetaceae, Pleurotaceae, Physalacriaceae, Polyporaceae, Schizophylaceae, Stereaceae and Tricholomataceae (Kirk, 2000) (Figs. 1, 2).

White rot fungi are well known for efficient production of various extracellular enzymes which play an important role in a variety of industrial applications (Bains et al., 2006; Rahi et al., 2009). They are capable of secreting enzymes which catalyse decomposition of a range of plant cell-wall polysaccharides, including cellulose, hemicellulose and lignin. Expression of these enzymes in fungi is induced by the substrate and the yield of enzymes depends on their growth and mode of cultivation (Rahi et al., 2018). Under nitrogen limiting conditions, these fungi have the ability to secrete extracellular non-specific enzyme complexes, such as lignin peroxidases (LiPs), manganese-dependent peroxidases (MnPs), versatile peroxidases (VPs) and laccasses through secondary metabolism (Wesenberg et al., 2003). The extracellular nature of these enzymes enables them to transform a variety of chemicals having structural similarity with lignin (Mansur et al., 2003). This makes the white rot fungi very attractive for different industrial and biotechnological applications, such as production of biofuel from plant biomass, biopulping, biobleaching and the degradation of recalcitrant pollutants. Besides enzymes, they also produce several metabolites with varied biological activities and therapeutic potential, such as polysaccharides, polyphenols, terpenoids, alkaloids, sterols and bioactive proteins.

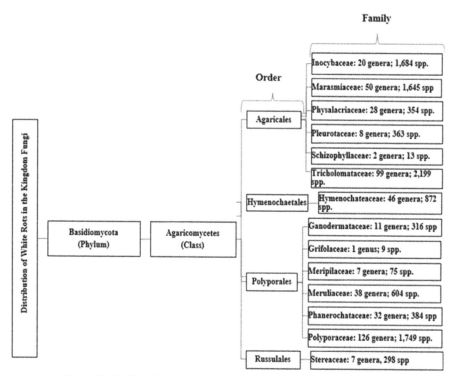

Fig. 1. The families of white rot fungi with respective generic and species tally.

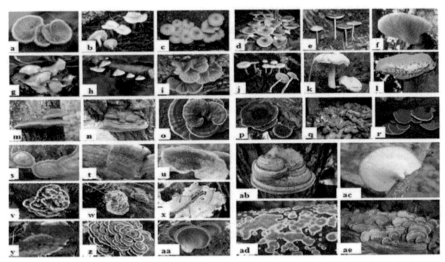

Fig. 2. Habit and habitat of important white rot fungi of the phylum Basidiomycota: a) *Crepidotus* sp.; b) *Marasmius* sp.; c) *Clitocybula* sp.; d) *Armillaria* sp.; e) *Hymenopellis* sp.; f) *Pleurotus* sp.; g) *Hohenbuehelia* sp.; h) *Auriculariopsis* sp.; i) *Schizophyllum* sp.; j) *Clitocybe* sp.; k) *Tricoholoma* sp.; l) *Hymenochateae* sp.; m) *Inonotus* sp.; n) *Phellinus* sp.; o) *Ganoderma* sp.; p) *Amauroderma* sp.; q) *Grifola* sp.; r) *Rigidoporus* sp.; s) *Physisporinus* sp.; t) *Gloeoporus* sp.; u) *Phlebia* sp.; v) *Bjerkander* sp.; w) *Antrodiella* sp.; x) *Ceriporiopsis* sp.; y) *Phanerochaete* sp.; z) *Trametes* sp.; aa) *Lentinus* sp.; ab) *Fomes* sp.; ac) *Polyporus* sp.; ad) *Aleurodiscus* sp., ae) *Stereum* sp.

This chapter focuses on the production of enzymes, industrial application (paper and textile industries) and bioremediation potential of enzymes produced by white rot fungi.

2. Production of Enzymes

2.1 Laccases

Laccases, belonging to the family of multi-copper oxidases (MCOs), are classified as benzenediol oxygen reductases (EC 1.10.3.2) and are also known as urushiol oxidases or *p*-diphenol oxidases (Giardina et al., 2010; Riva et al., 2016). Laccases contain three types of copper atoms, one of which is responsible for their characteristic blue colour. The enzymes lacking the blue copper atom are called yellow or white laccases. Laccases exist usually in monomeric form, but some are also present in homodimeric, heterodimeric and multimeric forms. The complete crystalline structure of laccase containing all four copper atoms in the active site has been published from *Cerrena maxima* and *Trametes versicolor* (Bertrand et al., 2002; Piontek et al., 2002). The structure of laccase consists of three cupredoxine-like domains and resembles that of ascorbate oxidase (Bertrand et al., 2002). Laccases are glycoproteins of white rot fungi generally have a molecular weight between 60–80 kDa (Hatakka, 2001). These enzymes have low substrate specificity and are capable of oxidising a large number of phenolic and non-phenolic molecules by using oxygen as electron acceptor while generating water as a byproduct. Wood rot Basidiomycetes causing white rot are most widely known to produce appreciable quantities of laccase. Almost all the species of white rot fungi are reported to produce varying degrees of laccase (Rahi et al., 2020). The first fungal laccase was reported by Bertrand et al. (1896) in the genus *Boletus*. A large number of fungi have been confirmed as laccase producers, among which white rot fungi are the most recognised ones (Table 1).

White rot fungi possess several laccase-encoding genes and secrete laccases as multiple isoforms (Palmieri et al., 2000; Hatakka, 2001). An interesting exception in the occurrence of laccase is in the widely studied white rot fungus, *P. chrysosporium*. In the sequenced genome of *Phanerochaete chrysosporium*, no

Table 1. Examples of laccase producing white rot fungi.

Mushroom	References
Cerrena unicolor	Michniewicz et al., 2003
Coprinopsis cinera	Arora and Sharma, 2010; Forootanfar and Faramarzi, 2015
Coriolopsis gallica	Reyes et al., 1999
Coriolopsis rigida	Gómez et al., 2005
Dichomitus squalens	Arora and Gill, 1999
Fomes fomentarius	Arora and Sharma, 2010; Forootanfar and Faramarzi, 2015
Funalia trogii	Ünyayar et al., 2005
Ganoderma lucidium	Postemsky et al., 2017
Irpex lacteus	Kasinath et al., 2003
Lentinus edodes	D'Annibale et al., 1999
Marasmius scorodonius	Risdianto et al., 2012
Panus rudis	Arora and Sharma, 2010; Forootanfar and Faramarzi, 2015
Phlebia fascicularia	Arora and Gill, 1999
Phlebia radiata	Arora and Sharma, 2010; Forootanfar and Faramarzi, 2015
Pleurotus eryngii	Camarero et al., 2004
Pleurotus oryzae	Carunchio et al., 2001
Pleurotus ostreatus	Hou et al., 2004; Palmieri et al., 2005
Pleurotus sajor-caju	Baborova, 2006; Arora and Sharma, 2010; Forootanfar and Faramarzi, 2015
Polyporus brumalis	Baborova, 2006; Arora and Sharma, 2010; Forootanfar and Faramarzi, 2015
Polyporus pinsitus	Claus et al., 2002
Pseudolagarobasidium acaciicola	Adak et al., 2016
Pycnoporus cinnabarinus	Baborova, 2006; Arora and Sharma, 2010; Forootanfar and Faramarzi, 2015
Trametes hirsuta	Abadulla et al., 2000; Moldes et al., 2003; Rodríguez Couto et al., 2004a; Domínguez et al., 2005
Trametes hirsuta	Campos et al., 2001
Trametes versicolor	Claus et al., 2002; Camarero et al., 2004
Trametes villosa	Baborova, 2006; Arora and Sharma, 2010; Forootanfar and Faramarzi, 2015

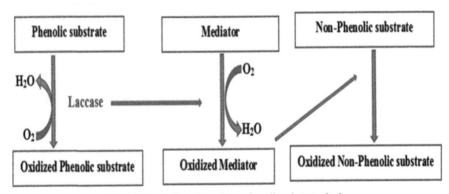

Fig. 3. Oxidation of phenolic and non-phenolic substrates by laccase.

close match to known laccase-encoding genes could be found (Kersten and Cullen, 2007). There are a few reports of laccase produced by *P. chrysosporium* (Srinivasan et al., 1995) but these results can be partly explained by unspecific nature of 2, 2'-azino-bis (3-ethylbenzthiazoline-6-sulfonic acid) (ABTS) oxidation reaction and the possibility of variation in strains. However, it seems evident that most of the *P. chrysosporium* strains do not produce laccase. Fungal laccase has been suggested to participate in morphogenesis, fungal plant-pathogen interaction, stress defence and detoxification of byproducts of lignin degradation (Thurston, 1994). Laccase production using white rot fungi can be induced by the addition of Cu^+ (Palmieri et al., 2000) and aromatic compounds, such as veratryl alcohol and 2,5-xylidine (Eggert et al., 1996). In some fungi (*C. subvermispora* and *Ganoderma lucidum*), laccase production increases in the presence of lignocellulosic material (Fukushima and Kirk, 1995). The catalysis of laccase occurs with reduction of one molecule of oxygen to water, which is accompanied by one electron oxidation of a wide range of aromatic compounds, including polyphenols, methoxy-substituted monophenols and aromatic amines (Bourbonnais and Paice, 1990; 1995). This oxidation results in generation of oxygen-centred free radical that can be converted into quinone in a second enzyme catalysed reaction. Laccase catalysis occurs in three steps: (1) type I Cu reduction by substrate; (2) electron transfer from type I Cu to the type II Cu and type III Cu trinuclear cluster; (3) reduction of oxygen to water at the trinuclear cluster. The laccase-mediated catalysis of non-phenolic substrates could be achieved with the help of mediators, such as 1-hydro-xybenzotriazole (HOBT), N-hydroxy-phthalimide (NHPI) and 2, 2-azinobis-3-ethylthiazoline-6-sulfonat (ABTS). The mediators are low molecular weight organic compounds, which get oxidised by laccase first to form highly active cation radicals that are capable of oxidising non-phenolic compounds which laccase alone cannot oxidise (Fig. 3).

2.2 *Lignin Peroxidase*

Lignin peroxidase (LiP, diaryl propane oxygenase) (EC 1.11.1.14) is a heme-containing monomeric enzyme known to catalyse the hydrogen peroxide-dependent oxidative degradation of lignin. It was first discovered in the extracellular medium of white rot fungus, *Phanerochaete chrysosporium* (Glenn et al., 1983). Lignin degradation has been extensively studied in wood rot fungi, especially white rot Basidiomycetes (Hatakka, 1994; Martinez et al., 2004; Wan and Li, 2012). The structure of LiP is characterised by its high redox potential with hydrogen peroxide enabling oxidation of non-phenolic aromatic compounds and by its long-range electron-transfer pathway enabling oxidation of polymers, like lignin. The enzyme has a globular structure (50 × 40 × 40 Å) with molecular weight ranging between 38–43 kDa. The globular structure of the enzyme contains eight major α-helices, eight minor helices and three short antiparallel β-sheets (Choinowski et al., 1999). It is segregated into proximal and distal domains by the heme. The heme is fixed completely in the enzyme and is made accessible only through two small channels. A variety of white rot fungi, such as *Pleurotus* and *Schizophyllum* are known to produce lignin peroxidases for a variety of biotechnological applications (Table 2). The catalytic cycle of lignin peroxidase involves three steps: the first reaction step is oxidation of the resting ferric enzyme [Fe (III)] by hydrogen peroxide (H_2O_2) as an electron acceptor, resulting in the formation of compound I oxo-ferryl intermediate; in the second step, the oxo-ferryl intermediate (deficient of 2e–) is reduced by a molecule of substrate, such as non-phenolic aromatic substrate (S), which donates one electron (1e–) to compound I to form the second intermediate, compound II (deficient of 1e–); the last step involves the subsequent donation of a second electron to compound II by the reduced substrate, thereby returning LiP to the resting ferric oxidation state, which indicates the completion of the oxidation cycle (Abdel-Hamid et al., 2013) (Fig. 4).

2.3 *Manganese Peroxidase*

Manganese peroxidase (MnP) (EC 1.11.1.13, Mn^{2+}: H_2O_2 oxidoreductases) is a monomeric protein with a molecular weight of 37 kDa. It belongs to the family of Oxidoreductases, which specifically oxidises Mn^{2+} to Mn^{3+}. The enzyme has five S-S bonds and Cys 341-Cys 348, which is located near the C-terminus of polypeptide

Table 2. Examples of lignin peroxidase (LiP) producing white rot fungi.

Mushroom	References
Bjerkandera sp.	Mester et al., 1996
Lentinula edodes	Bonnarme and Jeffries, 1990
Phanerochaete chrysosporium	Kang et al., 2004
Phanerochaete sordida	Hirai et al., 2005
Phellinus pini	Bonnarme and Jeffries, 1990
Phlebia ochraceofulva	Hatakka, 1994
Phlebia radiatar	Hatakka, 1994
Pleurotus ostreatus	Reddy et al., 2003
Pleurotus sajor-caju	Irshad and Asgher, 2011
Schizophyllum commune IBL-06	Asgher et al., 2012
Trametes suaveolens	Knežević et al., 2013
Trametes versicolor	Asgher et al., 2012

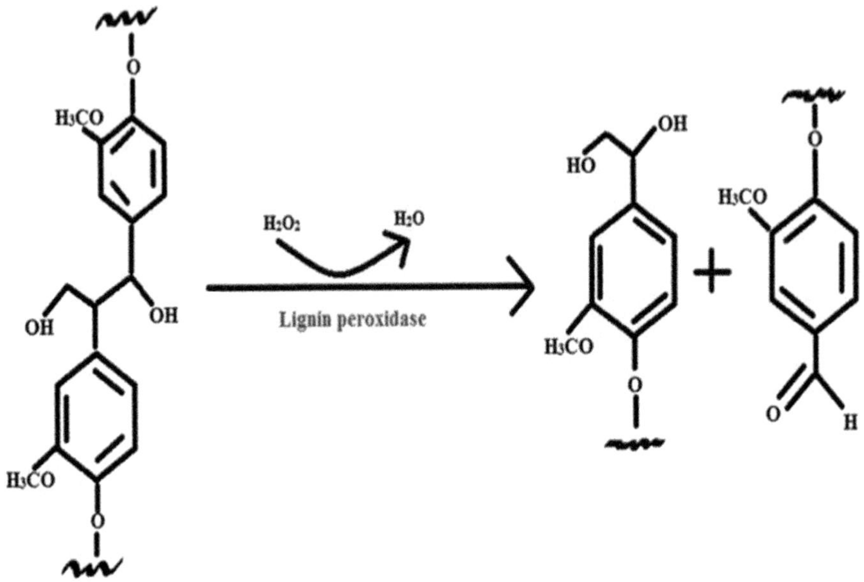

Fig. 4. Oxidative cleavage of β-1 linkage in lignin structure by lignin peroxidase (LiP).

chain. The formation of Mn (II) binding site and driving the C terminus part away from the foremost part of protein is caused by the fifth bond. The MnPs are more widespread than LiP (Hofrichter, 2002). The MnP production is reported to be produced by various white rot fungi, especially *Phanerochaetae chrysosporium, Pleurotus ostreatus* and *Trametes* spp. (Hofrichter, 2002; Elisashvili and Kachlishvili, 2009; Hatakka and Hammel, 2011) (Table 3). Indigenous ferric enzyme and H_2O_2 commence the catalytic cycle of MnP by forming compound I (Fe^{4+} oxo-porphyrin radical complex). The mono-chelated Mn^{2+} ions donates one electron to porphyrin, which is then intermediated into the compound II by the donation of one electron from Mn^{2+} to form Mn^{3+}. The chelated Mn^{3+} ions generated by MnP act as S-S charge transfer mediators, which allow the oxidation of various phenolic substrates, such as simple phenols, amines, phenolic lignin and several dyes (Fig. 5).

2.4 *Versatile Peroxidases*

Versatile peroxidases are class-II family peroxidases (EC 1.11.1.16) found in plants, fungi, and bacteria. These enzymes were first described in white rot fungus, *Pleurotus eryngii* followed by other white rot fungi, such as *Bjerkandera* sp., *Lentinus* sp., *Pleurotus* sp., *Trametes* sp. and so on (Table 4). Versatile peroxidases are monomeric glycosylated heme proteins with a molecular weight of 40–50 KDa and have four conserved disulphide bridges and two conserved calcium-binding sites. They operate through a central heme protoporphyrin and use hydrogen peroxide as an electron acceptor. Versatile peroxidases have two solvent access channels: wider channel for access by hydrogen peroxide and the narrow channel for access to manganese. These enzymes harbour multiple active sites and are capable of oxidising manganese

Table 3. Examples of manganese peroxidase (MnP) producing white rot fungi.

Mushroom	References
Bjerkandera sp.	Jarvinen et al., 2012
Ceriporiopsis subvermispora	Tanaka et al., 2009
Clitocybula dusenii	Nüske et al., 2002
Cerrena unicolor BBP6	Zhang et al., 2018
Dichomitus squalens	Hatakka, 1994
Ganoderma lucidum	Zhang et al., 2016
Ganoderma lucidum IBL-05	Bilal and Asgher, 2016
Ganoderma valesiacum	Nerud et al., 1991
Irpex flavus	Gill et al., 2002
Irpex lacteus F17	Baborova et al., 2006
Lentinus velutinus	Tekere et al., 2001
Nematoloma frowardii	Nüske et al., 2002
Phanerochaete chrysosporium	Singh et al., 2011
Phanerochaete sordida	Wang et al., 2011
Phlebia radiata	Hatakka, 1994
Physisporinus rivulosis	Jarvinen et al., 2012
Pleurotus ostreatus	Hatakka and Hammel, 2011
Polyporus sanguineus	Gill and Arora, 2003
Pycnoporus cinnabarinus	Nerud et al., 1991
Rigidoporus lignosus	Hatakka, 1994
Schizophyllum commune IBL-06	Irshad and Asgher, 2011
Stereum hirsutum	Nerud et al., 1991
Trametes sp. 48424	Zhang et al., 2016
Trametes villosa	Silva et al., 2014
Trametes cingulata	Tekere et al., 2001
Trametes elegans	Tekere et al., 2001
Trametes pocas	Tekere et al., 2001

ions and a variety of substrates (from hydroquinones to phenols and bulky lignin compounds) in different environmental conditions (Martinez et al., 2004). These enzymes are capable of oxidising the substrates even in the absence of mediators as compared to the oxidation carried out by lignin and manganese peroxidases, where the presence of mediators is necessary. Versatile peroxidases have a polyvalent catalytic site and the catalytic cycle for the oxidation of low molecular weight substrates, which are initiated at the heme centre by binding of hydrogen peroxide to the ferric state of heme to form iron peroxide complex (Compound I). This activated Compound I is then reduced to Compound II and finally to a resting form of enzyme accompanied with oxidation of two substrates (Busse et al., 2013; Ravichandaran and Sridhar, 2016) (Fig. 6).

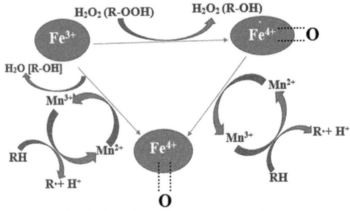

Fig. 5. The catalytic cycle of manganese peroxidase.

Table 4. Examples of versatile peroxidases producing white rot fungi.

Producer organism	References
Bjerkandera adusta	Davila-Vazquez et al., 2005
Galerina marginata	Kinnunen et al., 2017
Lentinus squarrosulus	Ravichandaran et al., 2019
Lentinus tigrinus	Zavarzina et al., 2018
Phanerochaete chrysosporim	Coconi-Linares, 2014
Phlebia radiata	Kinnunen et al., 2017
Pleurotus eryngii	Banci et al., 2003
Pleurotus ostreatus	Tsukihara et al., 2006
Pleurotus pulmonarius	Kinnunen et al., 2017
Trametes versicolor	Liu et al., 2004

3. Applications of Enzymes

The ligninolytic enzymes have gained more attention for various types of biotechnological applications, such as for degradation of toxic compounds, for generation of second generation alcohol, in pulp and paper, textile, food, pharmaceutical and in cosmetic industries. Owing to the higher redox potential (+800 mV) of fungal ligninolytic enzymes, they are implicated in several biotechnological applications, especially in degradation of lignin (Thurston, 1994). They have several applications owing to their oxidation ability towards a broad range of phenolic and non-phenolic compounds (Mohammadian et al., 2010). Other applications include bioremediation of industrial effluents, especially in pulp, paper, textile and petrochemical industries. They are also used in medical diagnostics, degradation of pollutants (e.g., herbicides, pesticides and some explosives in the soil) and also to clean water in many purification systems. They have applications in medical field to prepare certain drugs, such as anticancer drugs and serve as additives in cosmetics to minimise their toxic effects. They have enormous ability to remove

Versatile peroxidases

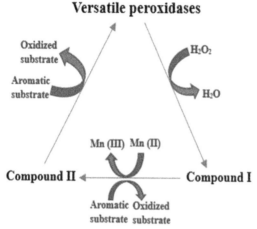

Fig. 6. The catalytic cycle of versatile peroxidases.

xenobiotics and produce polymeric products, thus facilitating their use for many bioremediation purposes (Couto and Herrera, 2006).

3.1 Pulp and Paper Industry

Paper manufacturing involves various pollution-causing chlorine-based delignification and bleaching procedures (Kuhad et al., 1997). These procedures result in the formation of various chlorinated aliphatic and aromatic compounds, which are carcinogenic, mutagenic and often toxic (Taspinar and Kolankaya, 1998). Therefore, an environment-friendly strategy for pre-treatment of wood using various enzymes from white rot fungi is employed for these purposes. The pulp- and paper-production technology is highly diverse and provides numerous opportunities for the application of microbial enzymes in processes like biopulping, biobleaching, de-inking, pitch removal, fibre grafting, paper coloration and bioremediation of effluents. Although many applications of enzymes in pulp and paper industries are still at the investigational and developmental stage, currently the most important application is in eco-friendly biobleaching of hard and soft wood pulps. In most instances, the enzymes employed are xylanases, laccases and rarely mannanases, which offer the potential alternative to conventional, environment-polluting chlorine and chlorine-based bleaching. During the pulp and paper production, it is necessary to separate the cellulose fibres from lignin. This is performed by using mechanical or chemical methods. Nowadays, prior to this conventional pulping method, the wood chips are treated with ligninolytic enzymes and the process is known as 'biopulping'. White rot fungal enzymes have been considered as potentially useful agents for biopulping because they reduce not only energy consumption but also the use of harmful chemicals, rendering the process environment-friendly. In addition, biopulping not only removes lignin but also some of the wood extractives, thereby reducing the pitch content and effluent toxicity (Ali and Sreekrishnan, 2001). Pretreatment of wood chips for mechanical and chemical pulping with white rot fungi or their enzymes has been developed by Mendonça et al. (2008). Laccases from white rot fungi can be applied in biopulping to partially degrade the lignin, thereby loosening the lignin structures effectively (Mendonça et al., 2008). Laccases from the white rot fungi have the ability to depolymerise lignin and delignify wood pulps (Virk et al., 2012). Laccase acts on small phenolic lignin fragments that react with the lignin polymer, thus leading to its easy degradation. A number of white rot fungi, such as *Ceriporiopsis subvermispora* and *Pleurotus*, are used in biopulping. Moreover, pretreatment of wood chips with ligninolytic enzymes increases the pulp strength, thus decresing the energy requirement for mechanical pulping. Laccases are also involved in reduction of the kappa number of pulp and improving the paper-making properties of pulp (Abadulla et al., 2000; Kuznetsov et al., 2001). Besides, the lignninolytic enzymes from white rot fungi are also used for bleaching of pulp and the process is called 'biobleaching'. Biobleaching of the pulp reduces the amount of chemical bleach required to obtain a desired brightness of pulp (Call and Call, 2005). *Bjerkandera* sp. BOS55, *Lentinus tigrinus, Phlebia radiata, Polyporus ciliatus* and *Stereum hirsutum* are found to be efficient biobleachers (Blanchette et al., 1992).

3.2 Textile Industry

The use of ligninolytic enzymes in the textile industry is growing at a faster rate. Besides decolourising the dyes released in the effluents of textile manufacturing industries, these enzymes are also used in the bleaching of cotton fabrics, which is an important step in textile manufacturing to modify as well as improve the surface texture of the fabrics (Setti et al., 1999; Shelke, 2001; Zille, 2005).

3.2.1 Cotton Bleaching

Cotton bleaching is generally carried out to give a white appearance to fabrics by decolourising the natural pigments. The most commonly used bleaching agent for this purpose is hydrogen peroxide, which is applied at alkaline pH and at boiling temperatures. This bleaching process however leads to decreased polymerisation and hence results in damage. Further, this process needs a huge amount of water to remove the bleaching agent from the fabrics, which again is a very cumbersome step. Use of laccase for the bleaching of cotton was first described by Tzanov et al. (2003). The replacement of hydrogen peroxide bleach with enzyme-mediated bleaching is considered a good alternative since it not only improves the quality of the cotton fabric, but also reduces the amount of water required in the bleaching process. According to Pereira et al. (2005), the laccase produced by *Trametes hirsua* is capable of improving the whiteness of cotton by oxidising the flavonoids.

3.2.2 Denim Bleaching

Use of laccases for denim bleaching in the textile industry is growing at a fast rate. Throughout the world, the light blue colour of the denim jeans is obtained via sodium hypochlorite bleaching. If not neutralised properly, sodium hypochlorite gives yellowness to the fabric, reduces the fabric strength and it is environmentally harmful owing to the increase in the COD and BOD levels of the effluents. Therefore, ligninolytic enzymes, mainly the laccases, produced by white rot fungi are considered as an ecofriendly alternative in the process of denim bleaching (Campos et al., 2001; Pazarogliu et al., 2005). Denilite I$_{TM}$, is the first industrial laccase and the first bleaching enzyme to be launched in 1996 by Novozyme (Novo Nordisk, Denmark). Other commercially available enzyme preparations which are useful in denim bleaching include DeniLiteII$_{TM}$ (Novozyme North America Inc., USA), Zylite (Zytex Pvt. Ltd., India), Bleach-cut 3S, (Chemicals Dyestuffs Ltd., Hong Kong), ECOSTONE®LCC 10 (AB Enzymes GmbH, Germany), Purizyme (Puridet Asia Ltd., Hong Kong), APCOZYME II-S (Apollo Chemical Company, LLC USA), Trilite II (Tri-Tex Co. Inc., Canada), Americos Laccase P and Americos Laccase LTC (Americos Industries Inc., India) and Hypozyme (Condor Speciality Products USA) (Rodriguez-Couto, 2012).

3.2.3 Shrink-proofing of Wool

Chlorination is conventionally used for shrink-proofing of wool and involves degradation of wool exo-cuticle, leading to formation of cysteic acid residues and protein losses. This conventional chlorination process however, has been replaced by treatment with proteinases, which are highly specific and cause very little environmental damage. The proteinase treatment however, has certain disadvantages leading to decrease in fibre strength and limited shrink resistance (Breier, 2000; Breier, 2002). Therefore, to overcome this problem, laccases are used as suitable alternatives to increase the shrink resistance of wool. In a study carried out by Lantto et al. (2004), it was observed that laccase was capable of activating the wool fibres in the presence of a suitable mediator. A patent has also been published for using laccase from *Trametes versicolor* to increase the shrink resistance of wool (Yoon, 1998).

3.3 Nanotechnology

Nanotechnology contributes to the development of efficient compact biosensors through controlled deposition or specific adsorption of biomolecules on different types of surfaces at nanoscale. Laccase catalyses oxidation of various aromatic compounds, especially phenols, which are organic pollutants in wastewaters. With specific function, laccases have a great impact on the development of biosensors to detect and remediate environmental pollutants or clinically relevant metabolites. Laccases are able to catalyse electron transfer reactions without additional cofactors. In biosensor technology, laccase is used because of its broad substrate-specificity, which allows the detection of a broad range of components, like phenolics and oxygen or azides. Biosensors that utilise laccase enzymes include electrodes to monitor phenols (e.g., catechols in tea or other products), polyphenolics in wine and lignins or phenols in wastewaters (Giovanelli and Ravasini, 1993; Palmore and Kim, 1999; Lanzellotto et al., 2014).

4. Bioremediation

4.1 Paper Industry Effluents

White rot fungi are the only microorganisms studied extensively for the degradation and decolourisation of lignin and its related monomers (Hofrichter, 2002). Use of enzymes produced by these fungi has great potential in tertiary treatment and removal of residual organic compounds in wastewater discharged from pulp and paper industries (Wu et al., 2005; Apiwattanapiwat et al., 2006; Da Re and Papinutti, 2011; Rajwar et al., 2017). White rot fungi, such as *Datronia* sp. (Chedchant et al., 2009), *Lentinus edodes* (Esposito et al., 1991; Wu et al., 2005), *Phanerochaete chrysosporium* (Zouari et al., 2002; Wu et al., 2005), *Phlebia radiata* (Lankinen et al., 1991), *Phlebia tremellosa* (Lankinen et al., 1991), *Pleurotus* spp., *Pleurotus citrinopileatus, Pleurotus platypus, Pleurotus sajor-caju* (Ragunathan and Swaminathan, 2004), *Steccherinum* sp. (Da Re and Papinutti, 2011), *Trametes versicolor* (Martin and Manzanares, 1994; Modi et al., 1998; Garg et al., 1999) and *Trichaptum* (Apiwattanapiwat et al., 2006) are reported to be effective in reducing the various pollutants existing in wastewaters of the pulp and paper industry.

4.2 Dyes of Textile Industry

The textile industry is a consumer of a large amount of water and chemicals, which are required in textile processing and thus its effluents are the main source of various harmful carcinogenic dyes (Banat et al., 1996). Dyeing is a low-yielding process and the amount of dye lost in the effluents could be sometimes as high as 50 per cent (Pierce, 1994; Pearce et al., 2003). Due to their complex synthetic chemical nature, dyes are resistant to fading on exposure to light, water and also to various chemicals, thus leading to difficulty in decolourising by common methods (Lin et al., 2010). The conventional treatment system used for dye decolourisation is based on chemical or physical methods, which are very expensive and demand a large quantity of chemicals as well as energy. Thus, alternate technologies, such as use of ligninolytic enzymes for dye degradation, have been studied recently. Laccases of white rot fungi are capable of oxidising a wide spectrum of pollutants and have received specific attention in degradation of dyes.

Table 5 details literature on different enzymes of white rot fungi that are involved in degradation of different dyes. These enzymes are capable of decolourising some of the azo dyes without direct cleavage of the azo bonds through a non-specific free-radical mechanism, which avoids the formation of toxic aromatic amines. Use of laccases in the textile industry is growing very fast because besides decolourising textile effluents, laccases are also used to bleach textiles, synthetic dyes and modify the surface of fabrics (Setti et al. 1999; Shelke, 2001; Zille, 2005). Laccase-based dye treatment provides a reasonable alternative for development of biotechnological processes for removal of industrial dye effluents at a large scale. It has also been reported that versatile peroxidase produced by white rot fungi is capable of efficiently oxidising high redox potential dyes (e.g., Reactive Black 5 and Reactive Blue 19) and substituting phenols and hydroquinones (Heinfling et al., 1998).

4.3 Xenobiotics

Contamination of soil, water and air by toxic chemicals is one the biggest environmental hazards, which requires immediate intervention for prevention

Table 5. Degradation of dyes by the enzymes produced by white rot fungi.

Mushroom	Enzymes	Nature of dye	References
Bjerkander adusta	Lip and MnP	Reactive orange, blue 15, black 5, violet 5, Remazol brilliant blue R	Heinfling et al., 1997; Novotny et al., 2001; Kornillowicz-Kowalska et al., 2014
Ceriporia metamorphosa	Laccase	Remazol brilliant blue R, Poly R-478	Novotny et al., 2001
Cerrena unicolor	Laccase	Safranine, Methylene blue, Azure blue and simulated textile effluent	Zhang et al., 2018
Dichomitus squalens	MnP and Laccase	Brilliant green, Cresol red, Crystal violet, Congo red and Orange II	Kim and Shoda, 1999; Gill et al., 2002
Fomes fomentarius	MnP	Congo red, Methylene blue and Malachite green	Jayasinghe et al., 2008
Ganoderma lucidium	Laccase	Congo red and Methylene blue	Jayasinghe et al., 2008
Irpex lacteus	Laccase, Lip and Mnp	Congo red, Methyl red, Napthol blue black, Remazol brilliant blue R and Bromophenol blue	Novotny et al., 2001
Lentinus tigrinus	Laccase and Mnp	Orange II, Reactive blue 38 and Remazol brilliant blue R	Moreira et al., 2000; Novotny et al., 2001
Naematoloma fasciculare	Laccase	Congo red	Jayasinghe et al., 2008
Phanerochaete chrysosporium	Laccase and LiP	Amido black 10 B, Malachite green and Nigrosin	Rani et al., 2014; Senthilkumar et al., 2014
Phlebia brevispora	Laccase, LiP and MnP	Brilliant green, Cresol red and Crystal violet	Gill et al., 2002
Pleurotus eryngii	Lip and MnP	Reactive violet, Reactive blue and Reactive black	Heinfling et al., 1998
Pleurotus ostreatus	MnP	Bromophenol blue, Bromocresol purple and Phenol red	Shrivastava et al., 2005
Pleurotus pulmonarius	Laccase and MnP	Congo red and Malachite green	Jayasinghe et al., 2008
Pleurotus sajor-caju	Laccase and MnP	Amaranth, New coccine, Orange G, Tartrazine and Indigo	Balan and Monteiro, 2001; Chagas and Durrant, 2001
Polyporus picipes	LiP and MnP	Triphenylmethane brilliant green and Azo Evans blue	Przystas et al., 2015
Pycnoporus sanguineus	Laccase	Bromophenol blue, Remazol brilliant blue R and Reactive blue 4	Zimbardi et al., 2016
Schizophyllum commune	Laccase	Solar brilliant red	Ashger et al., 2013

Table 5 contd....

... Table 5 contd.

Mushroom	Enzymes	Nature of dye	References
Stereum hirsutum	Laccase and MnP	Orange II, Reactive blue and Poly R-478	Moreira et al., 2000
Stereum ostrea	Laccase and MnP	Congo red	Jayasinghe et al., 2008
Trametes suaveolens	Laccase, LiP and MnP	Congo red and Methylene blue	Jayasinghe et al., 2008
Trametes versicolor	Laccase, Lip and Mnp	Malachite green, Bromothymol blue and Methyl red	Vrsanska et al., 2018

and conservation. Notified organic compounds released from industries and the extensive use of pesticides in agriculture constitute serious problems. Some of the carcinogenic and/or mutagenic compounds, such as polycyclic aromatic hydrocarbons (PAH), pentachlorophenols (PCP), polychlorinated biphenyls (PCB), dichlorodiphenyltrichloroethane (DDT), benzene, toluene, ethylbenzene, xylene (BTEX) and trinitrotoluene (TNT) are persistently present in the environment. The ability of white rot fungi to transform a wide variety of these hazardous chemical compounds has aroused interest towards adoption of strategies in bioremediation. Table 6 provides literature on white rot fungi involved in degradation of various xenobiotics.

The important white rot fungal genera participating in xenobiotic degradation include *Phanerochaete* spp., *Pleurotus* spp. and *Trametes* spp. Enzymatic treatment has become an alternative method for the removal of toxic xenobiotics from the

Table 6. Degradation of xenobiotics by enzymes of white rot fungi.

Organism	Pollutant	References
Bjerkandera sp. BOS 55	Anthracene and benzo(a)pyrene phenanthrene	Valentin et al., 2007
Ceriporiopsis subvermispora	Pentachlorophenol	Field et al., 1992
Chrysosporium lignosum	3,4-dichloroaniline, dieldrin and phenantherene	Morgan et al., 1991
Lentinus tigrinus	Polycyclic aromatic hydrocarbons (PAHs)	Valentin et al., 2006
Phanerochaete chrysosporium	14C-benzo(a)pyrene phenanthrene, Crystal violet, Azure blue and Pentachlorophenol	Bumpus, 1989
Phanerochaete sordida	Pentachlorophenol	Field et al., 1992
Pleurotus eryngii	Chlorinated biphenyls and PAHs	Rodriguez et al., 2004a
Pleurotus florida	Pyrene, benzo (a) anthracene and benzo (a) pyrene	Valentin et al., 2006
Pleurotus ostreatus	Pentachlorophenol	Field et al., 1992
Pleurotus ostreatus	Anthracene, pyrene, fluorine, dibenzothiophene and industrial dyes	Bezalel et al., 1996b
Trametes hirsuta	Pentachlorophenol	Field et al., 1992
Trametes versicolor	Anthracene and benzo(a)pyrene phenanthrene, synthetic dyes (Amaranth) and polysaccharide	Gavril and Hodson, 2007; Zhu et al., 2007

environment. The most prominent groups of enzymes utilised in xenobiotic bioremediation transformations are oxidoreductases (peroxidases, laccases and oxygenases) (Sharma et al., 2018). Oxidoreductases can detoxify compounds by catalysing oxidative coupling reactions using oxidising agents to support the reactions. Laccases (LACs) and CYP monooxygenases (P450s) use molecular oxygen as the electron acceptor, while the peroxidases use hydrogen peroxide to oxidise the substrates with both reactions resulting in the formation of water as a byproduct. Peroxidases are heme-containing proteins and the major types of peroxidases involved in detoxification processes in white rot fungi are manganese peroxidase (MnP), lignin peroxidase (LiP) and versatile peroxidases (VP) (Doddapaneni et al., 2005). Xenobiotic degrading enzymes are produced by diverse fungi, but maximum production is by white rot fungi.

5. Conclusion

In the recent past, significant progress has been made to understand the white rot Basidiomycetes and their enormous industrial potential. These fungi are a major source of various biotechnologically-valued ligninolytic enzymes, like laccases and peroxidases (LiP, MnP and Versatile peroxidase), which possess powerful oxidative capacity with broad substrate specificity. These enzymes are more effective in functionality compared to the synthetic or chemical compounds or strategies leading to development of sustainable technologies. The white rot fungi have been used in various fields, especially in the paper industry, textile industry, food industry, bioremediation of pollutants, degradation of xenobiotics and also in the field of nanobiotechnology. Although 90 per cent of the wood rot fungi belong to the category of white rots, a few of them have been studied and harnessed for their industrial potential, thus signifying the need for precise efforts in future for development of ecofriendly technologies to check the deterioration of our environment.

References

Abadulla, E., Tzanov, T., Costa, S., Robra, K.H., Cavaco-Paulo, A. and Gübitz, G.M. (2000). Decolorization and detoxification of textile dyes with a laccase from *Trametes hirsuta*. Appl. Environ. Microbiol., 66(8): 3357–3362.

Abdel-Hamid, A.M., Solbiati, J.O. and Cann, I.K. (2013). Insights into lignin degradation and its potential industrial applications. Adv. Appl. Microbiol., 82: 1–28.

Adak, A., Tiwari, R., Singh, S., Sharma, S. and Nain, L. (2016). Laccase production by a novel white-rot fungus *Pseudolagarobasidium acaciicola* LA 1 through solid-state fermentation of *Parthenium* biomass and its application in dyes decolourisation. Waste Biomass Valor., 7(6): 1427–1435.

Ali, M. and Sreekrishnan, T.R. (2001). Aquatic toxicity from pulp and paper mill effluents: A review. Adv. Enviorn. Res., 5(2): 175–196.

Apiwatanapiwat, W., Siriacha, P. and Vaithanomsat, P. (2006). Screening of fungi for decolourisation of wastewater from pulp and paper industry. Agric. Nat. Resour., 40(5): 215–221.

Arora, D.S. and Gill, P.K. (2001). Effects of various media and supplements on laccase production by some white rot fungi. Bioresour. Technol., 77(1): 89–91.

Arora, D.S. and Sharma, R.K. (2010). Ligninolytic fungal laccases and their biotechnological applications. Appl. Biochem. Biotechnol., 160(6): 1760–1788.

Asgher, M., Iqbal, H.M.N. and Asad, M.J. (2012). Kinetic characterisation of purified laccase produced from *Trametes versicolor* IBL-04 in solid state bio-processing of corncobs. BioResources, 7(1): 1171–1188.

Asgher, M., Yasmeen, Q. and Iqbal, H.M.N. (2013). Enhanced decolourisation of Solar brilliant red 80 textile dye by an indigenous white rot fungus *Schizophyllum commune* IBL-06. Saudi J. Biol. Sci., 20(4): 347–352.

Bains, R.K., Rahi, D.K. and Hoondal, G.S. (2006). Evaluation of wood degradation enzymes of some indigenous white rot fungi. J. Mycol. Pl. Pathol., 36: 161–164.

Balan, D.S. and Monteiro, R.T. (2001). Decolourisation of textile indigo dye by ligninolytic fungi. J. Biotechnol., 89(2-3): 141–145.

Banat, I.M., Nigam, P., Singh, D. and Marchant, R. (1996). Microbial decolourisation of textile-dyecontaining effluents: A review. Bioresour. Technol., 58(3): 217–227.

Banci, L., Camarero, S., Martínez, A.T., Martínez, M.J., Perez-Boada, M., Pierattelli, R. and Ruiz-Duenas, F.J. (2003). NMR study of manganese (II) binding by a new versatile peroxidase from the white-rot fungus *Pleurotus eryngii*. J. Biol. Inorg. Chem., 8(7): 751–760.

Bertrand, G. (1896). Simultaneous occurrence of laccase and tyrosinase in the juice of some mushrooms. CR Hebd. Séances Acad. Sci., 123: 463–465.

Bertrand, T., Jolivalt, C., Caminade, E., Joly, N., Mougin, C. and Briozzo, P. (2002). Purification and preliminary crystallographic study of *Trametes versicolor* laccase in its native form. Acta Crystallogr. D., 58(2): 319–321.

Bezalel, L., Hadar, Y. and Cerniglia, C.E. (1996b). Mineralisation of polycyclic aromatic hydrocarbons by the white rot fungus *Pleurotus ostreatus*. Appl. Environ. Microbiol., 62(1): 292–295.

Bilal, M. and Asgher, M. (2016). Enhanced catalytic potentiality of *Ganoderma lucidum* IBL-05 manganese peroxidase immobilised on sol-gel matrix. J. Mol. Catal. B - Enzym., 128: 82–93.

Blanchette, R.A., Burnes, T.A., Eerdmans, M.M. and Akhtar, M. (1992). Evaluating isolates of *Phanerochaete chrysosporium* and *Ceriporiopsis subvermispora* for use in biological pulping processes. Holzforschung, 46(2): 109–115.

Blanchette, R.A., Farrell, R.L. and Iverson, S. (1995). Pitch Degradation with White Rot Fungi. U.S. Patent # 5,476,790.

Bonnarme, P. and Jeffries, T.W. (1990). Selective production of extracellular peroxidases from *Phanerochaete chrysosporium* in an airlift bioreactor. J. Ferment. Bioeng., 70(3): 158–163.

Bourbonnais, R. and Paice, M.G. (1990). Oxidation of non-phenolic substrates: An expanded role for laccase in lignin biodegradation. FEBS Lett., 267(1): 99–102.

Bourbonnais, R., Paice, M.G., Reid, I.D., Lanthier, P. and Yaguchi, M. (1995). Lignin oxidation by laccase isozymes from *Trametes versicolor* and role of the mediator 2,2'-azinobis (3-ethylbenzthiazoline-6-sulfonate) in kraft lignin depolymerisation. Appl. Environ. Microbiol., 61(5): 1876–1880.

Breier, R. (2000). Lanazym-rein enzymatische Antifilzausrustung von Wolle-Von der Idee zur erfolgreichen Umsetzung in die Praxis. DWI Reports, 1: 49–62.

Breier, R. (2002). Enzymatische Antifilzausrüstung von Wolle. Textilveredlung, 37(1-2): 5–10.

Bumpus, J.A., Powers, R.H. and Sun, T. (1993). Biodegradation of DDE (1, 1-dichloro-2, 2-bis (4-chlorophenyl) ethene) by *Phanerochaete chrysosporium*. Mycol. Res., 97(1): 95–98.

Busse, N., Wagner, D., Kraume, M. and Czermak, P. (2013). Reaction kinetics of versatile peroxidase for the degradation of lignin compounds. Am. J. Biochem. Biotechnol., 9(4): 365–394.

Call, H.P. and Call, S. (2005). New generation of enzymatic delignification and bleaching. Pulp Paper-Canada, 106(1): 45–48.

Camarero, S., Garcıa, O., Vidal, T., Colom, J., del Rio, J.C. et al. (2004). Efficient bleaching of non-wood high-quality paper pulp using laccase-mediator system. Enzyme Microb. Technol., 35(2-3): 113–120.

Campos, R., Kandelbauer, A., Robra, K.H., Cavaco-Paulo, A. and Gübitz, G.M. (2001). Indigo degradation with purified laccases from *Trametes hirsuta* and *Sclerotium rolfsii*, J. Biotechnol., 89(2-3): 131–139.

Carunchio, F., Crescenzi, C., Girelli, A.M., Messina, A. and Tarola, A.M. (2001). Oxidation of ferulic acid by laccase: Identification of the products and inhibitory effects of some dipeptides. Talanta, 55(1): 189–200.

Chagas, E.P. and Durrant, L.R. (2001). Decolourisation of azo dyes by *Phanerochaete chrysosporium* and *Pleurotus sajorcaju*. Enz. Microb. Technol., 29(8-9): 473–477.

Chedchant, J., Petchoy, O., Vaithanomsat, P., Apiwatanapiwat, W., Kreetachat, T. and Chantranurak, S. (2009). *In*: Kasetsart University Annual Conference, Bangkok, March 17–20, 2009.

Choinowski, T., Blodig, W., Winterhalter, K.H. and Piontek, K. (1999). The crystal structure of lignin peroxidase at 1.70 Å resolution reveals a hydroxy group on the Cβ of tryptophan 171: A novel radical site formed during the redox cycle. J. Mol. Biol., 286(3): 809–827.

Claus, H., Faber, G. and König, H.J.A.M. (2002). Redox-mediated decolourisation of synthetic dyes by fungal laccases. Appl. Microbiol. Biotechnol., 59(6): 672–678.

Coconi-Linares, N., Magaña-Ortíz, D., Guzmán-Ortiz, D.A., Fernández, F., Loske, A.M. and Gómez-Lim, M.A. (2014). High-yield production of manganese peroxidase, lignin peroxidase, and versatile peroxidase in *Phanerochaete chrysosporium*. Appl. Microbial. Biotechnol., 98(22): 9283–9294.

Couto, S.R. and Toca Herrera, J.L. (2006). Industrial and biotechnological applications of laccases: A review. Biotechnol. Adv., 24(5): 500–513.

D'Annibale, A., Stazi, S.R., Vinciguerra, V., Di Mattia, E. and Sermanni, G.G. (1999). Characterisation of immobilised laccase from *Lentinula edodes* and its use in olive-mill wastewater treatment. Proc. Biochem., 34(6-7): 697–706.

Da Re, V. and Papinutti, L. (2011). Black liquor decolourisation by selected white-rot fungi. Appl. Biochem. Biotechnol., 165(2): 406–415.

Davila-Vazquez, G., Tinoco, R., Pickard, M.A. and Vazquez-Duhalt, R. (2005). Transformation of halogenated pesticides by versatile peroxidase from *Bjerkandera adusta*. Enz. Microb. Technol., 36(2-3): 223–231.

Doddapaneni, H., Chakraborty, R. and Yadav, J.S. (2005). Genome-wide structural and evolutionary analysis of the P450 monooxygenase genes (P450ome) in the white rot fungus *Phanerochaete chrysosporium*: Evidence for gene duplications and extensive gene clustering. BMC Genomics, 6(1): 92.

Domínguez, A., Couto, S.R. and Sanroman, M.A. (2005). Dye decolourisation by *Trametes hirsuta* immobilised into alginate beads. World J. Microb. Biotechnol., 21(4): 405–409.

Eggert, C., Temp, U. and Eriksson, K.E. (1996). The ligninolytic system of the white rot fungus *Pycnoporus cinnabarinus*: Purification and characterisation of the laccase. Appl. Environ. Microbiol., 62(4): 1151–1158.

Elisashvili, V. and Kachlishvili, E. (2009). Physiological regulation of laccase and manganese peroxidase production by white-rot Basidiomycetes. J. Biotechnol., 144(1): 37–42.

Esposito, E., Canhos, V.P. and Durán, N. (1991). Screening of lignin-degrading fungi for removal of color from Kraft mill wastewater with no additional extra carbon-source. Biotechnol. Lett., 13(8): 571–576.

Field, J.A., De Jong, E., Costa, G.F. and De Bont, J.A. (1992). Biodegradation of polycyclic aromatic hydrocarbons by new isolates of white rot fungi. Appl. Environ. Microbiol., 58(7): 2219–2226.

Forootanfar, H. and Faramarzi, M.A. (2015). Insights into laccase producing organisms, fermentation states, purification strategies, and biotechnological applications. Biotechnol. Prog., 31(6): 1443–1463.

Fukushima, Y. and Kirk, T.K. (1995). Laccase component of the *Ceriporiopsis subvermispora* lignin-degrading system. Appl. Environ. Microbiol., 61(3): 872–876.

Garg, S.K., Chandra, H. and Modi, D.R. (1999). Effect of glucose and urea supplements on decolourisation of bagasse and gunny bag-based pulp-paper mill effluent by *Trametes versicolor*. Indian J. Exp. Biol., 37: 302–304.

Gavril, M. and Hodson, P.V. (2007). Chemical evidence for the mechanism of the biodecoloration of Amaranth by *Trametes versicolor*. World J. Microbiol. Biotechnol., 23(1): 103.

Giardina, P., Faraco, V., Pezzella, C., Piscitelli, A., Vanhulle, S. and Sannia, G. (2010). Laccases: A never-ending story. Cell Mol. Life Sci., 67: 69–385.

Gill, P.K.D.A. and Arora, D. (2003). Effect of culture conditions on manganese peroxidase production and activity by some white rot fungi. J. Ind. Microbiol. Biotechnol., 30(1): 28–33.

Gill, P.K., Arora, D.S. and Chander, M. (2002). Biodecolourisation of azo and triphenylmethane dyes by *Dichomitus squalens* and *Phlebia* spp. J. Ind. Microbiol. Biot., 28(4): 201–203.

Giovanelli, G. and Ravasini, G. (1993). Apple juice stabilisation by combined enzyme-membrane filtration process. LWT Food Sci. Technol., 26(1): 1–7.

Glenn, J.K., Morgan, M.A., Mayfield, M.B., Kuwahara, M. and Gold, M.H. (1983). An extracellular H2O2-requiring enzyme preparation involved in lignin biodegradation by the white rot Basidiomycete *Phanerochaete chrysosporium.* Biochem. Biophy. Res. Com., 114(3): 1077–1083.

Gomez, J., Pazos, M., Couto, S.R. and Sanromán, M.Á. (2005). Chestnut shell and barley bran as potential substrates for laccase production by *Coriolopsis rigida* under solid-state conditions. J. Food Eng., 68(3): 315–319.

Hatakka, A. (1994). Lignin-modifying enzymes from selected white-rot fungi: Production and role from in lignin degradation. FEMS Microbial. Rev., 13(2-3): 125–135.

Hatakka, A. (2001). Biodegradation of lignin. pp. 129–180. *In*: Hofrichter, M. and Steinbuchel, A. (eds.). Biopolymers, vol. 1: Lignin, Humic Substances and Coal. Wiley-VCH, Weinheim.

Hatakka, A. and Hammel, K.E. (2011). Fungal biodegradation of lignocelluloses: Industrial applications. Mycota, 10: 319–340.

Heinfling, A.M.M.J.M.A.T.B.M.S.U., Martinez, M.J., Martinez, A.T., Bergbauer, M. and Szewzyk, U. (1998). Transformation of industrial dyes by manganese peroxidases from *Bjerkandera adusta* and *Pleurotus eryngii* in a manganese-independent reaction. Appl. Environ. Microbiol., 64(8): 2788–2793.

Heinfling, A., Bergbauer, M. and Szewzyk, U. (1997). Biodegradation of azo and phthalocyanine dyes by *Trametes versicolor* and *Bjerkandera adusta.* Appl. Microbiol. Biotechnol., 48(2): 261–266.

Hirai, H., Sugiura, M., Kawai, S. and Nishida, T. (2005). Characteristics of novel lignin peroxidases produced by white-rot fungus *Phanerochaete sordida* YK-624. FEMS Microbial. Lett., 246(1): 19–24.

Hofrichter, M. (2002). Lignin conversion by manganese peroxidase (MnP). Enz. Microb. Technol., 30(4): 454–466.

Hou, H., Zhou, J., Wang, J., Du, C. and Yan, B. (2004). Enhancement of laccase production by *Pleurotus ostreatus* and its use for the decolourisation of anthraquinone dye. Proc. Biochem., 39(11): 1415–1419.

Irshad, M. and Asgher, M. (2011). Production and optimisation of ligninolytic enzymes by white rot fungus *Schizophyllum commune* IBL-06 in solid state medium banana stalks. Afr. J. Biotechnol., 10(79): 18234–18242.

JarvinenJayasinghe, C., Imtiaj, A., Lee, G.W., Im, K.H., Hur, H. et al. (2008). Degradation of three aromatic dyes by white rot fungi and the production of ligninolytic enzymes. Mycobiology, 36(2): 114–120.

Kang, S.W., Park, Y.S., Lee, J.S., Hong, S.I. and Kim, S.W. (2004). Production of cellulases and hemicellulases by *Aspergillus niger* KK2 from lignocellulosic biomass. Bioresour. Technol., 91(2): 153–156.

Kasinath, A., Novotný, Č., Svobodová, K., Patel, K.C. and Šašek, V. (2003). Decolourisation of synthetic dyes by *Irpex lacteus* in liquid cultures and packed-bed bioreactor. Enz. Microb. Technol., 32(1): 167–173.

Kersten, P. and Cullen, D. (2007). Extracellular oxidative systems of the lignin-degrading Basidiomycete *Phanerochaete chrysosporium.* Fungal Genet. Biol., 44(2): 77–87.

Kim, S.J. and Shoda, M. (1999). Purification and characterisation of a novel peroxidase from *Geotrichum candidum* Dec 1 involved in decolourisation of dyes. Appl. Environ. Microbiol., 65(3): 1029–1035.

Kinnunen, A., Maijala, P., Jarvinen, P. and Hatakka, A. (2017). Improved efficiency in screening for lignin-modifying peroxidases and laccases of Basidiomycetes. Curr. Biotechnol., 6(2): 105–115.

Kirk, P.M. (2000). Species Fungorum (version January 2016), Species.

Knežević, A., Milovanović, I., Stajić, M. and Vukojević, J. (2013). Trametes suaveolens as ligninolytic enzyme producer. Zb. Matice Srp. Prir. Nauke., (124): 437–444.

Kuhad, R.C., Singh, A. and Eriksson, K.E.L. (1997). Microorganisms and enzymes involved in the degradation of plant fibre cell walls. pp. 45–125. *In*: Eriksson K.E.L. et al. (eds.). Biotechnology in the Pulp and Paper Industry, Springer, Berlin and Heidelberg.

Kuznetsov, B.A., Shumakovich, G.P., Koroleva, O.V. and Yaropolov, A.I. (2001). On applicability of laccase as label in the mediated and mediatorless electroimmunoassay: effect of distance on the direct electron transfer between laccase and electrode. Biosens. Bioelectron., 16(1-2): 73–84.

Lankinen, V.P., Inkeröinen, M.M., Pellinen, J. and Hatakka, A.I. (1991). The onset of lignin-modifying enzymes, decrease of AOX and color removal by white-rot fungi grown on bleach plant effluents. Water Sci. Technol., 24(3-4): 189–198.

Lantto, R., Schönberg, C., Buchert, J. and Heine, E. (2004). Effects of laccase-mediator combinations on wool. Text. Res. J., 74(8): 713–717.

Lanzellotto, C., Favero, G., Antonelli, M.L., Tortolini, C., Cannistraro, S., Coppari, E. and Mazzei, F. (2014). Nanostructured enzymatic biosensor based on fullerene and gold nanoparticles: Preparation, characterisation and analytical applications. Biosens. Bioelectron., 55: 430–437.

Lin, C.H., Sheu, G.T., Lin, Y.W., Yeh, C.S., Huang, Y.H. et al. (2010). A new immunomodulatory protein from *Ganoderma microsporum* inhibits epidermal growth factor mediated migration and invasion in A549 lung cancer cells. Proc. Biochem., 45(9): 1537–1542.

Liu, W., Chao, Y., Yang, X., Bao, H. and Qian, S. (2004). Biodecolourisation of azo, anthraquinonic and triphenylmethane dyes by white-rot fungi and a laccase-secreting engineered strain. J. Ind. Microbiol. Biotechnol., 31(3): 127–132.

Mansur, M., Arias, M.E., Copa-Patino, J.L., Flärdh, M. and González, A.E. (2003). The white-rot fungus *Pleurotus ostreatus* secretes laccase isozymes with different substrate specificities. Mycologia, 95(6): 1013–1020.

Martin, C. and Manzanares, P. (1994). A study of the decolourisation of straw soda-pulping effluents by *Trametes versicolor*. Bioresour. Technol., 47(3): 209–214.

Martinez, D., Larrondo, L.F., Putnam, N., Gelpke, M.D.S., Huang, K. et al. (2004). Genome sequence of the lignocellulose degrading fungus *Phanerochaete chrysosporium* strain RP78. Nat. Biotechnol., 22(6): 695–700.

Mendonça, R.T., Jara, J.F., González, V., Elissetche, J.P. and Freer, J. (2008). Evaluation of the white-rot fungi *Ganoderma australe* and *Ceriporiopsis subvermispora* in biotechnological applications. J. Ind. Microbiol. Biotechnol., 35(11): 1323–1330.

Mester, T., Pena, M. and Field, J.A. (1996). Nutrient regulation of extracellular peroxidases in the white rot fungus, *Bjerkandera* sp. strain BOS55. Appl. Microbial. Biotechnol., 44(6): 778–784.

Michniewicz, A., Ledakowicz, S., Jamroz, T., Jarosz-Wilkolazka, A. and Leonowicz, A. (2003). Decolouriaation of aqueous solution of dyes by the laccase complex from *Cerrena unicolor*. Biotechnol., 4: 194–203.

Modi, D.R., Chandra, H. and Garg, S.K. (1998). Decolourisation of bagasse-based paper mill effluent by the white-rot fungus *Trametes versicolor*. Bioresour. Technol., 66(1): 79–81.

Mohammadian, M., Fathi-Roudsari, M., Mollania, N., Badoei-Dalfard, A. and Khajeh, K. (2010). Enhanced expression of a recombinant bacterial laccase at low temperature and microaerobic conditions: Purification and biochemical characterisation. J. Ind. Microbiol. Biotechnol., 37(8): 863–869.

Moldes, D., Gallego, P.P., Couto, S.R. and Sanromán, A. (2003). Grape seeds: The best lignocellulosic waste to produce laccase by solid state cultures of *Trametes hirsuta*. Biotechnol. Lett., 25(6): 491–495.

Moreira, M.T., Mielgo, I., Feijoo, G. and Lema, J.M. (2000). Evaluation of different fungal strains in the decolourisation of synthetic dyes. Biotechnol. Lett., 22(18): 1499–1503.

Morgan, P., Lewis, S.T. and Watkinson, R.J. (1991). Comparison of abilities of white-rot fungi to mineralise selected xenobiotic compounds. Appl. Microbiol. Biotechnol., 34(5): 693–696.

Nerud, F., Zouchova, Z. and Mišurcová, Z. (1991). Ligninolytic properties of different white-rot fungi. Biotechnol. Lett., 13(9): 657–660.

Nüske, J., Scheibner, K., Dornberger, U., Ullrich, R. and Hofrichter, M. (2002). Large scale production of manganese-peroxidase using agaric white-rot fungi. Enz. Microb. Technol., 30(4): 556–561.

Palmieri, G., Giardina, P., Bianco, C., Fontanella, B. and Sannia, G. (2000). Copper induction of laccase isoenzymes in the ligninolytic fungus *Pleurotus ostreatus*. Appl. Environ. Microbiol., 66(3): 920–924.

Palmieri, G., Cennamo, G. and Sannia, G. (2005). Remazol Brilliant Blue R decolourisation by the fungus *Pleurotus ostreatus* and its oxidative enzymatic system. Enz. Microb. Technol., 36(1): 17–24.

Palmore, G.T.R. and Kim, H.H. (1999). Electro-enzymatic reduction of dioxygen to water in the cathode compartment of a biofuel cell. J. Electroanal. Chem., 464(1): 110–117.

Pearce, C.I., Lloyd, J.R. and Guthrie, J.T. (2003). The removal of colour from textile wastewater using whole bacterial cells: A review. Dyes Pigm., 58(3): 179–196.

Pereira, L., Bastos, C., Tzanov, T., Cavaco-Paulo, A. and Gübitz, G.M. (2005). Environmentally friendly bleaching of cotton using laccases. Environ. Chem. Lett., 3(2): 66–69.

Pierce, J. (1994). Colour in textile effluents—The origins of the problem. J. Soc. Dye. Colour., 110(4): 131–133.

Piontek, K., Antorini, M. and Choinowski, T. (2002). Crystal structure of a laccase from the fungus *Trametes versicolor* at 1.90—Å resolution containing a full complement of coppers. J. Biol. Chem., 277(40): 37663–37669.

Postemsky, P.D., Bidegain, M.A., González-Matute, R., Figlas, N.D. and Cubitto, M.A. (2017). Pilot-scale bioconversion of rice and sunflower agro-residues into medicinal mushrooms and laccase enzymes through solid-state fermentation with *Ganoderma lucidum*. Bioresour. Technol., 231: 85–93.

Przystas, W., Zablocka-Godlewska, E. and Grabinska-Sota, E. (2015). Efficacy of fungal decolorization of a mixture of dyes belonging to different classes. Braz. J. Microbiol., 46(2): 415–424.

Ragunathan, R. and Swaminathan, K. (2004). Biological treatment of a pulp and paper industry effluent by *Pleurotus* spp. World J. Microbiol. Biotechnol., 20(4): 389–393.

Rahi, D.K., Rahi, S., Pandey A.K. and Rajak, R.C. (2009). Enzymes from mushrooms and their industrial applications. pp. 136–184. *In*: Rai, M. (ed.). Advances in Fungal Biotechnology, I.K. International Publishing House Pvt. Ltd., New Delhi.

Rahi, D.K., Thakur, S. and Malik, D. (2018). Comparative qualitative profile of various extracellular enzymes produced by two indigenous fungus, *Lentinus cladopus* and *Pleurotus pulmonarius*. Int. J. Sci. Res. Biol. Sci., 5(3): 85–87.

Rahi, D.K., Kaur, A. and Kaur, M. (2020). Hyperproduction of laccase by an indigenous species of *PleurotusPulmonarius* under solid state fermentation and *in vitro* potential applications. Res. J. Biotechnol., 15(2): 117–128.

Rajwar, D., Paliwal, R. and Rai, J.P.N. (2017). Biodegradation of pulp and paper mill effluent by co-culturing ascomycetous fungi in repeated batch process. Environ. Monit. Assess., 189(9): 1–16.

Rani, B., Kumar, V., Singh, J., Bisht, S., Teotia, P. et al. (2014). Bioremediation of dyes by fungi isolated from contaminated dye effluent sites for bio-usability. Braz. J. Microbiol., 45(3): 1055–1063.

Ravichandran, A. and Sridhar, M. (2016). Versatile peroxidases: Super peroxidases with potential biotechnological applications—A mini review. J. Dairy Vet. Anim. Res., 4: 277–280.

Ravichandran, A., Rao, R.G., Thammaiah, V., Gopinath, S.M. and Sridhar, M. (2019). A versatile peroxidase from *Lentinus squarrosulus* towards enhanced delignification and *in vitro* digestibility of crop residues. BioResources, 14(3): 5132–5149.

Reddy, G.V., Babu, P.R., Komaraiah, P., Roy, K.R.R.M. and Kothari, I.L. (2003). Utilisation of banana waste for the production of lignolytic and cellulolytic enzymes by solid substrate fermentation using two *Pleurotus* species (*P. ostreatus* and *P. sajor-caju*). Process Biochem., 38(10): 1457–1462.

Reyes, P., Pickard, M.A. and Vazquez-Duhalt, R. (1999). Hydroxybenzotriazole increases the range of textile dyes decolorized by immobilised laccase. Biotechnol. Lett., 21(10): 875–880.

Risdianto, H., Sofianti, E., Suhardi, S.H. and Setiadi, T. (2012). Optimisation of laccase production using white rot fungi and agricultural wastes in solid-state fermentation. J. Eng. Technol. Sci., 44(2): 93–105.

Riva, S. (2016). Laccases: Blue enzymes for green chemistry. Tr. Biotechnol., 24: 219–226.

Rodríguez-Couto, S., Sanromán, M.A., Hofer, D. and Gübitz, G.M. (2004a). Production of laccase by *Trametes hirsuta* grown in an immersion bioreactor and its application in the decolorization of dyes from a leather factory. Eng. Life Sci., 4(3): 233–238.

Rodríguez-Couto, S. (2012). Laccases for denim bleaching: An eco-friendly alternative. Sigma, 1: 10–2.

Senthilkumar, S., Perumalsamy, M. and Prabhu, H.J. (2014). Decolourization potential of white-rot fungus Phanerochaete chrysosporium on synthetic dye bath effluent containing Amido black 10B. J. Saudi Chem. Soc., 18(6): 845–853.

Setti, L., Giuliani, S., Spinozzi, G. and Pifferi, P.G. (1999). Laccase catalyzed-oxidative coupling of 3-methyl 2-benzothiazolinone hydrazone and methoxyphenols. Enzyme Microb. Technol., 25(3-5): 285–289.

Sharma, B., Dangi, A.K. and Shukla, P. (2018). Contemporary enzyme based technologies for bioremediation: A review. J. Environ. Manag., 210: 10–22.

Shelke, V. (2001). Enzymatic decolourisation of denims: A novel approach. Colourage, 48(1): 25–26.

Shrivastava, R., Christian, V. and Vyas, B.R.M. (2005). Enzymatic decolourisation of sulfonphthalein dyes. Enzyme Microb. Technol., 36(2-3): 333–337.

Silva, M.L.C., de Souza, V.B., Santos, V.S., Kamida, H.M., de Vasconcellos-Neto, J.R.T. et al. (2014). Production of manganese peroxidase by *Trametes villosa* on unexpensive substrate and its application in the removal of lignin from agricultural wastes. Adv. Biosci. Biotechnol., 5(14): 1067.

Singh, D., Zeng, J. and Chen, S. (2011). Increasing manganese peroxidase productivity of *Phanerochaete chrysosporium* by optimising carbon sources and supplementing small molecules. Lett. Appl. Microbiol., 53(1): 120–123.

Srinivasan, C., Dsouza, T.M., Boominathan, K. and Reddy, C.A. (1995). Demonstration of laccase in the white rot Basidiomycete *Phanerochaete chrysosporium* BKM-F1767. Appl. Environ. Microbiol., 61(12): 4274–4277.

Tanaka, H., Koike, K., Itakura, S. and Enoki, A. (2009). Degradation of wood and enzyme production by *Ceriporiopsis subvermispora*. Enz. Microb. Technol., 45(5): 384–390.

Taşpınar, A. and Kolankaya, N. (1998). Optimisation of enzymatic chlorine removal from Kraft pulp. Bull. Environ. Contam. Toxcol., 61(1): 15–21.

Tekere, M., Zvauya, R. and Read, J.S. (2001). Ligninolytic enzyme production in selected sub-tropical white rot fungi under different culture conditions. J. Basic Microbiol., 41(2): 115–129.

Thurston, C.F. (1994). The structure and function of fungal laccases. Microbiology, 140(1): 19–26.

Tsukihara, T., Honda, Y., Sakai, R., Watanabe, T. and Watanabe, T. (2006). Exclusive overproduction of recombinant versatile peroxidase MnP2 by genetically modified white rot fungus, *Pleurotus ostreatus*. J. Biotechnol., 126(4): 431–439.

Tzanov, T., Silva, C.J., Zille, A., Oliveira, J. and Cavaco-Paulo, A. (2003). Effect of some process parameters in enzymatic dyeing of wool. Appl. Biochem. Biotechnol., 111(1): 1–13.

Ünyayar, A., Mazmanci, M.A., Ataçağ, H., Erkurt, E.A. and Coral, G. (2005). A Drimaren Blue X3LR dye decolourising enzyme from *Funalia trogii*: One step isolation and identification. Enzyme Microb. Technol., 36(1): 10–16.

Valentin, L., Feijoo, G., Moreira, M.T. and Lema, J.M. (2006). Biodegradation of polycyclic aromatic hydrocarbons in forest and salt marsh soils by white-rot fungi. Int. Biodeter. Biodegr., 58(1): 15–21.

Valentin, L., Lu-Chau, T.A., López, C., Feijoo, G., Moreira, M.T. and Lema, J.M. (2007). Biodegradation of dibenzothiophene, fluoranthene, pyrene and chrysene in a soil slurry reactor by the white-rot fungus *Bjerkandera* sp. BOS55. Proc. Biochem., 42(4): 641–648.

Virk, A.P., Sharma, P. and Capalash, N. (2012). Use of laccase in pulp and paper industry. Biotechnol. Prog., 28(1): 21–32.

Wan, C. and Li, Y. (2012). Fungal pretreatment of lignocellulosic biomass. Biotechnol. Adv., 30(6): 1447–1457.

Wang, J., Ogata, M., Hirai, H. and Kawagishi, H. (2011). Detoxification of aflatoxin B1 by manganese peroxidase from the white-rot fungus *Phanerochaete sordida* YK-624. FEMS Microb. Lett., 314(2): 164–169.

Wesenberg, D., Kyriakides, I. and Agathos, S.N. (2003). White-rot fungi and their enzymes for the treatment of industrial dye effluents. Biotechnol. Advn., 22(1-2): 161–187.

Wu, J., Xiao, Y.Z. and Yu, H.Q. (2005). Degradation of lignin in pulp mill wastewaters by white-rot fungi on biofilm. Bioresour. Technol., 96(12): 1357–1363.

Yoon, M.Y. (1998). Process for Improved Shrink Resistance in Wool. WO patent 98/27264.

Zavarzina, A.G., Lisov, A.V. and Leontievsky, A.A. (2018). The role of ligninolytic enzymes laccase and a versatile peroxidase of the white-rot fungus *Lentinus tigrinus* in biotransformation of soil humic matter: Comparative *in vivo* study. J. Geophys. Res., 123(9): 2727–2742.

Zhang, H., Zhang, S., He, F., Qin, X., Zhang, X. and Yang, Y. (2016). Characterisation of a manganese peroxidase from white-rot fungus *Trametes* sp. 48424 with strong ability of degrading different types of dyes and polycyclic aromatic hydrocarbons. J. Hazard. Mater., 320: 265–277.

Zhang, H., Zhang, J., Zhang, X. and Geng, A. (2018). Purification and characterisation of a novel manganese peroxidase from white-rot fungus *Cerrena unicolor* BBP6 and its application in dye decolorization and denim bleaching. Proc. Biochem., 66: 222–229.

Zhu, Y., Kaskel, S., Shi, J., Wage, T. and van Pée, K.H. (2007). Immobilisation of *Trametes versicolor* laccase on magnetically separable mesoporous silica spheres. Chem. Mater, 19(26): 6408–6413.

Zille, A. (2005). Laccase Reactions for Textile Applicationss. PhD Thesis, University of Minho, Braga.

Zimbardi, A.L., Camargo, P.F., Carli, S., Aquino Neto, S., Meleiro, L.P. et al. (2016). A high redox potential laccase from *Pycnoporus sanguineus* RP15: Potential application for dye decolourisation. Int. J. Mol. Sci., 17(5): 672.

Zouari, H., Labat, M. and Sayadi, S. (2002). Degradation of 4-chlorophenol by the white rot fungus *Phanerochaete chrysosporium* in free and immobilised cultures. Bioresour. Technol., 84(2): 145–150.

Pigments

13

Industrial Applications of Pigments from Macrofungi

Malika Suthar,[1] Ajay C Lagashetti,[1] Riikka Räisänen,[2] Paras N Singh,[1] Laurent Dufossé,[3] Seri C Robinson[4] and Sanjay K Singh[1,]*

1. INTRODUCTION

Macrofungi are fungi that produce visible, multi-cellular fruit bodies (sporocarps) either above the ground (epigeous) or below the ground (hypogeous) and generally belong to phylum Ascomycota and Basidiomycota, and a few are members of Zygomycota (Chang and Miles, 2004; Mueller et al., 2007). Macrofungi of the phylum Ascomycota include morels, truffles and cup fungi; whereas Basidiomycota includes all the agarics (mushrooms and toadstools), puffballs, bracket fungi, club fungi, coral fungi, stinkhorns, as well as chanterelles (Razak et al., 2014). Being one of the most diverse groups of organisms, macrofungi form an important part of the ecosystem due to the vital role they play in the ecosystem (Mueller et al., 2004; Zotti et al., 2013). By nature, most of the macrofungi are saprophytic or show symbiotic mycorrhizal association, while some of them are reported as plant pathogens (Mueller

[1] National Fungal Culture Collection of India (NFCCI), Biodiversity and Palaeobiology Group, Agharkar Research Institute, GG Agarkar Road, Pune- 411004, India.
[2] HELSUS Helsinki Institute of Sustainability Science, Craft Studies P.O. Box 8 (Siltavuorenpenger 10), 00014 University of Helsinki, Finland.
[3] Chimie et Biotechnologie des Produits Naturels (ChemBioProlab) & ESIROI Agroalimentaire, Université de la Réunion, 15 Avenue René Cassin, CS 92003, F-97744 Saint-Denis CEDEX, France.
[4] Department of Wood Science and Engineering, 119 Richardson Hall, Oregon State University, Corvallis, OR 97331, USA.
* Corresponding author: sksingh@aripune.org

et al., 2007; Razak et al., 2014; Singh et al., 2019). Saprophytic macrofungi are found to be involved in decomposition of dead and decaying matter and recycling of nutrients and symbiotic macrofungi shows a mutualistic association with both plants and animals, though some of the macrofungi live as parasites on living organisms (Piątek, 1999; Hou et al., 2012; Tang et al., 2015).

Since ancient times, macrofungi, especially the mushrooms, have drawn human attention and interest because of their nutritional and therapeutic properties which have proved beneficial to mankind (Chang and Miles, 2004; El Enshasy et al., 2013; Dutta and Acharya, 2014; Lu, 2014; Rahi and Malik, 2016; Badalyan and Zambonelli, 2019; Debnath et al., 2019; Winkelman, 2019). Many edible mushrooms belonging to genera *Agaricus, Auricularia, Cordyceps, Flammulina, Grifola, Hericium, Lentinus, Lactarius, Pleurotus, Tremella* and others have been used as food due to their nutritive properties (Chang and Miles, 2004; Boa, 2004; Bernas et al., 2006; Kalac et al., 2013; Badalyan and Zambonelli, 2019). Similarly, the medicinal properties of macrofungi (mainly belonging to Basidiomycota) have been known for ages and they have been used against different types of diseases or illnesses (Christensen, 1972; Wasser, 2010; Stamets and Zwickey, 2014). Mushrooms of the genera *Agaricus, Cordyceps, Ganoderma, Grifola, Inonotus, Lentinula, Lentinus, Schizophyllum, Trametes* and others have been used as medicine due to their therapeutic properties, such as antioxidant, antimicrobial, immunostimulatory, anti-inflammatory, anticancer and other activities (Ganeshpurkar et al., 2010; Hilszczańska, 2012; Pohleven et al., 2016; Badalyan and Zambonelli, 2019). Traditional Chinese medicine has also evolved with these mushrooms as an important ingredient in their ethnobotanical approaches towards disease therapy (Lee et al., 2012; Wu et al., 2019).

Macrofungi have been found to possess a large number of bioactive metabolites showing their potential applications in different industries, such as food, detergent, health and medicine. Numerous studies have shown that these macrofungi are a good source of industrially important bioactive compounds, like enzymes, hormones, antibiotics, terpenoids, alkaloids, acids, proteins, vitamins and pigments (Quang et al., 2006; Wasser, 2010; Camassola, 2013; Chang and Wasser, 2018; Agyare and Agana, 2019; Ho et al., 2020). Different types of enzymes, such as amylase, cellulase, protease, tyrosinase, laccase, lipase, urease, lignin peroxidase, manganese peroxidase and nitrate reductase from macrofungi or mushrooms were used in different industries, such as food and detergent among others (Nadim et al., 2015; Nguyen et al., 2018; Krupodorova et al., 2019). Mushrooms are also reported to be an ubiquitous source of important fatty acids, such as linoleic acid, palmitic acid, oleic acid, stearic acid and arachidic acids (Sande et al., 2019). Different types of medicinally important metabolites, for example antibiotics and anticancer drugs, are also extracted from mushrooms (Wasser and Weis, 1999; Schwan et al., 2012; Patel and Goyal, 2012; Valverde et al., 2015; Blagodatski et al., 2018; Agrawal and Dhanasekaran, 2019). Besides these, mushrooms have been reported to be a rich source of other nutritionally important metabolites, like proteins, carbohydrates and vitamins (Wani et al., 2010; Kakon et al., 2012; Kalac et al., 2013; Valverde et al., 2015).

In addition, pigment production is one of the charismatic traits of these macrofungi. Many 'beautifully coloured' mushrooms represent a potential source of pigments as they produce a high yield of pigments. Contrary to synthetic pigments, these macrofungal ones are versatile, biodegradable and eco-friendly; therefore, they are attracting greater public attention as well as a growing interest in industries across the world. Several studies have shown the pigment production potential of different macrofungi (Gill and Steglich, 1987; Zhou and Liu, 2010; Valisek and Cejpek, 2011; Singh, 2017; Lagashetti et al., 2019). Pigments of macrofungi are not only mere colours but also these are bioactive compounds exhibiting antioxidant, antimicrobial, anticancer, antiviral, antimalarial, cytotoxic, as well as other important activities. Due to these additional attributes, the macrofungal pigments display multifaceted applications in different industries, such as food, cosmetics, pharmaceuticals, textiles and others (Zhou and Liu, 2010; Valisek and Cejpek, 2011; Caro et al., 2017; Mukherjee et al., 2017; Lagashetti et al., 2019; Ramesh et al., 2019).

2. Large-scale Production Techniques

Among macrofungi described by taxonomists, about 200 of them are experimentally grown and 20 are commercially cultivated (Khanna and Sharma, 2014). The current world industry provides thousands of tons of *Agaricus bisporus*, *Agrocybe aegerita*, *Auricularia polytricha*, *Calocybe indica*, *Flammulina velutipes*, *Ganoderma lucidum*, *Lentinus edodes*, *Pleurotus* spp., *Volvariella volvacea* and others for food and medicinal markets. No doubt that new species will be domesticated in the future with new markets.

Up to now, large-scale cultivation of macrofungi is mostly based on traditional solid-growth media techniques. Cultivation requires the manufacture of composted substrate produced as a result of aerobic thermogenesis of different agricultural byproducts, like cereal straw, maize stalks, sugarcane bagasse or wood chips by a succession of microflora. The objective is mainly to produce whole fruit bodies.

With the aim of producing macrofungal pigments, such fruit bodies would be destroyed by pigment extraction and not sold. Only co-products would be valued (Fig. 1).

Future techniques could be based on:

a) solid-state fermentation, with a controlled environment (pH, temperature and agitation). Most researchers know that agitation may disrupt the pigment production from macrofungi. The separation of biomass and value-added compounds (i.e., pigments) from the solid medium is also an issue;

b) submerged fermentation, with also a controlled environment (pH, nutrients, temperature and agitation). Liquid growth of macrofungal hyphae is a major hurdle; however, separation of the biomass and/or value-added compounds (i.e., pigments) from the liquid medium is easy.

In our opinion, large-scale production of pigments from macrofungi is not economically viable on its own and value of the production should be also brought by additional co-products for health, for food, for industry (Fig. 2).

Large-scale cultivation of macrofungi

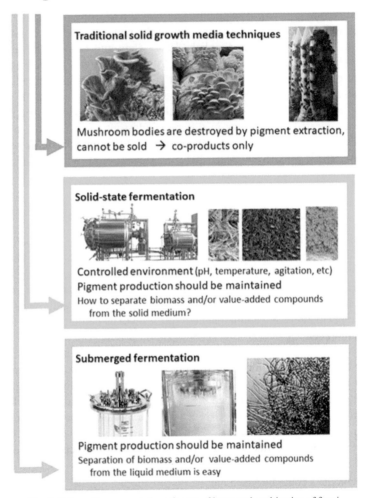

Fig. 1. Schematic representation of types of large-scale cultivation of fungi.

3. Pigments and the Textile Industry

3.1 Case Studies with Cortinarius Species

Natural compounds have a long history as colourants in the textile industry but after the synthetic dyes' revolution, their usage has been minimal. Synthetic dyes and pigments are produced in vast amounts with the global market value being 33.2 billion USD (2019) and which will rowing grow by 5.9 per cent until 2024—the growth being primarily due to the increasing demand of dyes in textiles, paints and coatings, construction and plastics (Textile Dyes Market Report, 2019). One of the main reasons for the decline in use of natural colourants in long-lasting applications, such as in textiles, due to the incomparable stability of synthetic dyes, which is an

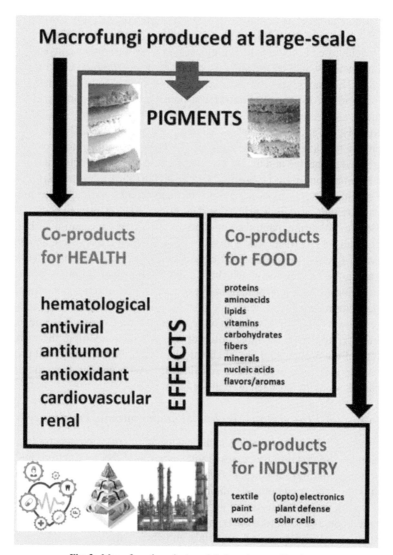

Fig. 2. Macrofungal products and their various applications.

important characteristic for efficiency in the dyeing process. However, in the current situation where sustainable solutions and circular economy principles are a major agenda, stability and inert nature of a compound pose a challenge in degradability. Another reason to search for more sustainable dyes for textiles is the observed toxicity and adverse effects that some synthetic dyes induce besides the hazards that dye containing wastewater can cause to photosynthesis and aquatic life (Hunger, 2003; Leme et al., 2015; Holkar et al., 2016; Oliveira et al., 2018; Tang et al., 2018; Räisänen et al., 2020; Umbuzeiro et al., 2020).

A great number of the recent research articles consider fungal colourants in food applications. This is obvious when looking at the stability and colour fastness of natural organic compounds, which, in most cases, are considerably lower than those of synthetic colourants. Bio-colourants are secondary metabolites of living organisms. They are seldom end products of biological processes but rather convert further to other compounds, which lead to their unstable character (Bell and Charlwood, 1980; Bell, 1981). This can be seen as a desired property for short-term material circulation and degradation processes required in a sustainable economy and for long-lasting applications, such as textiles.

Textile dyeing applications set several requirements for dyes. The most important properties of textile dyes are their affinity to the fibre, ability to produce strong bonding between the dye and the fibre, the stability of the dye structure and its ability to resist external conditions (Johnson, 1995; Hunger, 2003). Also, properties, such as the dye's light-fastness, toxicological profiles and tinctorial strength are important. Tinctorial strength of a dye primarily determines its cost-effectiveness (Hunger, 2003). Besides colouring, natural compounds are UV-protective, antimicrobial, water resistant and have electrical conductive properties, which impart them an additional value. Multi-functionality means efficiency in production, shorter processes and smaller energy requirements, which show a potential for innovations.

One major justification for natural dyes has been non-toxicity, but natural origin per se does not mean non-toxicity, nor sustainability. Textile producing and dyeing processes are very complex as they involve many different stages. It is true that dyes and dyeing create one of the most polluting stages, but in addition, there are many other steps of the process and use of chemicals, which influence the final properties of the textile, including toxicity and sustainability. With natural dyes, the usage of metal salts as mordant has been an issue of debate.

3.2 *Large-scale Production—Technical and Economic Feasibility*

When targeting industrial utilisation, technical feasibility of colourants for large-scale production and application processes is vital. Microbial dye production, in general, has enormous potential for a broad variety of dye structures, but more fundamental research is needed on compounds, their properties and scale up processes. Further, toxicity studies are necessary to be performed on each extracted colourant mixture and purified compound before choosing the targeted application. Investigations of different dyeing techniques and materials and their exposure to different testing conditions will finally answer the questions of feasibility and potentially, leading to further development of novel applications.

Macrofungi form an interesting source of raw material; also colourants (Orgiazzi et al., 2016) which has been recognised in many contexts. Macrofungi production for food purposes was nearly 9 million tonnes in 2018 (FAOSTAT, 2019). In the Northern Hemisphere, in Finland, the tradition of collecting fungal fruit bodies from forests through foraging networks is increasing and according to a latest report, there is a growing business in exploitation of all kinds of natural resources in the creation of value-added products and services; for example, in food, cosmetics and textile sectors (Honkanen, 2019; Rutanen, 2014). Control over the natural habitat conditions is also possible as it has been used successfully for producing eatable mushrooms, like truffles, *Tuber* species (Issakainen, 2013).

3.3 Pure Compound or a Mixture?

Water extracts of biological mass, which usually are complex mixtures of several chemical compounds, have traditionally been used as textile dyes. In most cases, these mixtures work well in small to medium-scale dyeing applications. However, when large-scale applications are involved, more attention needs to paid to additional impurities as optimisation of process parameters is important for a high-quality dyeing result and pro-environmental operations. Further, the dye as a mixture may more likely contain hazardous compounds. Purification seems necessary when looking at the contents of the Zero Discharge of Hazardous Chemicals (ZDHC) Manufactured Restricted Substance Lists (MRSL). The ZDHC MRSL goes beyond the traditional approaches to chemical restrictions, which only apply to finished products and it helps to protect consumers while minimising the possible impact of banned hazardous chemicals on the production workers, local communities, and the environment (ZDHC Foundation). Even though the pure compound would be the best solution, the question of purification needs further consideration when natural dyes are concerned because extraction and separation of individual substances are not straightforward processes. How can natural compounds, especially anthraquinones, be successfully and effortlessly separated from each other?

Extraction processes can be time consuming as they involve long and laborious steps. Often such steps combine various techniques, which have been selected on the basis of compound's solubility, volatility and stability. In this regard, the isolation of anthraquinones from the fungal material is relatively easy because anthraquinones are chemically stable, especially those of natural origin they contain plenty of OH-groups, rendering the compounds soluble in polar solvents, like water and alcohols. However, separation and purification of anthraquinones from each other sets challenges because anthraquinone structures have a strong tendency to interact with each other through π-π interactions as well as hydrogen bonding through hydroxyl and carboxyl groups forming self- and mixed-aggregates (i.e., dimers or greater clusters). In large-scale isolation, the problems with aggregation are even greater than in separation on an analytical scale. When targeting pro-environmental processes non-toxic solvents with the highest power of solvolysis are most desirable. Very often extraction solvents, such as water, alcohol and their mixtures are used (Diaz-Muñoz et al., 2018; Willemen et al., 2018). The purification process often utilises chromatographic techniques (Diaz-Muñoz et al., 2018), but liquid-liquid partition may be a more feasible alternative for large-scale separation and compound purification of complex mixtures (Hynninen and Räisänen, 2001; Räisänen, 2002; Poole, 2020). It seems that chromatographic methods have limited potential in separating samples containing several anthraquinone derivatives in very different ratios, as is the case with extracts of natural sources. Some of the derivatives overlap each other partially or completely. In addition, the adsorption of the compound on the solid support seems to be a problem in chromatography.

Even if the purified compound looks a reasonable target, the presence of non-colouring compounds in dye mixtures may play an important role in terms of other aspects, such as rate of solubility and stability of the colourant. It seems that the colour change (AE) value upon exposure to light is smaller for extracts than for

Compound	MW [g/mol] and formula	Chemical structure	Fibre and dyeing technique	Colour	LF
Emodin	270.0528 $C_{15}O_5H_{10}$		WO, direct dye		3
			WO, Al mordant dye		3
			WO, Fe mordant dye		2
			PA, direct dye		2
			PET, HT-disperse dye		7
Dermocybin	316.2650 $C_{16}O_7H_{12}$		WO, direct dye		2
			WO, Al mordant dye		1
			WO, Fe mordant dye		3
			PA, direct dye		1
			PET, HT-disperse dye		6
Dermorubin	344.0532 $C_{17}O_8H_{12}$	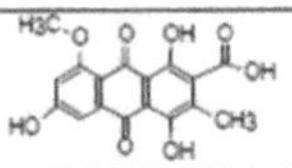	WO, acid dye		6
Mixture of anthraquinones	270.0528–478.1111 $C_{15}O_5H_{10}$– $C_{22}O_{12}H_{22}$		WO, direct dye		5
			WO, Al mordant dye		5
			PET, HT-disperse dye		6

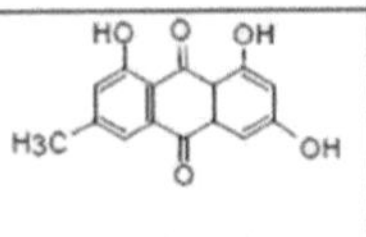

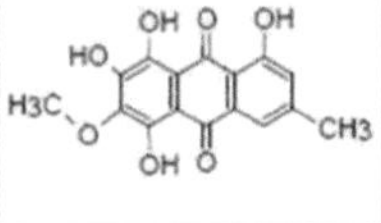

Fig. 3. Chemical structures of the anthraquinone dyes prepared from the fungi *Cortinarius sanguineus* and their usage in textile applications. LF = light-fastness according to ISO 105-B02 (Räisänen et al., 2001a, 2001b, 2002; Räisänen, 2002).

individual dye compounds, indicating a better colour stability to light for the total extract. This has been confirmed by studies where flavonoids from *Reseda luteola* and anthraquinones from *Rubia tinctorum* were used as dyes (Willemen et al., 2018). Actually, such conclusions could also be drawn, up to a point from our studies (Fig. 3). Willemen et al. in 2018 concluded that the use of a total plant extract is advantageous over the use of individual dye compounds in terms of rate of solubility, colour tone and stability towards light. Obviously, the production costs are also more favourable in the case of extracts. More research on the properties of pure dyes and mixtures needs to be performed to understand better the boundary conditions of dyeing processes versus dyed product qualities.

Numerous studies have reported the dyeing potential of different macrofungal pigments, like bright colours and good colour-fastness properties of natural anthraquinones, such as emodin, dermocybin, dermorubin and 5-chlorodermorubin obtained from *Cortinarius sanguineus* which have proved their significant potential as textile dyes (Räisänen et al., 2001a; 2001b; Räisänen, 2002; Räisänen et al., 2002). Similarly, another study has shown the dyeing potential of water-soluble pigments of *Cordyceps farinosa* (formerly known as *Isaria farinosa*) for dyeing cotton yarn (Velmurugan et al., 2010). An orange pigment from the mushrooms, such as *Amanita muscaria, Trametes versicolor* (formerly known as *Coriolus versicolor*) and *Ganoderma applanatum* has been used for dyeing silk and cotton fabrics. Likewise, dyes obtained from *Ganoderma lucidum* and *Pycnoporus sanguineus* have also been mentioned for their potential application in dyeing cotton and silk yarns (Karuppan et al., 2014). A study reports the use of orange dye obtained from fruit bodies of *Pycnoporus* species for dyeing pre-treated cotton and silk fabrics and suggested its potential application in the textile industry (Subramanian et al., 2014). High stability as well as good colour-fastness property of the green pigment, xylindein from *Chlorociboria aeruginosa*, suggest their possible potential application in the textile industry for dyeing different types of fabrics, such as bleached cotton, worsted wool, spun polyamide (nylon 6.6), spun polyester (Dacron 54), spun polyacrylic and garment fabrics (Weber et al., 2014; Hinsch et al., 2015).

3.4 Case Studies with Cortinarius Species

There are extensive studies of fungal anthraquinones and their applications as acid dyes for wool and polyamide (Räisänen et al., 2002), mordant dyes for wool and polyamide (Räisänen et al., 2001b) and disperse dyes for polyester in the high temperature and aqueous environment (Räisänen et al., 2001a). All these experiments showed that bio-based anthraquinone colourants, either pure single compounds or mixtures, result in thoroughly colouring of textile fibres and that their colour fastness properties are among the highest for natural dyes. For the thermoplastic fibers the colour fastness properties of pure dyes, emodin and dermocybin and the mixture of anthraquinones as disperse dyes, were excellent and comparable to the synthetic reference CI Disperse Red 60 (Fig. 3) (Räisänen et al., 2001a). Further, HT-disperse dyeing method with natural anthraquinones was successful without additional chemicals, with the dye uptake being clearly over 97 per cent, thereby indicating that the method was beneficial in terms of low emissions.

When targeting improved colouration of materials, focus can be laid on colourants, substrates and methods. It is known that textile-dyeing operations consume great amounts of water, especially during rinsing operations to remove the unreacted dye. To reduce water consumption and shorten the dyeing procedure, a waterless method has been developed (Montero et al., 2000). Supercritical carbon dioxide (SC-CO_2) dyeing can be applied as a single procedure where no rinsing is needed. Recently, waterless dyeing using supercritical carbon dioxide has gained more attention and studies of a natural dye for polyester (PET) and biodegradable polylactide (PLA) fabric colouration have been carried out. It seems that good HT-disperse dyeing results in an aqueous environment predict success in SC-CO_2 dyeing (Räisänen et al., 2020).

3.5 Discussion on Pigments in Textile Applications

The colour molecules with varying stability occur in nature as diverse mixtures, which lead to colours not as pure and bright as the synthetic ones. On the other hand, when purified dyes or dyes produced as a single component dye (e.g., through metabolic engineering of fungi) are used, bright colours can be obtained. It seems obvious that when increasing the usage of bio-based dyes, it is also important to debate the current mass production of colourants and introduce new ways to produce colours. From the food sector, it is known that consumers' adoption of new sources of commodities is not a straightforward and simple process. Even though colour is visible and present in everybody's closest environment, like clothing, people very seldom think about what is its origin (Yli-Heikkilä, 2020). Colour is an important factor to take into account when aiming towards greater sustainability in textile-dyeing operations but also because it is a crucial factor in designing and marketing. Designers have influence on the future of colour, how it is determined and used and what kind of meaning and impact are attached to it. Sustainability calls for a new mindset from everybody in the entire textile production and consumption chain. In addition, consumers' opinions and ideas on future aesthetics need to be considered and through co-creation of all actors, develop interventions towards greater responsibility in colourants and textile production as well as consumption.

3.6 Macrofungal Pigments and the Wood Industry

There is a long, if somewhat hidden, history of macrofungal pigments in wood products. The oldest recorded use is a small box from the church of Bad Gandersheim (originally from Sicily) and thought to have been made at the end of the 13th century (Robinson et al., 2016). Strips of blue-green wood peek out from a veneer stack, adding a bright strip of colour; otherwise it has brown decoration. Although crafters at the time did dye wood, the colours were not long lasting and have long since faded. This particular piece, and many others like it, have retained their unique blue-green colouration, even under exposure to UV light and humidity changes. For centuries, art historians thought this colour was due to an unknown dye; however, when the affected wood was viewed under a microscope, blue-green hyphae were seen, pointing to the colour originating from *Chlorociboria* species, which is likely either *C. aeruginascens* or *C. aeruginosa* (the two native European *Chlorociboria* species, both colloquially known as 'elf's cup') (Fig. 4a). The pigment produced by the genus *Chlorociboria* is known as xylindein—a quinone pigment that is yet to be completely synthesised.

In the middle of the fifteenth century, the blue-green colour of the elf's cup became integral to work done by the *maestri di legname* (masters of wood) (Robinson et al., 2016). The central perspective in tarsi featured everyday scenes—books, animals, landscapes, humans and relied heavily on elf's cup stained wood for areas like water, grass, robes, book bindings and to give an illusion of marble. Some of the most famous work from this time includes the Gubbio Studiolo by Giuliano da Maiano (made between 1478–1482), originally made for the Duke of Urbino and on display currently at the Metropolitan Museum of Art. Although intarsia works from this time frame are housed in museums across the world, most appear to have originated from either Italy or Germany and that from only a few guild families (Vega Gutierrez and Robinson, 2017). Elf's cup stained wood was popular until the late seventeenth century, when synthetic dyes progressed to a stage that allowed for longer colour permanence. These cheaper and stable dyes replaced the expensive and limited blue-green wood, and the technique and knowledge were mostly forgotten (Fig. 4b–e).

Although used far less often than elf's cup, 'brown oak' produced, when oak heartwood is colonised by *Fistulina hepatica* (beefsteak fungus), was also used in West European intarsia work. Brown oak veneers were used during the Italian Renaissance for backgrounds and to give a 'wood texture' to images meant to represent wooden objects. Although *F. hepatica* is a brown rotting fungus (fungi which tend to preferentially degrade cellulose and hemicellulose, leaving behind the brown-coloured lignin in wood), it also secretes polyphenolic deposits on the wood's parenchyma, leading to a uniform brown colouration.

Several well-known furniture and cabinet makers popularised brown oak in wooden furniture in the 1960s. Mies van der Rohe has several pieces housed

Fig. 4. (a) Elf's cup on a log in Ontario, Canada; (b–e) all of a bureau from South Germany, 1560–1570, housed in the Bilbao Fine Arts Museum, Spain; (f) illustration of beefsteak fungus, also known as 'poor man's beefsteak' (from Jan Kops, 1881; *Flora Batava*, vol. 16, Amsterdam); (g) zone lines on true sandal wood (*Santalum lanceolatum*); (h) zone lines on red maple (*Acer rubrum*); (i) spalted elm bowl by Mark Lindquist and Seri Robinson.

in the New National Gallery, Berlin, of brown oak office furniture (among them are a reception table, wardrobe, and writing desk). James Krenov made several cabinets from brown oak (along with other kinds of fungal-pigmented wood) in the late twentieth century. Brown oak is still a very popular wood for furniture today, especially in Europe, where it is known as 'English Brown Oak' (Fig. 4f).

The 1960s in North America showcased a huge resurgence in the popularity of *spalted* wood (wood coloured internally by fungi, a classification that both elf's cup stained wood and brown oak fall into) for handicrafts and art. This trend was spearheaded by the father-son duo, Melvin and Mark Lindquist, who were wood turners to primarily utilise zone-lined wood in their crafts and, later, sculpture. Zone lines, the winding (often) black lines on decaying wood, are composed of melanin, a high molecular weight pigment and often serve as demarcation lines between genetically distinct fungal species (Fig. 4g, h).

Numerous fungal species can make zone lines, most notably white rotting fungi (Basidiomycetes), such as *Bjerkandera adusta, Polyporus brumalis* and *Trametes versicolor* among many others. Some soft rotting fungi (Ascomycetes) can also make zone lines, the most utilised in the wood industry being *Xylaria polymorpha*, which makes zone lines as a response to low moisture conditions as well as genetic incompatibility between strains.

Zone-lined wood is used in numerous applications, ranging from flooring to furniture to panelling. It is most ubiquitous in the wood-turning world, where it is a well-known favourite of wood turners looking for visually distinct and challenging pieces. Spalted wood in wood-turned objects first gained popularity with articles, 'Spalted Wood: Rare Jewels from Death and Decay' (1977) and 'Turning Spalted Wood' (1978) in *Fine Woodworking*, by Mark Lindquist and was quickly adopted by other prominent wood-turners, such as David Ellsworth (Lindquist, 1978). Melvin and Mark Lindquist, who brought about the resurgence of spalted wood popularity in the late twentieth century, have several of their spalted turnings in the permanent collection at the Renwick Gallery of the Smithsonian American Art Museum (among many other museums) (Fig. 4i).

Today, spalted wood is both ubiquitous in the wood products industry and simultaneously invisible. For those familiar with the material it is a deeply sought after commodity, worthy of a high sticker price (Vega Gutierrez and Robinson, 2020). For those unfamiliar with its history and potential, it is as useful as firewood. Yet, as consumer and artistic knowledge about spalted wood grows, so too does scientific investment. Once a gift from the forest, spalted wood can now be manufactured in a lab by using known fungal strains or even by growing fungi in batch culture and harvesting their pigments for direct application on to wood (Weber et al., 2016). The extracted pigments continue to find new uses outside of the wood products industry, such as textile dyes (Vega Gutierrez et al., 2019) and photovoltaic thin films (Giesbers, 2019). However spalted wood in woodcraft remains a valuable, historic art form, with a deep and enduring popularity among woodworkers.

4. Health-related Applications

4.1 Antimicrobial Activities

A number of macrofungal genera belonging to Ascomycota, such as *Hypocrella, Cordyceps, Torrubiella* and others have been reported to possess antimicrobial pigments. Two antimicrobial pigments, hypocrellins A and B, have been extracted from parasitic fungus *Hypocrella bambusae*, among which hypocrellin A has shown good antifungal activity against *Candida albicans* and moderate antibacterial activity against *Pseudomonas aeruginosa, Staphylococcus aureus* and *Mycobacterium intracellulare*; whereas hypocrellin B has shown weak antimicrobial activity (Ma et al., 2004). Similarly, a yellow pigment, paecilosetin, obtained from *Cordyceps farinosa* (formerly known as *Paecilomyces farinosus*) was found to be active against the bacterium (*Bacillus subtilis*) and fungi (*Trichophyton mentagrophytes*, and *Cladosporium resinae*) (Lang et al., 2005). Pigments like torrubiellins A and B obtained from *Torrubiella* spp. BCC 28517 have shown significant antibacterial (*Bacillus cereus*) and antifungal (*Candida albicans*) activities and among these two, torrubiellin B has shown higher activity as compared to torrubiellin A (Isaka et al., 2012).

Besides, several species of the phylum Basidiomycota have reported a variety of pigments possessing antimicrobial potential. Pigments, such as peniophorins A and B, isolated from *Phanerochaete affinis* (formerly known as *Peniophora affinis*) have reported the antimicrobial activity specifically against Gram-positive cocci and *Proteus vulgaris* and among these, peniophorin B was found to be more active as compared to peniophorin A (Gerber et al., 1980). Pigments of *Punctularia atropurpurascens* (phlebiarubrone, 4'-hydroxyphlebiarubrone and 3/,4',4"-trihydroxyphlebiarubrone) have also displayed antimicrobial activity against some of the pathogenic bacteria (*Acinetobacter calcoaceticus, Bacillus brevis,* and *B. subtilis*) and fungi (*Mucor miehei, Nematospora coryli, Paecilomyces variotii,* and *Penicillium notatum*) (Anke, 1984). Many studies have shown that an orange pigment, cinnabarin, obtained from *Pycnoporus sanguineus*, has antibacterial potential against *B. cereus, Escherichia coli, E. faecalis, E. faecium, Klebsiella pneumoniae, Leuconostoc mesenteroides, L. plantarum, Psuedomonas aeruginosa, Salmonella typhi, Staphylococcus aureus* and several *Streptococcus* spp. (Smania Jr. et al., 1995a; 2003, Smania et al., 1998). Pulveraven A and vulpinic acid, isolated from fruit bodies of *Pulveroboletus ravenelii*, have displayed antibacterial activity (Duncan et al., 2003).

Two yellow pigments, such as 5-(3-chloro-4-hydroxybenzylidene) tetramic acid and methyl-3',5'-dichloro-4,4'-di-*O*-methylatromentate, extracted from *Leccinum pachyderme* (formerly known as *Chamonixia pachydermis*) and *Scleroderma* spp. respectively, have reported mild antibacterial activity against *B. subtilis* (Van Der Sar et al., 2005; Lang et al., 2006). Pigments of genus *Mycena* have been found to possess good antimicrobial potential. A red pigment, mycenaurin, obtained from *Mycena aurantiomarginata* has shown antibacterial activity against *Bacillus pumilus* (Jaeger and Spiteller, 2010). Another species of the genus *Mycena*, i.e., *Mycena haematopus* has been mentioned to produce different pigments, such as mycearubin A and mycearubins D-F, mycenaflavins A-D, haematopodin, and haematopodin B. Among these, mycenarubin D was found to be most effective against *Azovibrio restrictus, Azoarcus tolulyticus,* and *Azospirillum brasilense* while mycenaflavin A and haematopodin B were effective against *Azovibrio restrictus,* and *Azoarcus tolulyticus*; mycenaflavin B and mycenarubin E were effective only against *Azovibrio restrictus* (Lohmann et al., 2018).

Different pigments of the genera *Cortinarius* and *Dermocybe* (phallacinol, physcion and phlegmacin) have been mentioned to be potent antibacterial and antifungal agents (Gessler et al., 2013; Ramesh et al., 2019). Similarly, a yellow alkaloid, i.e., 6-hydroxyquinoline-8-carboxylic acid, isolated from fruit bodies of *Cortinarius subtortus*, has displayed inhibitory effect against phytopathogenic fungus, *Colletotrichum coccodes* (Teichert et al., 2008a). A recent study has reported weak antimicrobial activity of the nor-guanacastepene pigments (pyromyxones A, B, and D) of mushroom *Cortinarius pyromyxa* against bacteria (*B. subtilis* and *Aliivibrio fischeri*) and also against phytopathogenic fungi (*Botrytis cinerea, Septoria tritici* and *Phytophthora infestans*) (Lam et al., 2019). Basidiomycete *Quambalaria cyanescens* has been reported to produce three biologically active pigments (quambalarine A, quambalarine B, and mompain). Among them, quambalarine A was found to be effective against human pathogenic bacteria as well as fungi; quambalarine A was effective against fungi whereas mompain was effective only against bacteria (Stodulkova et al., 2015).

Natural melanin from different macrofungi has been found to possess promising biological activity. Melanin from *Schizophyllum commune* has antibacterial activity against *B. subtilis, E. coli, K. pneumoniae,* and *P. fluorescens* as well as antifungal activity against *Trichophyton simii* and *T. rubrum* (Arun et al., 2015). *Lachnum* sp. YM30 also exhibited anti-bacterial activity against *Vibrio parahaemolyticus* and *S. aureus*. Studies suggest that this antibacterial activity of melanin occurs by damaging the cell membrane and affecting bacterial membrane functioning (Xu et al., 2017). Pure melanin isolated from *Armillaria mellea* displayed antibacterial activity against *Bacillus cereus, B. subtilis, E. faecalis* and *P. aeruginosa* (Lopusiewicz et al., 2018c). Melanin obtained from *Auricularia auricula* showed antibacterial activity against *E. coli, P. aeruginosa* and *P. fluorescens* by inhibition of quorum-sensing (QS)-regulated biofilm formation (Bin et al., 2012). Similarly, melanin of the *Exidia nigricans* and *Scleroderma citrinum* has also been reported to have antibacterial activity against *E. faecalis* and *P. aeruginosa* (Lopusiewicz et al., 2018a; Lopusiewicz et al., 2018b).

Researchers have also shown the antibacterial potential of carotenoid pigments of certain xylotrophic Basidiomycetes (*Fistulina hepatica, Fomes fomentarius* and *Laetiporus sulphureus*) (Velygodska and Fedotov, 2016). An edible wood-rotting Basidiomycete *Laetiporus sulphureus* has also been mentioned to possess an antimicrobial pigment, laetiporic acid A (Popa et al., 2016; Ramesh et al., 2019). Besides this, a red pigment 7-acetyl-4-methylazulene-1-carbaldehyde isolated from *Lactarius deliciosus* has also exhibited moderate antibacterial activity against *S. aureus* (Tala et al., 2017). Also, pigments of *Tapinella atrotomentosa* (spiromentins B and C) have reported significant antibacterial activity against *Acinetobacter baumannii* and *E. coli* (Ben et al., 2018). Moreover, yellow-orange pigment, amitenone, from *Suillus bovinus* has also been described for its antibacterial ability (Ramesh et al., 2019).

4.2 *Antioxidant Activities*

Family Hymenochaetaceae has been found to be a rich source of a variety of antioxidant pigments. A variety of antioxidant pigments, which mainly include hispidin derivatives, such as inoscavins A-E, methylinoscavins A-D, phelligridins D-F, davallialactone, methydavallialactone, interfungins A-C have been reported from macrofungus *Hymenochaete xerantica* (formerly known as *Inonotus xeranticus*) by several researchers using superoxide, ABTS and DPPH free radical-scavenging assay (Kim et al., 1999; Lee et al., 2006a; 2006b, Lee et al., 2007b; Lee and Yun, 2006; Zhou and Liu, 2010). Likewise, yellow polyphenols (inonoblins A-C) along with phelligridin D (yellow) and phelligridins E-G (orange) extracted from *Inonotus obliquus* have displayed significant antioxidant activity against ABTS and DPPH radical but moderate activity against superoxide radical (Lee et al., 2007a). Another genus *Phellinus* has been described as an important source of antioxidant pigments. A yellow pigment extracted from *Phellinus igniarius* has been found to possess strong antioxidant activity with an IC_{50} value of 10.2 µM (Wang et al., 2005). Similarly, pigments, phelligridins H-J and davallialactone, isolated from Chinese medicinal mushroom *Phellinus igniarius* have shown antioxidant activity by inhibiting rat liver microsomal lipid peroxidation (Wang et al., 2007b). Pigments extracted from *Phellinus* sp. KACC93057P (phellinins A-C, and hispidin) exhibited the potent antioxidant activity in a concentration-dependent manner when tested with DPPH, ABTS and superoxide free radical-scavenging assay (Lee et al., 2009a; Lee et al., 2010). Certain yellow-orange pigments of *Phellinus ellipsoideus* (formerly known as *Fomitiporia ellipsoidea*), which include inoscavin C, inonoblin B and phelligridin K, have shown potent ABTS radical-scavenging ability (Zan et al., 2015).

Besides these, many other macrofungi have been described to produce a diverse class of pigments showing good antioxidant potential. A new radical scavenger, thelephorin A (grey), has been isolated from the mushroom *Thelephora vialis* and its antioxidant activity has been confirmed by DPPH assay (Tsukamoto et al., 2002). The styrylpyrone pigments (hispidin and bisnoryangonin) of *Gymnopilus junonius* (formerly known as *Gymnopilus spectabilis*) have displayed significant and promising antioxidant activity when tested with the ABTS, DPPH and superoxide free radical-scavenging assay (Lee et al., 2008). A brown pigment, clitocybin A, isolated from the culture broth of *Hygrophoropsis aurantiaca* (formerly known as *Clitocybe aurantiaca*) reported the free radical, scavenging activity against DPPH, ABTS and superoxide radicals (Kim et al., 2008). Likewise, (iso)-quinoline alkaloids (6-hydroxyquinoline-8-carboxylic acid and 7-hydroxy-1-oxo-1,2-dihydroisoquinoline-5-carboxylic acid) extracted from fruit bodies of *Cortinarius subtortus* have shown moderate antioxidant activity by DPPH radical-scavenging method (Teichert et al., 2008a). Antioxidant potential of anthraquinone pigments (rufoolivacin, rufoolivacin C, rufoolivacin D, leucorufoolivacin) isolated from the edible macrofungus, *Cortinarius purpurascens* have been confirmed by DPPH radical-scavenging assay (Bai et al., 2013). Three yellow pigments (cyathusals A-C) obtained from a fermented mushroom *Cyathus stercoreus* have shown antioxidant activity when evaluated with DPPH and ABTS free radical-scavenging assay (Kang et al., 2007).

Melanin, isolated from *Hypoxylon archeri*, has shown antioxidant activity by significantly inhibiting oxidation of TNB (5-thio-2-nitrobenzoic acid) caused by HOCl and hydrogen peroxide. Hence, this melanin has proven to be an efficient scavenger of peroxide free radicals (Wu et al., 2008). Similarly, melanin from *Ophiocordyceps sinensis* has also proved to be an effective DPPH scavenger as compared to BHT (butylated hydroxyl toluene) and alpha-tocopherol, the commercial antioxidants. It has also displayed much better ferrous ion-chelating ability than that of water extract, making it a strong ferrous ion chelator (Dong et al., 2012). Melanin extracted from *Lachnum singerianum* has inhibited lipid peroxidation and slows down the ageing process of this Basidiomycete, suggesting that this pigment could be used as a potential anti-ageing drug (Lu et al., 2014). Natural melanin extracted from *Inonotus hispidus* has also shown strong antioxidant activity (Arun et al., 2015). Melanin pigment from *Schizophyllum commune* has exhibited concentration-dependent free radical-scavenging activity of DPPH. The scavenging activity has increased from 87 per cent to 96 per cent when the melanin concentration was raised from 10 μg to 50 μg (Lopusiewicz et al., 2018a). Likewise, melanin obtained from *Scleroderma citrinum* has been reported for its significant antioxidant activity (Lopusiewicz et al., 2018b).

Carotenoid pigments of *Laetiporus sulphureus*, *Fomes fomentarius*, and *Fistulina hepatica* have also been found to possess good antioxidant potential (Velygodska and Fedotov, 2016). Moreover, yellow pigment BNT (1,1'–binaphthalene–4,4'–5,5'–tetrol), isolated from the ascomycetous fungus, *Hypoxylon fuscum*, has shown very good antioxidant activity when tested with DPPH radical-scavenging assay (Quang et al., 2004a). Two violet pigments, i.e., spiromentins B and C, obtained from *Tapinella atrotomentosa* have also reported remarkable antioxidant activity upon screening by DPPH and oxygen radical absorbance capacity (ORAC) assay (Ben et al., 2018).

4.3 Anticancer Activities

The anticancer property of several macrofungal pigments has been studied by several investigators. Most of the pigments of Basidiomycetes have shown the anticancer potential. Pigments of mushroom, *Phellinus linetus* (meshimakobnol A, meshimakobnol B, phellifuropyranone A, and phelligridin G) have reported *in vitro* anti-proliferative activity against human lung cancer cells and mouse melanoma cells (Nagatsu et al., 2004; Kojima et al., 2008). Similarly, yellow pigments (phelligridin C and phelligridin D) isolated from *Phellinus igniarius* have shown *in vitro* selective cytotoxicity against a human lung adenocarcinoma cell line (A549) and hepatocarcinoma cell line (Bel7402) (Mo et al., 2004); whereas phelligridin J was found to be cytotoxic against four cancer cell lines (A2708, A549, Bel-7402, HCT-8) (Wang et al., 2007b). Moreover, brown pigments, meshimakobnols A and B isolated from *Phellinus igniarius*, have effectively inhibited HepG2 cells (IC_{50} values 19.2 ± 2.0, and 16.7 ± 1.6 μM respectively) and Lu cells (IC_{50} values 35.2 ± 3.4 and 21.7 ± 2.8 μM respectively) (Thanh et al., 2017). Likewise, pigments of *Albatrellus confluens*, such as albatrellin as well as aurovertins P and S, have also reported anticancer activity. Pigment albatrellin exhibited cytotoxic activity against human lung carcinoma cells (HepG2) with an IC_{50} value of 1.55 μg/ml (Yang et al., 2008); whereas yellow pigments, aurovertins P and S, displayed moderated cytotoxicity against five human cancer cell lines, which include lung cancer (A-549), breast cancer (SK-BR-3), pancreatic cancer (PANC-1), human myeloid leukemia (HL-60) and hepatocellular carcinoma (SMMC-7721) cell lines (Guo et al., 2013).

Also, pigments obtained from bracket fungus, *Pycnoporus cinnabarinus*, have reported varying degrees of antitumor activity against murine leukemia cell line (P388). Cinnabarin showed significant antitumor activity with an IC_{50} value of 13 μM at 1 mg/ml followed by pycnoporin, cinnabarinic acid and tramesanguin (Dias and Urban, 2009). A stereaceous Basidiomycete has also produced two yellow polyene pigments [(3Z,5E,7E,9E,11E,13Z,15E,17E)-18-methyl-19-oxoicosa-

3,5,7,9,11,13,15,17-octaenoic acid and (3E,5Z,7E,9E,11E,13E,15Z,17E,19E)-20-methyl-21-oxodocosa-3,5,7,9,11,13,15,17,19-nonaenoic acid], showing anticancer potential against HUVEC (human umbilical vein epithelial cells), K-562 leukemia cells and HeLa cells (Schwenk et al., 2014). Melanin from *S. commune* was also screened against cancer cell lines by MTT assay and it has shown anticancer activity against human epidermoid larynx carcinoma cell line (HEP-2). This *in vitro* anti-cell proliferation effect was concentration-dependent. Melanin at 60 μg was reported to inhibit cell viability by 53 per cent (Arun et al., 2015). A study has mentioned that quambalarine B and mompain extracted from *Quambalaria cyanescens* have slowed down cell division in various cancer cell lines (HeLa, HEK 293, HTC 116, and A549) in a concentration-dependent manner by selectively targeting the mitochondria (Stodulkova et al., 2015).

Besides Basidiomycetes, many ascomycetous macrofungi have reported anticancer activity against different types of cancer cells. Yellow pigments (paecilosetin and farinosone B) isolated from *Cordyceps farinosa* (formerly known as *Paecilomyces farinosus*) have displayed anticancer activity against murine leukemic P388 cell line (IC_{50} values of 3.1 and 1.1 μg/ml, respectively) (Lang et al., 2005). Pigment shiraiarin, from the Ascomycete *Shiraia bambusicola*, has exhibited anticancer activity against human breast cancer cell lines (MCF-7, MCF-7b, and MDA-MB-435); whereas hypocrellin D has shown anticancer activity against different cancer cell lines (Bel-7721, A-549, and Anip-973) (Fang et al., 2006; Cai et al., 2008). Different pigments of the genus *Torrubiella* (*Torrubiella longissima* and *Torrubiella* spp. BCC 28517), such as torrubiellins A-B, torrubiellone A and torrubiellone E have reported cytotoxicity against different human cancer cell lines MCF-7 (breast cancer), NCI-H187 (lung cancer) and KB (oral cavity cancer) (Isaka et al., 2010, 2012, 2014). Aurovertins B, D, and M, discovered from *Calcarisporium arbuscula*, have also exhibited significant cytotoxic activities against triple-negative breast cancer cell lines (MDA-MB-231) (Zhao et al., 2016).

4.4 *Enzyme-inhibition Activities*

Yellow-red pigments (betulinans A-B) isolated from *Lenzites betulinus* have shown lipid peroxidation inhibition in rat liver microsomes with IC_{50} values of 0.46 and 2.88 μg/ml, respectively (Lee et al., 1996). Similarly, three yellow pigments (orirubenones A-C), showing extracellular hyaluronan-degradation inhibitory activity in human skin fibroblasts (Detroit 551), have been extracted from the mushroom, *Tricholoma orirubens* (Kawagishi et al., 2004). In one of the studies, researchers screened the inhibitory activity of 15 azaphilones from mushrooms of Xylariaceae family, which includes daldinins C, E, and F from *Hypoxylon fuscum*, entonaemin A and rubiginosins A-C from *H. rubiginosum*, rutilins A-B from *H. rutilum*, multiformin D from *Jackrogersella multiformis* (formerly known as *H. multiforme*), cohaerins A-B from *Jackrogersella cohaerens* (formerly known as *H. cohaerens*), and sassafrins A-C from *Creosphaeria sassafras* and identified these azaphilones as potential inhibitors of nitric oxide production with the help of cell-based system (RAW 264.7 cells). Finally, they have concluded that the strongest inhibitory activity has been observed in dimeric azaphilones (Quang et al., 2006d).

A study has reported cholinesterase inhibitory activity of yellow-green pigment, brunnein A, obtained from *Cortinarius brunneus* (Teichert et al., 2007). Likewise, pigments of several species of the genus *Phellinus* have shown inhibitory activity against different enzymes. For instance, yellow phenylpropanoid-derived polyketides (squarrosidine and pinillidine) isolated from *Pholiota squarrosa* and *Phellinus pini* have shown potent xanthine oxidase inhibitory activity (Wangun and Hertweck, 2007), orange pigments (phelligridins H and I) extracted from medicinal mushroom *Phellinus igniarius* have inhibited protein tyrosine phosphatase 1B (PTP1B) (Wang et al., 2007b); whereas pigments davallialactone, hypholomine B, and ellagic acid isolated from *Phellinus linetus* have exhibited inhibitory activity against human recombinant aldose reductase and rat lens aldose reductase with IC_{50} values of 0.56, 1.28, 1.37 µM and, 0.33, 0.82, 0.63 µM, respectively (Lee et al., 2008). Furthermore, a yellow clitocybin D, extracted from *Hygrophoropsis aurantiaca* (formerly known as *Clitocybe aurantiaca*), displayed the significant human neutrophil elastase (HNE) inhibitory activity with an IC_{50} value of 17.8 µM (Kim et al., 2009). A recent study has reported two important neuraminidase inhibitors (comazaphilone D and rubiginosin A) from the fruit bodies of *Glaziella splendens*. Both of them have inhibited significantly three types of neuraminidases from recombinant rvH1N1, H3N2, and H5N1 influenza A viruses (Kim et al., 2019).

4.5 Antiviral Activities

Red pigment cinnabarin, isolated from *Pycnoporus sanguineus*, has shown progressive reduction of rabies virus infection on mouse neuroblastoma cell (NA) culture (Smania A. Jr. et al., 2003; Ramesh et al., 2019). A pale-yellow pigment concentricolide, extracted from *Daldinia concentrica*, has also displayed antiviral activity against HIV-1 with EC_{50} value 0.31 µg/ml (Qin et al., 2006; Zhou and Liu, 2010). Similarly, pigments phelligridin F and phelligridin G, extracted from the fruit bodies of *Phellinus igniarius*, have been found to inhibit neuraminidases from recombinant influenza viruses (rvH1N1, H3N2, and H5N1), with IC_{50} values ranging from $0.7 \sim 8.1$ µM, which suggests their possible potential application as an antiviral drug (Kim et al., 2016). Along with these, genus *Suillus* has reported many antiviral pigments. Yellow pigment, flazin, isolated from *Suillus granulatus*, has displayed anti-HIV activity with an EC_{50} value of 2.36 µM, and therapeutic index 12.1 (Dong et al., 2007; Zhou and Liu, 2010). Likewise, amitenone from *Suillus bovinus* has been reported to possess antiviral activity (Ramesh et al., 2019).

4.6 Cytotoxic Activities

A study has confirmed the cytotoxic activity of the pigment of *Punctularia atropurpurascens* (phlebiarubrone and 4',4"-dihydroxyphlebiarubrone and 3/,4',4"-trihydroxyphlebiarubrone) using plant growth-regulating activity tests with *Lepidium sativum* and *Setaria italica* (Anke, 1984). Researchers have studied the toxicity of the cinnabarin obtained from *Pycnoporus sanguineus* and they have found that cinnabarin had no effect on mouse neuroblastoma cells at a concentration of 0.31 mg/ml. Moreover, a high concentration of cinnabarine also had no toxic effects on animals (Smania Jr. et al., 2003; Ramesh et al., 2019). Another pigment, xylariamide A, extracted from *Xylaria* species, has also exhibited cytotoxic activity when tested with brine shrimp (*Artemia salina*) lethality assay (Davis, 2005; Zhou and Liu, 2010).

4.7 Anti-malarial Activities

Yellow pigment, torrubiellone A, extracted from spider pathogenic fungus *Torrubiella* spp. BCC 2165, has shown weak antimalarial activity against *Plasmodium falciparum*, with an IC_{50} value of 8.1 µM (Isaka et al., 2010). Torrubiellin B, obtained from *Torrubiella* spp. BCC 28517, exhibited significant biological activity against *Plasmodium falciparum* (Isaka et al., 2012). Similarly, pigment torrubiellone E of *Torrubiella longissima* has reported antimalarial activity with an IC_{50} value of 3.2 µg/ml (Isaka et al., 2014).

4.8 Anti-Leishmanial Activities

Certain red pigments, such as hypocrellins A and B, isolated from *Hypocrella bambusae*, have shown anti-leishmanial activity against parasites, *Leishmania donovani* (Ma et al., 2004).

4.9 Anti-diabetic Activity

Researchers have isolated a red-brown pigment, fomentariol, from plant pathogenic macrofungus *Fomes fomentarius*, and its anti-diabetic potential has been evaluated (Maljuric et al., 2018).

4.10 Sensitizers

Many researchers have studied the photochemical properties of hypocrellins, such as hypocrellin A and B, isolated from parasitic fungus *Hypocrella bambusae*, and proposed their potential use in valuable photosensitisers for photodynamic therapy (PDT) in cancer or viral infection (Zhenjun and Lown, 1990; Zhenjun, 1995).

4.11 Cytochrome P450 Inhibitors

Pulvinic acid derivatives, such as atromentic acid, variegatic acid and xerocomic acid obtained from *Boletus calopus* and *Suillus bovinus*, have shown non-specific inhibitory effects against all four cytochrome P450 (CYP) (1A2, 2C9, 2D6, and 3A4) (Huang et al., 2009; Zhou and Liu, 2010).

5. Other Industrial Applications

5.1 Food Colourants

Researchers have extracted laetiporic acids from edible fruit bodies of *Laetiporus sulphureus* and mentioned their promising use as food colourants (Davoli et al., 2005). Yellow-red pigments, obtained from different species of *Cortinarius*, such as emodin, physcion, dermocybin, dermocybin-1-β-D-glycopyranoside, dermorubin, flavomannin, have also been reported as potent food-grade colourants (Caro et al., 2012; Ramesh et al., 2019).

5.2 Solar Cells

One of the studies has reported dye extracts of *Cortinarius sanguineus* as a source of sensitising dyes in dye-sensitised solar cells. In their study, they found that the device sensitised with dye extract of *C. sanguineus* was most efficient in the photon-to-current conversion process (Zalas et al., 2015).

6. Photo-defence/Light Barrier Properties

Purified natural melanin, obtained from *Exidia nigricans* and *Scleroderma citrinum*, has shown better light barrier properties compared to raw melanin (Lopusiewicz et al., 2018a, 2018b). A recent study has reported for the first time the fungal pigments of Basidiomycetes (*Inonotus obliquus*, *Cortinarius croceus*, and *Tapinella atrotomentosa*) as phototoxic metabolites and also suggested their potential role in photochemical defence (Siewert et al., 2019).

6.1 Opto Electronics

The genus *Chlorociboria* produces a blue-green pigment known as xylindein, which forms a durable porous film layer on various surfaces. Xylindein has moderate conductive properties in this form, which led researchers to study its (opto) electronic properties. Xylindein was found to have high electron mobility and was photostable, while the pigment is currently being used to make prototype organic semiconductors (Giesbers et al., 2018; Giesbers et al., 2019) (Fig. 5).

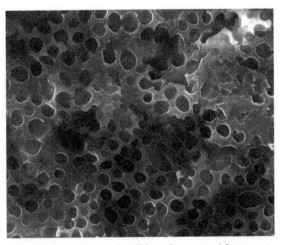

Fig. 5. The porous film of xylindein, made by *Chlorociboria* spp. (photo courtesy Dr. Sarath Vega Gutierrez).

6.2 Insecticidal Activity

Yellow polyene pigments [(3Z,5E,7E,9E,11E,13Z,15E,17E)-18-methyl-19-oxoicosa-3,5,7,9,11,13,15,17-octaenoic acid and (3E,5Z,7E,9E,11E,13E,15Z,17E,19E)-20-methyl-21-oxodocosa-3,5,7,9,11,13,15,17,19-nonaenoic acid] of a stereaceous Basidiomycete have confirmed their anti-insecticidal activity against *Drosophila melanogaster* (Schwenk et al., 2014).

6.3 Bioremediation Activities

Melanin pigment from the rhizomorphs of four *Armillaria* species (*A. ostoyae*, *A. calvescens*, *A. gemina* and *A. sinapina*) were studied and showed adsorption of cations of Al, Zn, Fe, Cu and Pb, such that these ions were found to be 50–100 times more focussed on rhizomorphs than in soil (Rizzo et al., 1992). In this way, melanin in fungi is useful in bioremediation of the environment because the pigment binds to radioactive wastes in polluted locales and reduces their spread (Pombeiro-Sponchiado et al., 2017).

7. Conclusion

Fungal pigments have a vast variety of structures and thus many interesting properties which open up possibilities for historically known and re-discovered applications as well as novel inventions. The exploration of the possibilities of fungal pigments has merely been started and yet there is a lot to discover. Besides colouring, fungal compounds have UV-protective, antimicrobial, water resistance and electrically conductive properties, among others, which give additional value to those compounds when applied to consumer commodities. It is also very important to understand the properties of individual compounds and their mixtures to ensure safe and sustainable products, as natural origin alone is not a guarantee of safety nor sustainability. Efforts should also be conducted by the scientific community to ensure large-scale production of the pigmented macrofungi, in solid state fermentation or submerged fermentation. Full valorisation of all biochemicals contained in macrofungal biomass is also an economic issue.

Considering their immense potential, diversity of fungi needs to be systematically explored and studied from various habitats and ecological niches in the world. Bioprospecting of colours/pigments produced from these fungi need to be explored for various commercial applications in different industries and their sustainable utilisation in future.

Acknowledgements

MS, AL, PNS and SKS thank Director ARI for providing the facilities. MS and AL also thank CSIR, New Delhi (India) for providing research fellowships (JRF and SRF) and S.P. Pune University, Pune (India) for allowing students to register for Ph.D degree. LD was supported by grants from the *Conseil Régional de La Réunion*, Reunion Island, France (grant numbers 'COLORMAR RUN110356' and '*Arômes et colorants microbiens* RUN003261'). RR was supported by the Strategic Research Council at the Academy of Finland (grant # 327178, 2019).

References

Agrawal, D.C. and Dhanasekaran, M. (2019). Medicinal Mushrooms: Recent Progress in Research and Development, Springer, Singapore, p. 419.

Agyare, C. and Agana, T.A. (2019). Bioactive metabolites from Basidiomycetes. pp. 230–253. *In*: Sridhar, K.R. and Deshmukh, S.K. (eds.). Advances in Macrofungi: Diversity, Ecology and Biotechnology, CRC Press, Boca Raton.

Anke, H., Casser, I., Herrfnann, R. and Steglich, W. (1984). New terphenylquinones from mycelial cultures of *Punctularia atropurpurascens* (Basidiomycetes). Zeitschriftfür Naturforschung C, 39(7-8): 695–698.

Arun, G., Eyini, M. and Gunasekaran, P. (2015). Characterisation and biological activities of extracellular melanin produced by *Schizophyllum commune* (Fries). Ind. J. Exp. Biol., 53(6): 380–387.

Badalyan, S.M. and Zambonelli, A. (2019). Biotechnological exploitation of macrofungi for the production of food, pharmaceuticals, and cosmeceuticals. pp. 199–230. *In*: Sridhar, K.R. and Deshmukh, S.K. (eds.). Advances in Macrofungi: Diversity, Ecology and Biotechnology, CRC Press, Boca Raton.

Bai, M.S., Wang, C., Zong, S.C., Lei, M. and Gao, J.M. (2013). Antioxidant polyketide phenolic metabolites from the edible mushroom *Cortinarius purpurascens*. Food Chem., 141(4): 3424–3427.

Bell, E. and Charlwood, B. (1980). Secondary plant products. Encyclopedia of Plant Physiology (New Series), vol. 8, Springer-Verlag, Heidelberg, p. 674.

Bell, E. (1981). The physiological role(s) of secondary (natural) products. pp. 1–19. *In*: Conn, E. (ed.). The Biochemistry of Plants: A Comprehensive Treatise, Secondary Plant Products, vol. 7, Academic Press, London.

Ben, Z., Dekany, M., Kovacs, B., Csupor-Loffler, B., Zomborszki, Z.P. et al. (2018). Bioactivity-guided isolation of antimicrobial and antioxidant metabolites from the mushroom *Tapinella atrotomentosa*. Molecules, 23(5): 1082. 10.3390/molecules23051082.

Bernas, E., Jaworska, G. and Lisiewska, Z. (2006). Edible mushrooms as a source of valuable nutritive constituents. Acta Scientiarum Polonorum Technologia Alimentaria, 5(1): 5–20.

Blagodatski, A., Yatsunskaya, M., Mikhailova, V., Tiasto, V., Kagansky, A. and Katanaev, V.L. (2018). Medicinal mushrooms as an attractive new source of natural compounds for future cancer therapy. Oncotarget, 9(49): 29259–29274.

Boa, E.R. (2004). Non-wood forest products # 17. Wild Edible Fungi: A Global Overview of their use and Importance to People, Food and Agriculture Organisation, United Nations, Rome, p. 148.

Cai, Y., Din, Y., Ta, G. and Lia, X. (2008). Production of 1, 5-dihydroxy-3-methoxy-7-methylanthracene-9, 10-dione by submerged culture of *Shiraia bambusicola*. J. Microbiol. Biotechnol., 18(2): 322–327.

Camassola, M. (2013). Mushrooms—The incredible factory for enzymes and metabolites productions. Ferment. Technol., 2(1): 1000e117.

Caro, Y., Anamale, L., Fouillaud, M., Laurent, P., Petit, T. and Dufossé, L. (2012). Natural hydroxyanthraquinoid pigments as potent food-grade colourants: An overview. Nat. Prod. Bioprospect., 2(5): 174–193.

Caro, Y., Venkatachalam, M., Lebeau, J., Fouillaud, M. and Dufossé, L. (2017). Pigments and colorants from filamentous fungi. pp. 499–568. *In*: Merillon, J.-M. and Ramawat, K.G. (eds.). Fungal Metabolites. Springer International, Cham, Switzerland.

Chang, S.T. and Miles, P.G. (2004). Mushrooms: Cultivation, Nutritional Value, Medicinal Effect and Environmental Impact, CRC press, USA, p. 451.

Chang, S.T. and Wasser, S.P. (2018). Current and future research trends in agricultural and biomedical applications of medicinal mushrooms and mushroom products. Int. J. Med. Mush., 20(12): 1121–1133.

Christensen, C.M. (1972). Common Edible Mushrooms. Minneapolis, The University of Minnesota Press, USA, p. 124.

Davis, R.A. (2005). Isolation and structure elucidation of the new fungal metabolite (–)- xylarimide A. J. Nat. Prod., 68(5): 769–772.

Davoli, P., Mucci, A., Schenetti, L. and Weber, R.W.S. (2005). Laetiporic acids, a family of non-carotenoid polyene pigments from fruit-bodies and liquid cultures of *Laetiporus sulphureus* (Polyporales fungi). Phytochem., 66(7): 817–823.

Debnath, S., Debnath, B., Das, P. and Saha, A.K. (2019). Review on an ethnomedicinal practices of wild mushrooms by the local tribes of India. J. Appl. Pharmaceut. Sci., 9(8): 144–156.

De Oliveira, G.A.R., Leme, D.M., De Lapuente, J., Brito, L.B., Porredón, C. et al. (2018). A test battery for assessing the ecotoxic effects of textile dyes. Chemico-Biological Interactions, 291: 171–179.

Dias, D.A. and Urban, S. (2009). HPLC and NMR studies of phenoxazone alkaloids from *Pycnoporus cinnabarinus*. Nat. Prod. Com., 4(4): 489–498.

Diaz-Muñoz, G., Miranda, I.L., Sartori, S.K., De Rezende, D.C. and Diaz, M.A. (2018). Anthraquinones: An overview. Stud. Nat. Prod. Chem., 58: 313–338.

Dong, C. and Yao, Y. (2012). Isolation, characterisation of melanin derived from *Ophiocordyceps sinensis*, an entomogenous fungus endemic to the Tibetan Plateau. J. Biosci. Bioeng., 113(4): 474–479.

Dong, Z.J., Wang, F., Wang, R.R., Yang, L.M., Zheng, Y.T. and Liu, J.K. (2007). Chemical constituents of the fruiting bodies from the Basidiomycete *Suillus granulatus*. Chin. Trad. Herbal Drugs, 38: 17–19.

Duncan, C.J., Cuendet, M., Fronczek, F.R., Pezzuto, J.M., Mehta, R.G. et al. (2003). Chemical and biological investigation of the fungus *Pulveroboletus ravenelii*. J. Nat. Prod., 66(1): 103–107.

Dutta, A.K. and Acharya, K. (2014). Traditional and ethno-medicinal knowledge of mushrooms in West Bengal, India. Asian J. Pharmaceut. Clin. Res., 7(4): 36–41.

El Enshasy, H., Elsayed, E.A., Aziz, R. and Wadaan, M.A. (2013). Mushrooms and truffles: Historical biofactories for complementary medicine in Africa and in the Middle East. Evid. Based Complem. Alt. Med., article ID: 620451.10.1155/2013/620451.

Fang, L.Z., Shao, H.J., Yang, W.Q. and Liu, J.K. (2006). Two new azulene pigments from the fruiting bodies of the Basidiomycete *Lactarius hatsudake*. Helvetica Chimica Acta, 89(7): 1463–1466.

FAOSTAT 2019. Food and Agriculture Organization of the United Nations (FAO). Mushrooms. http://www.fao.org/faostat/, (accessed on May 2, 2020).

Ganeshpurkar, A., Rai, G. and Jain, A.P. (2010). Medicinal mushrooms: Towards a new horizon. Pharmacogn. Rev., 4(8): 127–135.

Gerber, N.N., Shaw, S.A. and Lechevalier, H.A. (1980). Structures and antimicrobial activity of peniophorin A and B, two polyacetylenic antibiotics from *Peniophora affinis* Burt. Antimicrob. Ag. Chemother., 17(4): 636–641.

Gessler, N.N., Egorova, A.S. and Belozerskaya, T.A. (2013). Fungal anthraquinones. Applied Biochemistry and Microbiology, 49(2): 85–99.

Giesbers, G., Van Schenck, J., Gutierrez, S.V., Robinson, S. and Ostroverkhova, O. (2018). Fungi-derived pigments for sustainable organic (Opto) electronics. MRS Adv., 3(59): 3459–3464.

Giesbers, G., Van Schenck, J., Quinn, A., Van Court, R., Vega Gutierrez, S.M. et al. (2019). Xylindein: Naturally produced fungal compound for sustainable (Opto) electronics. ACS Omega, 4(8): 13309–13318.

Gill, M. and Steglich, W. (1987). Pigments of fungi (Macromycetes). Progress in the Chemistry of Organic Natural Product, Springer-Verlag/Wien, p. 297.

Goyal, S., Ramawat, K.G. and Merillon, J.M. (2016). Different shades of fungal metabolites: An overview. pp. 1–29. *In*: Mérillon, J.M. and Ramawat, K. (eds.). Fungal Metabolites, Reference Series in Phytochemistry, Springer, Cham, Switzerland.

Guo, H., Feng, T., Li, Z.H. and Liu, J.K. (2013). Ten new aurovertins from cultures of the Basidiomycete *Albatrellus confluens*. Nat. Prod. Bioprospect., 3(1): 8–13.

Gutierrez, P.V. and Robinson, S.C. (2020). Complexity of biodegradation patterns in spalted wood and its influence on the perception of US woodturners. Eur. J. Wood Prod., 78(1): 173–183.

Hilszczańska, D. (2012). Medicinal properties of macrofungi. For. Res. Pap., 73(4): 347–353.

Hinsch, E.M., Chen, H.L., Weber, G. and Robinson, S.C. (2015). Colourfastness of extracted wood-staining fungal pigments on fabrics: A new potential for textile dyes. J. Tex. Apparel. Technol. Manage., 9(3): 1–11.

Ho, L.H., Zulkifli, N.A. and Tan, T.C. (2020). Edible mushroom: Nutritional properties, potential nutraceutical values, and its utilisation in food product development. *In*: An Introduction to Mushroom, Intech Open. https://cdn.intechopen.com/pdfs/71993.pdf.

Holkar, C.R., Jadhav, A.J., Pinjari, D.V., Mahamuni, N.M. and Pandit, A.B. (2016). A critical review on textile wastewater treatments: Possible approaches. J. Environ. Manage., 182: 351–366.

Honkanen, M. (2019). Sector Report for the Natural Products Sector for 2019, Publications of the Ministry of Economic Affairs and Employment, Sector Reports 2019: 32. Helsinki: Ministry of Economic Affairs and Employment.http://urn.fi/URN:ISBN:978-952-327-428-0 (accessed on June 24, 2020).

Hou, W., Lian, B., Dong, H., Jiang, H. and Wu, X. (2012). Distinguishing ectomycorrhizal and saprophytic fungi using carbon and nitrogen isotopic compositions. Geosci. Front., 3(3): 351–356.

Huang, Y.T., Onose, J.I., Abe, N. and Yoshikawa, K. (2009). *In vitro* inhibitory effects of pulvinic acid derivatives isolated from Chinese edible mushrooms, *Boletus calopus* and *Suillus bovinus*, on cytochrome P450 activity. Biosc. Biotechnol. Biochem., 73(4): 855–860.

Hunger, K. (ed.). (2003). Industrial Dyes: Chemistry, Properties, Applications, Wiley, Weinheim, p. 659.

Hynninen, P.H. and Räisänen, R. (2001). Stepwise pH-gradient elution for the preparative separation of natural anthraquinones by multiple liquid-liquid partition. Zeitschriftfür Naturforschung C, 56(9-10): 719–725.

Isaka, M., Chinthanom, P., Supothina, S., Tobwor, P. and Hywel-Jones, N.L. (2010). Pyridone and tetramic acid alkaloids from the spider pathogenic fungus *Torrubiella* sp. BCC 2165. J. Nat. Prod., 73(12): 2057–2060.

Isaka, M., Palasarn, S., Tobwor, P., Boonruangprapa, T. and Tasanathai, K. (2012). Bioactive anthraquinone dimers from the leafhopper pathogenic fungus *Torrubiella* sp. BCC 28517. J. Antibiot., 65(11): 571–574.

Isaka, M., Haritakun, R., Intereya, K., Thanakitpipattana, D. and Hywel-Jones, N.L. (2014). Torrubiellone E, an antimalarial *N*-hydroxypyridone alkaloid from the spider pathogenic fungus *Torrubiella longissima* BCC 2022. Nat. Prod. Comm., 9(5): 627–628.

Issakainen, J. (2013). *Sienten viljely* [cultivating fungi]. pp. 355–362. *In*: Sientenbiologia [Biology of Fungi], Gaudeamus, Helsinki. http: //hdl.handle.net/10138/313222 (accessed on June 25, 2020).

Jaeger, R.J. and Spiteller, P. (2010). Mycenaaurin A, an antibacterial polyene pigment from the fruiting bodies of *Mycena aurantiomarginata*. J. Nat. Prod., 73(8): 1350–1354.

Johnson, A. (1995). The theory of colouration of textiles. 2nd ed., Society of Dyers and Colourists, Bradford, p. 552.

Kakon, A.J., Choudhury, M.B.K. and Saha, S. (2012). Mushroom is an ideal food supplement. J. Dhaka Nat. Med. Coll. Hosp., 18(1): 58–62.

Kalač, P. (2013). A review of chemical composition and nutritional value of wild-growing and cultivated mushrooms. J. Sci. Food Agric., 93(2): 209–218.

Kang, H.S., Jun, E.M., Park, S.H., Heo, S.J., Lee, T.S. et al. (2007). Cyathusals A, B, and C, antioxidants from the fermented mushroom *Cyathus stercoreus*. J. Nat. Prod., 70(6): 1043–1045.

Karuppan, P., Subramanian, C.S., Velusamy, K., Sadasivam, M., Mahalingam, P. et al. (2014). Prospective aspects of myco-chrome as promising future textiles. pp. 412–416. *In*: Proceedings of 8th International Conference on Mushroom Biology and Mushroom Products (ICMBMP8), New Delhi, November 19–22, 2014, vol. I and II, ICAR-Directorate of Mushroom Research, New Delhi.

Kawagishi, H., Tonomura, Y., Yoshida, H., Sakai, S. and Inoue, S. (2004). Orirubenones A, B, and C, novel hyaluronan-degradation inhibitors from the mushroom *Tricholoma orirubens*. Tetrahedron 60(33): 7049–7052.

Khanna, P.K. and Sharma, S. (2014). Production of mushrooms. pp. 509–555. *In*: Panesar, P.S. and Marwaha, S.S. (eds.). Biotechnology in Agriculture and Food Processing: Opportunities and Challenges. CRC Press, Boca Raton, USA.

Kim, J.P., Yun, B.S., Shim, Y.K. and Yoo, I.D. (1999). Inoscavin A, a new free radical scavenger from the mushroom *Inonotus xeranticus*. Tetrahed. Lett., 40(36): 6643–6644.

Kim, J.Y., Kim, D.W., Hwang, B.S., Woo, E.E., Lee, Y.J. et al. (2016). Neuraminidase inhibitors from the fruiting body of *Phellinus igniarius*. Mycobiology, 44(2): 117–120.

Kim, J.Y., Woo, E.E., Ha, L.S., Ki, D.W., Lee, I.K. and Yun, B.S. (2019). Neuraminidase inhibitors from the fruiting body of *Glaziella splendens*. Mycobiology, 47(2): 256–260.

Kim, Y.H., Cho, S.M., Hyun, J.W., Ryoo, I.J., Choo, S.J. et al. (2008). A new antioxidant, clitocybin A, from the culture broth of *Clitocybe aurantiaca*. J. Antibiot., 61(9): 573–576.

Kim, Y.H., Ryoo, I.J., Choo, S.J., Xu, G.H., Lee, S. et al. (2009). Clitocybin D, a novel human neutrophil elastase inhibitor from the culture broth of *Clitocybe aurantiaca*. J. Microbiol. Biotechnol., 19(10): 1139–1141.

Kojima, K., Ohno, T., Inoue, M., Mizukami, H. and Nagatsu, A. (2008). Phellifuropyranone A: A new furopyranone compound isolated from fruit bodies of wild *Phellinus linteus*. Chem. Pharmaceut. Bull., 56(2): 173–175.

Krupodorova, T., Ivanova, T. and Barshteyn, V. (2019). Screening of extracellular enzymatic activity of macrofungi. J. Microbiol. Biotechnol. Food Sci., 3(4): 315–318.

Lagashetti, A.C., Dufossé, L., Singh, S.K. and Singh, P.N. (2019). Fungal pigments and their prospects in different industries. Microorganisms, 7(12): 604. 10.3390/microorganisms7120604.

Lam, Y.T., Palfner, G., Lima, C., Porzel, A., Brandt, W. et al. (2019). Nor-guanacastepene pigments from the Chilean mushroom *Cortinarius pyromyxa*. Phytochem., 165: 112048. 10.1016/j.phytochem.2019.05.021.

Lang, G., Blunt, J.W., Cummings, N.J., Cole, A.L. and Munro, M.H. (2005). Paecilosetin, a new bioactive fungal metabolite from a New Zealand isolate of *Paecilomyces farinosus*. J. Nat. Prod., 68(5): 810–811.

Lang, G., Cole, A.L., Blunt, J.W. and Munro, M.H. (2006). An unusual oxalylatedtetramic acid from the New Zealand Basidiomycete *Chamonixia pachydermis*. J. Nat. Prod., 69(1): 151–153.

Lee, I.K., Yun, B.S., Cho, S.M., Kim, W.G., Kim, J.P. et al. (1996). Betulinans A and B, two Benzoquinone compounds from *Lenzites betulina*. J. Nat. Prod., 59(11): 1090–1092.

Lee, I.K. and Yun, B.S. (2006). Hispidinanalogs from the mushroom *Inonotus xeranticus* and their free radical scavenging activity. Bioorg. Med. Chem. Lett., 16(9): 2376–2379.

Lee, I.K., Seok, S.J., Kim, W.K. and Yun, B.S. (2006a). Hispidin derivatives from the mushroom *Inonotus xeranticus* and their antioxidant activity. J. Nat. Prod., 69(2): 299–301.

Lee, I.K., Jung, J.Y., Seok, S.J., Kim, W.G. and Yun, B.S. (2006b). Free radical scavengers from the medicinal mushroom *Inonotus xeranticus* and their proposed biogenesis. Bioorg. Med. Chem. Lett., 16(21): 5621–5624.

Lee, I.K., Kim, Y.S., Jang, Y.W., Jung, J.Y. and Yun, B.S. (2007a). New antioxidant polyphenols from the medicinal mushroom *Inonotus obliquus*. Bioorg. Med. Chem. Lett., 17(24): 6678–6681.

Lee, I.K., Kim, Y.S., Seok, S.J. and Yun, B.S. (2007b). Inoscavin E, a free radical scavenger from the fruiting bodies of *Inonotus xeranticus*. J. Antibiot., 60(12): 745–747.

Lee, I.K., Seo, G.S., Jeon, N.B., Kang, H.W. and Yun, B.S. (2009a). Phellinins A1 and A2, new styrylpyrones from the culture broth of *Phellinus* sp. KACC93057P: I. Fermentation, taxonomy, isolation and biological properties. J. Antibiot., 62(11): 631–634.

Lee, I.K., Jung, J.Y., Kim, Y.H. and Yun, B.S. (2010). Phellinins B and C, new Styrylpyrones from the culture broth of *Phellinus* sp. J. Antibiot., 63(5): 263–266.

Lee, K.H., Morris-Natschke, S.L., Yang, X., Huang, R., Zhou, T. et al. (2012). Recent progress of research on medicinal mushrooms, foods, and other herbal products used in traditional Chinese medicine. J. Trad. Compl. Med., 2(2): 1–12.

Lee, Y.S., Kang, Y.H., Jung, J.Y., Kang, I.J., Han, S.N. et al. (2008). Inhibitory constituents of aldose reductase in the fruiting body of *Phellinus linteus*. Biol. Pharmaceut. Bull., 31(4): 765–768.

Leme, D.M., Primo, F.L., Gobo, G.G., da Costa, C.R.V., Tedesco, A.C. and de Oliveira, D.P. (2015). Genotoxicity assessment of reactive and disperse textile dyes using human dermal equivalent (3D cell culture system). J. Toxicol. Environ. Health (Part A), 78(7): 466–480.

Lindquist, M. (1977). Spalted wood—Rare jewels from death and decay. Fine Woodworking, 7: 50–53.

Lindquist, M. (1978). Turning spalted wood. Fine Woodworking, 2(1): 50–53.

Lohmann, J.S., Wagner, S., von Nussbaum, M., Pulte, A., Steglich, W. and Spiteller, P. (2018). Mycenaflavin A, B, C, and D: Pyrroloquinoline alkaloids from the fruiting bodies of the mushroom *Mycena haematopus*. Chemistry-A Eur. J., 24(34): 8609–8614.

Lopusiewicz, L. (2018a). Isolation, characterisation and biological activity of melanin from *Exidia nigricans*. World Sci. News, 91: 111–129.

Lopusiewicz, L. (2018b). *Scleroderma citrinum* melanin: Isolation, purification, spectroscopic studies with characterisation of antioxidant, antibacterial and light barrier properties. World Sci. News, 94(2): 115–130.

Lopusiewicz, L. (2018c). The isolation, purification and analysis of the melanin pigment extracted from *Armillaria mellea* rhizomorphs. World Sci. News, 100: 135–153.

Lu, D. (2014). Ancient Chinese people's knowledge of macrofungi as medicinal material during the period from 581 to 979 AD. Int. J. Med. Mush., 16(2): 189–204.

Lu, Y., Ye, M., Song, S., Li, L., Shaikh, F. and Li, J. (2014). Isolation, purification and anti-aging activity of melanin from *Lachnum singerianum*. Appl. Biochem. Biotechnol., 174(2): 762–771.

Ma, G., Khan, S.I., Jacob, M.R., Tekwani, B.L., Li, Z. et al. (2004). Antimicrobial and antileishmanial activities of hypocrellins A and B. Antimicrob. Ag. Chemother., 48(11): 4450–4452.

Maljuric, N., Golubovic, J., Ravnikar, M., Zigon, D., Strukelj, B. and Otaševic, B. (2018). Isolation and determination of Fomentariol: Novel potential antidiabetic drug from fungal material. J. Anal. Met. Chem., 2018: 2434691. 10.1155/2018/2434691.

Mo, S., Wang, S., Zhou, G., Yang, Y., Li, Y. et al. (2004). Phelligridins C–F: Cytotoxic pyrano [4, 3-c] [2] benzopyran-1, 6-dione, and Furo [3, 2-c] pyran-4-one derivatives from the fungus *Phellinus igniarius*. J. Nat. Prod., 67(5): 823–828.

Montero, G.A., Smith, C.B., Hendrix, W.A. and Butcher, D.L. (2000). Supercritical fluid technology in textile processing: An overview. Ind. Eng. Chem. Res., 39(12): 4806–4812.

Mueller, G.M., Foster, M. and Bills, G.F. (eds.). (2004). Biodiversity of Fungi Inventory and Monitoring Methods, Academic Press, Burlington, p. 777.

Mueller, G.M., Schmit, J.P., Leacock, P.R., Buyck, B., Cifuentes, J. et al. (2007). Global diversity and distribution of macrofungi. Biodiver. Conserv., 16(1): 37–48.

Mukherjee, G., Mishra, T. and Deshmukh, S.K. (2017). Fungal pigments: An overview. pp. 525–541. *In*: Satyanarayana, T., Deshmukh, S. and Johri, B. (eds.). Developments in Fungal Biology and Applied Mycology, Springer, Singapore.

Nadim, M., Deshaware, S., Saidi, N., Abd-Elhakeem, M.A., Ojamo, H. and Shamekh, S. (2015). Extracellular enzymatic activity of *Tuber maculatum* and *Tuber aestivum* mycelia. Adv. Microbiol., 5(7): 523–530.

Nagatsu, A., Itoh, S., Tanaka, R., Kato, S., Haruna, M. et al. (2004). Identification of novel substituted fused aromatic compounds, meshimakobnol A and B, from natural *Phellinus linteus* fruit body. Tetrahed. Lett., 45(30): 5931–5933.

Nguyen, K.A., Wikee, S. and Lumyong, S. (2018). Brief review: Lignocellulolytic enzymes from polypores for efficient utilisation of biomass. Mycosphere, 9(6): 1073–1088.

Orgiazzi, A., Bardgett, R., Barrios, E., Behan-Pelletier, V., Briones, M. et al. (2016). Global Soil Biodiversity Atlas, European Commission, Publications Office of the European Union, Luxembourg. https://doi.org/10.2788/2613 (accessed on June 24, 2020).

Patel, S. and Goyal, A. (2012). Recent developments in mushrooms as anti-cancer therapeutics: A review. 3 Biotech, 2(1): 1–15.

Piątek, M. (1999). Parasitic macrofungi (Basidiomycetes) on fruit shrubs and trees in the Tarnów town (S. Poland). Acta Mycol., 34(2): 329–344.

Pohleven, J., Korošec, T. and Gregori, A. (2016). Medicinal mushrooms, *MycoMedica*, Kranjska Gora, Slovenia, p. 54.

Pombeiro-Sponchiado, S.R., Sousa, G.S., Andrade, J.C., Lisboa, H.F. and Goncalves, R.C.R. (2017). Production of melanin pigment by fungi and its biotechnological applications. pp. 47–75. *In*: InTech. http://dx.doi.org/10.5772/67375.

Poole, C.F. (ed.). (2020). Liquid-phase extraction. Handbooks in Separation Science, Elsevier, Amsterdam, p. 816.

Popa, G., Cornea, C.P., Luta, G., Gherghina, E., Israel- Roming, F. et al. (2016). Antioxidant and antimicrobial properties of *Laetiporus sulphureus* (Bull.) Murrill. AgroLife Sci. J., 5(1): 168–173.

Qin, X.D., Dong, Z.J., Liu, J.K., Yang, L.M., Wang, R.R. et al. (2006). Concentricolide, an anti-HIV agent from the Ascomycete *Daldinia concentrica*. Helvetica Chimica Acta, 89(1): 127–133.

Quang, D.N., Hashimoto, T., Tanaka, M., Stadler, M. and Asakawa, Y. (2004a). Cyclic azaphilones daldinina E and F from the Ascomycete fungus *Hypoxylon fuscum* (Xylariaceae). Phytochem., 65(4): 469–473.

Quang, D.N., Hashimoto, T. and Asakawa, Y. (2006). Inedible mushrooms: A good source of biologically active substances. The Chemical Record, 6(2): 79–99.

Quang, D.N., Harinantenaina, L., Nishizawa, T., Hashimoto, T., Kohchi, C. et al. (2006d). Inhibition of nitric oxide production in RAW 264.7 cells by Azaphilones from Xylariaceous fungi. Biol. Pharmaceut. Bull., 29(1): 34–37.

Rahi, D.K. and Malik, D. (2016). Diversity of mushrooms and their metabolites of nutraceutical and therapeutic significance. J. Mycol., 2016: 7654123. 10.1155/2016/7654123.

Räisänen, R., Nousiainen, P. and Hynninen, P.H. (2001a). Emodin and dermocybin natural anthraquinones as high-temperature disperse dyes for polyester and polyamide. Tex. Res. J., 71(10): 922–927.

Räisänen, R., Nousiainen, P. and Hynninen, P.H. (2001b). Emodin and dermocybin natural anthraquinones as mordant dyes for wool and polyamide. Tex. Res. J., 71(11): 1016–1022.

Räisänen, R. (2002). Anthraquinones from the fungus *Dermocybe sanguinea* as textile dyes. Ph.D. thesis, University of Helsinki, Finland, pp. 1–107. http://urn.fi/URN:ISBN:978-952-10-5928-5 (accessed on June 24, 2020).

Räisänen, R., Nousiainen, P. and Hynninen, P.H. (2002). Dermorubin and 5-chlorodermorubin natural anthraquinone carboxylic acids as dyes for wool. Tex. Res. J., 72(11): 973–976.

Räisänen, R., Montero, G.A. and Freeman, H.S. (2020). BioColour—Bio-based colourants for sustainable material markets: a fungal-based anthraquinone for PLA and PET in supercritical carbon dioxide (SC-CO2) dyeing. Proceedings in AIC2020 Natural Colours—Digital Colours Conference, October, 2020, Avignon, France. https://aic2020.org/, in press.

Ramesh, C., Vinithkumar, N.V., Kirubagaran, R., Venil, C.K. and Dufossé, L. (2019). Multifaceted applications of microbial pigments: Current knowledge, challenges, and future directions for public health implications. Microorganisms, 7(7): 186. 10.3390/microorganisms7070186.

Razaq, A., Shahzad, S., Ali, H. and Noor, A. (2014). New reported species of macro fungi from Pakistan. J. Agri-Food Appl. Sci., 2(3): 67–71.

Rizzo, D.M., Blanchette, R.A. and Palmer, M.A. (1992). Biosorption of metal ions by *Armillaria rhizomorphs*. Can. J. Bot., 70(8): 1515–1520.

Robinson, S.C., Michaelsen, H. and Robinson, J.C. (2016). Spalted Wood: The History, Science, and Art of a Unique Material, Schiffer Publishing, Atglen, p. 288.

Rutanen, J. (2014). Luonnontuotteet monipuolistuvissa arvoverkoissa—luonnontuotealan toimintaohjelma, 2020. [Natural Products in Versatile Value Nets—Plan of Action 2020 for Natural Products' Sector], Mikkeli: University of Helsinki Ruralia institute. http://hdl.handle.net/10138/229380 (accessed on June 26, 2020).

Sande, D., Oliveira, G.P., Moura, M.A.F.E., Martins, B.A., Lima, M.T.N.S. and Takahashi, J.A. (2019). Edible mushrooms as a ubiquitous source of essential fatty acids. Food Res. Int., 125: 108524. 10.1016/j.foodres.2019.108524.

Schwan, W.R. (2012). Mushrooms: An untapped reservoir for nutraceutical antibacterial applications and antibacterial compounds. Curr. Top. Nutraceut. Res., 10(1): 75–82.

Schwenk, D., Nett, M., Dahse, H.M., Horn, U., Blanchette, R.A. and Hoffmeister, D. (2014). Injury-induced biosynthesis of methyl-branched polyene pigments in a white-rotting Basidiomycete. J. Nat. Prod., 77(12): 2658–2663.

Siewert, B., Vrabl, P., Hammerle, F., Bingger, I. and Stuppner, H. (2019). A convenient workflow to spot photosensitisers revealed photo-activity in Basidiomycetes. RSC Adv., 9(8): 4545–4552.

Singh, O.V. (ed.). (2017). Bio-pigmentation and Biotechnological Implementations, Wiley-Blackwell, Hoboken, New Jersey, p. 291.

Singh, R.P., Kashyap, A.S., Pal, A., Singh, P. and Tripathi, N.N. (2019). Macrofungal diversity of north-eastern part of Uttar Pradesh (India). Int. J. Curr. Microbiol. Appl. Sci., 8(2): 823–838.

Smania, A. Jr., Delle Monache, F., Smania, E.F.A., Gil, M.L., Benchetrit, L.C. and Cruz, F.S. (1995a). Antibacterial activity of a substance produced by the fungus *Pycnoporus sanguineus* (Fr.) Murr. J. Ethnopharmacol., 45(3): 177–181.

Smania, A. Jr., Marques, C.J.S., Smania, E.F.A., Zanetti, C.R., Carobrez, S.G. et al. (2003). Toxicity and antiviral activity of cinnabarin obtained from *Pycnoporus sanguineus* (Fr.) Murr. Phytother. Res., 17(9): 1069–1072.

Smania, E.D.F.A., Smania, A. Jr. and Loguercio-Leite, C. (1998). Cinnabarin synthesis by *Pycnoporus sanguineus* strains and antimicrobial activity against bacteria from food products. Rev. Microbiol., 29(4): 317–320.

Stamets, P. and Zwickey, H. (2014). Medicinal mushrooms: Ancient remedies meet modern science. Integr. Med., 13(1): 46–47.

Stodulkova, E., Cisarova, I., Kolarik, M., Chudickova, M., Novak, P. et al. (2015). Biologically active metabolites produced by the Basidiomycete *Quambalaria cyanescens*. PLoS ONE, 10(2): e0118913.

Subramanian, C.S., Velusamy, K. and Karuppan, P. (2014). Orange dye from *Pycnoporus* sp. for textile industries. pp. 456–460. *In*: Proceedings of 8th International Conference on Mushroom Biology and Mushroom Products (ICMBMP8), New Delhi, India, 19–22 November, 2014, Vol. I and II, ICAR–Directorate of Mushroom Research.

Tala, M.F., Qin, J., Ndongo, J.T. and Laatsch, H. (2017). New azulene-type sesquiterpenoids from the fruiting bodies of *Lactarius deliciosus*. Nat. Prod. Bioprospect., 7(3): 269–273.

Tang, A.Y., Lo, C.K. and Kan, C.W. (2018). Textile dyes and human health: A systematic and citation network analysis review. Coloration Technol., 134(4): 245–257.

Tang, X., Mi, F., Zhang, Y., He, X., Cao, Y. et al. (2015). Diversity, population genetics and evolution of macrofungi associated with animals. Mycology, 6(2): 94–109.

Teichert, A., Schmidt, J., Porzel, A., Arnold, N. and Wessjohann, L. (2007). Brunneins A-C, β-carboline alkaloids from *Cortinarius brunneus*. J. Nat. Prod., 70(9): 1529–1531.

Teichert, A., Schmidt, J., Porzel, A., Arnold, N. and Wessjohann, L. (2008a). (Iso)-quinoline alkaloids from fungal fruiting bodies of *Cortinarius subtortus*. J. Nat. Prod., 71(6): 1092–1094.

Textile Dyes Market Report. (2019). Global Forecast to 2024, Hadapsar-Pune: Markets and Markets Research Private Ltd. https://www.marketsandmarkets.com/Market-Reports/textile-dye-market-226167405.html (accessed on June 24, 2020).

Thanh, N.T., Tuan, N.N., Kuo, P.C., Dung, D.M., Phuong, D.L. et al. (2017). Chemical constituents from the fruiting bodies of *Phellinus igniarius*. Nat. Prod. Res., 32(20): 2392–2397.

Tsukamoto, S., Macabalang, A.D., Abe, T., Hirota, H. and Ohta, T. (2002). Thelephorin A: A new radical scavenger from the mushroom *Thelephora vialis*. Tetrahedron, 58(6): 1103–1105.

Umbuzeiro, G., Albuquerque, A., Morales, D., Szymczyk, M. and Freeman, H.S. (2020). Strategy for the evaluation of mutagenicity of natural dyes: Application to the BioColour project. SETAC Sci. Con., 2020, Dublin: Europe, p. 337.

Valverde, M.E., Hernández-Pérez, T. and Paredes-López, O. (2015). Edible mushrooms: Improving human health and promoting quality of life. Int. J. Microbiol., 2015: 376387. 10.1155/2015/376387.

Van Der Sar, S.A., Blunt, J.W., Cole, A.L., Din, L.B. and Munro, M.H. (2005). Dichlorinated pulvinic acid derivative from a Malaysian *Scleroderma* sp. J. Nat. Prod., 68(12): 1799–1801.

Vega Gutierrez, P.T. and Robinson, S.C. (2017). Determining the presence of spalted wood in Spanish Marquetry woodworks of the 1500s through the 1800s. Coatings, 7(11): 188.

Vega Gutierrez, S.M., He, Y., Cao, Y., Stone, D., Walsh, Z., Malhotra, R., Chen, H.L., Chang, C.H. and Robinson, S.C. (2019). Feasibility and surface evaluation of the pigment from *Scytalidium cuboideum* for inkjet printing on textiles. Coatings, 9(4): 266.

Velisek, J. and Cejpek, K. (2011). Pigments of higher fungi—A review. Czech J. Food Sci., 29(2): 87–102.

Velmurugan, P., Lee, Y.H., Nanthakumar, K., Kamala-Kannan, S., Dufossé, L. et al. (2010). Water-soluble red pigments from *Isaria farinosa* and structural characterisation of the main coloured component. J. Basic Microbiol., 50(6): 581–590.

Velygodska, A.K. and Fedotov, O.V. (2016). The production and analysis of carotenoid preparations from some strains of xylotrophic Basidiomycetes, Visnyk of Dnipropetrovsk University. Biology, Ecology, 24(2): 290–294.

Wang, Y., Wang, S.J., Mo, S.Y., Li, S., Yang, Y.C. and Shi, J.G. (2005). Phelligridimer A, a highly oxygenated and unsaturated 26-membered macrocyclic metabolite with antioxidant activity from the fungus *Phellinus igniarius*. Org. Lett., 7(21): 4733–4736.

Wang, Y., Shang, X.Y., Wang, S.J., Mo, S.Y., Li, S. et al. (2007b). Structures, biogenesis, and biological activities of pyrano [4, 3-c] isochromen-4-one derivatives from the fungus *Phellinus igniarius*. J. Nat. Prod., 70(2): 296–299.

Wangun, H.V.K. and Hertweck, C. (2007). Squarrosidine and pinillidine: 3, 3'-fused bis (styrylpyrones) from *Pholiota squarrosa* and *Phellinus pini*. Eur. J. Org. Chem., 2007(20): 3292–3295.

Wani, B.A., Bodha, R.H. and Wani, A.H. (2010). Nutritional and medicinal importance of mushrooms. J. Med. Pl. Res., 4(24): 2598–2604.

Wasser, S.P. and Weis, A.L. (1999). Therapeutic effects of substances occurring in higher Basidiomycetes mushrooms: A modern perspective. Critical Reviews™ in Immunology, 19(1): 65–96.

Wasser, S.P. (2010). Medicinal mushroom science: History, current status, future trends, and unsolved problems. Int. J. Med. Mush., 12(1): 1–16.

Weber, G., Chen, H.L., Hinsch, E., Freitas, S. and Robinson, S. (2014). Pigments extracted from the wood-staining fungi *Chlorociboria aeruginosa, Scytalidium cuboideum*, and *S. ganodermophthorum* show potential for use as textile dyes. Colouration Technol., 130(6): 445–452.

Weber, G.L., Boonloed, A., Naas, K.M., Koesdjojo, M.T., Remcho, V.T. and Robinson, S.C. (2016). A method to stimulate production of extracellular pigments from wood-degrading fungi using a water carrier. Curr. Res. Environ. Appl. Mycol., 6(4): 218–230.

Willemen, H., van den Meijdenberg, G.J., van Beek, T.A. and Derksen, G.C. (2019). Comparison of madder (*Rubia tinctorum* L.) and weld (*Reseda luteola* L.) total extracts and their individual dye compounds with regard to their dyeing behaviour, colour, and stability towards light. Colouration Technol., 135(1): 40–47.

Winkelman, M. (2019). Introduction: Evidence for entheogen use in prehistory and world religions. J. Psyched. Stu., 3(2): 43–62.

Wu, F., Zhou, L.W., Yang, Z.L., Bau, T., Li, T.H. and Dai, Y.C. (2019). Resource diversity of Chinese macrofungi: Edible, medicinal and poisonous species. Fungal Diversity, 98: 1–76.

Wu, Y., Shan, L., Yang, S. and Ma, A. (2008). Identification and antioxidant activity of melanin isolated from *Hypoxylon archeri*, a comparison fungus of *Tremella fuciformis*. J. Basic Microbiol., 48(3): 217–221.

Xu, C., Li, J., Shi, F., Yang, L. and Ye, M. (2017). Antibacterial activity and a membrane damage mechanism of *Lachnum* YM30 melanin against *Vibrio parahaemolyticus* and *Staphylococcus aureus*. Food Control, 73: 1445–1451.

Yang, X.L., Qin, C., Wang, F., Dong, Z.J. and Liu, J.K. (2008). A new meroterpenoid pigment from the Basidiomycete *Albatrellus confluens*. Chem. Biodiver., 5(3): 484–489.

Yli-Heikkilä, E. (2020). Clothing, colours and sustainability—Consumers' views of textile consumption and the origin and the features of dyes [in Finnish], Master thesis, University of Helsinki.

Zalas, M., Gierczyk, B., Bogacki, H. and Schroeder, G. (2015). The *Cortinarius* fungi dyes as sensitisers in dye-sensitised solar cells. Int. J. Photoener., 2015: 653740. 10.1155/2015/653740.

Zan, L.F., Bao, H.Y., Bau, T. and Li, Y. (2015). A new antioxidant pyrano [4, 3-c][2] benzopyran-1, 6-dione derivative from the medicinal mushroom *Fomitiporia ellipsoidea*. Nat. Prod. Comm., 10(2): 315–316.

ZDHC Foundation [Online]. ZDHC Manufacturing Restricted Substances List. https://mrsl.roadmaptozero. com/ (accessed on June 24, 2020).

Zhao, H., Wu, R., Ma, L.F., Wo, L.K., Hu, Y.Y. et al. (2016). Aurovertin-type polyketides from *Calcarisporium arbuscula* with potent cytotoxic activities against triple-negative breast cancer. Helvetica Chimica Acta, 99(7): 543–546.

Zhenjun, D. and Lown, J.W. (1990). Hypocrellins and their use in photosensitisation. Photochem. Photobiol., 52(3): 609–616.

Zhenjun, D. (1995). Novel therapeutic and diagnostic applications of hypocrellins and hypericins. Photochem. Photobiol., 61(6): 529–539.

Zhou, Z.Y. and Liu, J.K. (2010). Pigments of fungi (Macromycetes). Nat. Prod. Rep., 27(11): 1531–1570.

Zotti, M., Persiani, A.M., Ambrosio, E., Vizzini, A., Venturella, G. et al. (2013). Macrofungi as ecosystem resources: Conservation versus exploitation. Pl. Biosyst., 147(1): 219–225.

14

Colorful Macrofungi and their Pigment Structures

Ajay C Lagashetti,[1] *Riikka Räisänen,*[2] *Seri C Robinson,*[3] *Sanjay K Singh*[1] *and Laurent Dufossé*[4,*]

1. INTRODUCTION

The fungi kingdom comprises a large group organisms, which include both micro- and macrofungi. Microfungi refer to fungi that are generally not visible to the naked eye, whereas macrofungi refer to fungi that form visible fruit bodies that can be picked by hand. Fruit bodies of macrofungi are either epigeous (above the ground) or hypogeous (below the ground) (Chang and Miles, 1992; Bhandari and Jha, 2017). Macrofungi are fleshy and include mushrooms or toadstools, jelly fungi, gilled fungi, bracket fungi, stink fungi, coral fungi, bird nest fungi, puffballs and truffles (Kinge et al., 2017). Members of macrofungi mostly belong to the phylum Basidiomycota or Ascomycota and very few are Zygomycota. Most of the macrofungi produce sexual reproductive structures in the form of visible, fleshy and colloidal fruit bodies, while some produce visible asexual structures, like sclerotia (Mueller et al., 2007; Tang et al., 2015; Lallawmsanga et al., 2018). They may live as saprophytes on dead and decaying organic matter or as parasites on living organisms, while some show symbiotic association with plants and animals (Piatek, 1999; Hou et al., 2012; Tang et al., 2015).

[1] National Fungal Culture Collection of India (NFCCI), Biodiversity and Palaeobiology Group, Agharkar Research Institute, G.G. Agarkar Road, Pune-411004, India.

[2] HELSUS Helsinki Institute of Sustainability Science, Craft Studies P.O. Box 8 (Siltavuorenpenger 10), 00014 University of Helsinki, Finland.

[3] Department of Wood Science and Engineering, 119 Richardson Hall, Oregon State University, Corvallis, OR 97331, USA.

[4] Chimie et Biotechnologie des Produits Naturels (ChemBioProLab) & ESIROI Agroalimentaire, Université de la Réunion, 15 Avenue René Cassin, CS 92003, F-97744 Saint-Denis CEDEX, France.

* Corresponding author: laurent.dufosse@univ-reunion.fr

As per an initial estimation, there are a total of about 1.5 million species of fungi on the earth and among them, 1,40,000 are mushroom species, but recent molecular methods and sequencing of the environmental samples have increased this number to vary between 2.2–3.8 million species. So, as per the assumption adopted in 2001, the total estimated macrofungal species would be between 220,000–380,000. This means that we know only 3.7–6.4 per cent of them out of the total species (Hawksworth, 2019) while the rest of the species are still unknown and unexplored. Ainsworth in 2008 mentioned that among the total described 1,00,000 fungal species, only 6,000 develop visible fruit bodies and sclerotia.

Macrofungi play many important roles in the natural ecosystem. Most importantly, they maintain the ecological balance in the environment (Borovicka et al., 2010; Gates et al., 2011). Most of the macrofungi act as decomposers, which cause biodegradation of organic matter, especially of ligno-cellulolytic materials and recycle the nutrients in the ecosystem by continuing the food chain (Redhead, 1997; Conceicao et al., 2018). Many macrofungi are involved in the process of mineralisation or rock dissolution and thereby produce the soil and maintain soil fertility (Dighton, 2019). These macrofungi are also engaged in bioconversion (degradation of organic matter and production of valuable byproducts) and biotransformation (use of macrofungal enzymes for the production of valuable byproducts from specific substrates) activity (Conceicao et al., 2018). Some of the macrofungi show mutualistic behaviour and live symbiotically with many plants and animals. From plants and animals, they obtain food, shelter and also protection against microbial infections; in return, they provide the essential nutrients and sometimes protection (Tang et al., 2015; Dighton, 2019; Ye et al., 2019). Some of the macrofungi may harm plants and animals through parasitism (Sinclair and Lyon, 2005; Gedminas et al., 2015; Shang et al., 2015; Ye et al., 2019).

These macrofungi were found to be beneficial for humanity. Several macrofungi have been used since ancient times for different purposes, such as food or medicine (Stamets and Zwickey, 2014; Valverde et al., 2015; Singh, 2017; Badalyan and Zambonelli, 2019). They have been reported as potential producers of a variety of industrially important products, such as enzymes, pigments, antibiotics, hormones, pheromones, toxins and so on (Redhead, 1997; Camassola, 2013; Goyal et al., 2017; Hyde et al., 2019). Many of these macrofungi have medicinal importance. Mushrooms are reported to have numerous bioactive substances which perform several biological activities, such as antimicrobial, anticancer, cytotoxic, antiviral, antioxidant, antidiabetic, anti-HIV and others (Jin-Ming, 2006; Ganeshpurkar et al., 2010; Yarlagadda et al., 2013; De Silva et al., 2013; Glamoclija et al., 2015; Valverde et al., 2015; Fernando et al., 2018; Cheong et al., 2018).

2. Pigment-Producing Macrofungi and their Pigments

Many macrofungi are reported across the world for their pigment-production potential (Zhou and Liu, 2010; Velisek and Cejpek, 2011; Caro et al., 2017; Mukherjee et al., 2017; Hyde et al., 2019; Lagashetti et al., 2019; Ramesh et al., 2019; Sen et al., 2019). Pigments of macrofungi makes them very colourful, attractive and useful to humans. Several researchers have discovered plentiful pigments of diverse chemical classes from different macrofungi (Ascomycetes and Basidiomycetes) (Table 1) (Fig. 1).

Table 1. Pigments produced by macrofungi and their characteristic features.

Macrofungus	Nature of pigment	References
Agaricus bisporus	2-hydroxy-4-iminocyclohexa-2,5-dienone (pink-red), melanin (black), 4-glutaminyhydroxylbenzene melanin (black)	Margalith, 1992; Hanson, 2008a; Hanson, 2008b; Ramesh et al., 2019
Agaricus xanthodermus	Agaricone (yellow)	Hanson, 2008a; Hanson, 2008b
Albatrellus caeruleoporus	Grifolinone B (purple)	Quang et al., 2006b
Albatrellus confluens	Grifolinone B (purple), grifolinone C (red), albatrellin (red), aurovertins B-C (yellow), aurovertin E (yellow), aurovertins I-S (yellow)	Wang et al., 2005; Yang et al., 2008; Zhou et al., 2009; Guo et al., 2013
Albatrellus cristatus	Cristatomentin (green)	Koch et al., 2010
Albatrellus flettii	Grifolinone B (purple), albatrellin (red), 16-hydroxyalbatrellin (violet-blue)	Koch and Steglich, 2007
Albatrellus ovinus	Albatrellin D (yellow), albatrellin E (yellow), albatrellin F (yellow)	Liu et al., 2013
Albatrellus spp.	Grifolinone B (purple), albatrellin (red)	Velisek and Cejpek, 2011
Aleuria aurantia	Aleuriaxanthin (red)	Margalith, 1992; Ramesh et al., 2019
Amanita muscaria	Muscapurpurin (purple), amavadin (blue), muscarufin (red-brown), muscoflavin (yellow), uncharacterized (orange), muscaurins I–VII (orange)	Bayer and Kneifel, 1972; Muso, 1976; Li and Oberlies, 2005; Hanson, 2008b; Duran et al., 2009; Karuppan et al., 2014
Amanita pantherina	Hispidin (yellow-brown), bisnoryangonin (yellow)	Repke et al., 1978
Armillaria cepistipes	Melanin (black)	Ribera et al., 2019
Arthrographis cuboidea	Uncharacterized (pink)	Robinson et al., 2012b
Atheliachaete sanguinea (formerly known as *Phanerochaete sanguine*)	Xylerythrin (black), 5-*O*-methylxylerythrin (black), peniophorinin (dark brown), penioflavin (yellow)	Abrahamsson and Innes, 1966; Gripenberg, 1970; Gripenberg, 1978; Velisek and Cejpek, 2011
Auricularia auricula	Melanin (black)	Prados-Rosales et al., 2015; Zou and Ma, 2018
Austroboletus gracilis	Badione A (chocolate brown), austrogracilins A-B (yellow)	Bartsch et al., 2004
Boletinus cavipes	Atromentic acid (orange-red), xerocomic acid (red-orange), variegatic acid (orange), variegatorubin (red), cavipetins A-E	Besl and Bresinsky, 1997
Boletinus paluster	Xerocomic acid (red-orange), variegatic acid (orange), variegatorubin (red)	Besl and Bresinsky, 1997
Boletopsis leucomelaena	Cycloleucomelone (brown)	Jagers et al., 1987; Steglich et al., 2000
Boletus curtisii	Curtisin (pale yellow), 9-deoxycurtisin (pale yellow)	Brockelmann et al., 2004
Boletus laetissimus	Boletocrocin A (orange), Boletocrocin B (orange)	Kahner et al., 1998

Species	Pigment/Compound	References
Boletus subalpinus (formerly known as *Gastroboletus subalpinus*)	Variegatic acid (orange)	Besl and Bresinsky, 1997
Calcarisporium arbuscula	Aurovertins A-E (yellow), aurovertin M (yellow), aurovertins T-U (yellow)	Mulheirn et al., 1974; Guo et al., 2013; Zhao et al., 2016
Caloboletus calopus (formerly known as *Boletus calopus*)	Atromentic acid (orange-red), xerocomic acid (red-orange), variegatic acid (orange)	Huang et al., 2009; Zhou and Liu, 2010
Calostoma cinnabarinum	Calolstomal (orange)	Gruber and Steglich, 2007; Zhou and Liu, 2010
Cantharellus cibarius	α-carotene, β-carotene (orange), lycopene (bright red), γ- and δ-isomers of carotene	Velisek and Cejpek, 2011
Cantharellus cinnabarinus	Canthaxanthin (red), β-carotene (orange)	Haxo, 1950; Velisek and Cejpek, 2011
Cerioporus squamosus (formerly known as *Polyporus squamosus*)	Melanin (black)	Tudor, 2013; Tudor et al., 2013
Chalciporus piperatus	Sclerocitrin (bright yellow), chalcitrin (yellow), norbadione A (golden yellow), variegatic acid (orange), variegatorubin (red)	Winner et al., 2004; Velisek and Cejpek, 2011
Chamonixia caespitosa	Chamonixin (blue), gyrocyanin (blue), gyroporin (blue)	Gill and Steglich, 1987; Feling et al., 2001; Steglich et al., 1977
Chlorociboria aeruginascens	Xylindein (green), xylindein quinol (yellow)	Robinson et al., 2012a; Hinsch, 2015
Chlorociboria aeruginosa	Xylindein (green)	Hinsch, 2015; Hinsch et al., 2015; Hernandez et al., 2016
Chroogomphus helveticus	Xerocomic acid (red-orange), boviquinone 3 (yellow), boviquinone 4 (yellow-orange)	Besl and Bresinsky, 1997
Chroogomphus rutilus	Xerocomic acid (red-orange), atromentic acid (orange-red), boviquinone 3-4 (yellow-orange), diboviquinone-3,4, diboviquinone-4,4, methylenediboviquinone-3,3, methylenediboviquinone-3,4, methylenediboviquinone-4,4	Besl and Bresinsky, 1997
Chroogomphus tomentosus	Xerocomic acid (red-orange), boviquinones	Besl and Bresinsky, 1997; Ramesh et al., 2019
Cordyceps bifusispora	Cordycepoid A (yellow)	Lu et al., 2013; Ramesh et al., 2019
Cordyceps farinosa	Paecilosetin (pale yellow), farinosone B (yellow), an anthraquinone derivative (red), anthraquinonoid (red)	Lang et al., 2005; Velmurugan et al., 2010; Ramesh et al., 2019
Cordyceps sp.	Cordycepin (yellow)	Tuli et al., 2014; Ramesh et al., 2019
Cortinarius atrovirens	(3R)-atrochrysone (green)	Gill and Steglich, 1987; Gill, 1994, Velisek and Cejpek, 2011

Table 1 Contd. ...

...Table 1 Contd.

Macrofungus	Nature of pigment	References
Cortinarius infractus	Canthin-6-one (yellow), infractin A (blue), infractin B (blue)	Gill, 1996; Velisek and Cejpek, 2011
Cortinarius scaurus	Scaurin A, scaurin B	Velisek and Cejpek, 2011
Cortinarius abnormis	Hispidin (yellow-brown), bisnoryangonin (yellow)	Gill, 1994
Cortinarius alienates (formerly known as *Dermocybe alienata*)	Hypericin (red), skyrin (orange)	Gill, 1995
Cortinarius atrovirens	Atrovirin B (green-brown)	Antonowitz et al., 1994; Gill, 1995
Cortinarius austrovenetus (formerly known as *Dermocybe austroveneta*)	Austrovenetin (yellow-green), protohypericin (violet), hypericin (red), skyrin (orange), hypericin-like (red)	Margalith, 1992; Gill, 1995; Beattie et al., 2010; Ramesh et al., 2019
Cortinarius basirubescens	Emodin (yellow), physcion (yellow), erythroglaucin (yellow), (1S,3S)-austrocortirubin (red), (1S,3S)-austrocortilutein (orange-yellow), (1S,3R)-austrocortilutein (orange-yellow), (S)-torosachrysone (yellow), 6-methylxanthopurpurin-3-O-methyl ether (orange-yellow)	Beattie et al., 2010
Cortinarius brunneus	Brunnein A (yellow-green), brunneins B–C (yellow), 3-(7-hydroxy-9H-β-carboline-1-yl)propanoic acid (yellow-brown), 1-(1-β-glucopyranosyl)-3-(methoxymethyl)-1H-indole (yellow-brown), 1-(1-β-glucopyranosyl)-1H-indole-3-carbaldehyde (yellow-brown), N-1-β-glucopyranosyl-3-(carboxymethyl)-1H-indole (yellow-brown), N-1-β-glucopyranosyl-3-(2-methoxy-2-oxoethyl)-1H-indole (yellow-brown), 1-(1-β-glucopyranosyl)-1H-indole-3-acetic acid (yellow-brown), methyl 1-(1-β-glucopyranosyl)-1H-indole-3-acetate (yellow-brown)	Teichert et al., 2007; Teichert et al., 2008b; Zhou and Liu, 2010
Cortinarius canarius (formerly known as *Dermocybe canaria*)	Physcion (yellow), 4-aminophyscion (yellow), erythroglaucin (red), dermocanarin-1(yellow), dermocanarin-2 (orange-yellow), dermocanarin-3 (yellow)	Gill, 1995; Gill and Gimenez, 1995; Beattie et al., 2010
Cortinarius cardinalis (formerly known as *Dermocybe cardinalis*)	Cardinalins 2-3 (yellow), cardinalin 4 (deep red), cardinalins 5-6 (deep red)	Gill and Yu, 1994; Gill, 1995
Cortinarius cinnabarinus	Fallacinol (6-O-methoxycitreorosein) (dark orange)	Gill and Steglich, 1987; Gill, 1994; Velisek and Cejpek, 2011
Cortinarius cinnamomeoluteus	Flavomannin-6,6'-di-O-methyl ether (bright yellow)	Gill and Steglich, 1987; Gill, 1994; Velisek and Cejpek, 2011
Cortinarius citrinus	Flavomannin-6,6'-di-O-methyl ether (bright yellow), (3S)-torosachrysone-8-O-methyl ether (yellow)	Gill and Steglich, 1987; Gill, 1994; Velisek and Cejpek, 2011

Cortinarius croceus	Flavomannin-6,6'-di-*O*-methyl ether (bright yellow)	Gill and Steglich, 1987; Gill, 1994; Velisek and Cejpek, 2011
Cortinarius fulmineus	(3*S*)-torosachrysone-8-*O*-methyl ether (yellow)	Gill and Steglich, 1987; Gill, 1994; Velisek and Cejpek, 2011
Cortinarius icterinoides (formerly known as *Dermocybe icterinoides*)	Icterinoidins A-B (yellow-brown), icterinoidin C (red-brown), atrovirin B (green-brown), hypericin (purple red), skyrin (red-orange)	Antonowitz et al., 1994; Gill, 1995
Cortinarius odoratus	(3*R*)-atrochrysone (green)	Gill and Steglich, 1987; Gill, 1994; Velisek and Cejpek, 2011
Cortinarius odorifer	Phlegmacin (yellow), anhydrophlegmacin (yellow)	Steglich and Topper-Petersen, 1972; Zhou and Liu, 2010
Cortinarius persplendidus (formerly known as *Dermocybe splendida*)	Austrocortilutein (yellow), 1-deoxyaustrocortilutein (yellow), austrocortirubin (red), 1-deoxyaustrocortirubin (red), torosachrysone (yellow)	Gill, 1995; Beattie et al., 2010
Cortinarius purpurascens	Rufoolivacin A (red), rufoolivacins C-D (red), leucorufoolivacin (yellow), citreorosein 6,8-dimethyl ether (orange), verbindung cr11 (yellow), verbindung cr60 (dark brown), physcion (yellow), 1-hydroxy-3-methyl-2-isopropanyl-6,8-dimethoxy anthraquinone (orange)	Bai et al., 2013
Cortinarius pyromyxa	Pyromyxones A-D (yellow)	Lam et al., 2019
Cortinarius rufo-olivaceus	Rufoolivacin A (red), rufoolivacin B (red)	Zhang et al., 2009
Cortinarius sanguineus (formerly known as *Dermocybe sanguinea*)	Emodin (yellow), physcion (yellow), dermocybin (red-purple), dermorubin (red), 5-chlorodermorubin (red), dermoglaucin (brick red), flavomannin (bright yellow), dermolutein (orange), dermocybin-1-β-D glycopyranoside (red)	Räisänen et al., 2001a; Räisänen, 2002; Räisänen et al., 2002; Caro et al., 2012; Gessler et al., 2013; Zalas et al., 2015; Räisänen, 2019; Ramesh et al., 2019
Cortinarius semisanguineus	Emodin (yellow), Emodin-1-β-D-glucopyranoside (yellow), physcion (yellow), dermocybin (red-purple), dermocybin-1-β-D glycopyranoside (red), dermorubin (red), 5-chlorodermorubin (red), erythroglaucin (red), dermoglaucin (brick red), dermolutein (orange), 5-chlorodermolutein (orange), endocrocin (orange-red), flavomannin (bright yellow), dermoxanthone (bright yellow)	Gill, 1999; Velisek and Cejpek, 2011; Zalas et al., 2015; Räisänen, 2019; Ramesh et al., 2019
Cortinarius sinapicolor	Hydroxyphlegmacinquinone (yellow), (3*S*,3'*S*,*P*)-anhydrophlegmacin-9,10-quinone 8'-*O*-methyl ether (orange), (2'*S*,3*S*,3'*S*,*P*)-2'-hydroxyanhydrophlegmacin-9,10-quinone 8'-*O*-methyl ether (orange), dermocanarin-4 (orange), sinapiquinone (red), sinapicolone (red)	Gill, 1995; Gill et al., 1998; Elsworth et al., 1999; Beattie et al., 2010

Table 1 Contd. ...

Macrofungus	Nature of pigment	References
Cortinarius sp.	Phlegmacin (yellow)	Gessler et al., 2013
Cortinarius splendens	(3*S*)-torosachrysone-8-*O*-methyl ether (yellow)	Gill and Steglich, 1987; Gill, 1994; Velisek and Cejpek, 2011
Cortinarius subtortus	6-Hydroxyquinoline-8-carboxylic acid (yellow), 7-hydroxy-1-oxo-1,2-dihydroisoquinoline-5-carboxylic acid (light brown)	Teichert et al., 2008a
Cortinarius vitiosus	Physcion (yellow), dermocybin (red-purple), dermocybin-1-β-D glycopyranoside (red), dermorubin (red), 5-chlorodermorubin (red), dermolutein (orange), 5-chlorodermolutein (orange), dermoglaucin (brick red), endocrocin (orange-red), 5,7-dichloroendocrocin (orange-red)	Räisänen, 2019
Cyathus stercoreus	Cyathusals A–C (yellow), pulvinatal (yellow)	Kang et al., 2007
Daldinia concentrica	Daldinol (dark brown), daldinins A–C (green-olivaceous-isabelline), Daldinals A–C (yellow), BNT (1,1′-Binaphthalene-4,4′-5,5′-tetrol) (yellow), concentricolide (pale yellow), (17β,20*R*,22*E*,24*R*)-19-norergosta-1,3,5,7,9,14,22-heptaene (pale yellow), (17β,20*R*,22*E*,24*R*)-1-methyl-19-norergosta-1,3,5,7,9,14,22-heptaene (pale green), 8–methoxy–1–napthol, 2–hydroxy–5–methylchromone	Hashimoto et al., 1994; Qin and Liu, 2004; Qin et al., 2006; Caro et al., 2017
Daldinia bambusicola	Daldinol (dark brown), BNT (1,1′-Binaphthalene-4,4′-5,5′-tetrol) (yellow), 8–methoxy–1–napthol, 2–hydroxy–5–methylchromone	Caro et al., 2017
Daldinia caldariorum	Daldinol (dark brown), BNT (1,1′-Binaphthalene-4,4′-5,5′-tetrol) (yellow), 8–methoxy–1–napthol, 2–hydroxy–5–methylchromone	Caro et al., 2017
Daldinia childiae	Daldinol (dark brown), BNT (1,1′-Binaphthalene-4,4′-5,5′-tetrol) (yellow), 8–methoxy–1–napthol, 2–hydroxy–5–methylchromone	Caro et al., 2017
Daldinia clavata	Daldinol (dark brown), BNT (1,1′-Binaphthalene-4,4′-5,5′-tetrol) (yellow), 8–methoxy–1–napthol, 2–hydroxy–5–methylchromone	Caro et al., 2017
Daldinia eschscholzii	Daldinol (dark brown), daldinals A–C (yellow), BNT (1,1′-Binaphthalene-4,4′-5,5′-tetrol) (yellow), 8–methoxy–1–napthol, 2–hydroxy–5–methylchromone	Caro et al., 2017
Daldinia fissa	Daldinol (dark brown), BNT (1,1′-Binaphthalene-4,4′-5,5′-tetrol) (yellow), 8–methoxy–1–napthol, 2–hydroxy–5–methylchromone	Caro et al., 2017
Daldinia grandis	Daldinol (dark brown), BNT (1,1′-Binaphthalene-4,4′-5,5′-tetrol) (yellow), 8–methoxy–1–napthol, 2–hydroxy–5–methylchromone	Caro et al., 2017

Species	Pigment/Compound	Reference
Daldinia lloydi	Daldinol (dark brown), BNT (1,1′-Binaphthalene-4,4′-5,5′-tetrol) (yellow), 8–methoxy–1–napthol, 2–hydroxy–5–methylchromone	Caro et al., 2017
Daldinia loculata	Daldinol (dark brown), BNT (1,1′-Binaphthalene-4,4′-5,5′-tetrol) (yellow), 8–methoxy–1–napthol, 2–hydroxy–5–methylchromone	Caro et al., 2017
Daldinia petriniae	Daldinol (dark brown), BNT (1,1′-Binaphthalene-4,4′-5,5′-tetrol) (yellow), 8–methoxy–1–napthol, 2–hydroxy–5–methylchromone	Caro et al., 2017
Daldinia singularis	Daldinol (dark brown), BNT (1,1′-Binaphthalene-4,4′-5,5′-tetrol) (yellow), 8–methoxy–1–napthol, 2–hydroxy–5–methylchromone	Caro et al., 2017
Dermocybe sp.	Phlegmacin (yellow)	Gessler et al., 2013
Dermocybe sp.	Austrocolorin A_1 (green), austrocolorin B_1 (yellow-green), Phallacinol (dark orange)	Beattie et al., 2004; Gessler et al., 2013; Ramesh et al., 2019
Entonaema liquescens (formerly known as *Glaziella splendens*)	Glaziellin A (yellow), entonaemins A-B (yellow), rubiginosin A (orange-brown), comazaphilone D (yellow)	Kim et al., 2019
Epichloe bertonii (formerly known as *Hyperdermium bertonii*)	Skyrin (orange-red)	Caro et al., 2017
Exidia nigricans	Melanin (black)	Lopusiewicz et al., 2018
Fistulina hepatica	Mixture of carotenoids	Velygodska and Fedotov, 2016
Fomes fomentarius	Melanin (black), fomentariol (red-brown), a mixture of carotenoids	Velisek and Cejpek, 2011; Tudor, 2013; Tudor et al., 2013; Velygodska and Fedotov, 2016
Ganoderma applanatum	Uncharacterized (orange)	Karuppan et al., 2014
Ganoderma lucidum	Uncharacterized	Karuppan et al., 2014
Gastroboletus ruber	Xerocomic acid (red-orange), variegatic acid (orange), variegatorubin (red)	Besl and Bresinsky, 1997
Gastroboletus turbinatus	Xerocomic acid (red-orange), variegatic acid (orange)	Besl and Bresinsky, 1997
Gastroboletus valdivianus	Atromentic acid (orange-red), xerocomic acid (red-orange), variegatic acid (orange)	Besl and Bresinsky, 1997
Gastroboletus xerocomoides	Variegatic acid (orange), xerocomic acid (red-orange)	Besl and Bresinsky, 1997
Gliocephalotrichum simplex	Melanin (black)	Jalmi et al., 2012
Gomphidius glutinosus	Gomphidic acid (yellow), xerocomic acid (red-orange), atromentic acid (orange-red), gomphilactone (red)	Steglich et al., 1969; Knight and Pattenden, 1976; Besl and Bresinsky, 1997

Table 1 Contd. ...

...Table 1 Contd.

Macrofungus	Nature of pigment	References
Gomphidius maculatus	Gomphidic acid (yellow), xerocomic acid (red-orange), atromentic acid (orange-red), gomphilactone (red)	Besl and Bresinsky, 1997
Gomphidius roseus	Atromentic acid (orange-red)	Besl and Bresinsky, 1997
Gomphidius subroseus	Xerocomic acid (red-orange), variegatic acid (orange), variegatorubin (red)	Besl and Bresinsky, 1997
Gymnopilus junonius (formerly known as *Gymnopilus spectabilis*)	Hispidin (yellow-brown), bisnoryangonin (yellow)	Lee et al., 2008
Gymnopilus punctifolius	Hispidin (yellow-brown), bisnoryangonin (yellow)	Repke et al., 1978
Gyroporus cyanescens	Gyroporin (blue)	Hanson, 2008b
Hapalopilus rutilans (formerly known as *Hapalopilus nidulans* = *Polyporus nidulans*)	Polyporic acid (dark red)	Hanson, 2008b; Velisek and Cejpek, 2011
Hydnellum spp.	Aurantiacin (red)	Velisek and Cejpek, 2011
Hygrocybe conica	Muscaflavin (yellow), hygroaurins (yellow)	Muso, 1976; Li and Oberties, 2005; Velisek and Cejpek, 2011
Hygrophoropsis aurantiaca	Clitocybin A (pale brown), clitocybin D (yellow)	Kim et al., 2008; Kim et al., 2009
Hygrophorus agathosmus (formerly known as *Hygrophorus hyacinthinus*)	Brunnein A (yellow-green)	Teichert et al., 2008c
Hygrophorus eburneus	Harmane (yellow), norharmane (yellow)	Teichert et al., 2008c
Hymenochaete xerantica (formerly known as *Inonotus xeranticus*)	Inoscavin A (yellow), inoscavin B (orange), inoscavins C-E (yellow), methylinoscavins A-D (yellow), phelligridin D (yellow), phelligridin F (orange), davallialactone (yellow), methyldavallialactone (yellow), interfungins A-C (yellow), hispidin (yellow-brown)	Kim et al., 1999; Lee et al., 2006a; 2006b; Lee et al., 2007b; Lee and Yun, 2006; Zhou and Liu, 2010
Hypholoma fassiculare	Hypholomines A-B (yellow)	Fiasson et al., 1977
Hypholoma spp.	Hispidin (yellow-brown), bisnoryangonin (yellow), β-keto ester (red)	Gill, 1994; Velisek and Cejpek, 2011
Hypocrella bambusae	Hypocrellin A (blood-red), hypocrellin B (black-red)	Ma et al., 2004
Hypoxylon fragiforme	Hypoxyxylerone (green), fragiformins A–B, rutilin C (red), rutilin D (yellow), fragirubrins A–E (yellow), lenormandin F (yellow), mitorubrin (yellow), (+)-mitorubrinol (yellow), mitorubrinol-acetate (yellow-orange), and (+)-mitorubrinic acid (yellow)	Steglich et al., 1974; Edwards et al., 1991; Caro et al., 2017; Surup et al., 2018

Hypoxylon fulvo-sulphureum	(+)-6''-hydroxymitorubrinol acetate (yellow),(+)-mitorubrinol acetate (yellow), (+)-6''-hydroxymitorubrinol (yellow), (+)-mitorubrinol (yellow)	Sir et al., 2015; Caro et al., 2017
Hypoxylon fuscum	Daldinins A–C (green-olivaceous-isabelline), BNT (1,1´-Binaphthalene–4,4´–5,5´–tetrol) (yellow), daldinins E-F	Quang et al., 2004a; Quang et al., 2006d; Caro et al., 2017
Hypoxylon howeanum	Mitorubrin (yellow), (+)-mitorubrinol (yellow), mitorubrinol-acetate (yellow-orange), and (+)-mitorubrinic acid (yellow)	Caro et al., 2017
Hypoxylon jaklitschii	Lenormandins A–G (yellow)	Caro et al., 2017
Hypoxylon lechatii	Vermelhotin (orange-red), hypoxyvermelhotin A (orange), hypoxyvermelhotins B–C (yellow)	Kuhnert et al., 2014; Caro et al., 2017
Hypoxylon lenormandii	Lenormandins A–G (yellow)	Caro et al., 2017
Hypoxylon rickii	Rickenyl B (red), rickenyl D (brown)	Caro et al., 2017
Hypoxylon rubiginosum	Mitorubrin (yellow-orange), rubiginosins A–C (orange-brown), hypomiltin (yellowish-green), entonaemin A (yellow)	Quang et al., 2004b; Stadler et al., 2008; Caro et al., 2017
Hypoxylon rutilum	Rutilins A-B (dark red), rubiginosins A-B (orange-brown)	Quang et al., 2005b
Hypoxylon sclerophaeum	Hypoxylone (orange)	Caro et al., 2017
Imleria badia (formerly known as *Xerocomus badius*)	Badione A (chocolate brown), norbadione A (golden yellow)	Steffan and Steglich, 1984; Aumann et al., 1989; Velisek and Cejpek, 2011
Inocutis rheades	Rheadinin (yellow), phellinin A (orange), hypholomines A-B (yellow), hispidin (yellow-brown), bisnoryangonin (yellow)	Olennikov et al., 2017
Inonotus hispidus	Melanin (black), hispidin (yellow-brown), hispolon (yellow), uncharacterized (yellow)	Ali et al., 1996; Robinson et al., 2012b; Tudor, 2013; Tudor et al., 2013; Robinson et al., 2014
Inonotus obliquus	Inonoblins A-C (yellow), phelligridin D (yellow), phelligridins E-G (orange), melanin (black)	Babitskaya et al., 2002; Lee et al., 2007a
Jackrogersella cohaerens (formerly known as *Annulohypoxylon cohaerens/Hypoxylon cohaerens*)	Cohaerin A (yellow), cohaerins B-F (yellow)	Quang et al., 2005a; Quang et al., 2006c; Caro et al., 2017
Lactarius blennius	Blennione (green)	Spiteller and Steglich, 2002; Velisek and Cejpek, 2011

Table 1 Contd. ...

...Table 1 Contd.

Macrofungus	Nature of pigment	References
Lactarius deliciosus	Lactarazulene (blue), lactaroviolin (red-violet), 7-acetyl-4-methylazulene-1-carbaldehyde (red), 7-(1-hydroxyethyl)-4-methyl-1-azulenecarboxaldehyde (purple), 7-(1,2-dihydroxy-1-methylethyl)-4-methylazulene-1-carbaldehyde (red), 7-acetyl-4-methylazulene-1-carboxylic acid (purple), dihydroazulen-1-ol (orange), 7-isopropenyl-4-methyl-azulene-1-carboxylic acid (purple), 15-hydroxy-3,6-dihydrolactarazulene (orange), 15-hydroxy-6,7-dihydrolactarazulene (orange), quinizarine (orange-red-brown), acetylazulene (red), dihydroxyazulene (red)	Yang et al., 2006; Rai, 2009; Zhou and Li, 2010; Velisek and Cejpek, 2011; Tala et al., 2017; Tian et al., 2019; Ramesh et al., 2019
Lactarius fuliginosus	4-methoxy-2-(3-methylbut-2-enyl) phenol	Velisek and Cejpek, 2011
Lactarius hatsudake	Lactarioline A (blue), lactarioline B (red), lactaroviolin (red-violet), 7-(1-hydroxy-1-methylethyl)-4-methylazulene-1-carbaldehyde (red-purple), 4-methyl-7-(1-methylethyl)azulene-1-carboxylic acid (purple), 4-methyl-7-(1-methylethyl)azulene-1-carbaldehyde (red-purple), 1-[(15E)-buten-17-one]-4-methyl-7-isopropylazulene (green), 4-methyl-7-isopropylazulene-1-carboxylic acid (red), 1-formyl-4-methyl-7-isopropyl azulene (red), 1-formyl-4-methyl-7-(1-hydroxy-1-methylethyl) azulene (red)	Fang et al., 2006; Fang et al., 2007; Xu et al., 2010; Zhou and Liu, 2010
Lactarius indigo	Lactaroviolin (red-violet), stearoyldeterrol (blue), 1-hydroxymethyl-4-methyl-7-(1-methylethenyl)azulene stearate (brilliant blue), 1-stearoyloxymethylene-4-methyl-7-isopropenylazulene (blue)	Nelsen, 2010; Harmon et al., 1980; Velisek and Cejpek, 2011; Ramesh et al., 2019
Lactarius lilacinus	Lilacinone (red)	Spiteller et al., 2003; Velisek and Cejpek, 2011
Lactarius picinus	4-methoxy-2-(3-methylbut-2-enyl)phenol	Velisek and Cejpek, 2011
Lactarius sanguifluus	1,3,5,7(11),9-pentaenyl-14-guaianal (red)	Velisek and Cejpek, 2011
Lactarius scrobiculatus	Lactarane, secolactarane sesquiterpenoids	Velisek and Cejpek, 2011
Lactarius sp.	Azulenes (blue)	Caro et al., 2017
Lactarius turpis	Necatarone, Necatarone dehydrodimer, 10-deoxydehydrodimer	Velisek and Cejpek, 2011
Lactarius uvidus	Uvidin A, uvidin B, drimenol	De Bernardi et al., 1980; Velisek and Cejpek, 2011
Laetiporus sulphureus	Laetiporic acid A (orange), 2-dehydro-3-deoxylaetiporic acid A (orange), Laetiporic acids B-C (orange), a mixture of carotenoids	Weber et al., 2004; Davoli et al., 2005; Popa et al., 2016; Velygodska and Fedotov, 2016
Leccinum pachyderme (formerly known as *Chamonixia pachydermis*)	Pachydermin (yellow), 5-(3-chloro-4-hydroxybenzylidene) tetramic acid (yellow)	Lang et al., 2006

Species	Pigment/Compound	References
Lentinus brumalis (formerly known as *Polyporus brumalis*)	Melanin (black)	Tudor, 2013; Tudor et al., 2013
Lenzites betulina	Betulinan A (yellow), betulinan B (red)	Lee et al., 1996
Leucocoprinus birnbaumii	Birnbaumins A-B (yellow)	Bartsch et al., 2005
Macrolepiota albuminosa (formerly known as *Termitomyces albuminosus*)	Melanin (black)	De Souza et al., 2018
Melanogaster broomeanus	Melanocrocin (orange)	Aulinger et al., 2001
Mycena aurantiomarginata	Mycenaaurin A (red)	Jaeger and Spiteller, 2010; Ramesh et al., 2019
Mycena haematopus	Haematopodin (reddish-pink), haematopodin B (red), sanguinolentaquinone (red), mycenarubins A-B (red), mycenarubins D-F (red), mycenaflavins A-C (yellow), mycenaflavin D (purple)	Baumann et al., 1993; Hopmann and Steglich, 1996; Peters et al., 2008; Velisek and Cejpek, 2011; Lohmann et al., 2018
Mycena pelianthina	Pelianthinarubins A-B (red), mycenarubins A & D (red)	Pulte et al., 2016
Mycena rosea	Mycearubins A-B (red)	Peters and Spiteller, 2007a; Zhou and Liu, 2010; Velisek and Cejpek, 2011
Mycena sanguinolenta	Sanguinone A (blue), sanguinone B (blue), sanguinolentaquinone (red), decarboxydehydrosanguinone A (yellow)	Peters and Spiteller, 2007b; Velisek and Cejpek, 2011
Ophiocordyceps unilateralis	Erythrostominone (red), 3,5,8–TMON (red), deoxyerythrostominone (red), deoxyerythrostominol (red), 4–O–methyl erythrostominone (red), epierythrostominol (red)	Caro et al., 2017
Oudemansiella melanotricha	Xerulin (orange-yellow), dihydroxerulin (orange-yellow), xerulinic acid (orange-yellow)	Kuhnt et al., 1990; Velisek and Cejpek, 2011
Oxyporus populinus	Melanin (black)	Tudor, 2013; Tudor et al., 2014
Paxillus involutus	Involutin (brown), involutone (bright yellow), chamonixin (blue), gyroporin (blue), 4-(3,4-dihydroxyphenyl)-2-(4-hydroxyphenyl)-2-(2-pyrrolidon-5-yl)-4-cyclopentene-1,3-dione (deep yellow), (4Z)-5-hydroxy-2-(3,4-dihydroxyphenyl)-5-(4-hydroxyphenyl)-2,4-pentadien-4-olide (deep yellow)	Feling et al., 2001; Antkowiak et al., 2003; Mikolajczyk and Antkowiak, 2009; Hanson, 2008b; Zhou and Liu, 2010; Braesel et al., 2015
Phanerochaete affinis (formerly known as *Peniophora affinis*)	Peniophorin A (pink-red), peniophorin B (brown-black)	Gerber et al., 1980
Phellinus ellipsoideus (formerly known as *Fomitiporia ellipsoidea*)	Hispidin (yellow-brown), phelligridin K (orange), inoscavin C (yellow), inonoblin B (yellow)	Zan et al., 2015

Table 1 Contd. ...

...Table 1 Contd.

Macrofungus	Nature of pigment	References
Phellinus igniarius	Phelligridins C-D (yellow), phelligridins E-I (orange), phelligridin J (yellow), phelligridimer A (yellow), meshimakobnols A-B (brown), hispidin (yellow-brown), hispolon (yellow), uncharacterized phenolic pigment (dark brown)	Kirk et al., 1975; Mo et al., 2003a; 2003b; Mo et al., 2004; Wang et al., 2005; Wang et al., 2007b; Velisek and Cejpek 2011; Kim et al., 2016; Thanh et al., 2017
Phellinus linetus	Meshimakobnols A-B (brown), phellifuropyranone A (orange), phelligridin G (orange), phelligridimer A (yellow), davallialactone (yellow), methyldavallialactone (yellow), inoscavin A (yellow), hispidin (yellow-brown), interfungin A (yellow), hypholomine B (yellow), ellagic acid (yellow)	Nagatsu et al., 2004; Wang et al., 2005; Kojima et al., 2008; Lee et al., 2008; Zhou and Liu, 2010
Phellinus pini	Pinillidine (yellow)	Wangun and Hertweck, 2007
Phellinus ribis	Phelliribsin A (orange)	Kubo et al., 2014
Phellinus spp.	Phellinin A (orange), Phellinins B-C (yellow), hispidin (yellow-brown)	Lee et al., 2009a; Lee et al., 2009b; Lee et al., 2010
Pholiota squarrosa	Squarrosidine (yellow)	Wangun and Hertweck, 2007
Pholiota squarroso-adiposa	Hispidin (yellow-brown), bisnoryangonin (yellow)	Brady and Benedict, 1972; Gill, 1994
Phycomyces blakesleeanus	β-carotene (yellow-orange)	Caro et al., 2017
Pisolithus arrhizus	Norbadione A (golden yellow), badione A (chocolate brown)	Winner et al., 2004; Velisek and Cejpek, 2011
Pleurotus citrinopileatus	Melanin (black)	Vallimayil and Eyini, 2013
Pleurotus citrinopileatus (formerly known as *Pleurotus cornucopiae* var. *citrinopileatus*)	Uncharacterized (yellow)	Shirasaka et al., 2012
Pleurotus cystidiosus	Melanin (black)	Selvakumar et al., 2008
Pleurotus djamor	Melanin (black)	Vallimayil and Eyini, 2013
Pleurotus ostreatus	Mevinolin (lovastatin) (red)	Parthasarathy and Sathiyabama, 2015; Ramesh et al., 2019
Pulveroboletus cramesinus	Vulpinic acid (yellow)	Duncan et al., 2003
Pulveroboletus lignicola	Variegatic acid (orange), variegatorubin (red), xerocomic acid (red-orange)	Duncan et al., 2003
Pulveroboletus ravenelii	Pulverolide (yellow), pulveraven A (yellow), pulveraven B (orange), vulpinic acid (yellow)	Duncan et al., 2003; Zhang et al. 2006; Zhou and Liu, 2010

Punctularia atropurpurascens	Phlebiarubrone (red), 4'-hydroxyphlebiarubrone (violet), 4',4''-dihydroxyphlebiarubrone (violet), 3,4',4'',4'''-trihydroxyphlebiarubrone (violet)	Anke, 1984
Pycnoporus cinnabarinus	Cinnabarin (red), cinnabarinic acid (orange), tramesanguin (red-orange), pycnoporin (red-orange)	Sullivan and Henry, 1971; Dias and Urban, 2009; Velisek and Cejpek, 2011
Pycnoporus coccineus	Cinnabarin (red), cinnabarinic acid (orange), tramesanguin (red-orange)	Sullivan and Henry, 1971
Pycnoporus sanguineus	Cinnabarin (red), cinnabarinic acid (orange), tramesanguin (red-orange)	Sullivan and Henry, 1971; Smania Jr. et al., 2003; Karuppan et al., 2014; Ramesh et al., 2019
Pycnoporus sp.	Uncharacterised (orange)	Subramanian et al., 2014
Quambalaria cyanescens	Quambalarine A (brown-cinnamon), quambalarine B (deep violet), mompain (red)	Stodulkova et al., 2015
Rhizopogon pumilionum	Rhizopogone (red), 2-acetoxyrhizopogone	Lang et al., 2009
Rubroshiraia bambusae	Hypocrellin A (blood-red), hypocrellin B (black-red)	Dai et al., 2019
Russula ochroleuca	Ochroleucin A (red), ochroleucin B (yellow)	Sontag et al., 2006
Sanghuangporus baumii	Uncharacterised (yellow)	Heo et al., 2018
Sarcodon leucopus	sarcoviolin α (violet)	Cali et al., 2004; Velisek and Cejpek,2011
Schizophyllum commune	Melanin (black-brown), Indigo (blue), indirubin (red), schizocommunin (orange), isatin (orange-red), tryptanthrin (yellow)	Miles et al., 1956; Epstein and Miles, 1966; Hosoe et al., 1999; Arun et al., 2015
Scleroderma citrinum	Sclerocitrin (bright yellow), norbadione A (golden yellow), badione A (chocolate brown), xerocomic acid (red-orange)	Winner et al., 2004; Velisek and Cejpek, 2011
Scleroderma sp.	Methyl-3',5'-dichloro-4,4'-di-*O*-methylatromentate (yellow)	van der Sar et al., 2005; Zhou and Liu, 2010
Serpula lacrymans	Variegatic acid (orange), xerocomic acid (red-orange), isoxerocomic acid (orange-red), atromentic acid (orange-red), xerocomorubin (red)	Hanson, 2008b; Tauber et al., 2016
Shiraia bambusicola	Shiraiarin (red), hypocrellin A (blood-red), hypocrellins B-C (black-red), hypocrellin D (orange-red)	Kishi et al., 1991; Fang et al., 2006; Cai et al., 2008
Stephanospora caroticolor	Stephanosporin (bright orange), 2-chloro-4-nitrophenol (yellow-green)	Lang et al., 2001; Velisek and Cejpek, 2011
Stereaceous Basidiomycete	(3Z,5E,7E,9E,11E,13Z,15E,17E)-18-methyl-19-oxoicosa-3,5,7,9,11,13,15,17-octaenoic acid (yellow), (3E,5Z,7E,9E,11E,13E,15Z,17E,19E)-20-methyl-21-oxodocosa-3,5,7,9,11,13,15,17,19-nonaenoic acid (yellow)	Schwenk et al., 2014
Suillus acidus	Grevillin D (orange-red)	Besl and Bresinsky, 1997
Suillus aeruginascens	Grevillins B-C (orange-red)	Nelsen, 2010

Table 1 Cond. ...

...Table 1 Contd.

Macrofungus	Nature of pigment	References
Suillus albidipes	Grevillin D (orange-red)	Nelsen, 2010
Suillus amabilis	Grevillins B-C (orange-red), variegatic acid (orange)	Besl and Bresinsky, 1997
Suillus americanus	Grevillins A-D (orange-red), suillin (brown)	Besl and Bresinsky, 1997; Nelsen, 2010
Suillus bellinii	Grevillins A-D (orange-red), suillin (brown)	Besl and Bresinsky, 1997
Suillus bovinus	Atromentin (brown), atromentic acid (orange-red), xerocomic acid (red-orange), variegatic acid (orange), variegatorubin (red), methyl bovinate (orange), boviquinone-4 (yellow-orange), amitenone (yellow-orange)	Besl and Bresinsky, 1997; Muhlbauer et al., 1998; Besl et al., 2008; Huang et al., 2009; Velisek and Cejpek, 2011; Ramesh et al., 2019
Suillus bresadolae	Grevillins A-C (orange-red)	Nelsen, 2010
Suillus brevipes	Grevillins B-D (orange-red)	Besl and Bresinsky, 1997
Suillus clintonianus	Grevillins B-C (orange-red), cyclovariegatin (red), thelephoric acid (violet)	Besl and Bresinsky, 1997
Suillus collinitus	Grevillins B-D (orange-red), xerocomic acid (red-orange), variegatic acid (orange), suillin (brown), bovilactone-4,4 (brown)	Besl and Bresinsky, 1997; Nelsen, 2010
Suillus flavidus	Grevillin D (orange-red)	Besl and Bresinsky, 1997
Suillus granulatus	Grevillins B-D (orange-red), suillin (brown), flazin (yellow)	Besl et al., 1974; Dong et al., 2007; Velisek and Cejpek, 2011
Suillus grevillei	Grevillins A-C (orange-red), variegatic acid (orange), 3',4',4-trihydroxypulvinone, cyclovariegatin (red), thelephoric acid (violet), aurantricholide B (orange), pyrandione (orange), furanones (orange)	Besl and Bresinsky, 1997; Nelsen, 2010; Velisek and Cejpek, 2011; Ramesh et al., 2019
Suillus grisellus	Xerocomic acid (red-orange), variegatic acid (orange), variegatorubin (red)	Besl and Bresinsky, 1997
Suillus hirtellus	Xerocomic acid (red-orange), variegatic acid (orange), variegatorubin (red)	Besl and Bresinsky, 1997
Suillus lakei	Grevillins B-C (orange-red)	Nelsen, 2010
Suillus leptopus	Grevillin D (orange-red)	Besl and Bresinsky, 1997; Nelsen, 2010
Suillus luteus	Grevillins A-C (orange-red), atromentic acid (orange-red), xerocomic acid (red-orange), variegatic acid (orange)	Besl et al., 1974; Besl and Bresinsky, 1997; Velisek and Cejpek, 2011
Suillus neoalbidipes	Grevillin D (orange-red)	Besl and Bresinsky, 1997
Suillus pictus	Grevillins B-C (orange-red)	Nelsen, 2010
Suillus placidus	Grevillins A-D (orange-red), atromentic acid (orange-red), xerocomic acid (red-orange), variegatic acid (orange)	Besl and Bresinsky, 199; Nelsen, 2010

Suillus plorans	Atromentic acid (orange-red), xerocomic acid (red-orange), variegatic acid (orange), variegatorubin (red)	Besl and Bresinsky, 1997
Suillus pseudobrevipes	Grevillin D (orange-red)	Besl and Bresinsky, 1997; Nelsen, 2010
Suillus punctatipes	Grevillins A-D (orange-red)	Besl and Bresinsky, 1997; Nelsen, 2010
Suillus punctipes	Variegatic acid (orange), variegatorubin (red)	Besl and Bresinsky, 1997
Suillus riparius	Grevillin D (orange-red)	Besl and Bresinsky, 1997; Nelsen, 2010
Suillus serotinus	Xerocomic acid (red-orange), variegatic acid (orange), gyroporin (blue), chamonixin (blue), involutin (brown)	Besl and Bresinsky, 1997
Suillus sibiricus	Grevillins B-D (orange-red)	Besl and Bresinsky, 1997; Nelsen, 2010
Suillus spectabilis	Xerocomic acid (red-orange), variegatic acid (orange), variegatorubin (red)	Besl and Bresinsky, 1997
Suillus spraguei	Grevillins B-D (orange-red), boviquinone-4 (yellow-orange), variegatic acid (orange)	Besl and Bresinsky, 1997
Suillus tomentosus	Xerocomic acid (red-orange), variegatic acid (orange), variegatorubin (red)	Besl and Bresinsky, 1997
Suillus tridenticus	Tridentoquinone (red), tridentorubin (red), grevillins A-C (orange-red), cyclovariegatin (red), thelephoric acid (violet	Besl and Bresinsky, 1997; Lang et al., 2008; Velisek and Cejpek, 2011
Suillus umbonatus	Grevillin D (orange-red)	Besl and Bresinsky, 1997; Nelsen, 2010
Suillus variegatus	Atromentic acid (orange-red), variegatic acid (orange), variegatorubin (red)	Besl and Bresinsky, 1997
Suillus viscidus	Xerocomic acid (red-orange), variegatic acid (orange), variegatorubin (red), chamonixin (blue)	Besl and Bresinsky, 1997
Tapinella atrotomentosa	Flavomentins A-D (orange-yellow),spiromentins A-D (violet), xerocomic acid (red-orange), atromentic acid (orange-red), osmundalactone (purple), 5-hydroxy-hex-2-en-4-olide (purple)	Gill and Steglich, 1987; Liu, 2006; Velisek and Cejpek, 2011; Beni et al. 2018
Tapinella panuoides	Flavomentins A-D (orange-yellow)	Gill and Steglich, 1987; Liu, 2006; Velisek and Cejpek, 2011
Terana coerulea (formerly known as *Corticium coeruleum*)	Corticins A-B (blue)	Hanson, 2008b
Thelebolus microsporus	β-carotene (orange)	Duarte et al., 2019; Singh et al., 2014
Thelephora aurantiotincta	Thelephantins A-D (grey), thelephantin E (red-brown), thelephantin F (grey), thelephantin G (green-brown), thelephantin H (grey), thelephorin A (grey), atromentin (brown)	Quang et al., 2003a; 2003b; Quang et al., 2006a

Table 1 Contd. ...

...*Table 1 Contd.*

Macrofungus	Nature of pigment	References
Thelephora ganbajun	Ganbajunin A (orange)	Hu et al., 2001
Thelephora terrestris	Terrestrin B (reddish-violet), terrestrin C (red), terrestrin D (greyish violet), terrestrin E (greyish grey), terrestrin F (blue), terrestrin G (grey), thelephantin F (grey), thelephantin H (grey)	Radulovic et al., 2005
Thelephora vialis	Thelephorin A (grey)	Tsukamoto et al., 2002
Torrubiella longissima	Torrubiellones A-B (yellow), torrubiellone E (pale yellow)	Isaka et al., 2014
Torrubiella spp.	Torrubiellones A-D (yellow)	Isaka et al., 2010
Torrubiella spp.	Torrubiellins A-B (dark brown), emodin (yellow), chrysophanol (orange-red)	Isaka et al., 2012
Trametes versicolor	Melanin (black), uncharacterised (orange)	Robinson et al., 2012b; Tudor, 2013; Tudor et al., 2013; Karuppan et al., 2014
Tricholoma sulphureum	(3S)-torosachrysone-8-O-methyl ether (yellow)	Gill and Steglich, 1987; Gill, 1994, Velisek and Cejpek, 2011
Tricholoma aurantium	Aurantricholone (orange-red), aurantricholides A-B (yellow)	Klostermeyer et al., 2000; Hanson, 2008b; Velisek and Cejpek, 2011
Tricholoma equestre	(3S)-torosachrysone-8-O-methyl ether (yellow)	Gill and Steglich, 1987; Gill, 1994; Velisek and Cejpek, 2011
Tricholoma orirubens	Orirubenones A-G (yellow)	Kawagishi et al., 2004; Sakai et al., 2005
Tricholoma terreum	Terreumols A-D (yellow)	Yin et al., 2013
Tuber melanosporum	Melanin (black)	Harki et al., 1997; Butler and Day, 1998
Xylaria euglossa	Phlegmacin A 8,8'-di-O-methyl ether (green-yellow), (S)-torosachrysone-8-O-methyl ether (yellow), emodin-6,8-di-O-methyl ether (yellow)	Wang et al., 2005
Xylaria polymorpha	Melanin (black)	Robinson et al., 2012b; Tudor, 2013; Tudor et al., 2013
Xylaria sp.	(−)-Xylariamide A (yellow)	Davis, 2005; Zhou and Liu, 2010

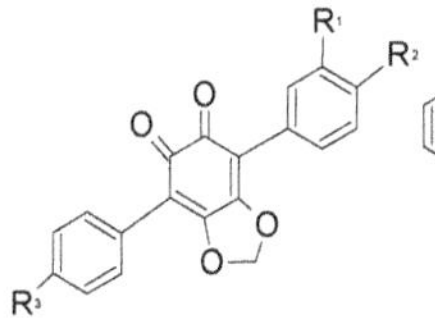

Phlebiarubrone
$(R_1 = H, R_2 = H, R_3 = H)$
4'-hydroxyphlebiarubrone
$(R_1 = H, R_2 = OH, R_3 = H)$

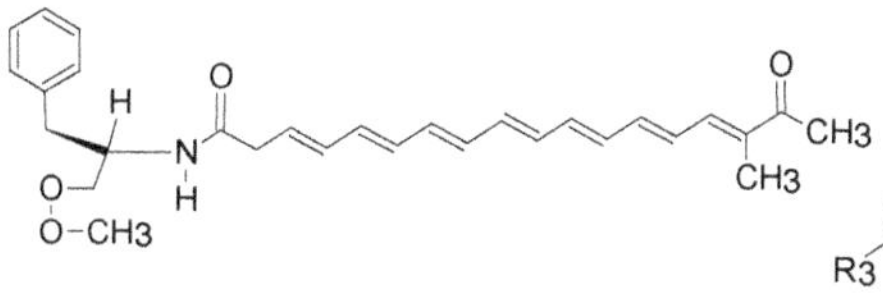

Melanocrocin

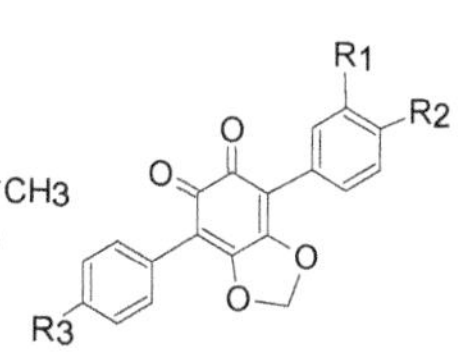

4',4''-dihydroxyphlebiarubrone
$(R_1 = H, R_2 = OH, R_3 = OH)$
3/,4',4''-trihydroxyphlebiarubrone
$(R_1 = OH, R_2 = OH, R_3 = OH)$

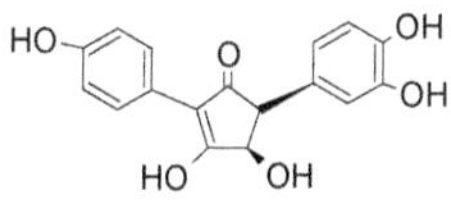

Involutin

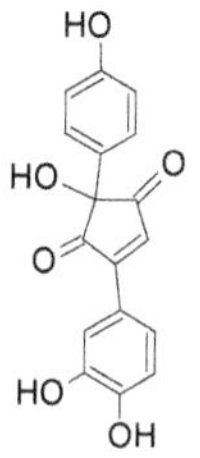

Involutone

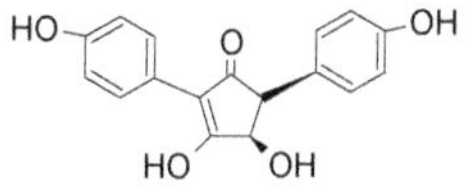

Chamonixin

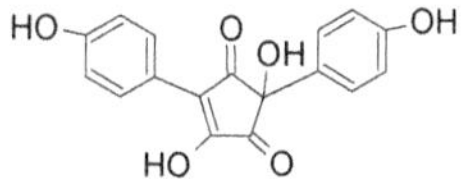

Gyroporin

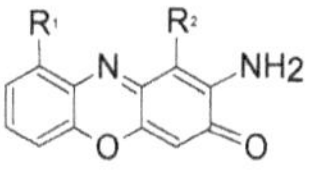

Cinnabarinic acid
$(R_1 = COOH, R_2 =$
$COOH)$**Cinnabarin**
$(R_1 = CH_2OH, R_2 =$
$COOH)$**Tramesanguin**
$(R_1 = COOH, R_2 = CHO)$

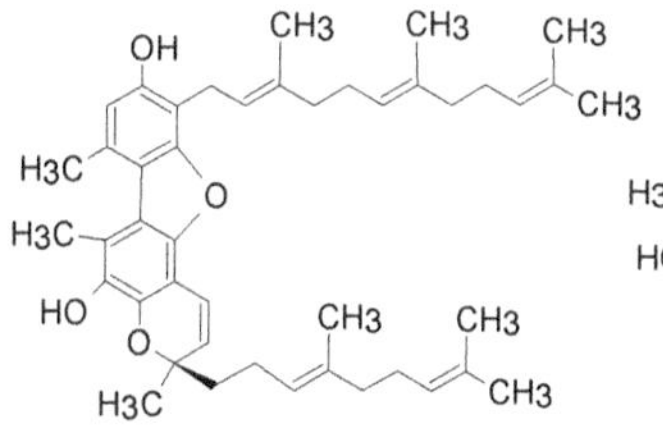

Albatrellin D

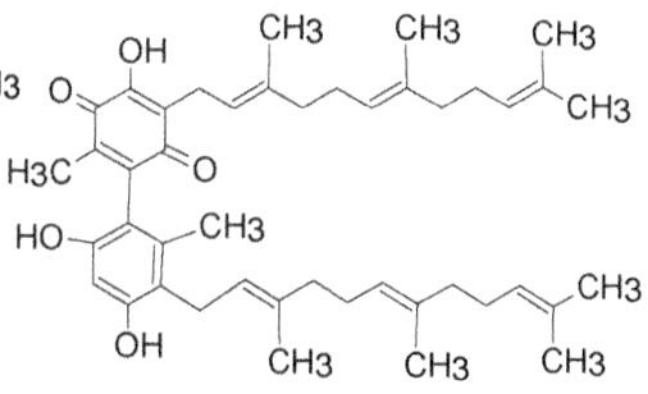

Albatrellin E

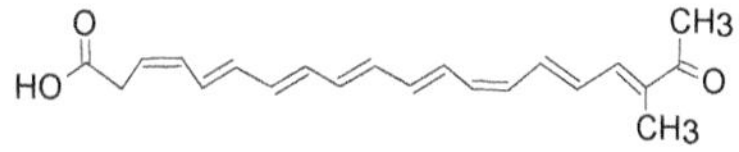

Albatrellin F

(3Z,5E,7E,9E,11E,13Z,15E,17E)-18-methyl-19-oxoicosa-3,5,7,9,11,13,15,17-octaenoic acid

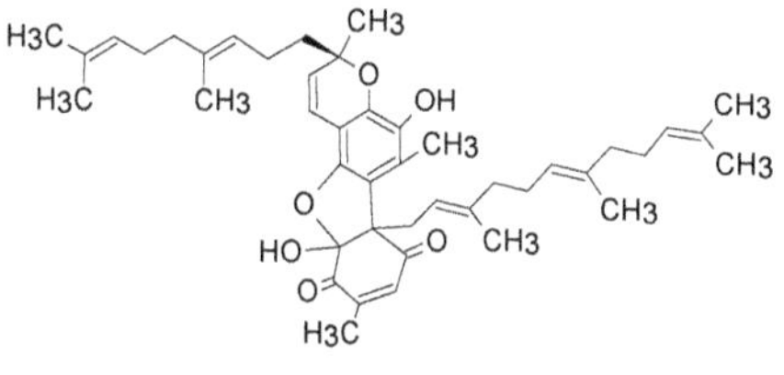

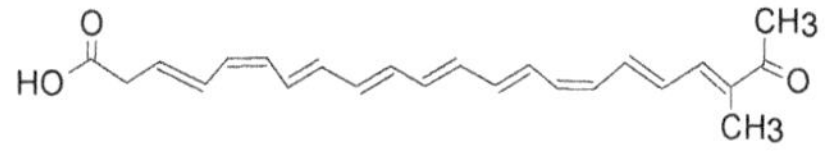

(3E,5Z,7E,9E,11E,13E,15Z,17E,19E)-20-methyl-21-oxodocosa-3,5,7,9,11,13,15,17,19-nonaenoic acid

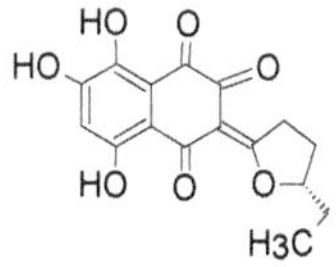

Quambalarine A

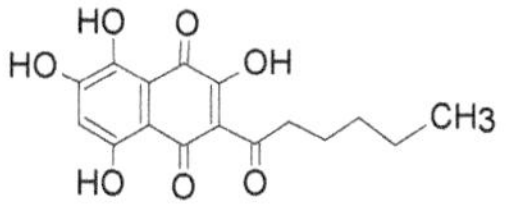

Quambalarine B

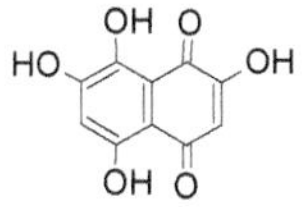

Mompain

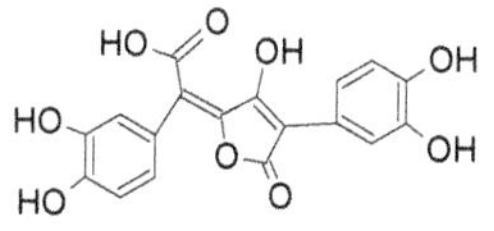

Variegatic acid

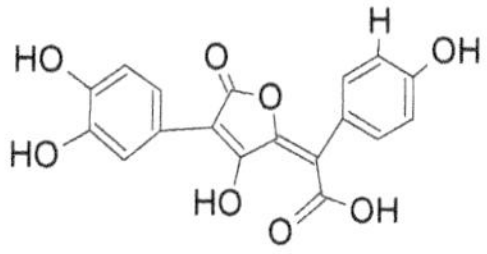

Xerocomic acid

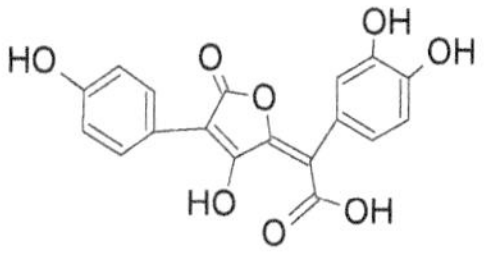

Isoxerocomic acid

Fig. 1 Contd. ...

...Fig. 1 Contd.

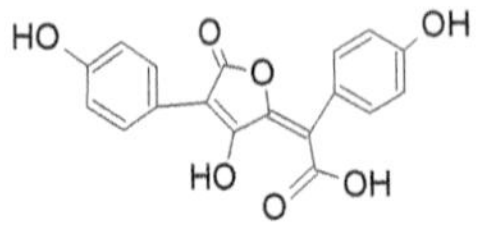

Atromentic acid

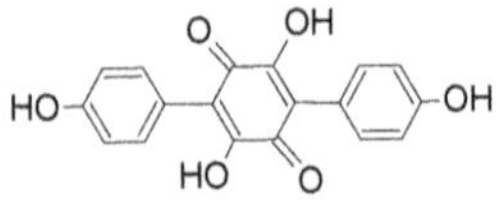

Atromentin

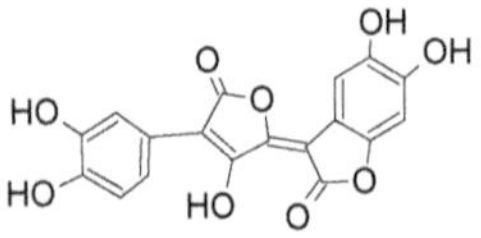

Variegatorubin

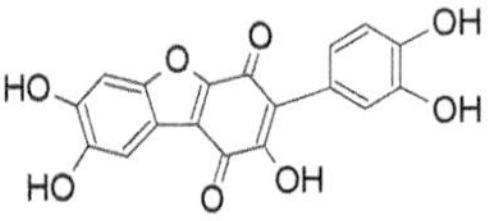

Cyclovariegatin

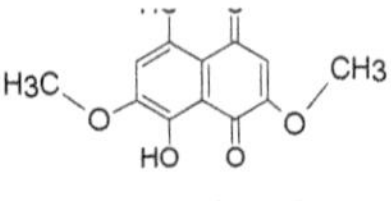

Draconin red

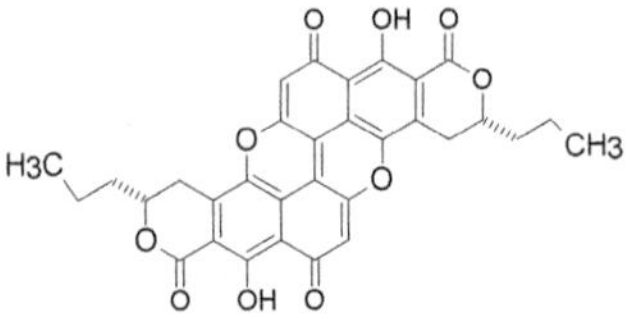

Xylindein

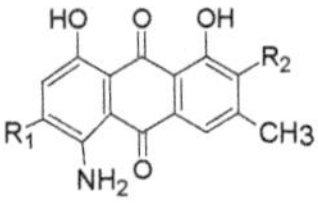

Anthraquinone derivative

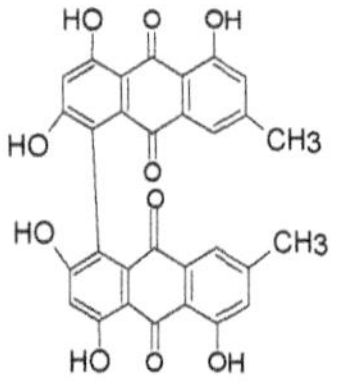

Skyrin

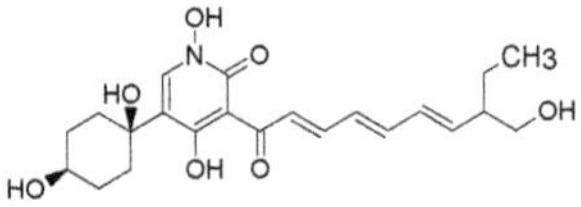

Torrubiellone A

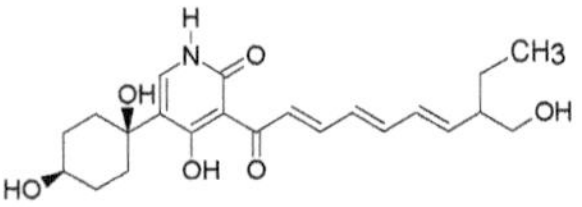

Torrubiellone B

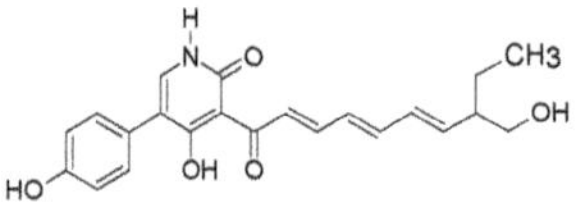

Torrubiellone C

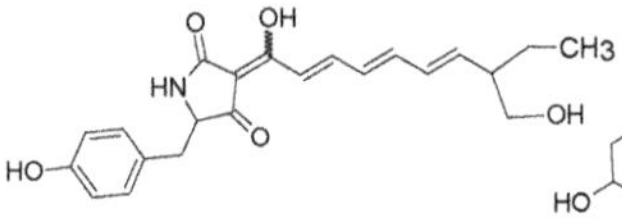

Torrubiellone D

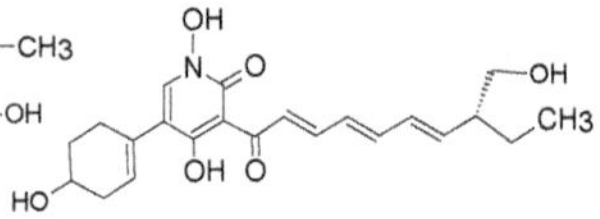

Torrubiellone E

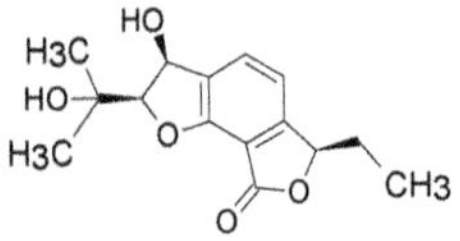

Daldinin A

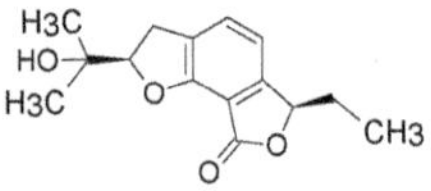

Daldinin B

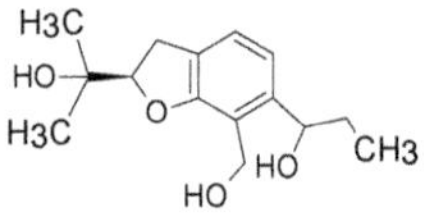

Daldinin C

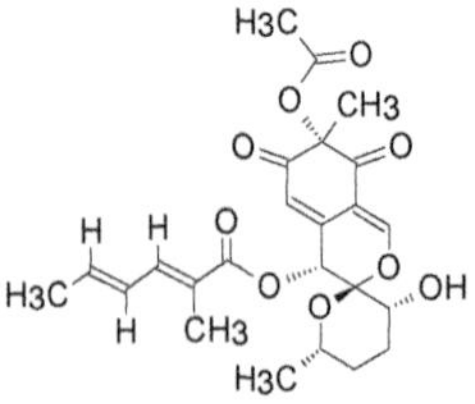

Daldinin E

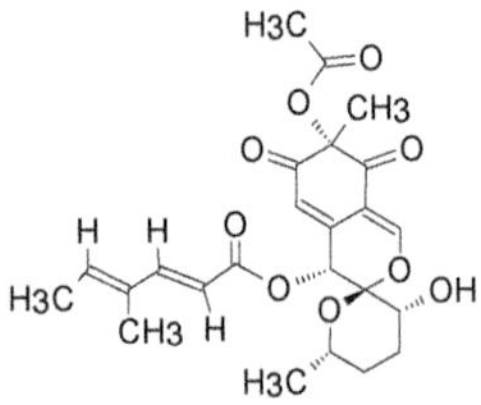

Daldinin F

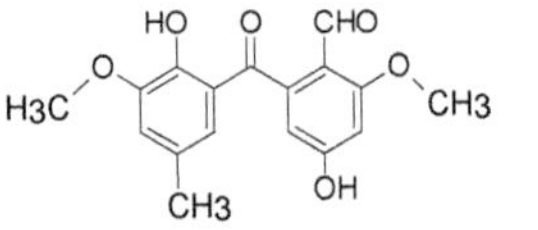

Daldinal A

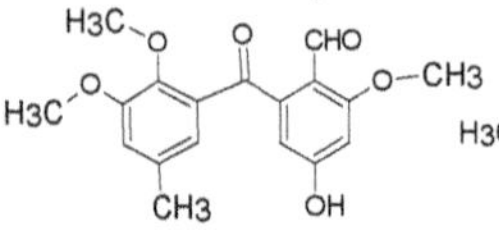

Daldinal B

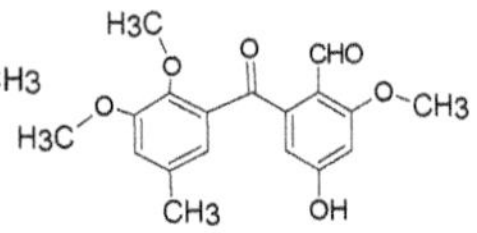

Daldinal C

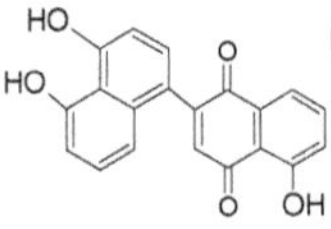

Hypoxylone

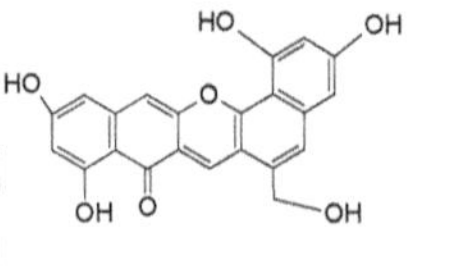

Hypoxyxylerone

Fig. 1 Contd. ...

...Fig. 1 Contd.

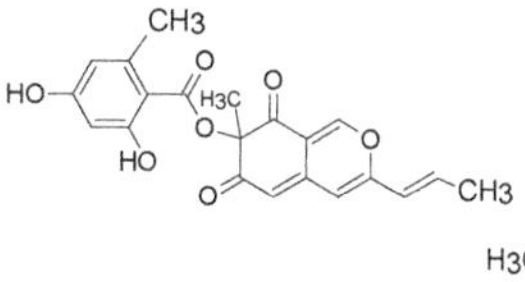

Mitrorubrin

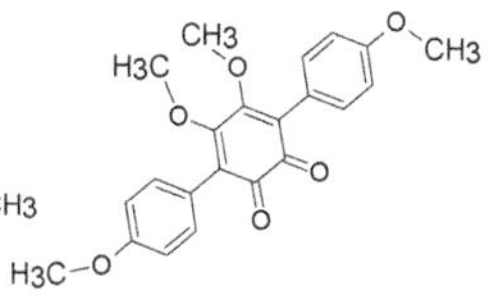

Rickenyl B

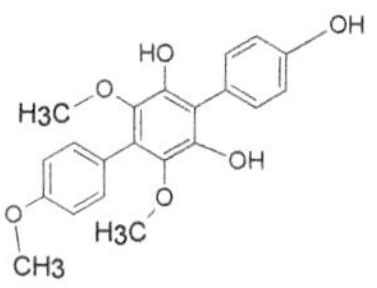

Rickenyl D

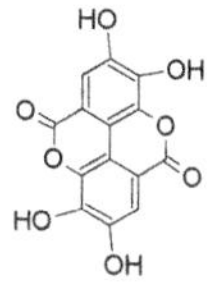

Ellagic acid

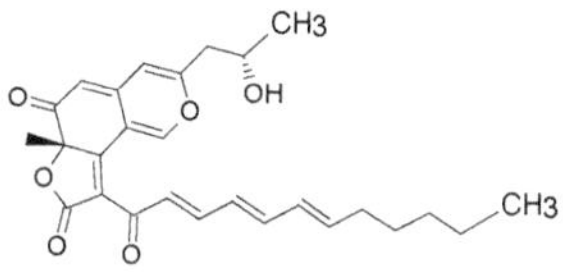

Lenormandin A

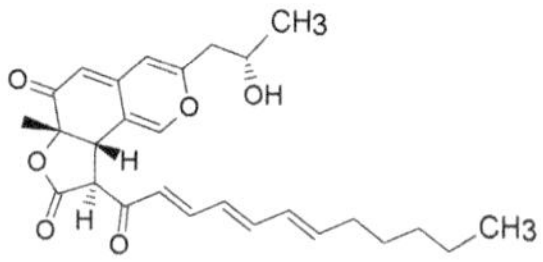

Lenormandin B

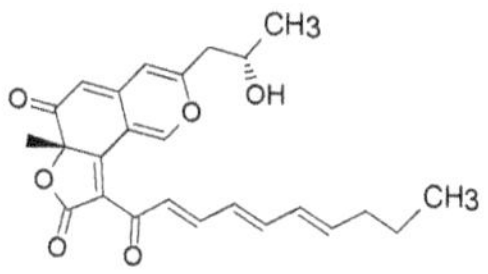

Lenormandin C

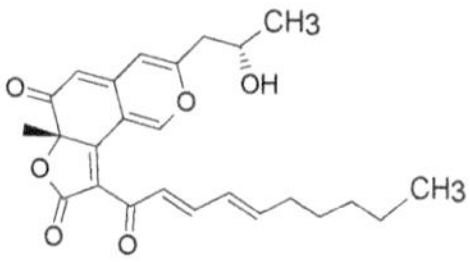

Lenormandin D

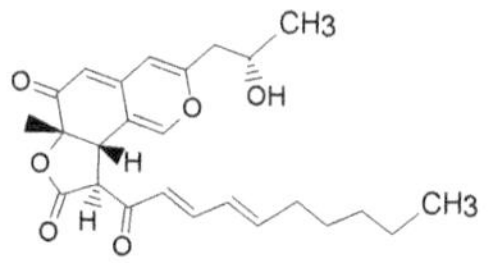

Lenormandin E

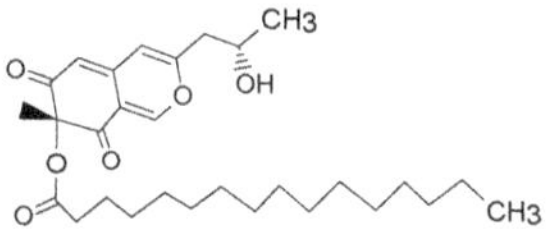

Lenormandin F

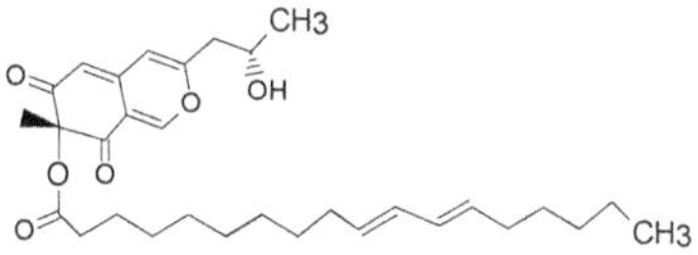

Lenormandin G

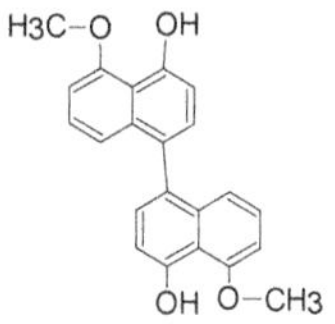

Daldinol

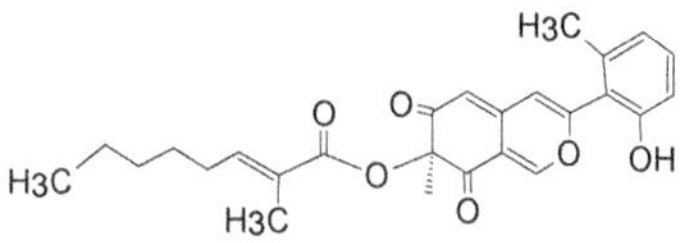

Cohaerin A

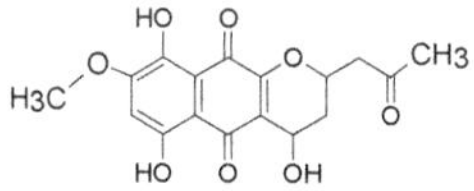

Erythrostominone

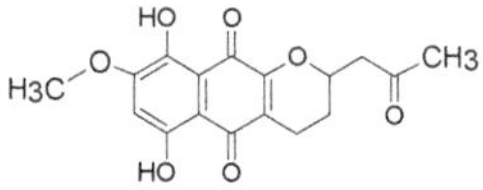

Deoxyerythrostominone

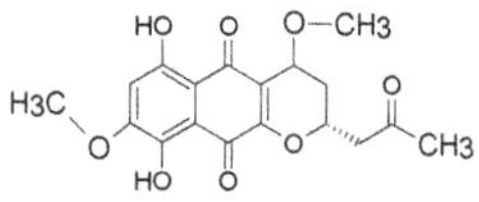

4-*O*-methyl erythrostominone

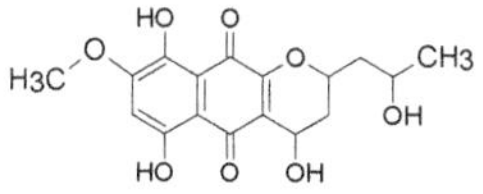

Epierythrostominol

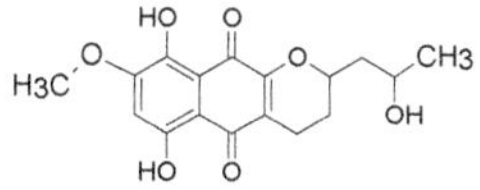

Deoxyerythrostominol

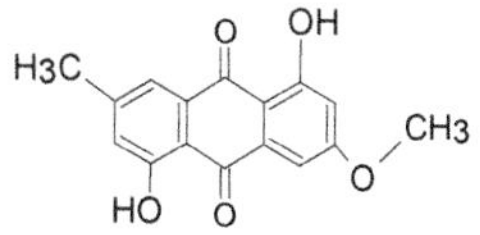

Shiraiarin

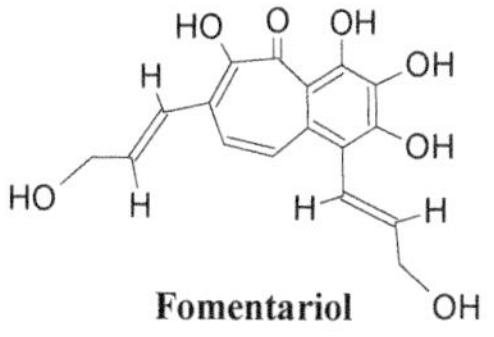

Fomentariol

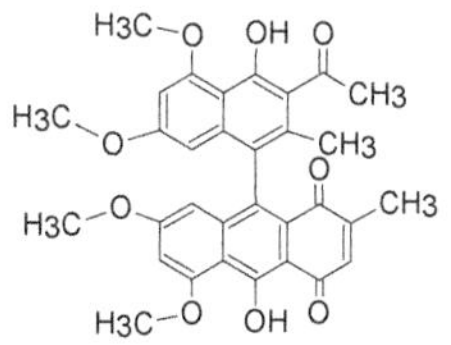

Rufoolivacin A

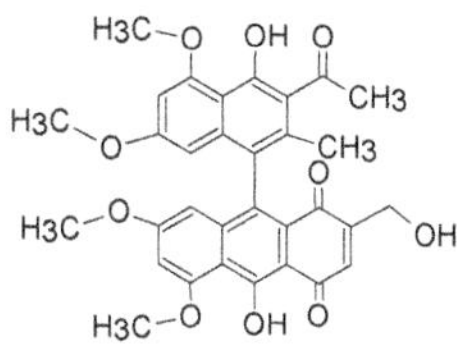

Rufoolivacin B

Fig. 1 Contd. ...

...Fig. 1 Contd.

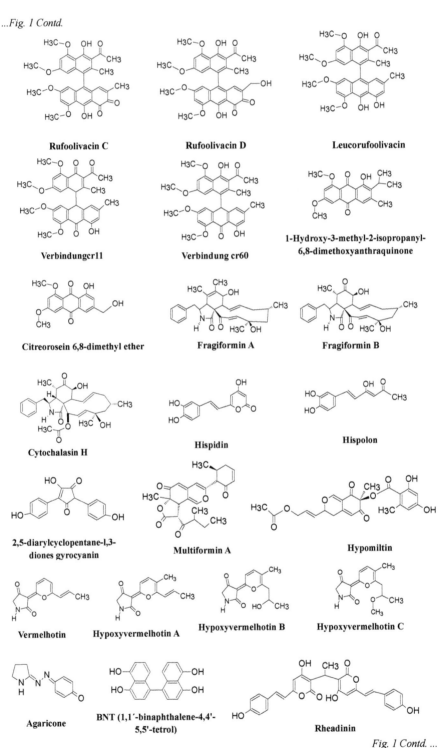

Rufoolivacin C

Rufoolivacin D

Leucorufoolivacin

Verbindungcr11

Verbindung cr60

1-Hydroxy-3-methyl-2-isopropanyl-6,8-dimethoxyanthraquinone

Citreorosein 6,8-dimethyl ether

Fragiformin A

Fragiformin B

Cytochalasin H

Hispidin

Hispolon

2,5-diarylcyclopentane-1,3-diones gyrocyanin

Multiformin A

Hypomiltin

Vermelhotin

Hypoxyvermelhotin A

Hypoxyvermelhotin B

Hypoxyvermelhotin C

Agaricone

BNT (1,1′-binaphthalene-4,4′-5,5′-tetrol)

Rheadinin

Fig. 1 Contd. ...

...Fig. 1 Contd.

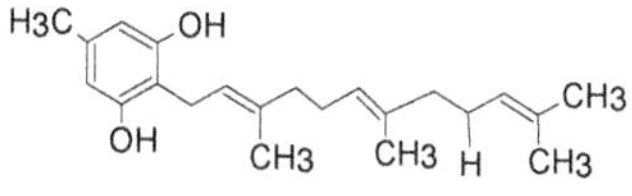

Grifolin

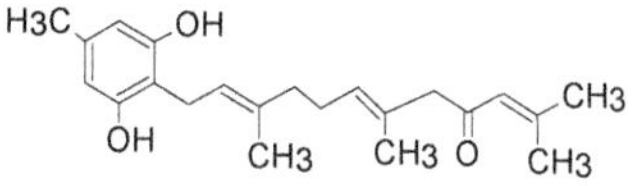

Grifolinone A

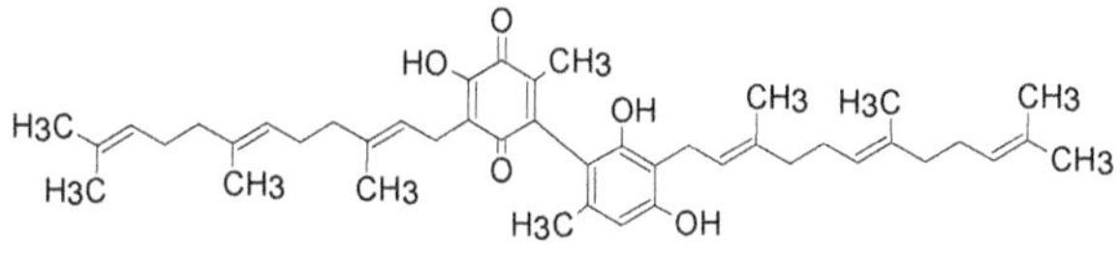

Grifolinone C

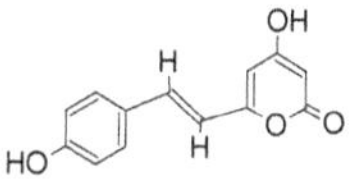

Bisnoryangonin

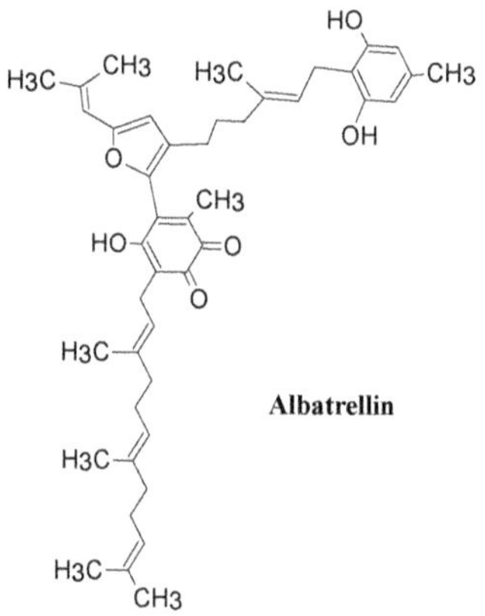

Albatrellin

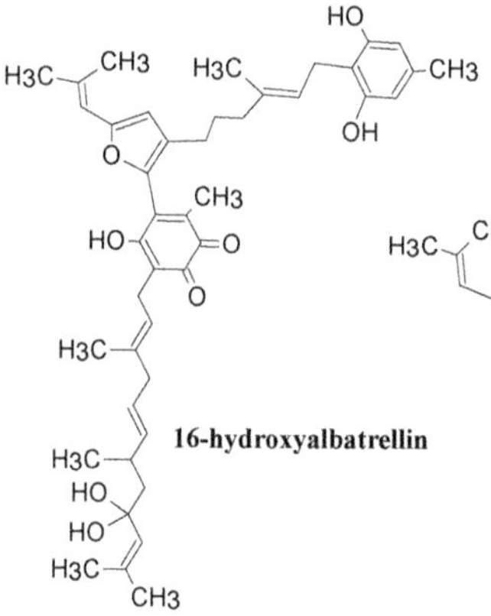

16-hydroxyalbatrellin

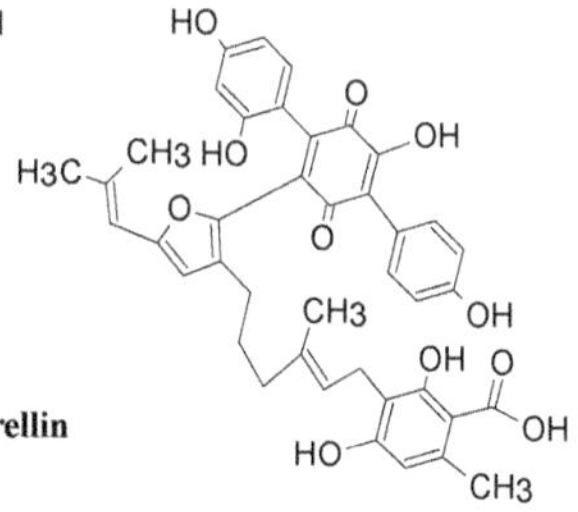

Cristatomentin

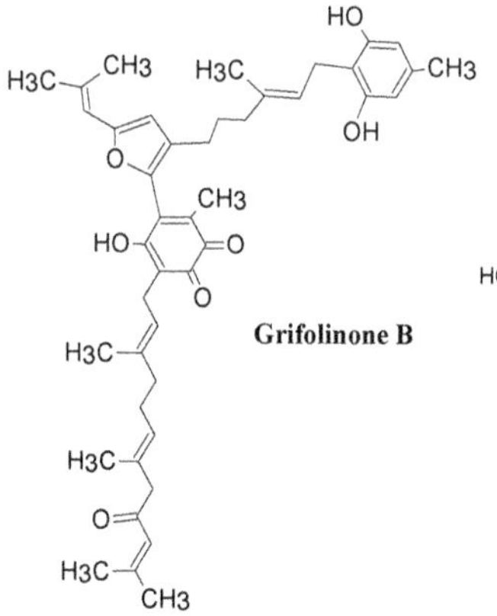

Grifolinone B

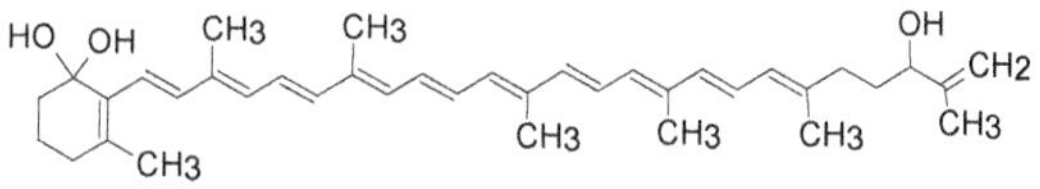

Aleuriaxanthin

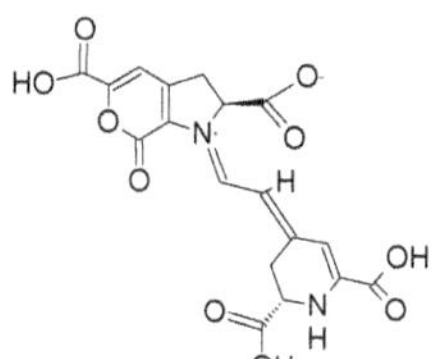

Muscapurpurin

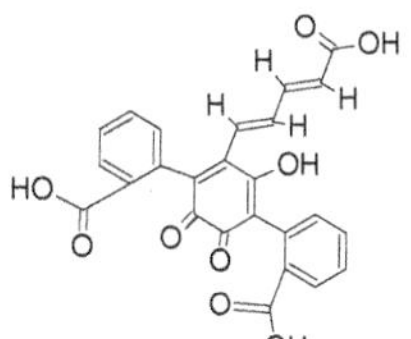

Muscarufin

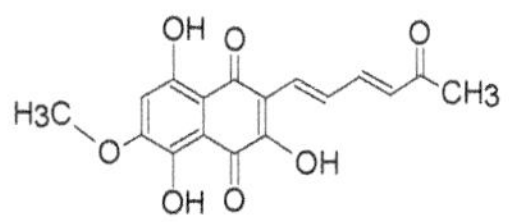

3,5,8-TMON (3,5,8-trihydroxy-6-methoxy-2-
(5-oxohexa-1,3-dienyl)-1,4-naphthoquinone)

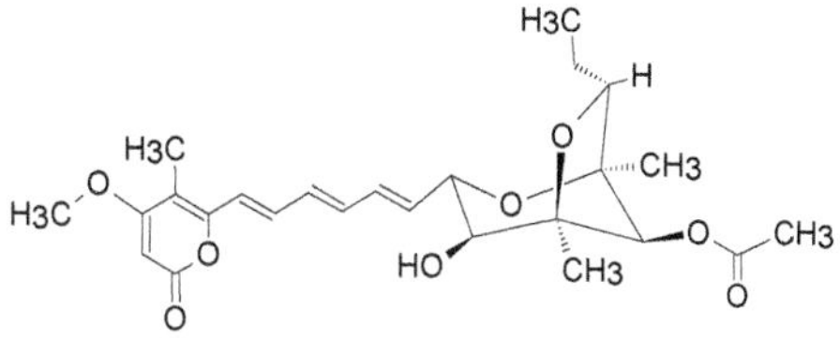

Aurovertin B

Fig. 1 Contd. ...

...*Fig. 1 Contd.*

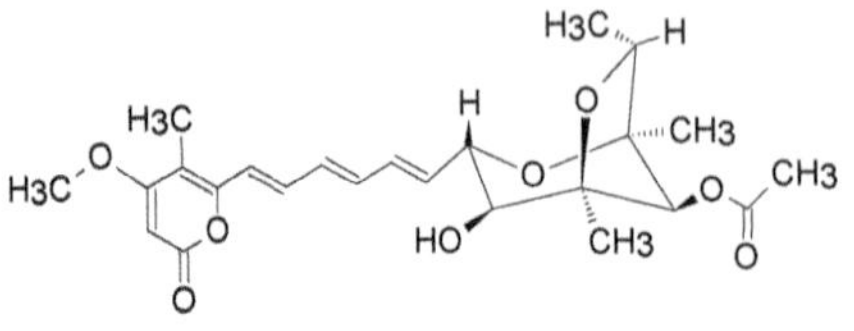

Aurovertin C

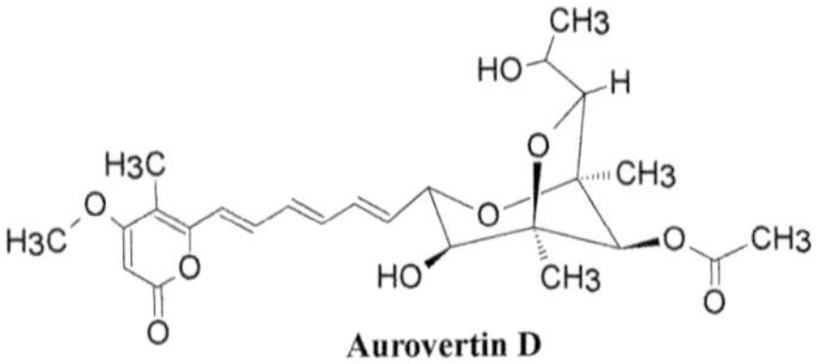

Aurovertin D

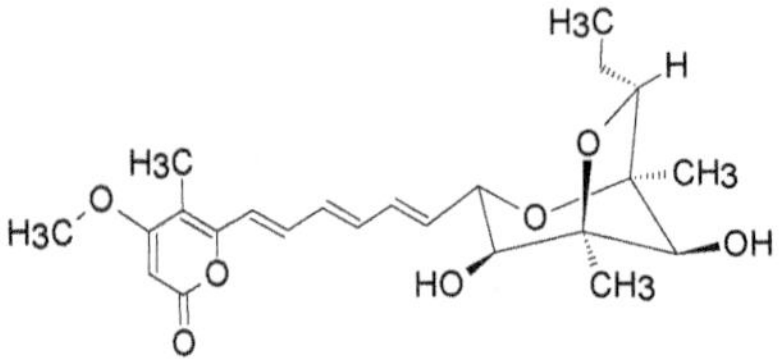

Aurovertin E

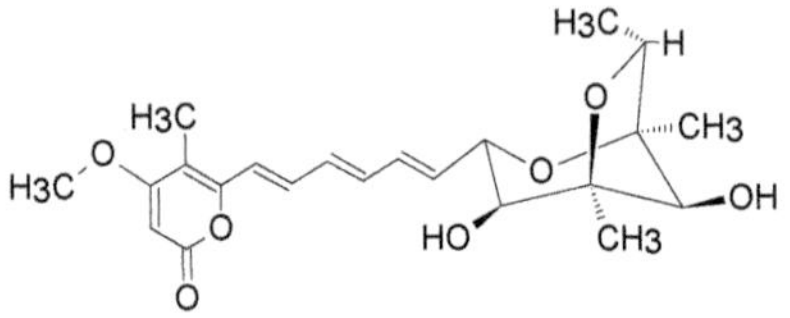

Aurovertin I

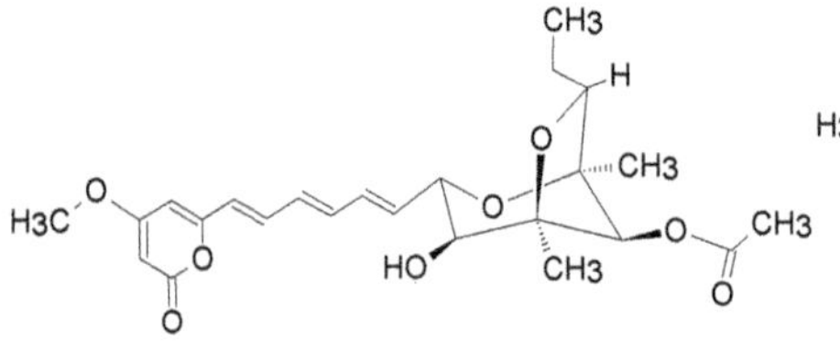

Aurovertin J

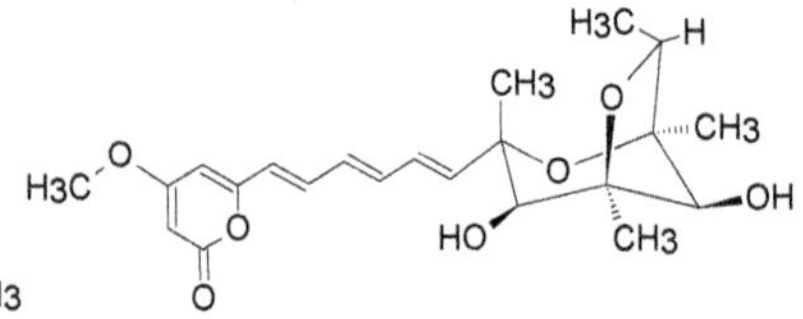

Aurovertin K

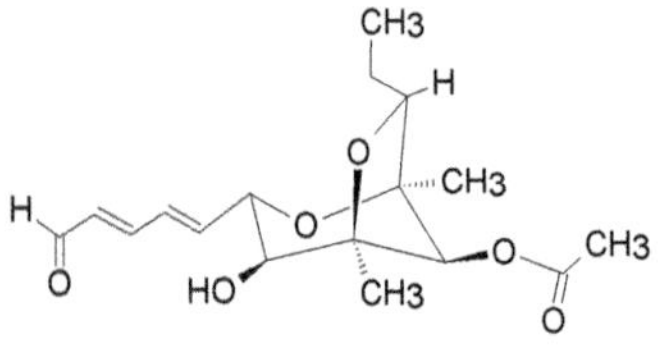

Aurovertin L

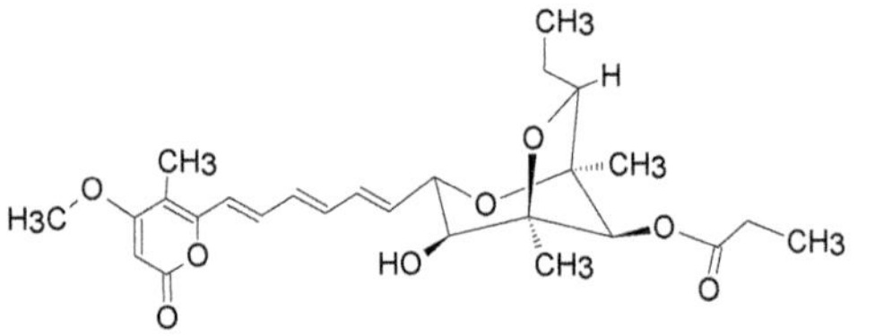

Aurovertin M

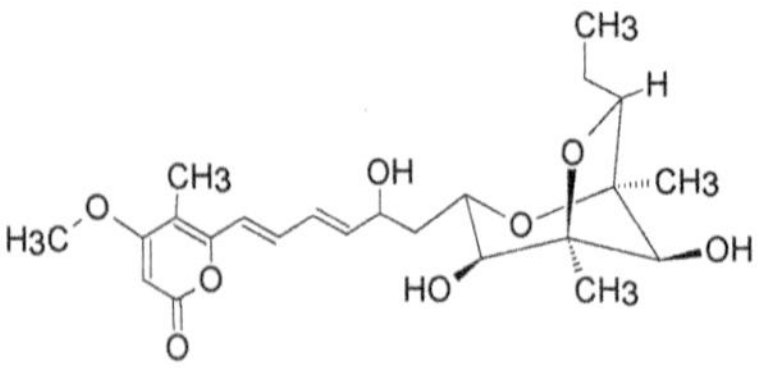

Aurovertin N

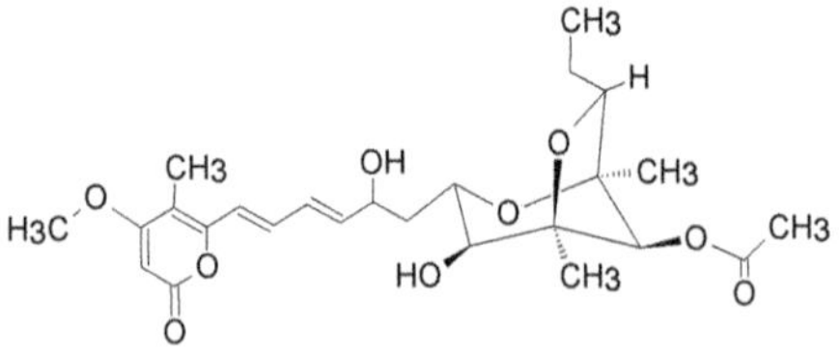

Aurovertin O

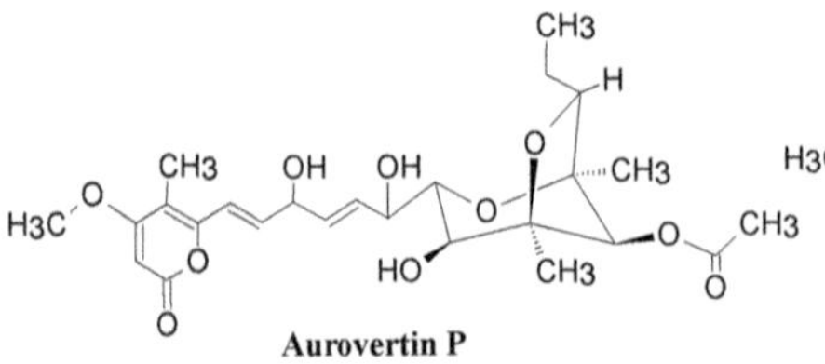

Aurovertin P

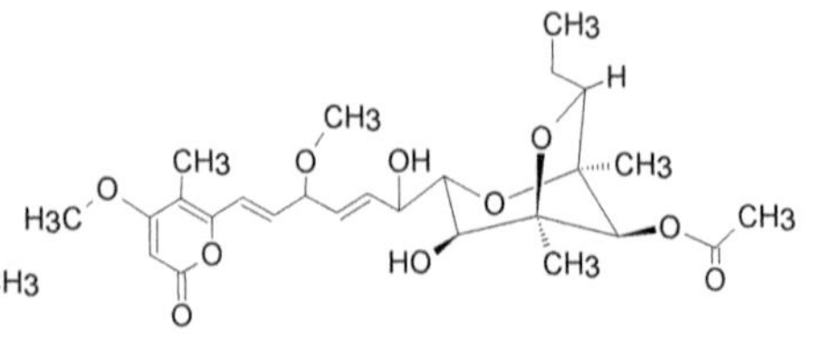

Aurovertin Q

Fig. 1 Contd. ...

...Fig. 1 Contd.

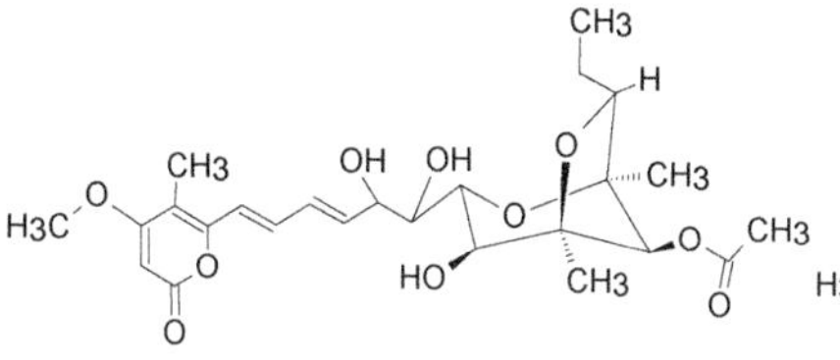

Aurovertin R

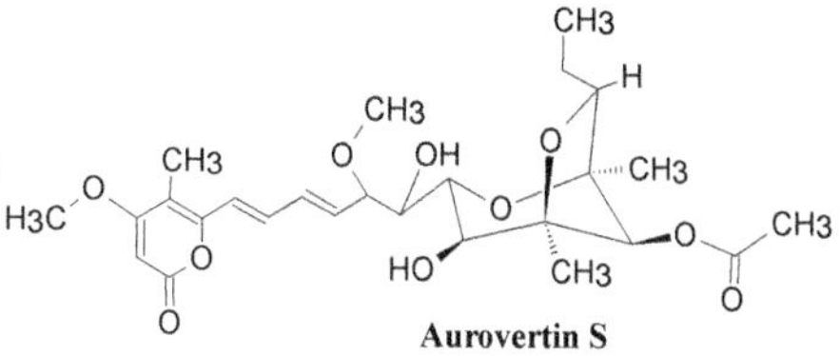

Aurovertin S

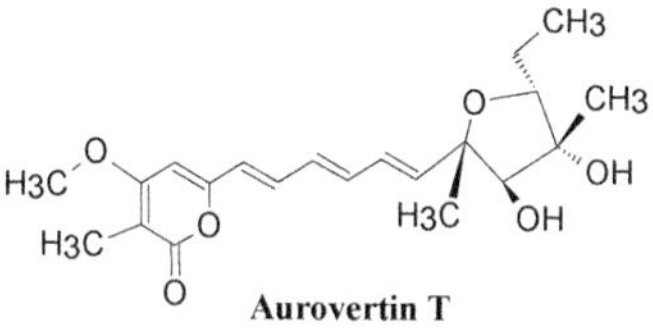

Aurovertin T

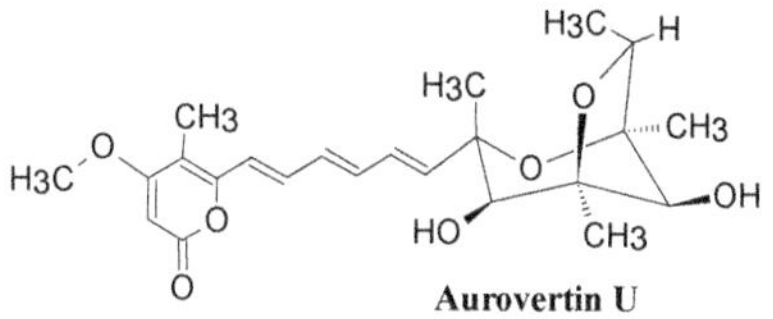

Aurovertin U

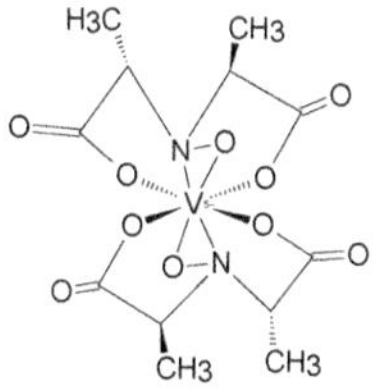

Amavadin

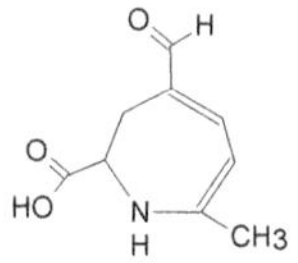

Muscoflavin

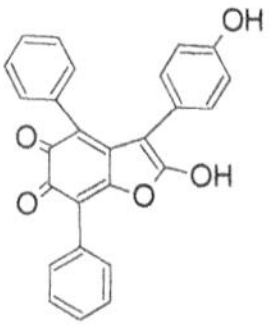

Xylerythrin

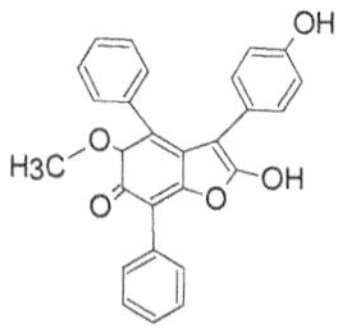

5-*O*-methylxylerythrin

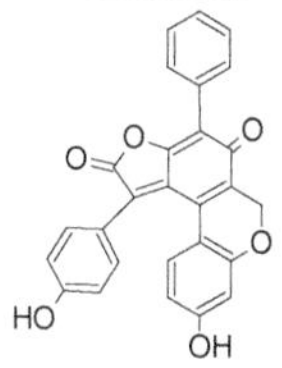

Peniophorinin

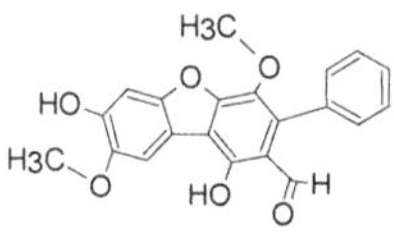

Penioflavin

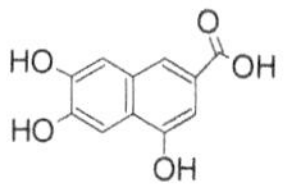

Austrogracilin A

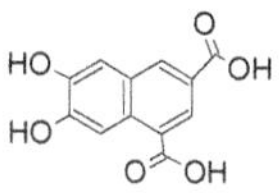

Austrogracilin B

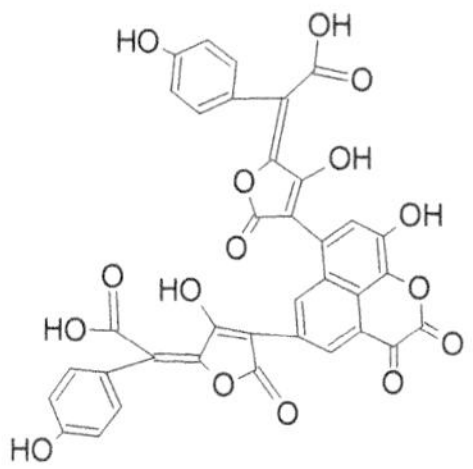

Badione A

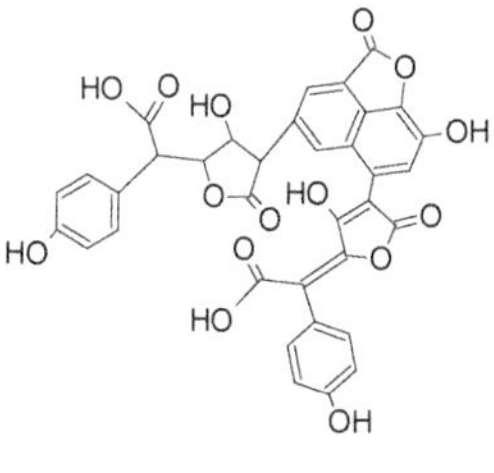

Norbadione A

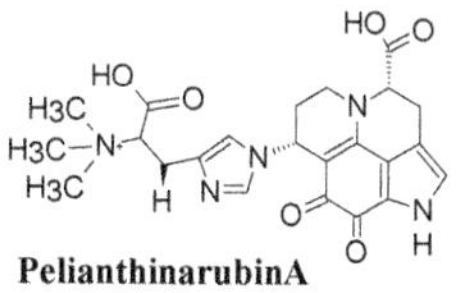

PelianthinarubinA

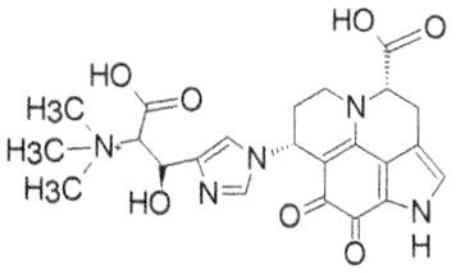

Pelianthinarubin B

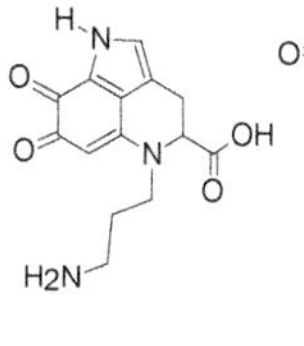

Mycenarubin A

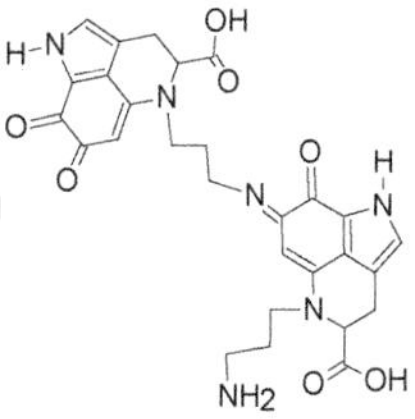

Mycenarubin B

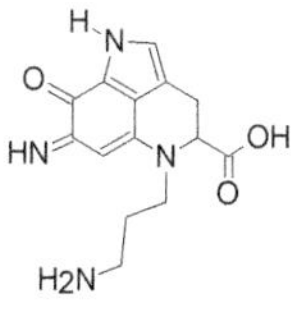

Mycenarubin D

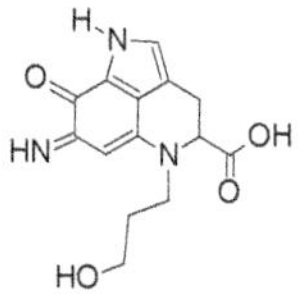

Mycenarubin E

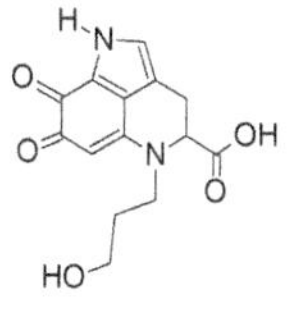

Mycenarubin F

Fig. 1 Contd. ...

...Fig. 1 Contd.

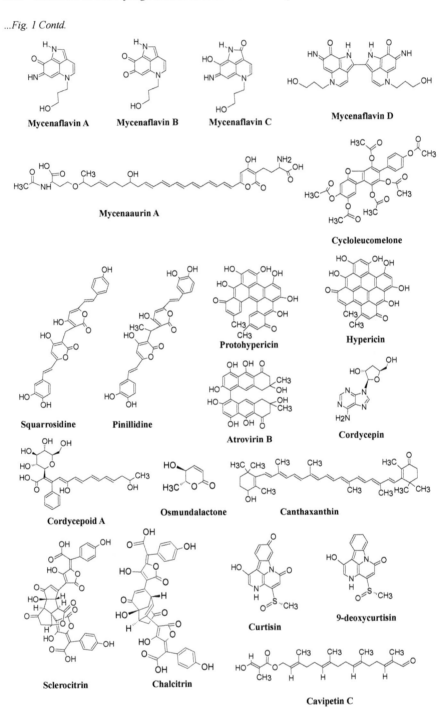

Mycenaflavin A

Mycenaflavin B

Mycenaflavin C

Mycenaflavin D

Mycenaaurin A

Cycloleucomelone

Squarrosidine

Pinillidine

Protohypericin

Hypericin

Atrovirin B

Cordycepin

Cordycepoid A

Osmundalactone

Canthaxanthin

Sclerocitrin

Chalcitrin

Curtisin

9-deoxycurtisin

Cavipetin C

Fig. 1 Contd. ...

...Fig. 1 Contd.

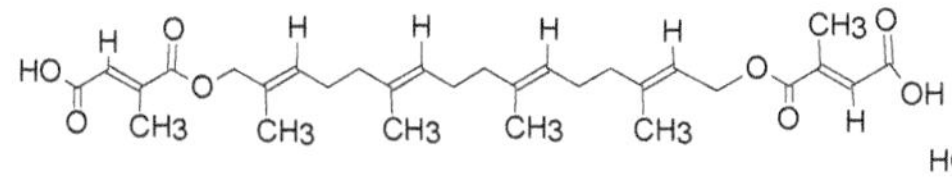

Cavipetin A

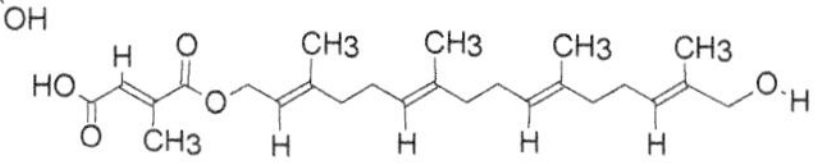

Cavipetin D

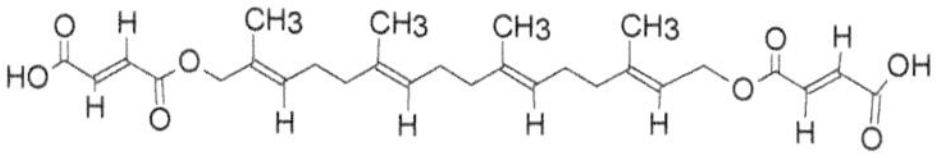

Cavipetin B

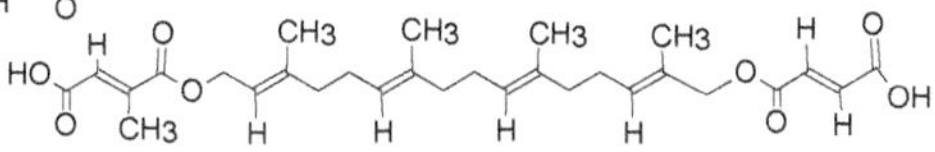

Cavipetin E

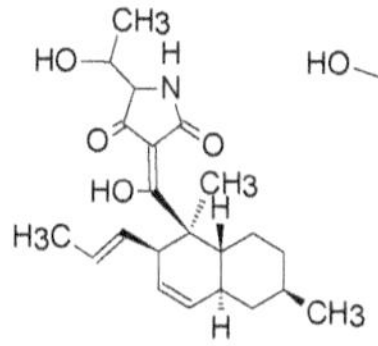

Paecilosetin

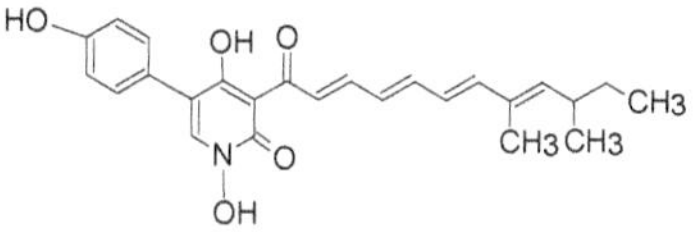

Farinosone B

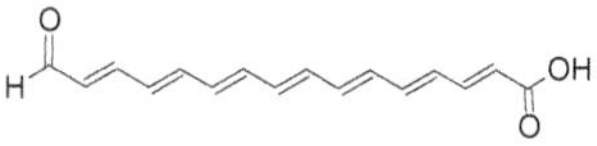

Calolstomal

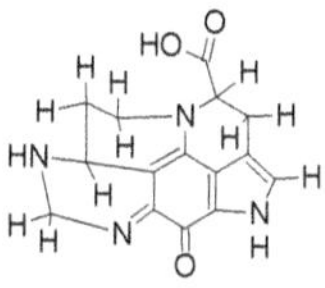

Sanguinone A

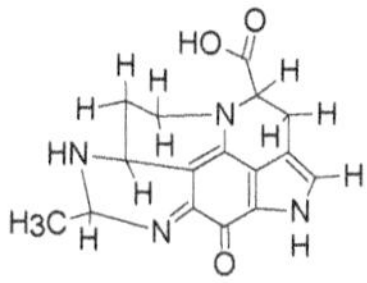

Sanguinone B

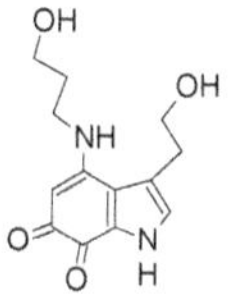

Sanguinolentaquinone

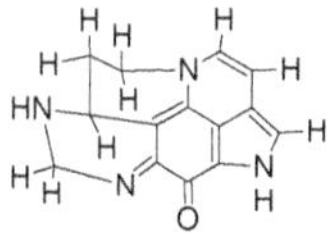

Decarboxydehydrosanguinone A

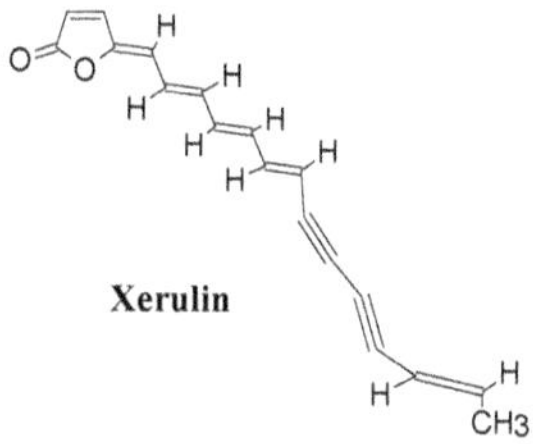

Xerulin

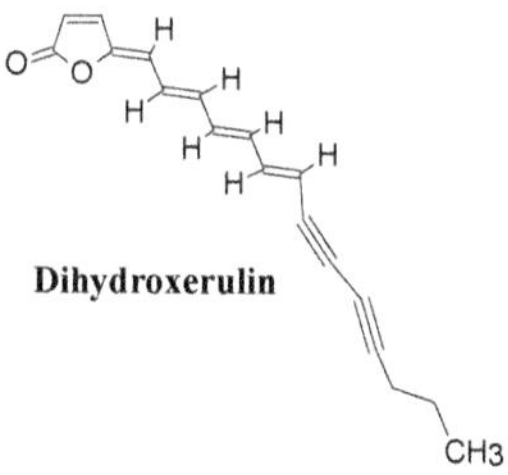

Dihydroxerulin

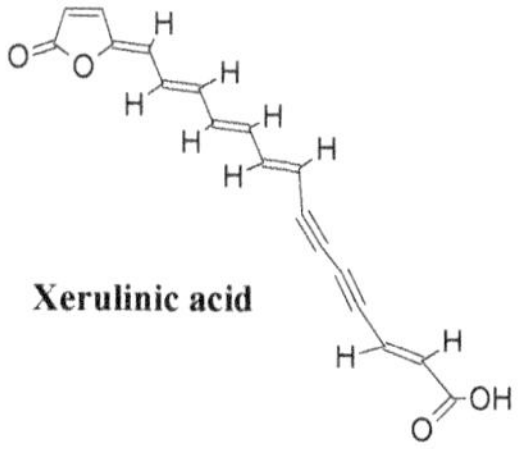

Xerulinic acid

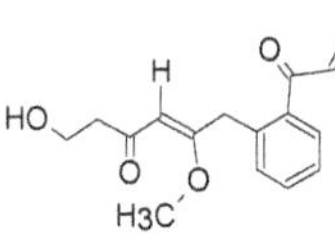

Peniophorin A

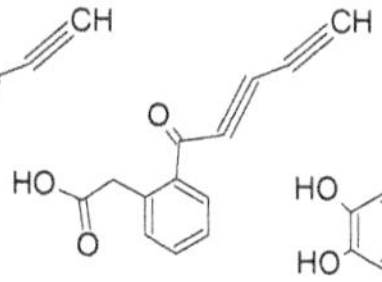

Peniophorin B

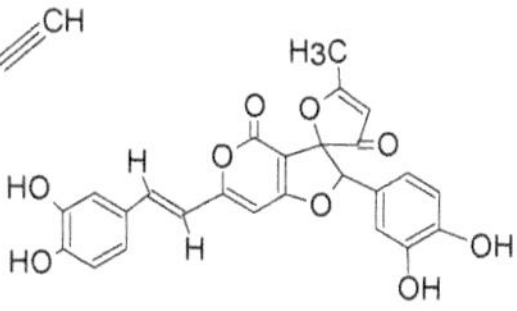

Inoscavin A

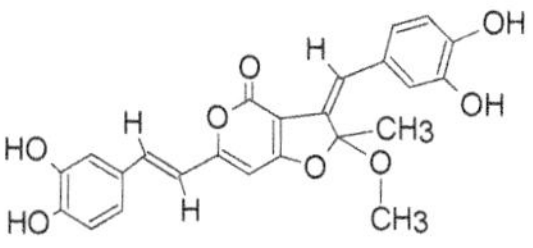

Inoscavin B

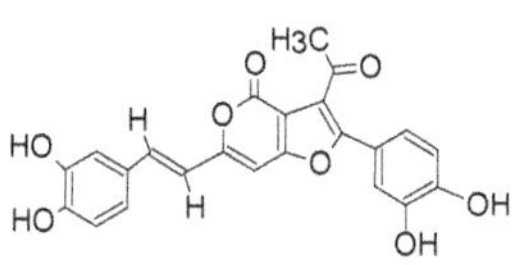

Inoscavin C

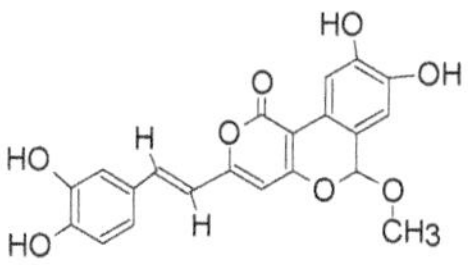

Inoscavin D

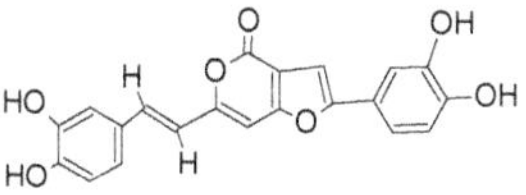

Inoscavin E

Fig. 1 Contd. ...

...Fig. 1 Contd.

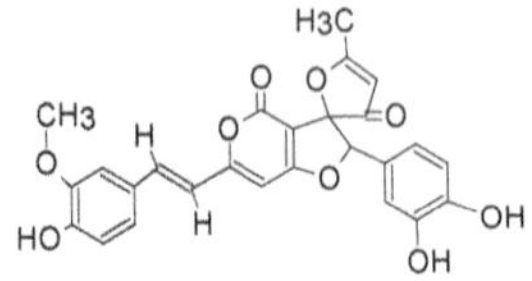

Methylinoscavin A

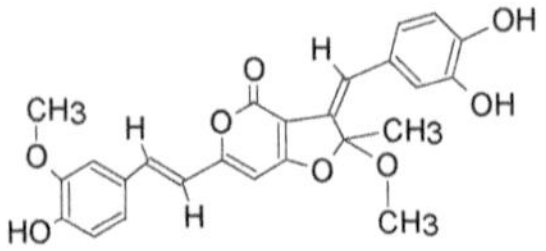

Methylinoscavin B

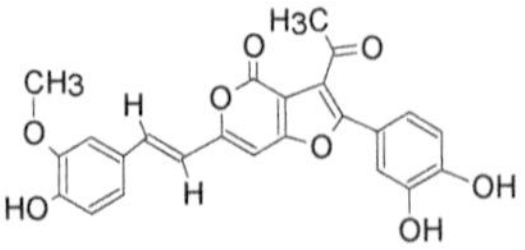

Methylinoscavin C

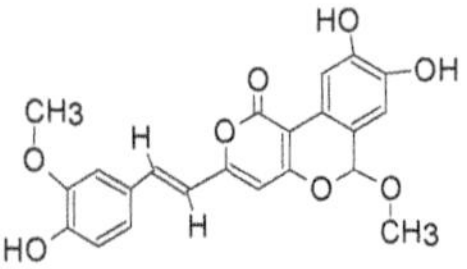

Methylinoscavin D

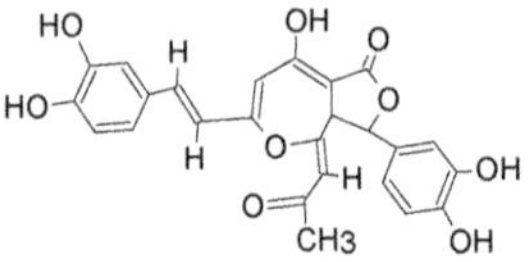

Davallialactone

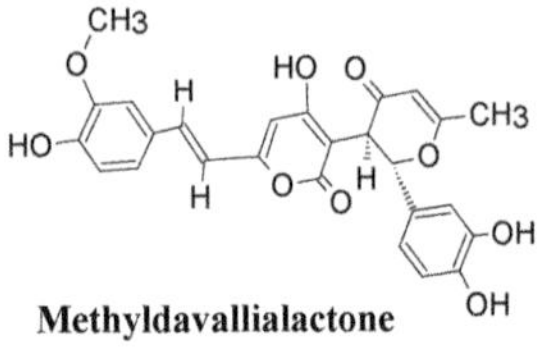

Methyldavallialactone

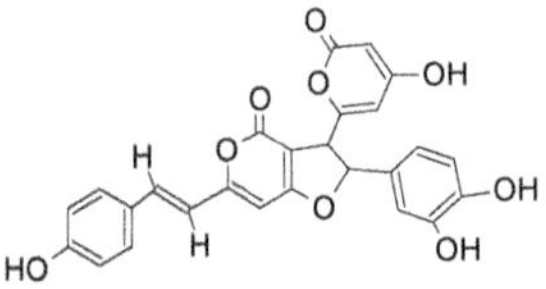

Hypholomine A

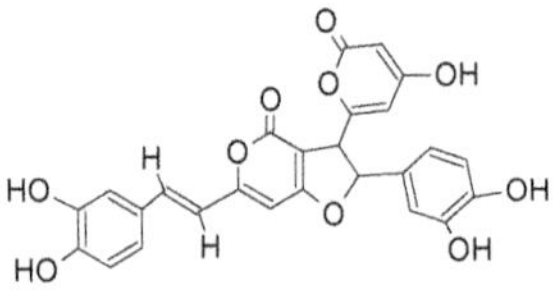

Hypholomine B

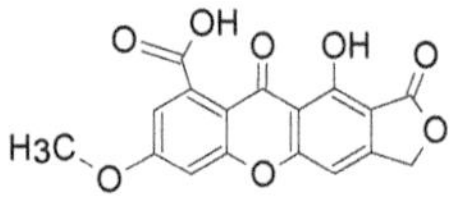

Dermoxanthone

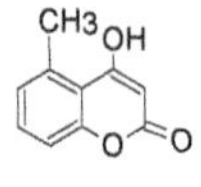

2–hydroxy–5–methylchromone

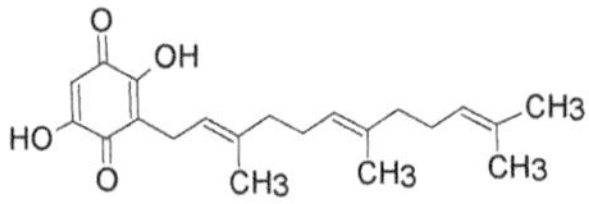

Boviquinone 3

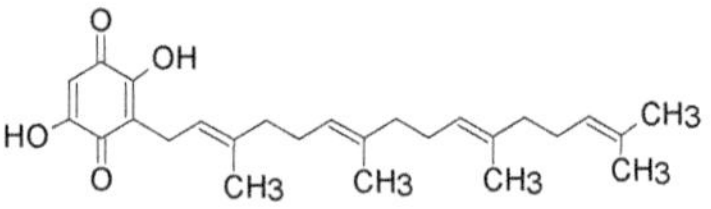

Boviquinone 4

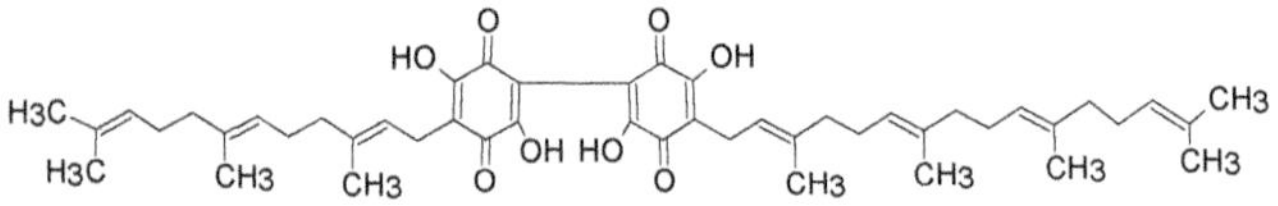

Diboviquinone-3,4

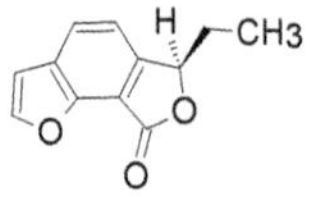

Concentricolide

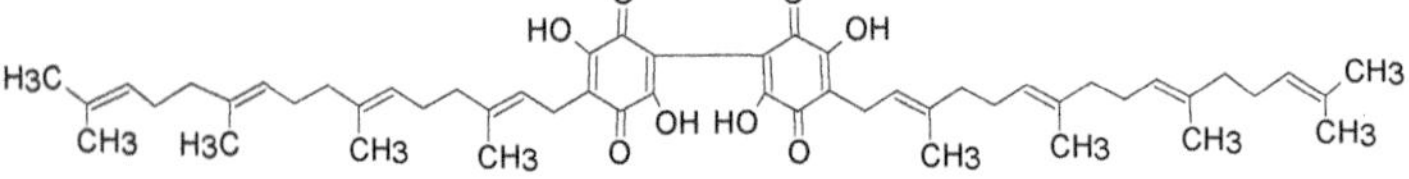

Diboviquinone-4,4

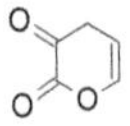

Pyrandione

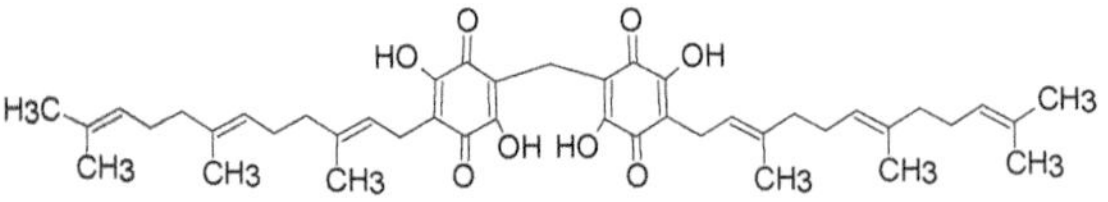

Methylenediboviquinone-3,3

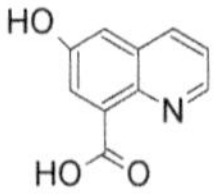

6-Hydroxyquinoline-8-carboxylic acid

Fig. 1 Contd. ...

...Fig. 1 Contd.

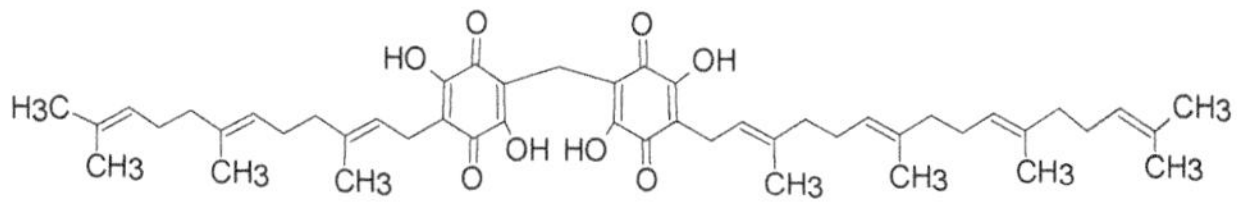

Methylenediboviquinone-3,4

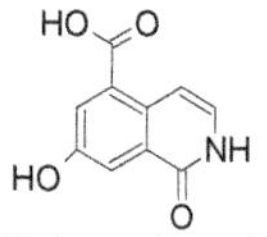

7-Hydroxy-1-oxo-1,2-dihydroisoquinoline-5-carboxylic acid

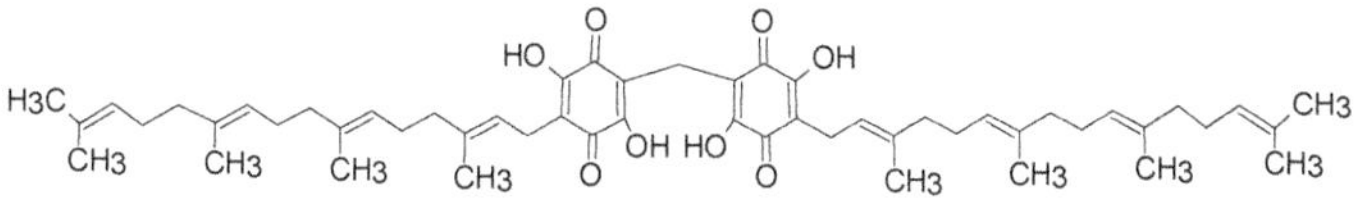

Methylenediboviquinone-4,4

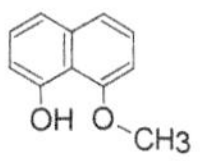

8–Methoxy–1–napthol

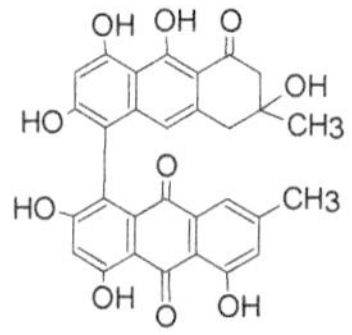

Icterinoidin A

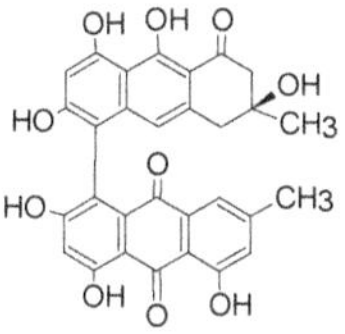

IcterinoidinB

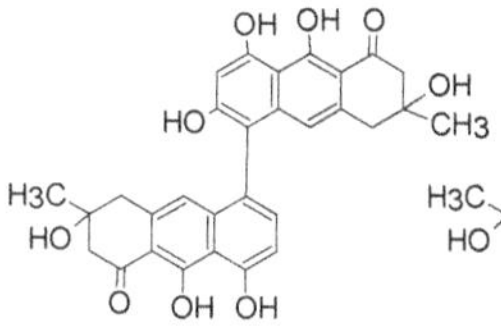

IcterinoidinC

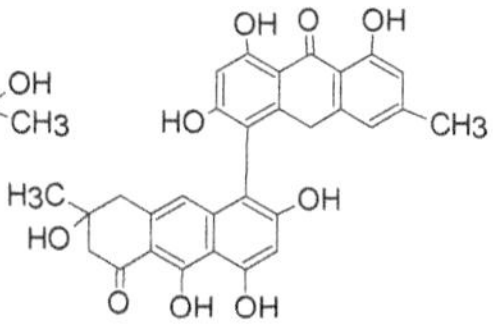

Austrovenetin

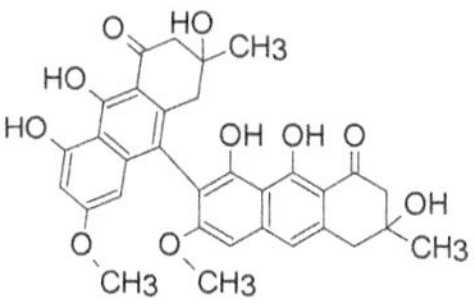

Phlegmacin

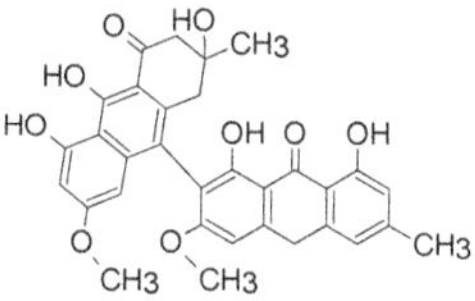

Anhydrophlegmacin

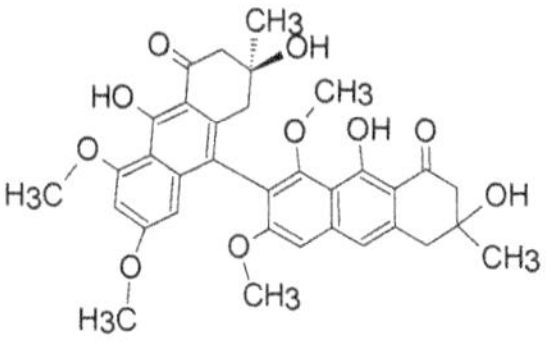

Phlegmacin A 8,8'-di-*O*-methyl ether

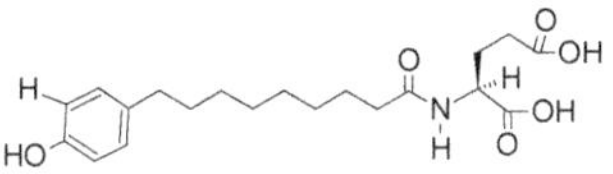

Scaurin A

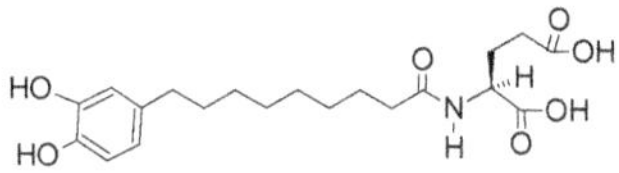

Scaurin B

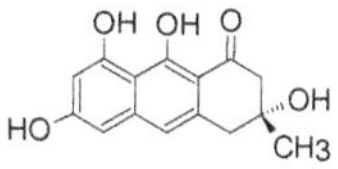

(3*R*)-atrochrysone

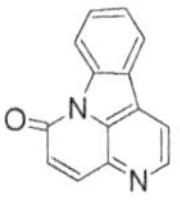

Canthin-6-one

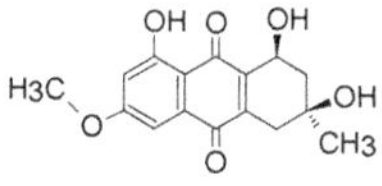

(1*S*,3*S*)-austrocortilutein

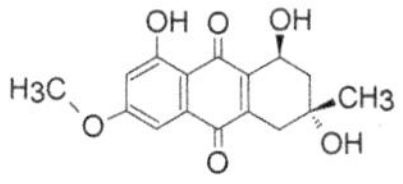

(1*S*,3*R*)-austrocortilutein

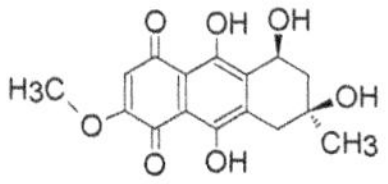

(1*S*,3*S*)-austrocortirubin

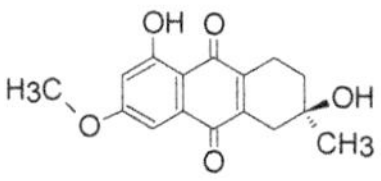

1-deoxyaustrocortilutein

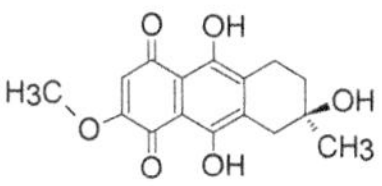

1-deoxyaustrocortirubin

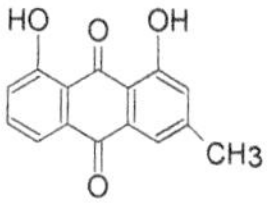

Chrysophanol

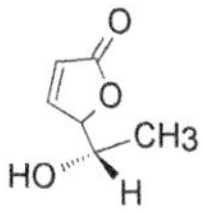

5-hydroxy-hex-2-en-4-olide

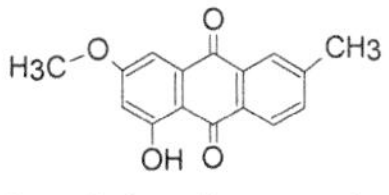

6-methylxanthopurpurin-3-*O*-methyl ether

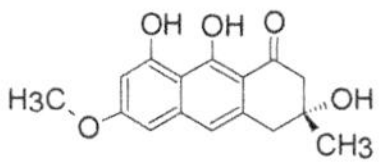

(3*R*)-torosachrysone

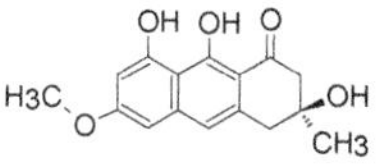

(*S*)-torosachrysone

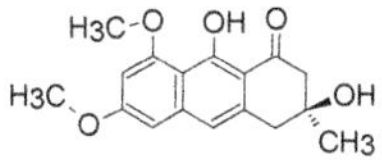

(3*S*)-torosachrysone-8-*O*-methyl ether

Fig. 1 Contd. ...

...Fig. 1 Contd.

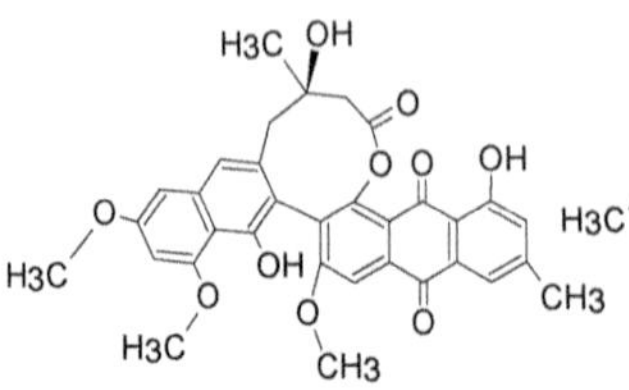

Dermocanarin 1

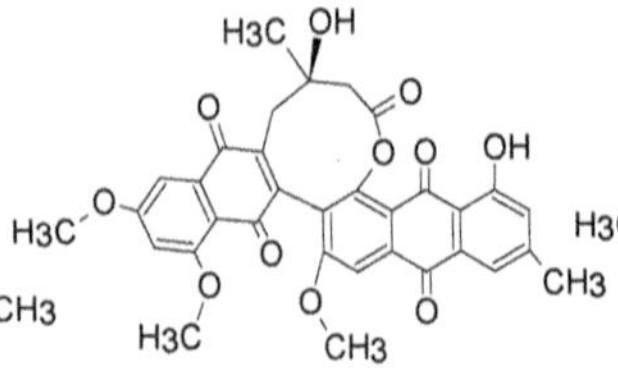

Dermocanarin 2

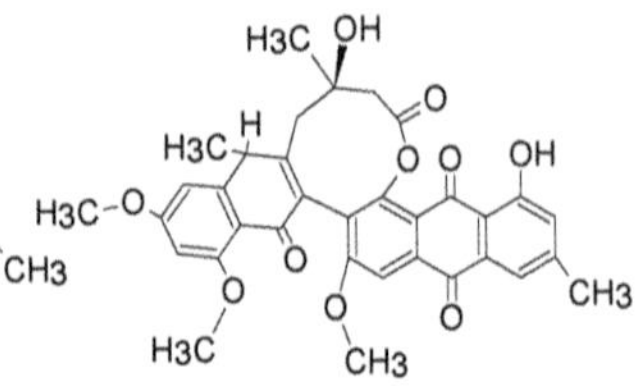

Dermocanarin 3

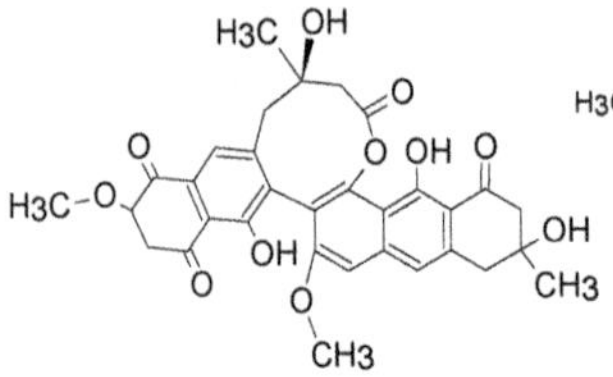

Dermocanarin 4

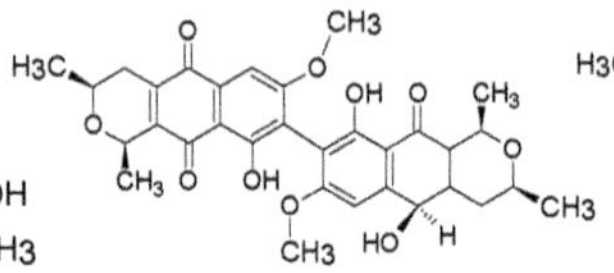

Cardinalin 2

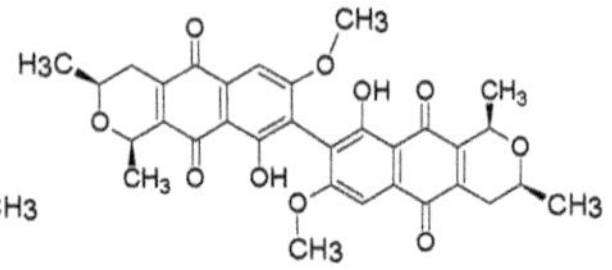

Cardinalin 3

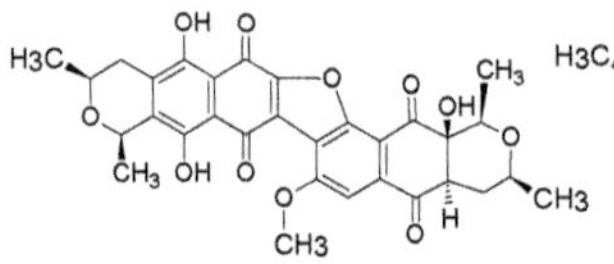

Cardinalin 4

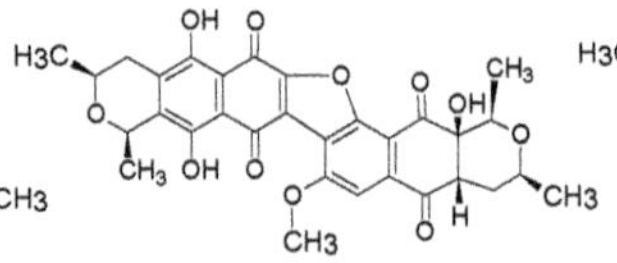

Cardinalin 5

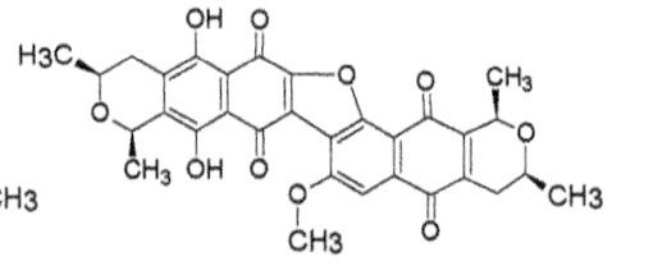

Cardinalin 6

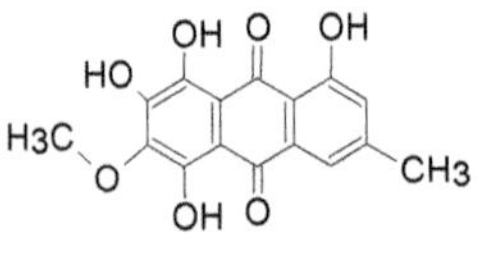

Dermocybin

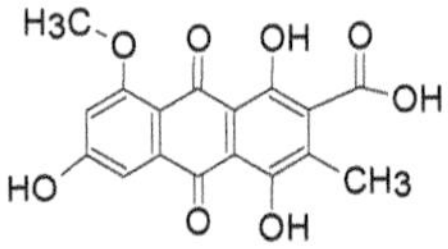

Dermorubin

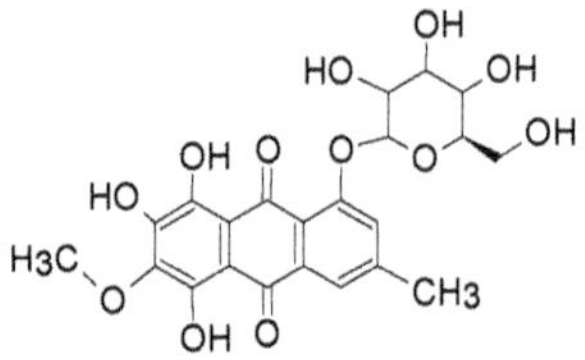

Dermocybin-1-β-D glycopyranoside

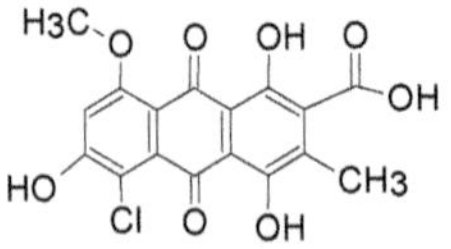

5-chlorodermorubin

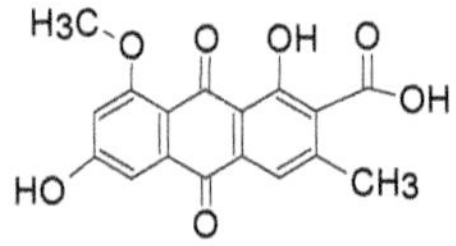

Dermolutein

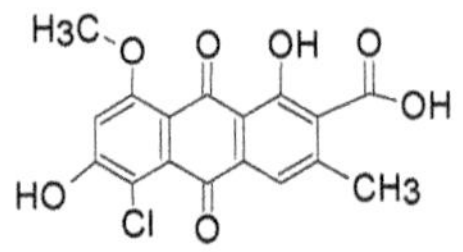

5-chlorodermolutein

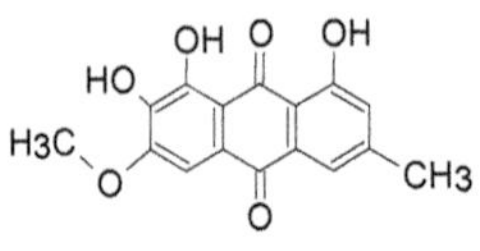

Dermoglaucin

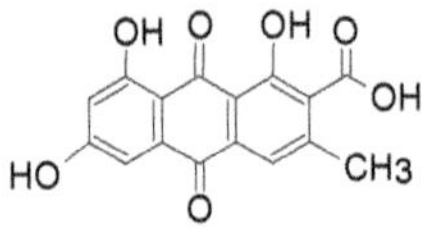

Endocrocin

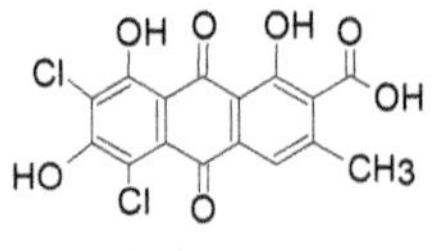

5,7-dichloroendocrocin

Fig. 1 Contd. ...

...*Fig. 1 Contd.*

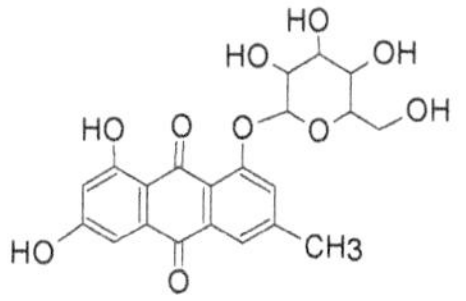

Emodin-1-β-D-glucopyranoside

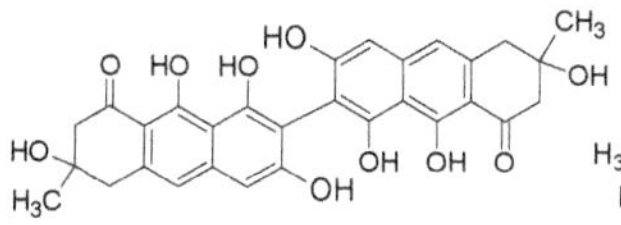

Flavomannin

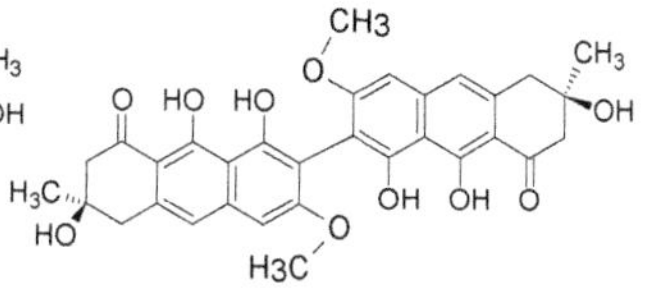

Flavomannin-6,6'-di-*O*-methyl ether

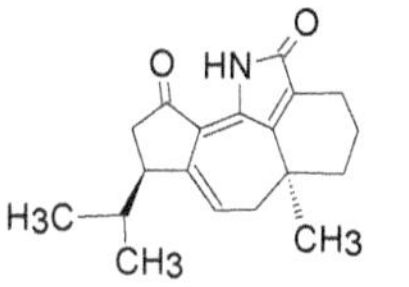

Pyromyxone A

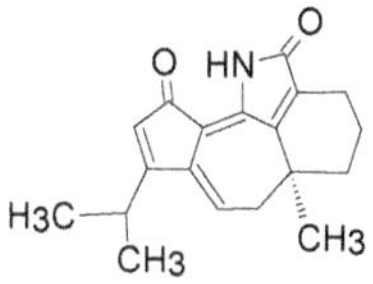

Pyromyxone B

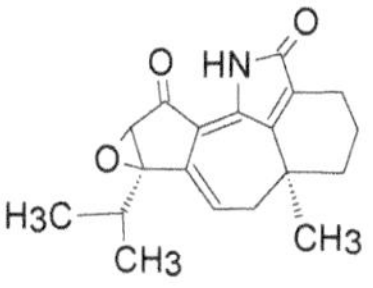

Pyromyxone C

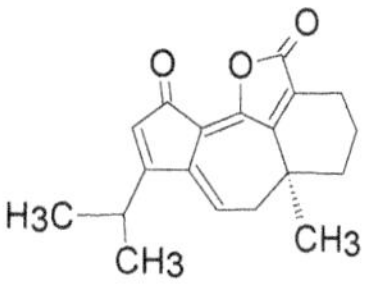

Pyromyxone D

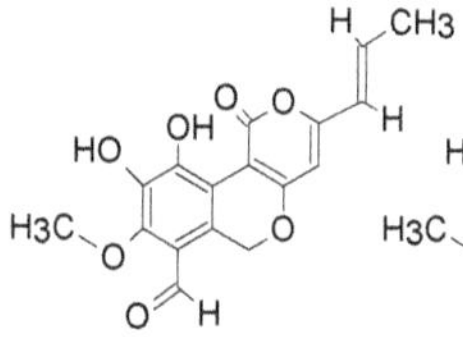

Cyathusal A

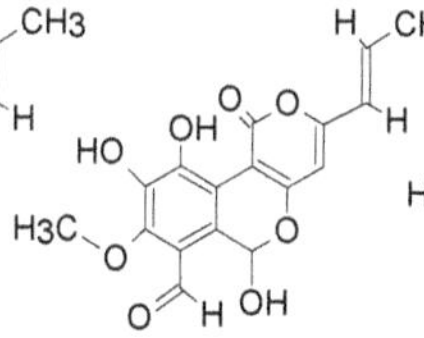

Cyathusal B

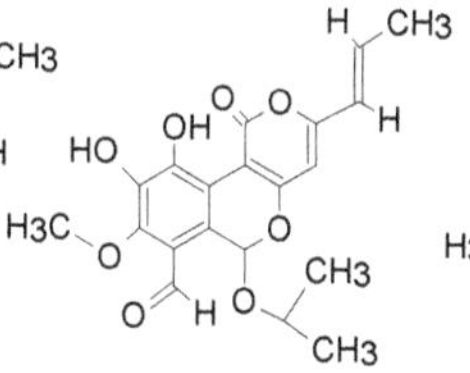

Cyathusal C

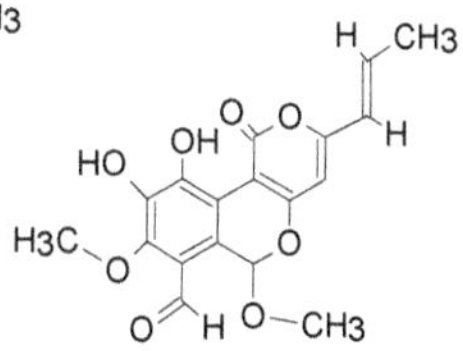

Pulvinatal

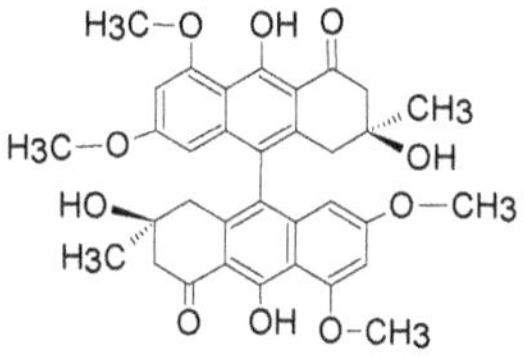

Austrocolorin A₁

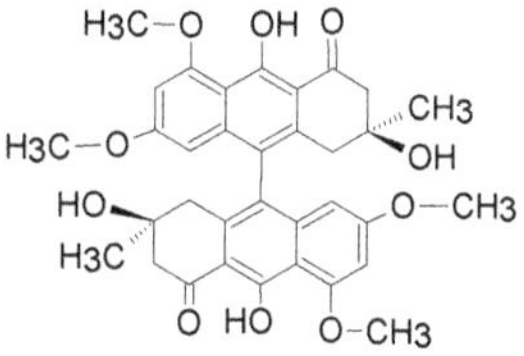

Austrocolorin B₁

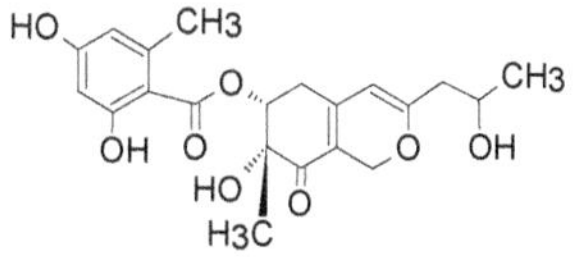

Glaziellin A

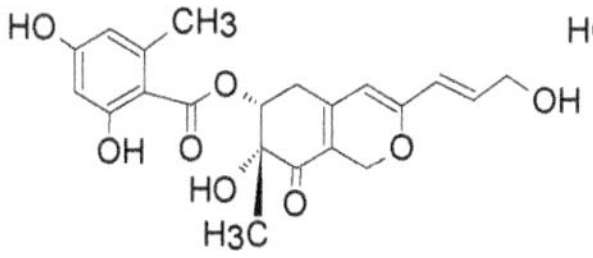

Entonaemin A

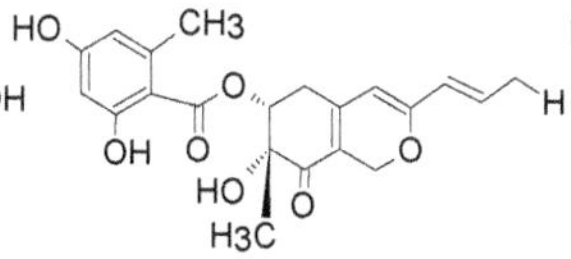

Comazaphilone D

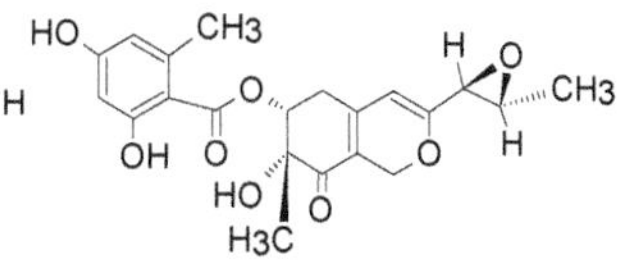

Entonaemin B

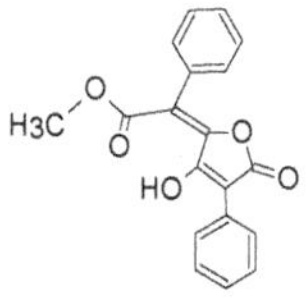

Vulpinic acid

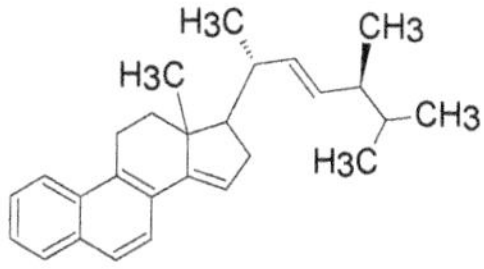

(17β,20R,22E,24R)-19-norergosta-1,3,5,7,9,14,22-heptaene

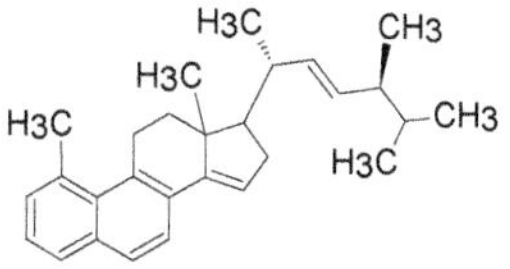

(17β,20R,22E,24R)-1-methyl-19-norergosta-1,3,5,7,9,14,22-heptaene

Fig. 1 Contd. ...

...Fig. 1 Contd.

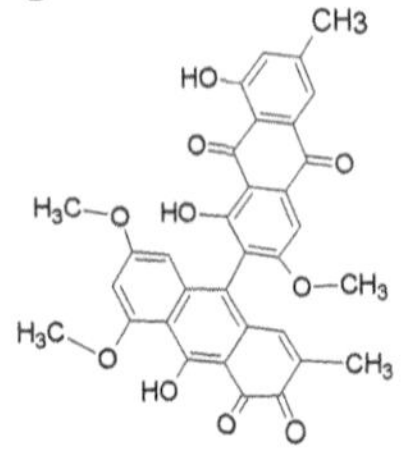

Sinapiquinone

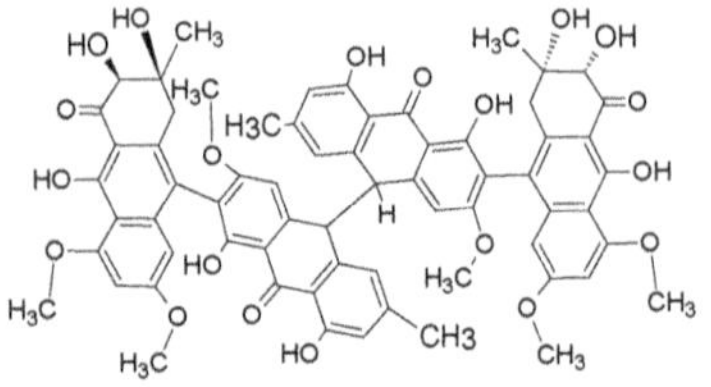

Sinapicolone

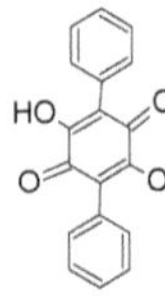

Polyporic acid

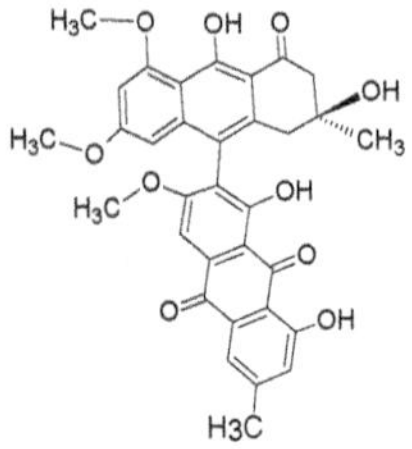

(3S,3'S,P)-anhydrophlegmacin-
9,10-quinone 8'-O-methyl ether

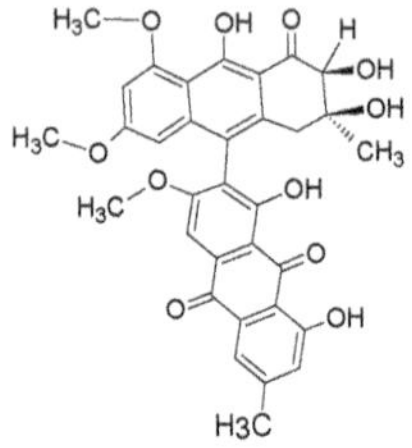

(2'S,3S,3'S,P)-2'-
hydroxyanhydrophlegmacin-
9,10-quinone 8'-O-methyl ether

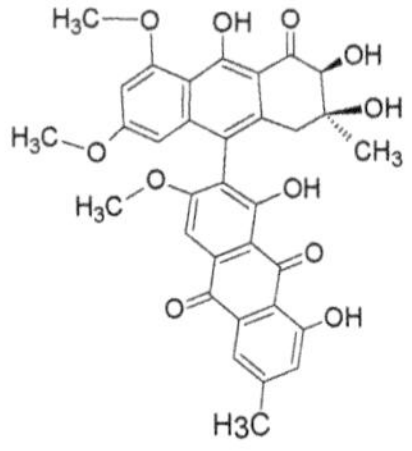

Hydroxyphlegmacinquinone

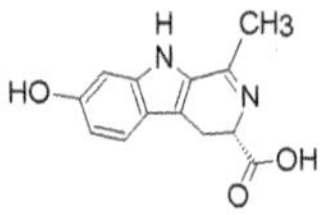

Brunnein A

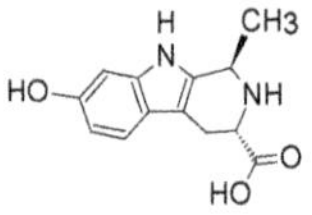

Brunnein B

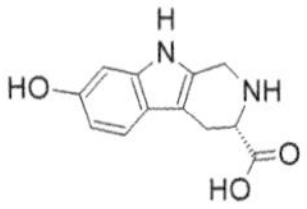

Brunnein C

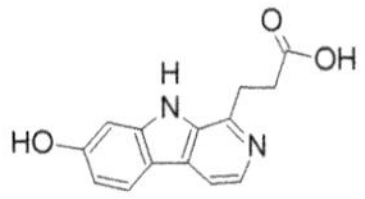

3-(7-Hydroxy-9H-β-carboline-1-
yl)propanoic acid

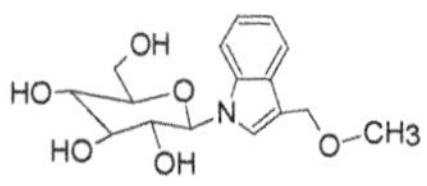

1-(1-β-glucopyranosyl)-3-
(methoxymethyl)-1H-
indole

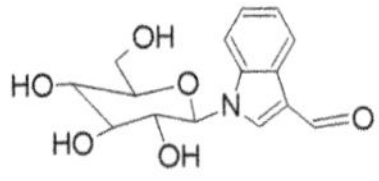

1-(1-β-glucopyranosyl)-
1H-indole-3-
carbaldehyde

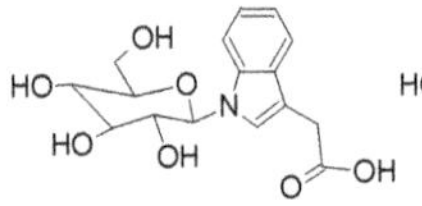

1-(1-β-glucopyranosyl)-
1H-indole-3-acetic acid

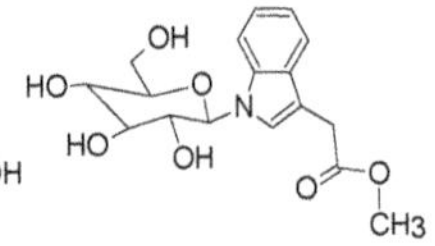

Methyl 1-(1-β-
glucopyranosyl)-1H-
indole-3-acetate

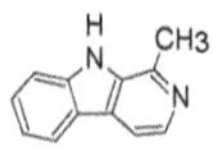

Harmane

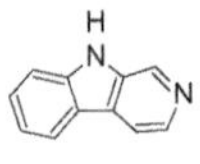

Nonharmane

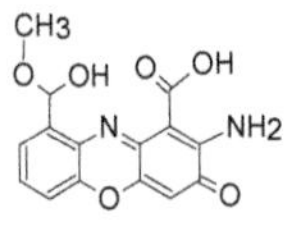

Pycnoporin

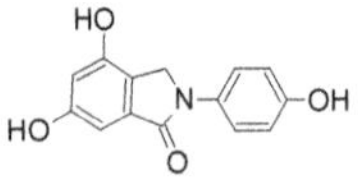

Clitocybin A

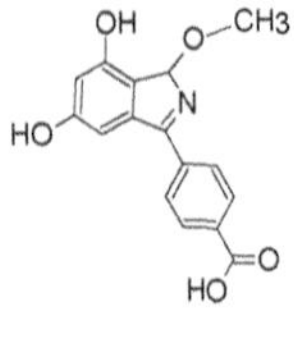

Clitocybin D

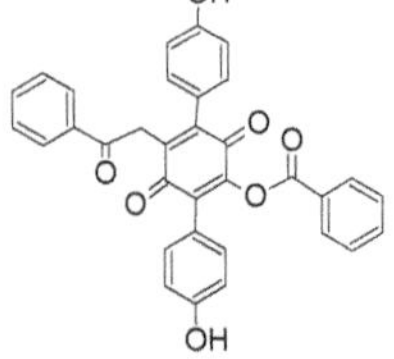

Aurantiacin

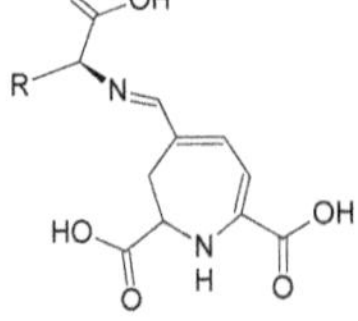

Hygroaurins
(R=aminoacidresidue)

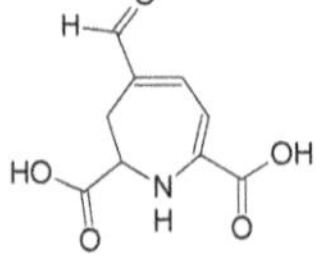

Muscaflavin

Fig. 1 Contd. ...

...Fig. 1 Contd.

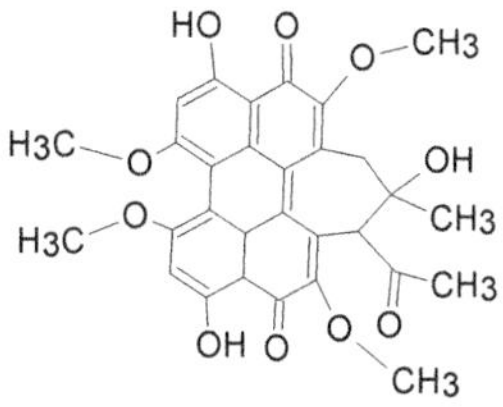

Hypocrellin A

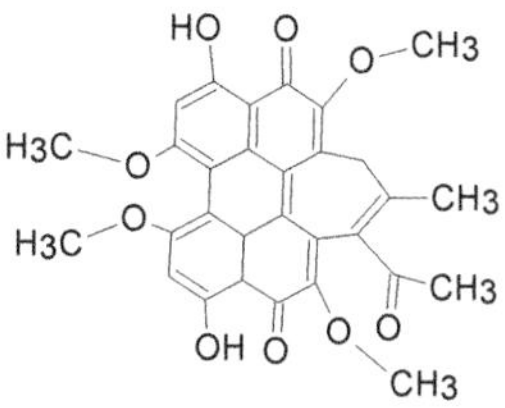

Hypocrellin B

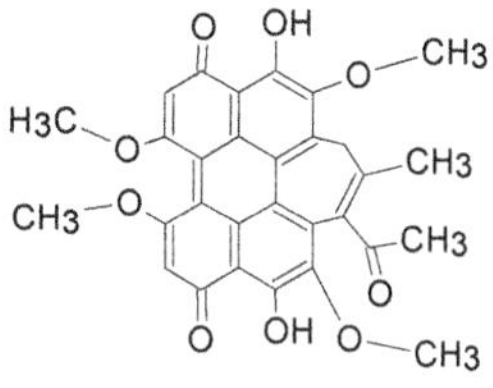

Hypocrellin C

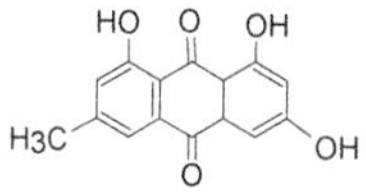

Emodin

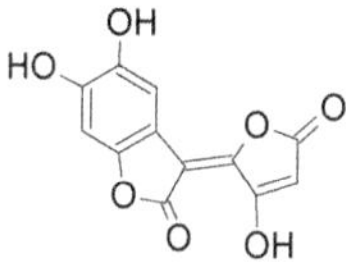

Gomphilactone

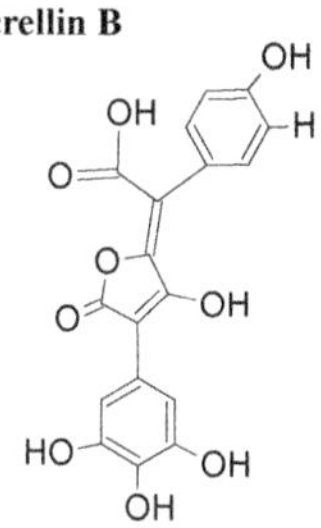

Gomphidic acid

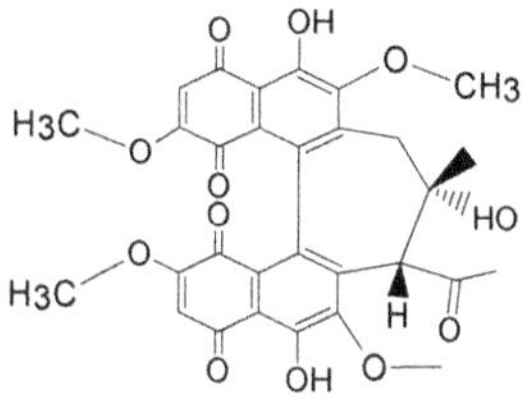

Erythroglaucin

Hypocrellin D

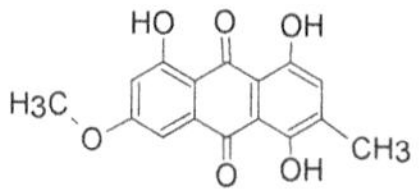

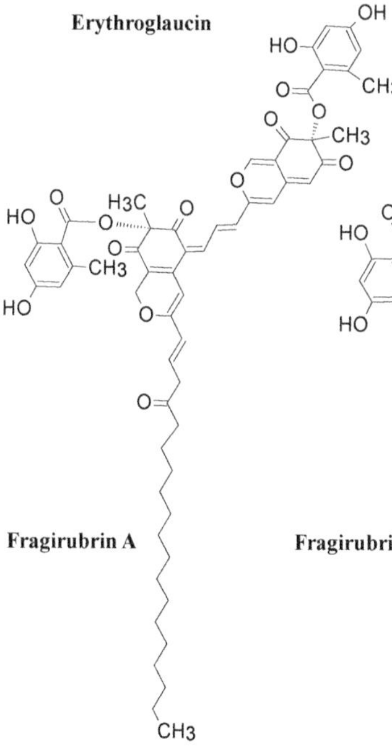

Fragirubrin A

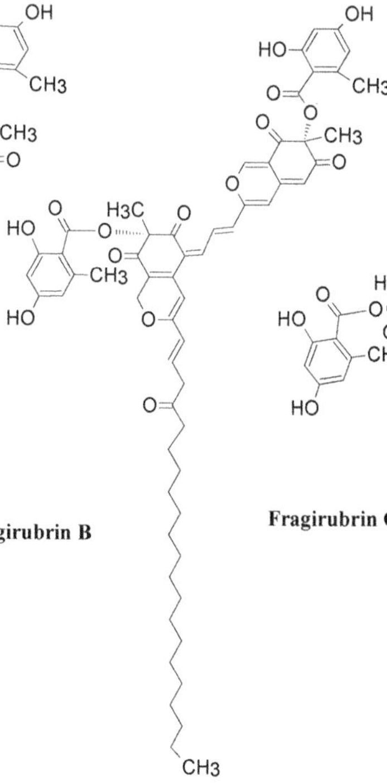

Fragirubrin B

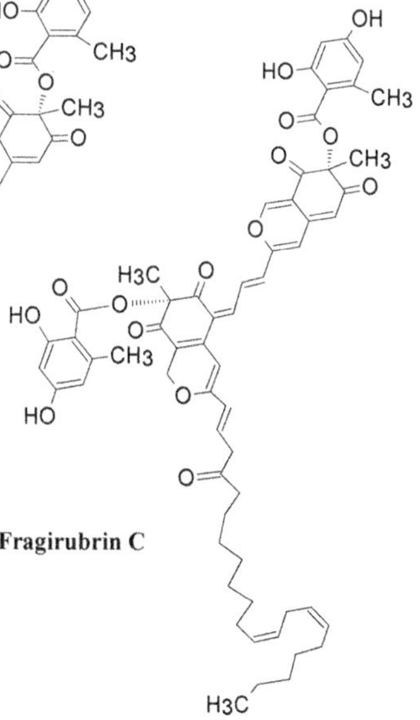

Fragirubrin C

Fig. 1 Contd. ...

...Fig. 1 Contd.

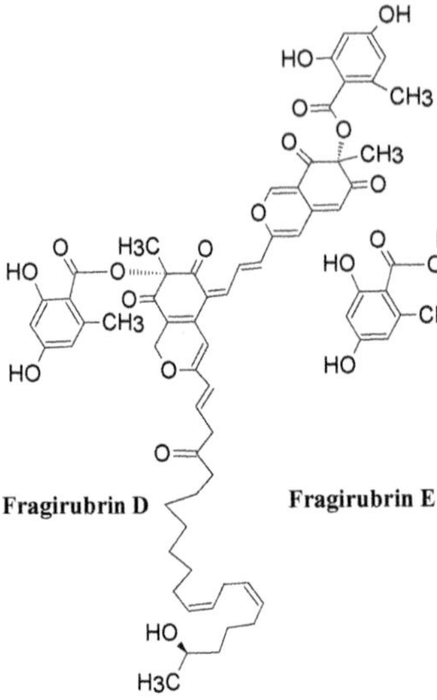

Fragirubrin D

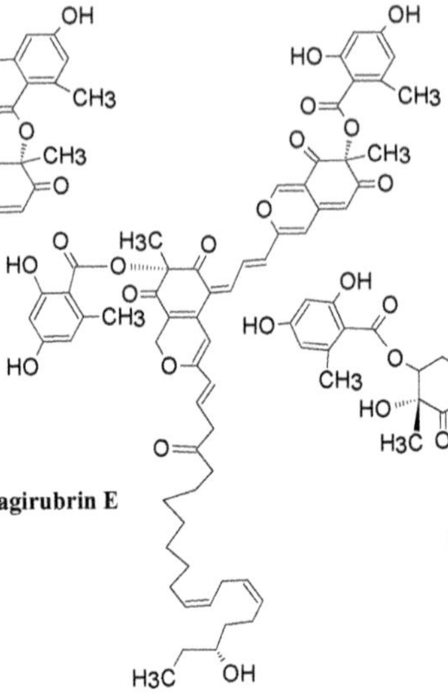

Fragirubrin E

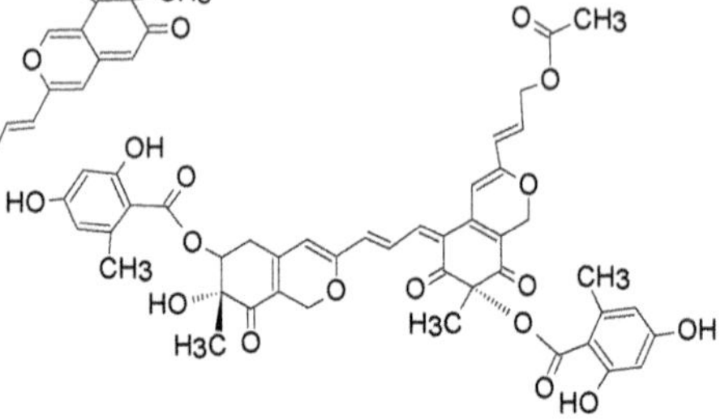

Rutilin A

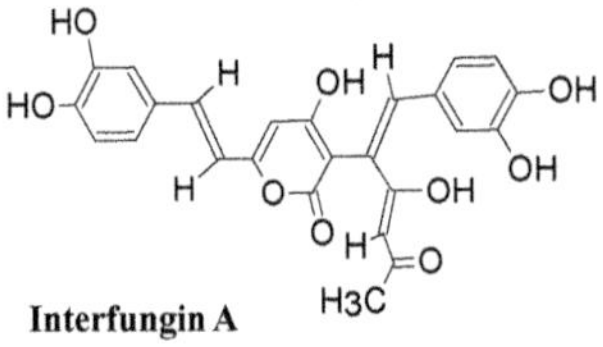

Interfungin A

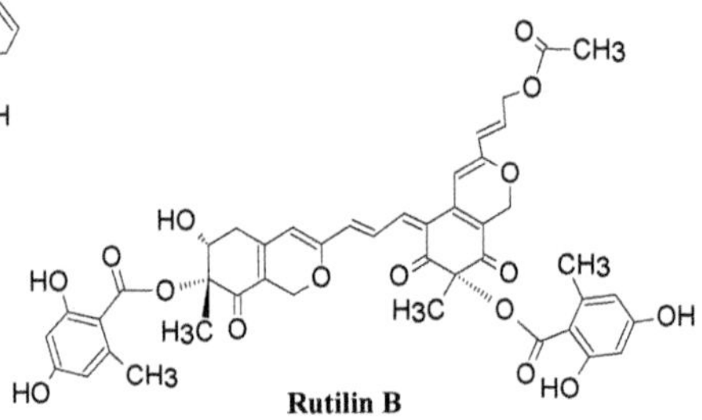

Rutilin B

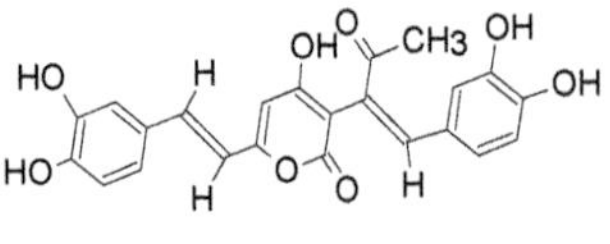

Interfungin B

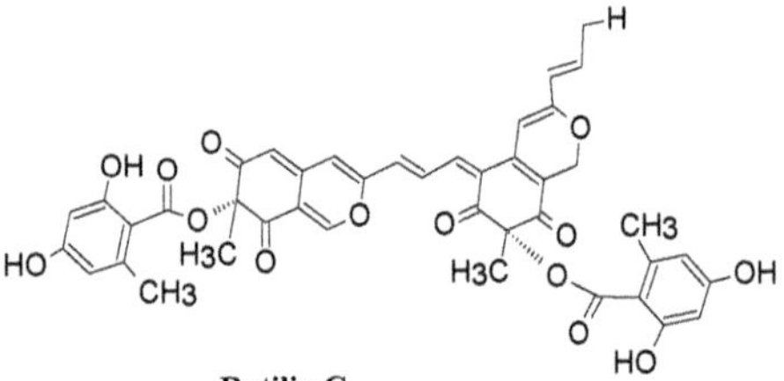

Rutilin C

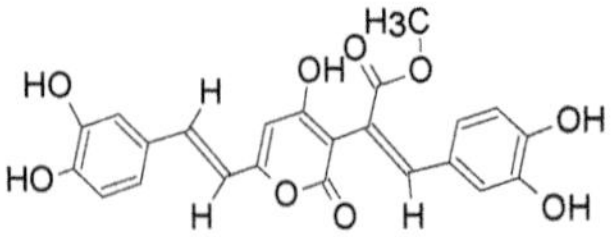

Interfungin C

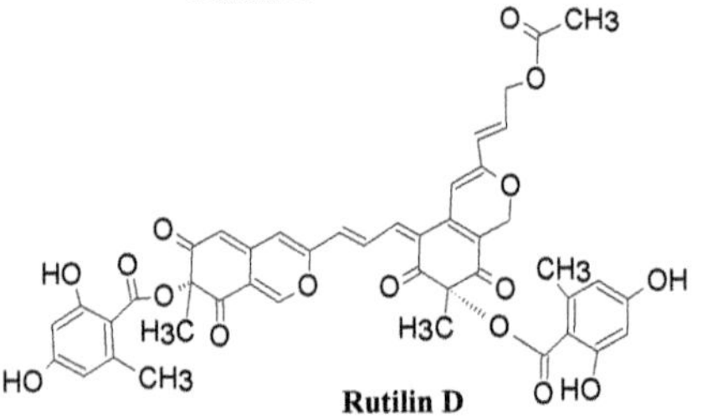

Rutilin D

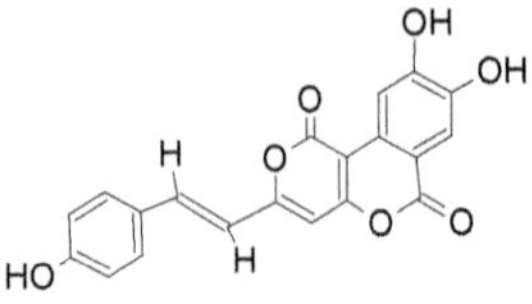

Phelligridin C

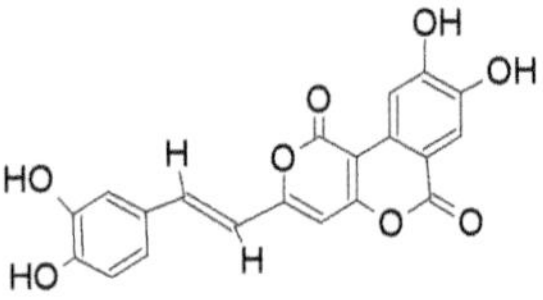

Phelligridin D

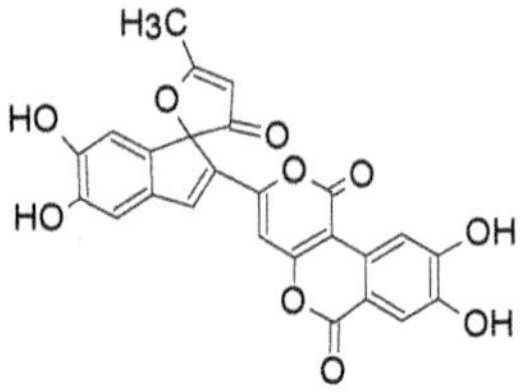

Phelligridin E

Fig. 1 Contd. ...

...Fig. 1 Contd.

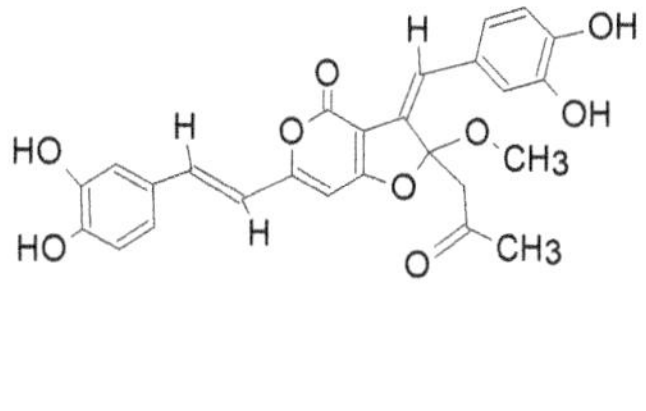

Phelligridin F

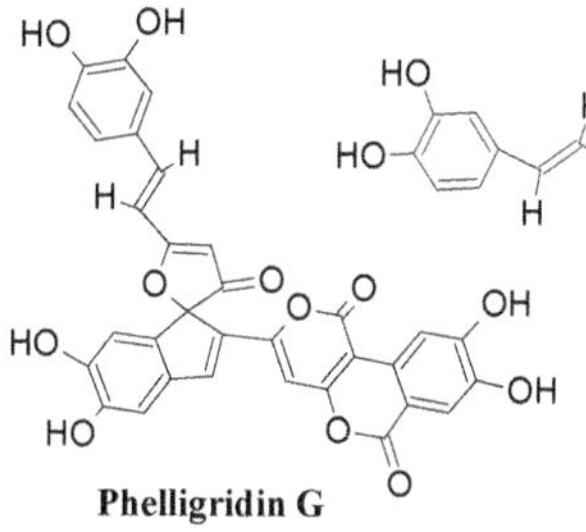

Phelligridin G

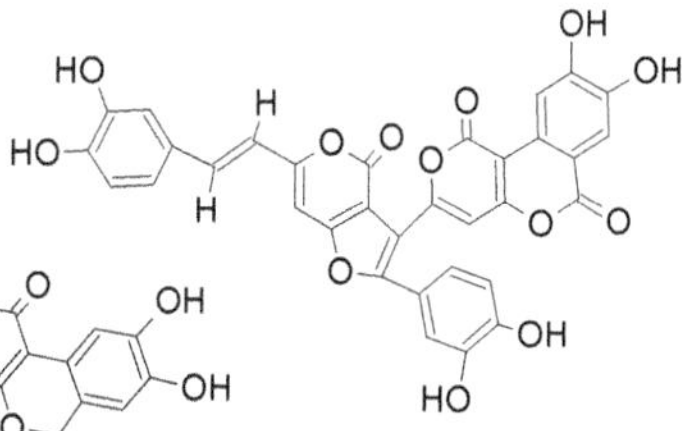

Phelligridin H

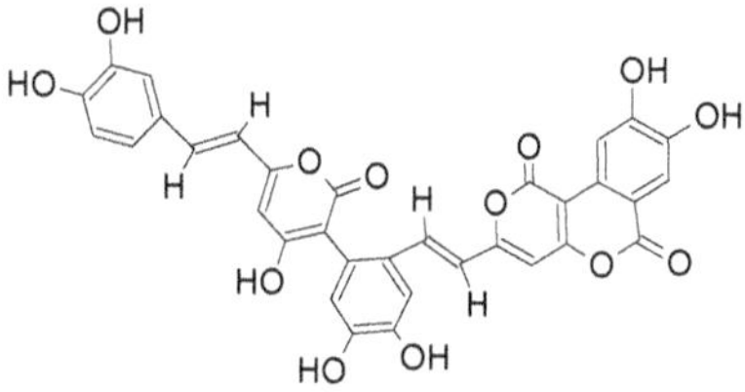

Phelligridin I

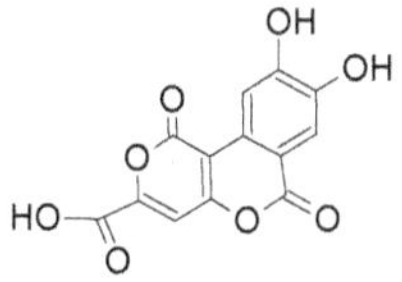

Phelligridin J

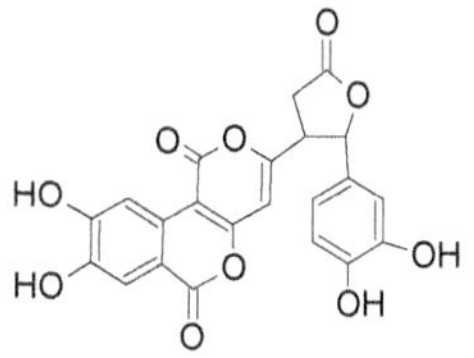

Phelligridin K

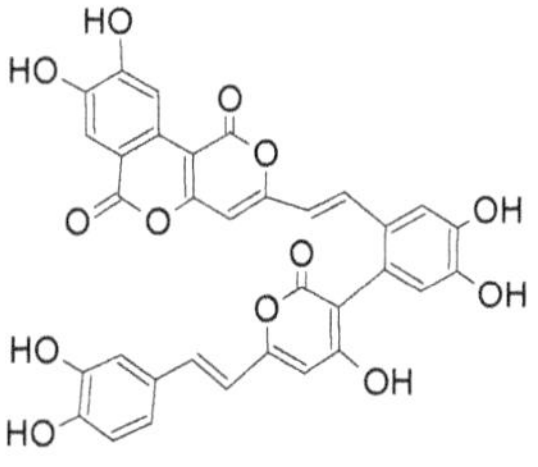

Inonoblin A

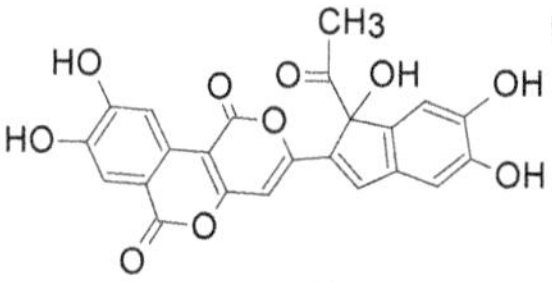

Inonoblin B

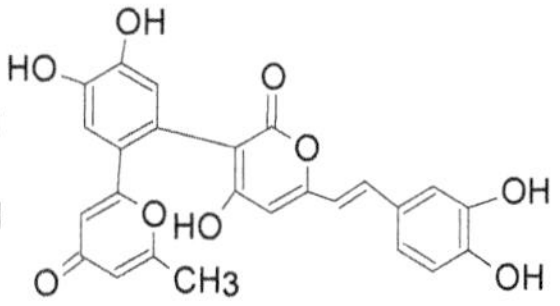

Inonoblin C

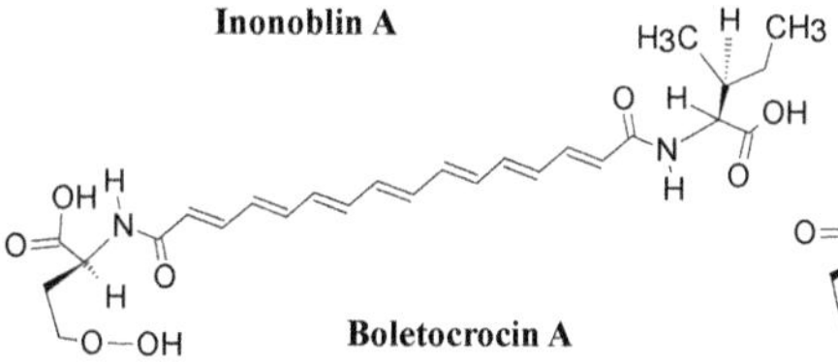

Boletocrocin A

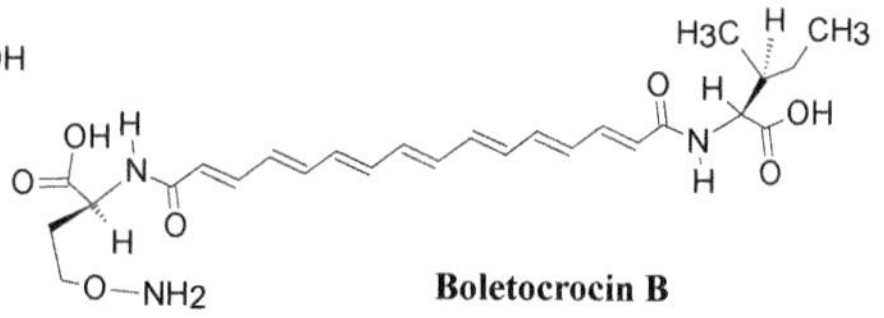

Boletocrocin B

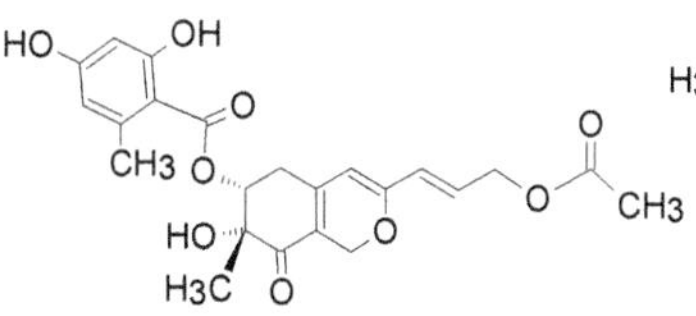

Rubiginosin A

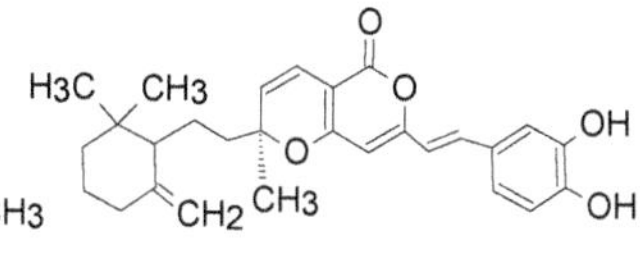

Phellinin A

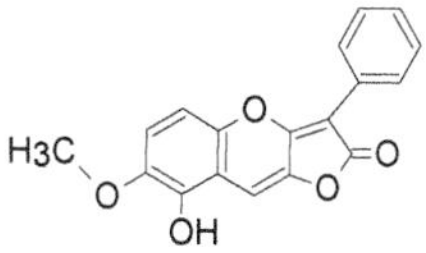

Pulverolide

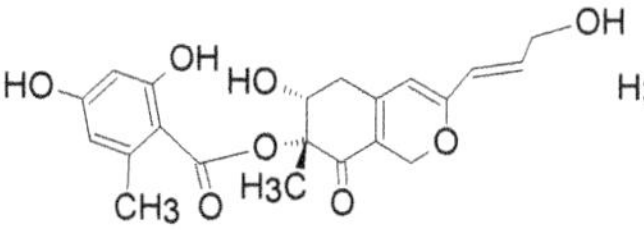

Rubiginosin B

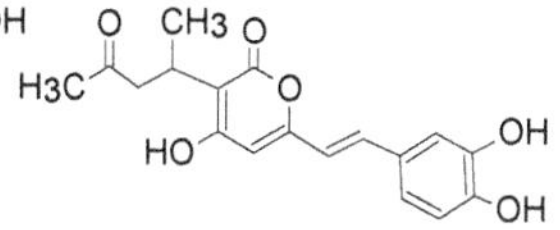

Phellinin B

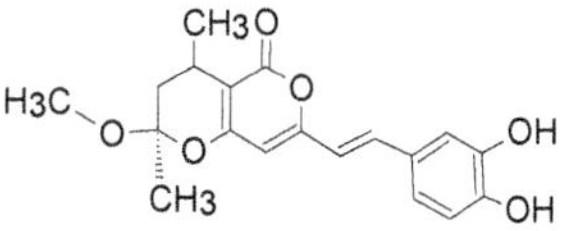

Phellinin C

Fig. 1 Contd. ...

...Fig. 1 Contd.

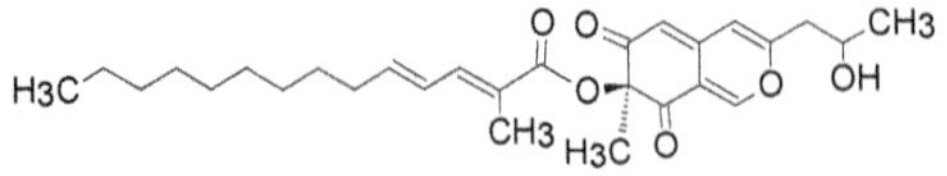

Rubiginosin C

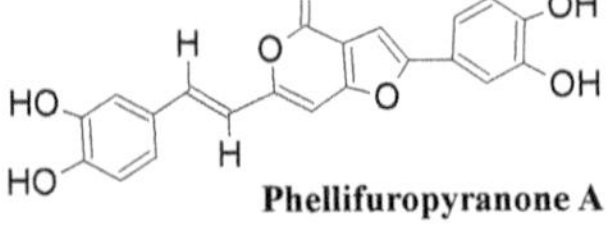

Phellifuropyranone A

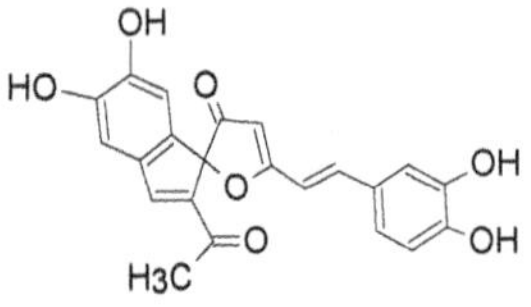

Phelliribsin A

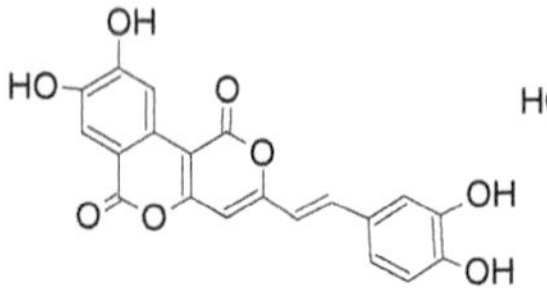

Meshimakobnol A

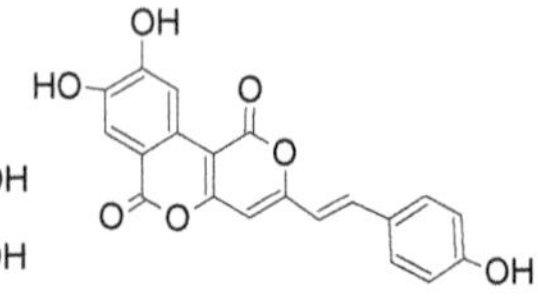

Meshimakobnol B

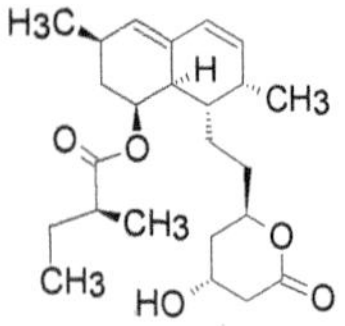

Mevinolin (lovastatin)

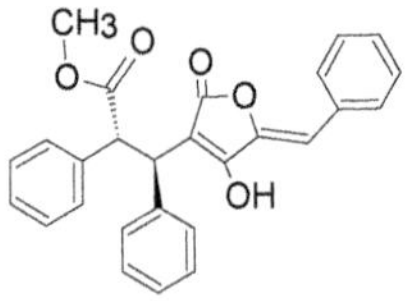

Pulveraven A

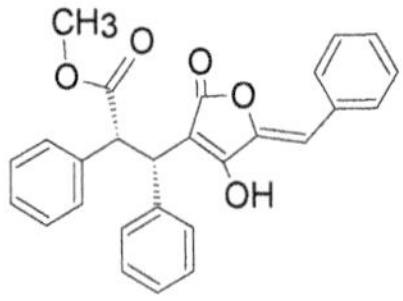

Pulveraven B

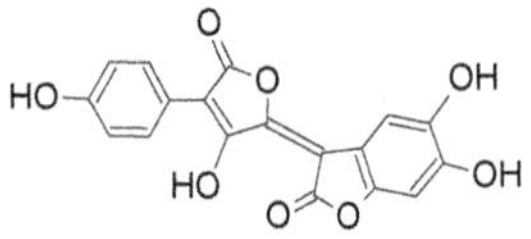

Xerocomorubin

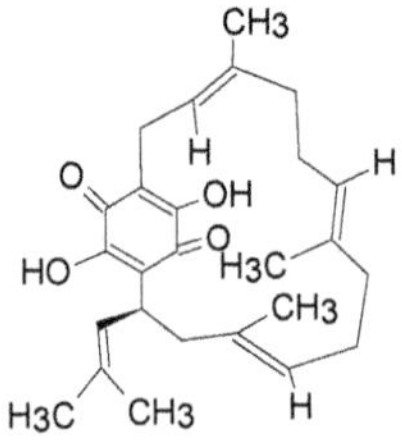

Rhizopogone

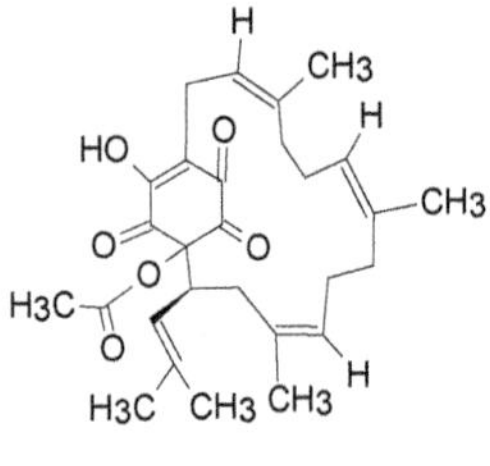

2-acetoxyrhizopogone

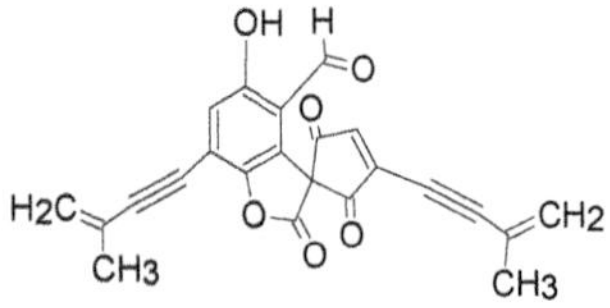

Ochroleucin A

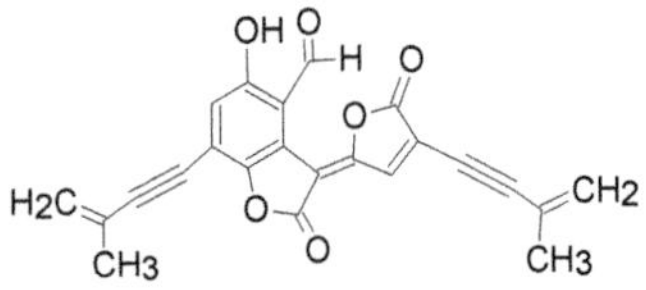

Ochroleucin B

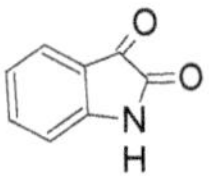

Isatin

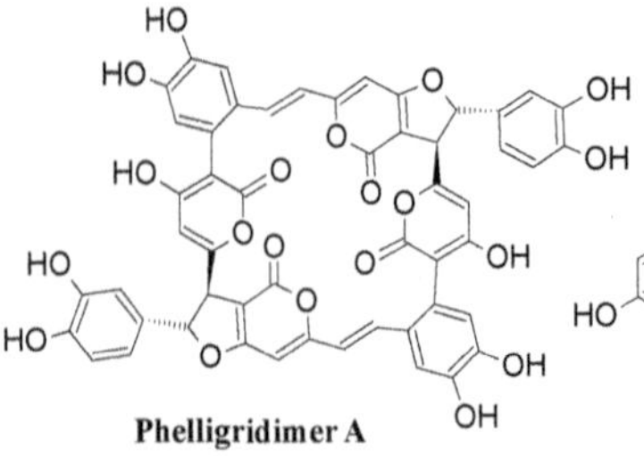

Phelligridimer A

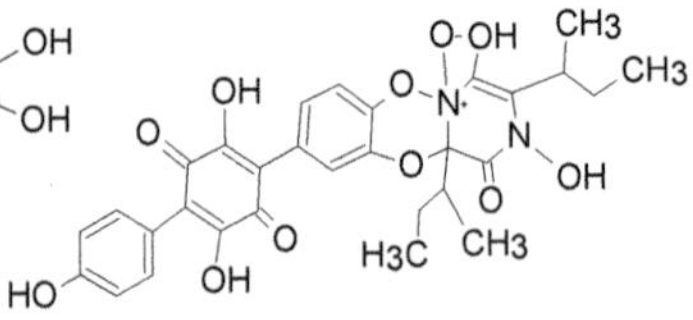

Sarcoviolin α

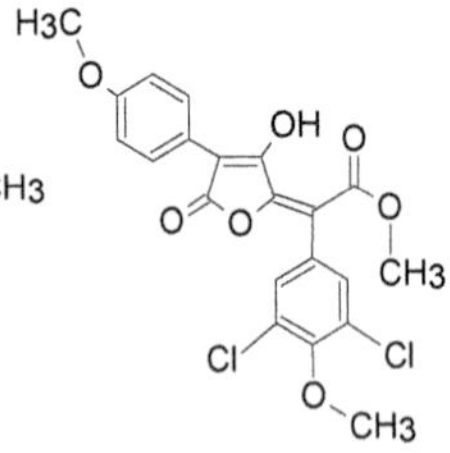

Methyl-3',5'-dichloro-4,4'-
di-O-methylatromentate

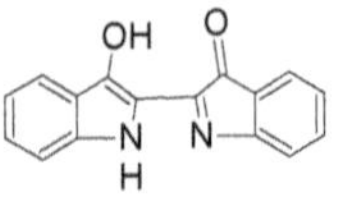

Indigo

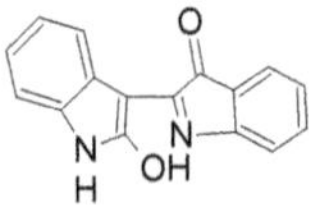

Indirubin

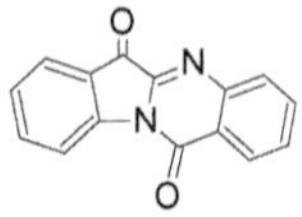

Tryptanthrin

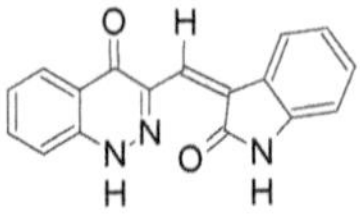

Schizocommunin

Fig. 1 Contd. ...

...Fig. 1 Contd.

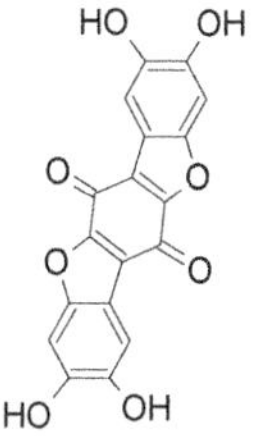

Thelephoric acid 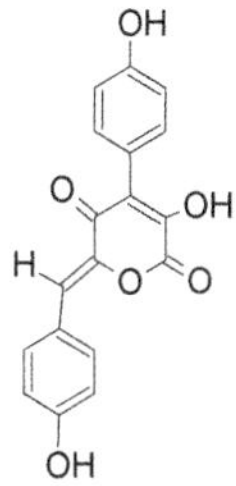   **Grevillin A** 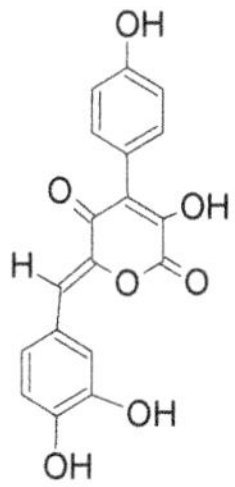   **Grevillin B** 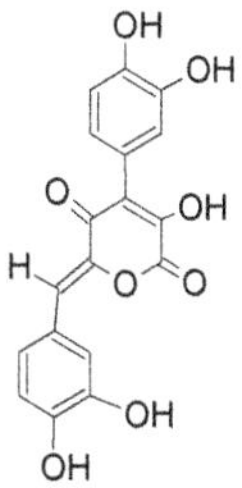   **Grevillin C** 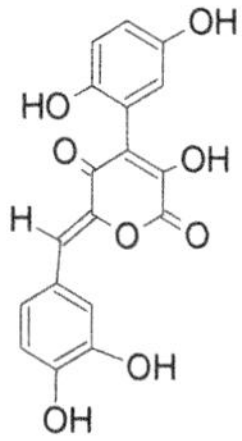   **Grevillin D**

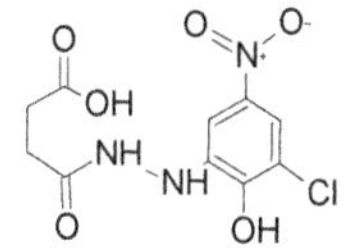

Stephanosporin 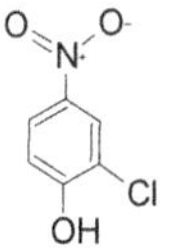   **2-chloro-4-nitrophenol** 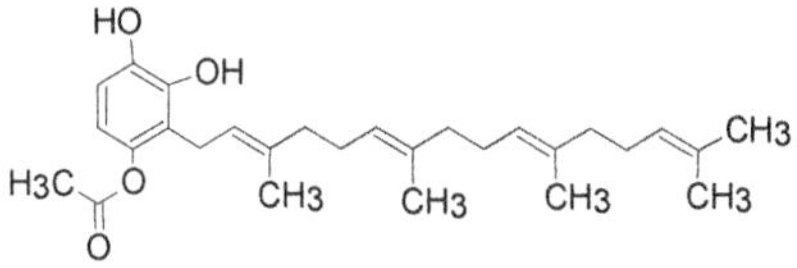   **Suillin**

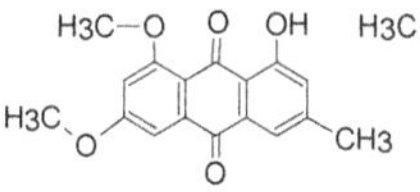

Emodin-6,8-di-*O*-methyl ether 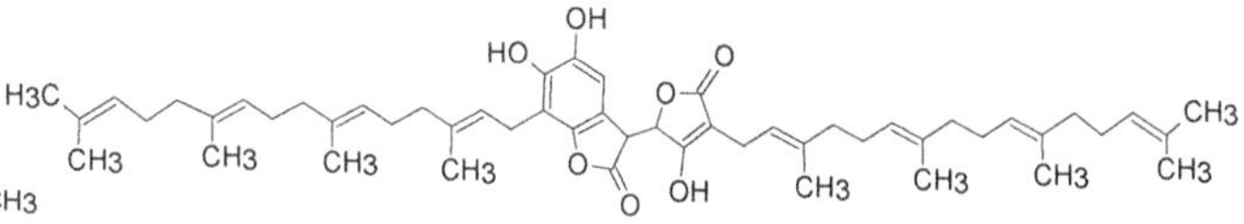   **Bovilactone-4,4**

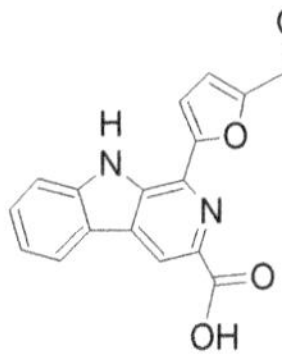

Flazin 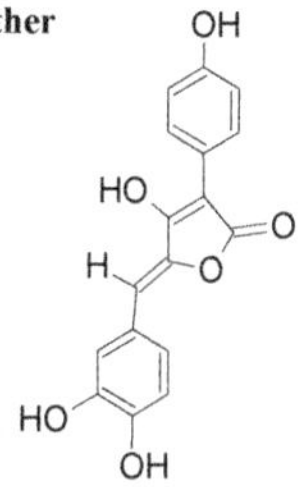   **3',4',4-trihydroxypulvinone** 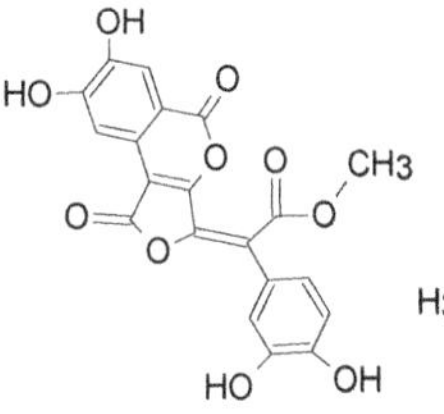   **Methyl bovinate** 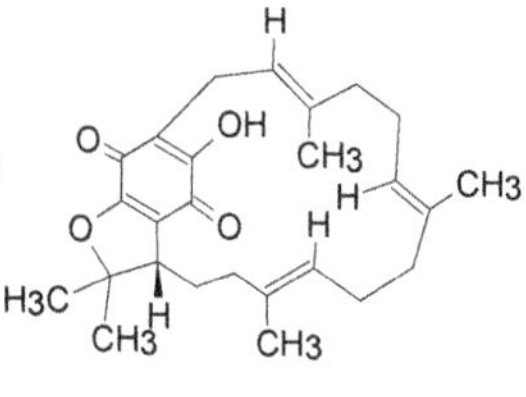   **Tridentoquinone**

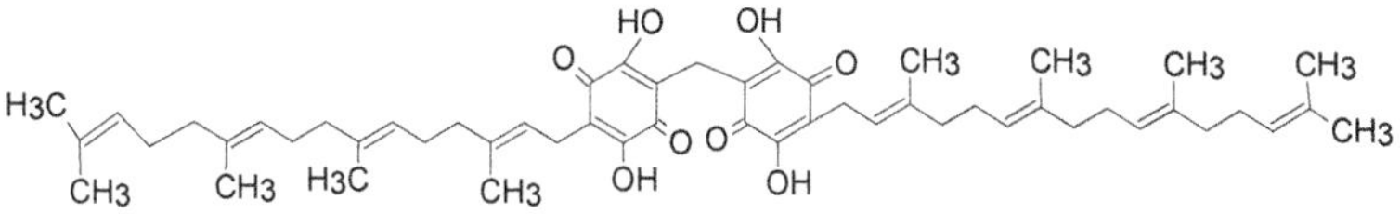

Amitenone

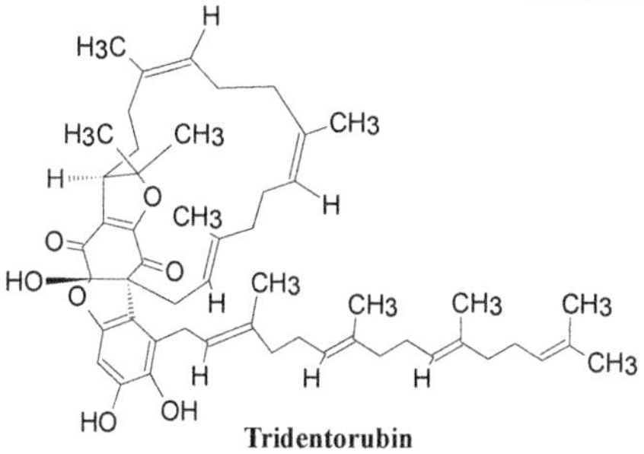

Tridentorubin 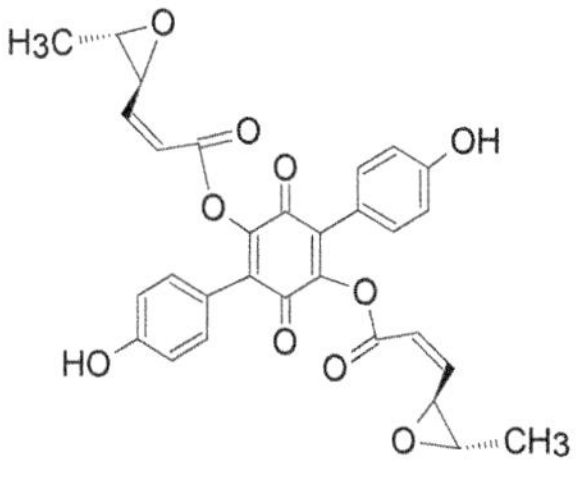   **Flavomentin A**

Fig. 1 Contd. ...

...Fig. 1 Contd.

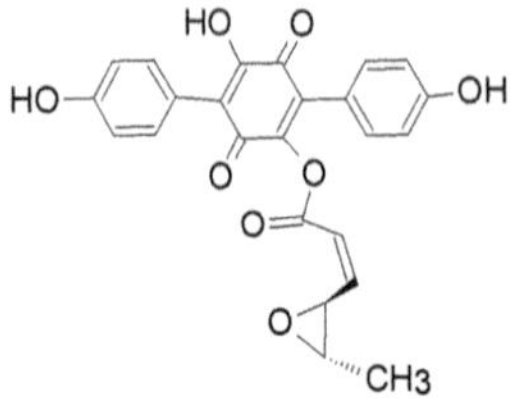

Flavomentin B

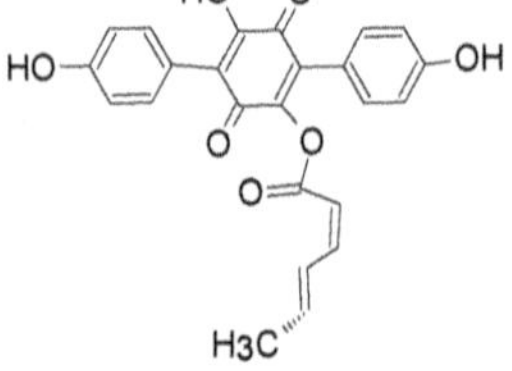

Flavomentin C

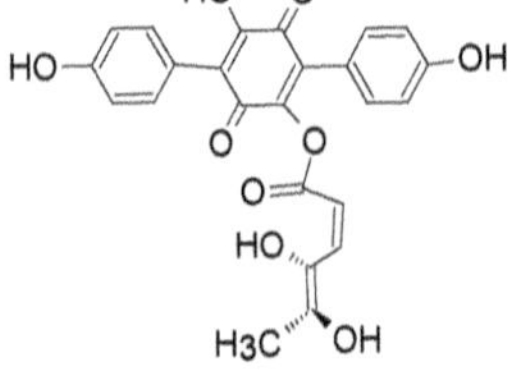

Flavomentin D

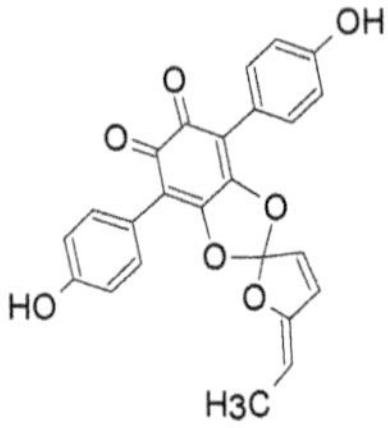

Spiromentin A

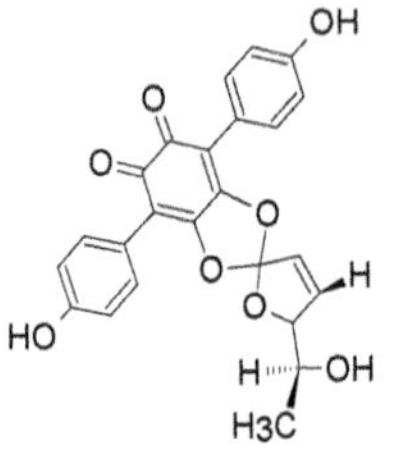

Spiromentin B

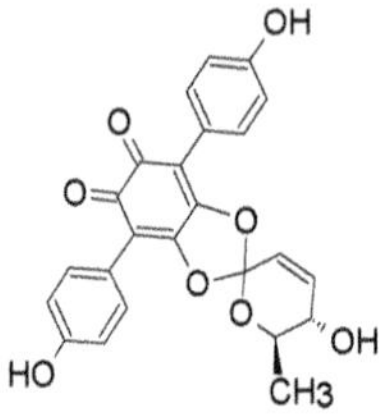

Spiromentin C

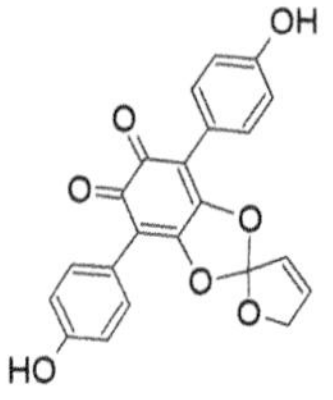

Spiromentin D

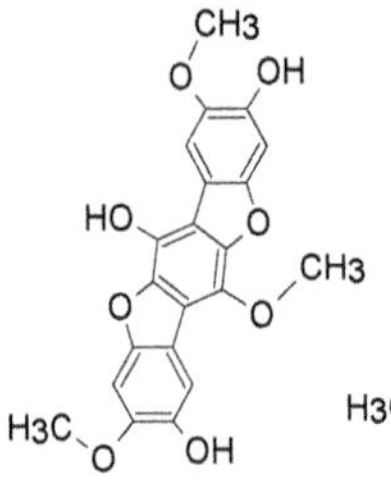

Corticin A

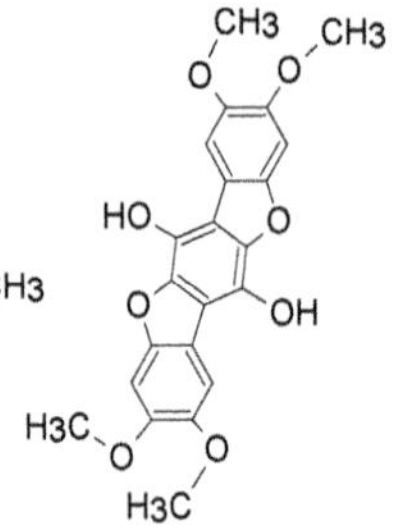

Corticin B

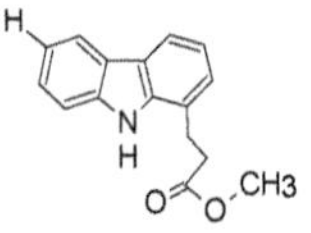

Infractin A

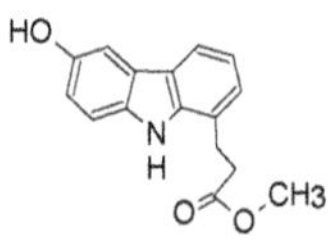

Infractin B

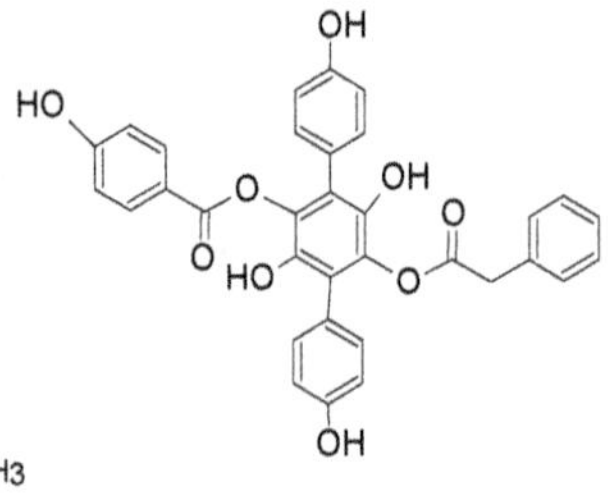

Thelephorin A

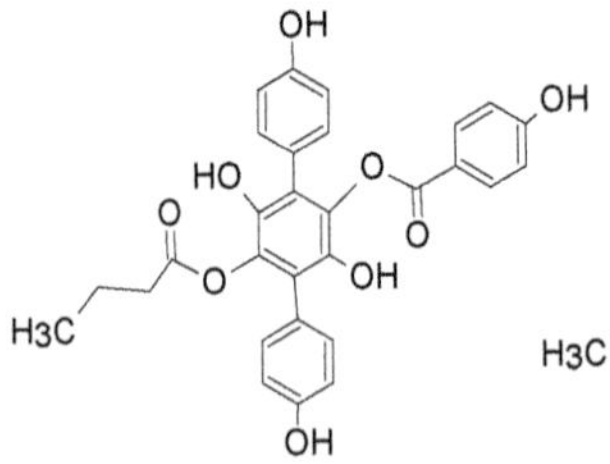

Thelephantin A

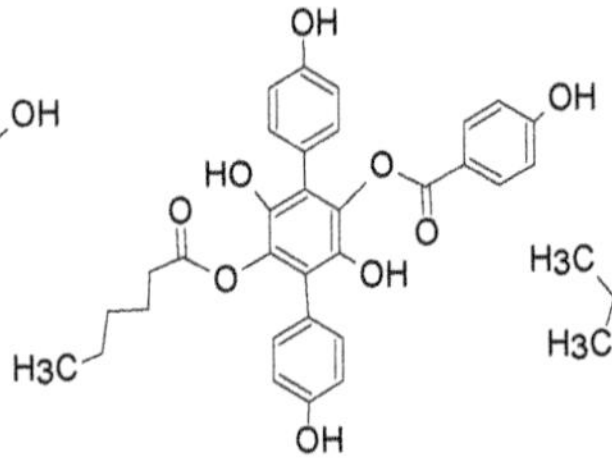

Thelephantin B

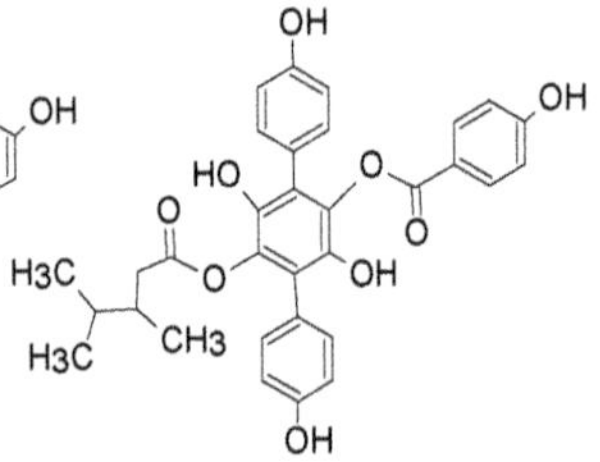

Thelephantin C

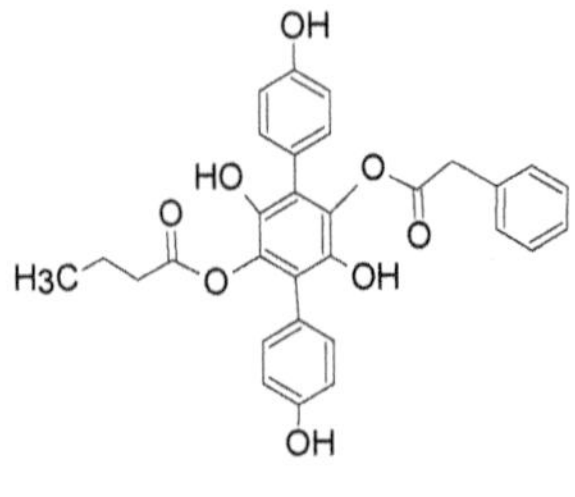

Thelephantin D

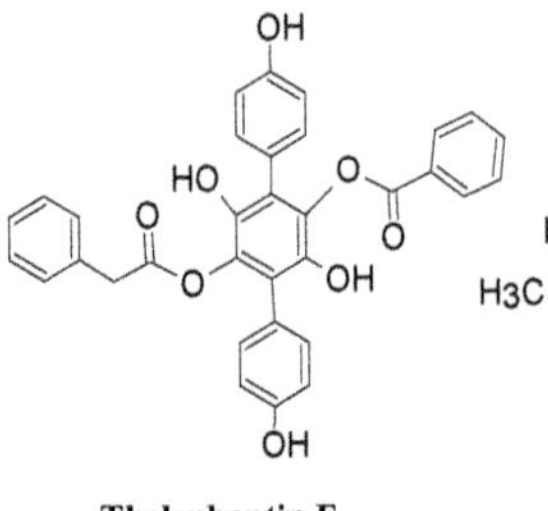

Thelephantin E

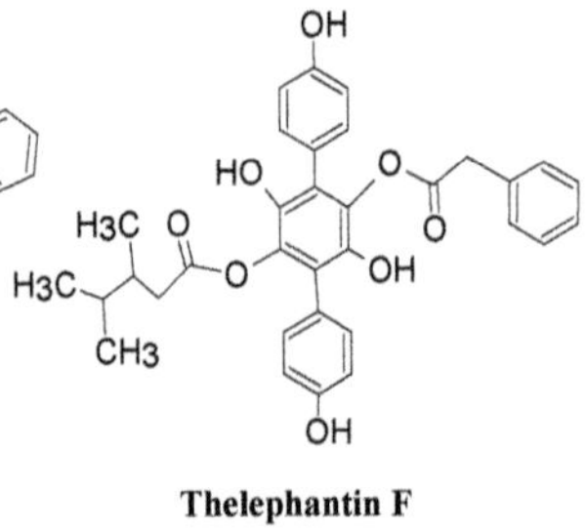

Thelephantin F

Fig. 1 Contd. ...

...Fig. 1 Contd.

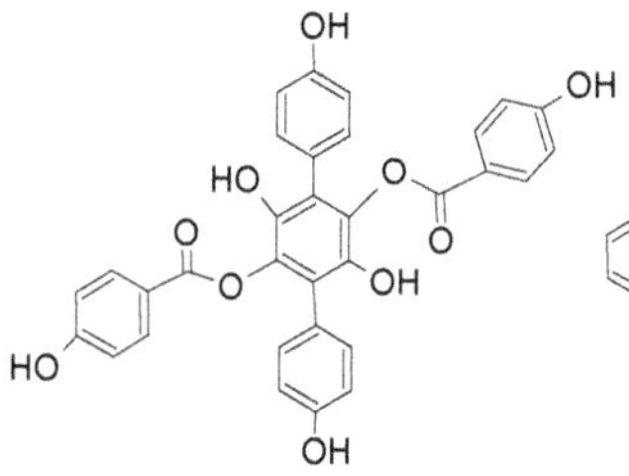

Thelephantin G

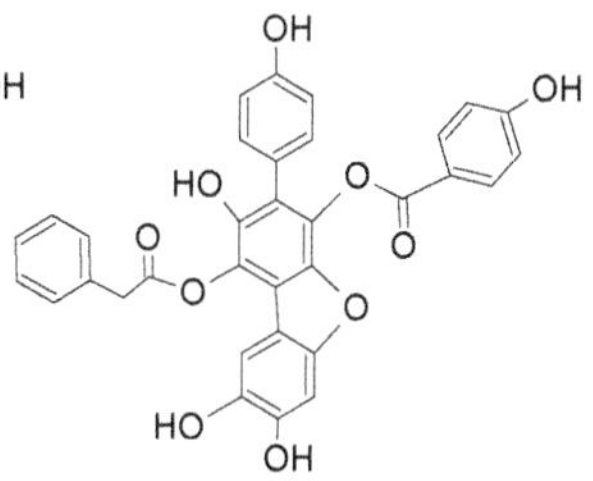

Thelephantin H

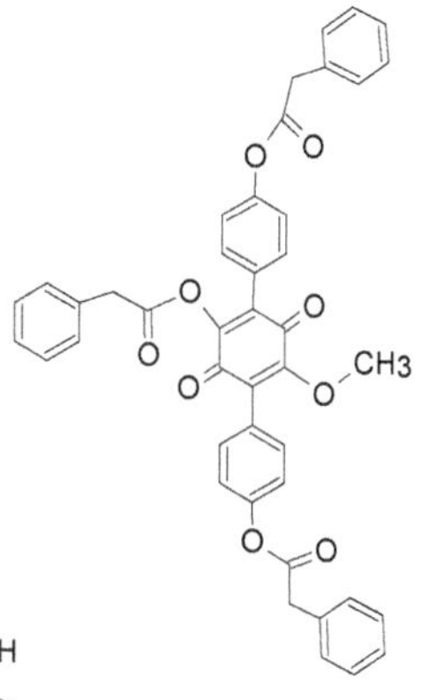

Ganbajunin A

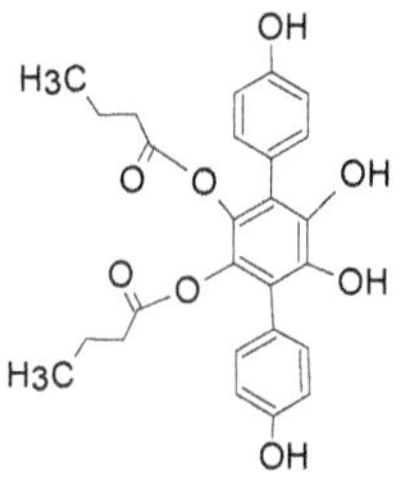

Terrestrin B

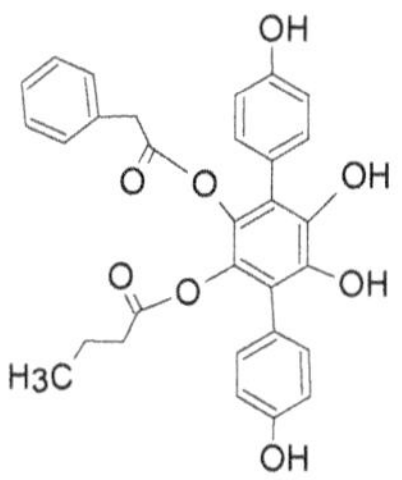

Terrestrin C

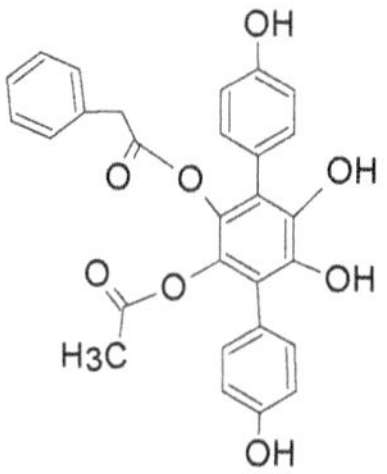

Terrestrin D

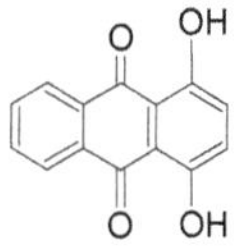

Quinizarine

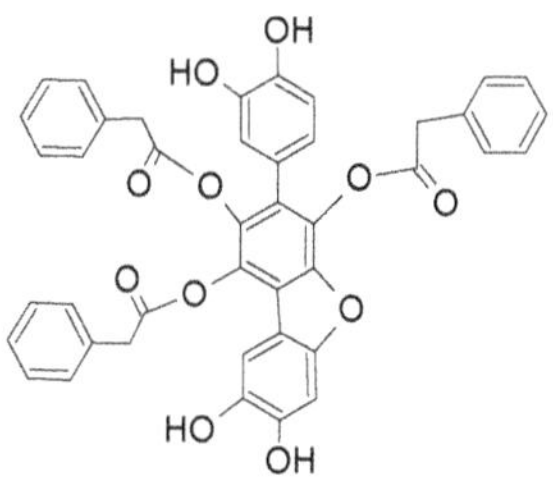

Terrestrin E

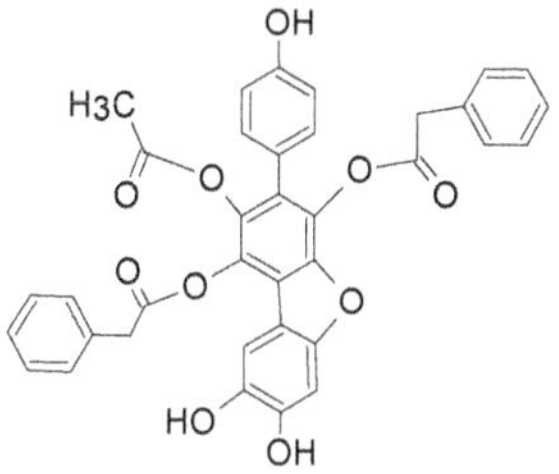

Terrestrin F

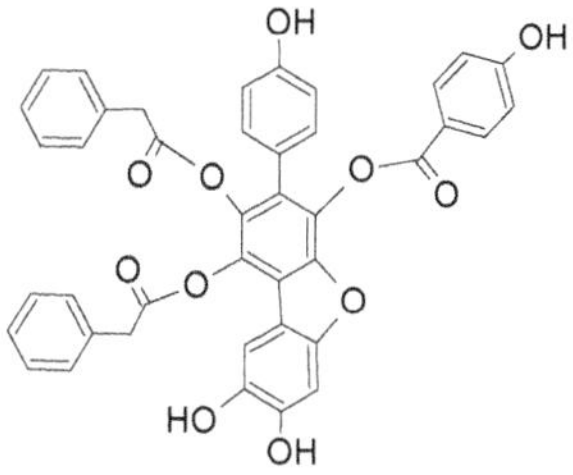

Terrestrin G

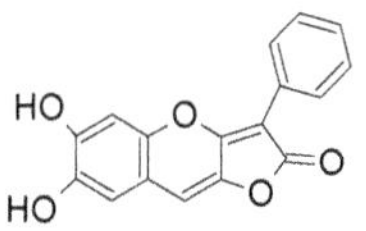

Aurantricholide A

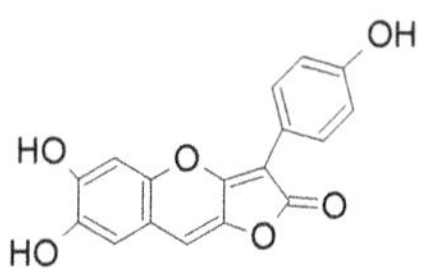

Aurantricholide B

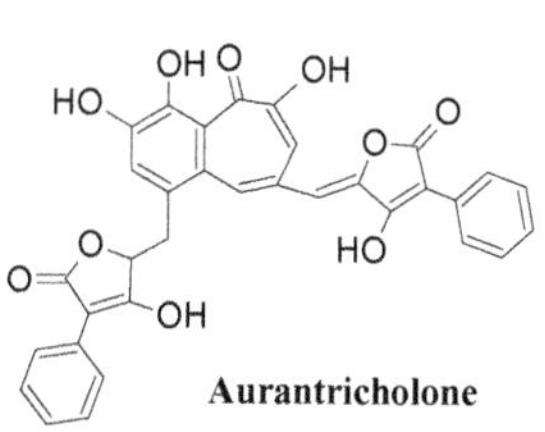

Aurantricholone

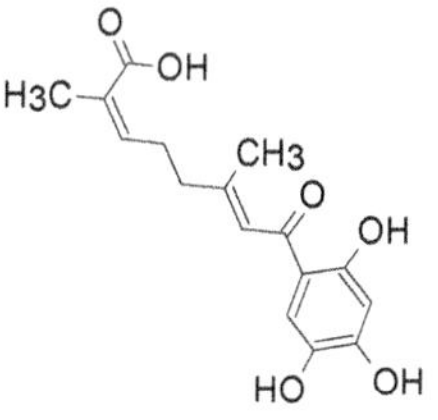

Orirubenone A

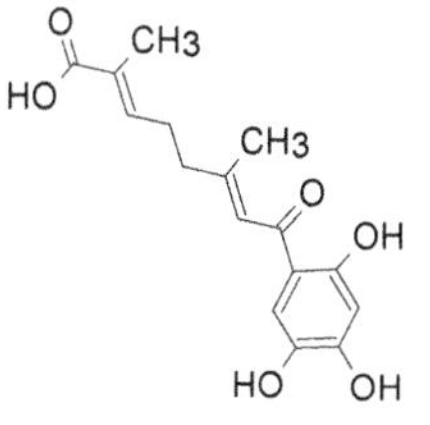

Orirubenone B

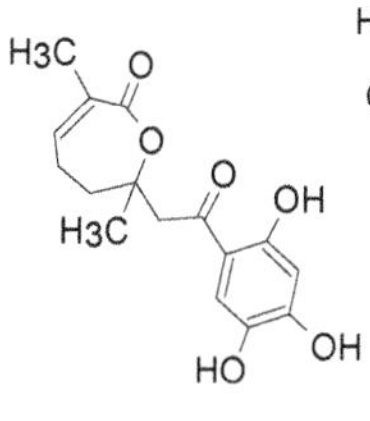

Orirubenone C

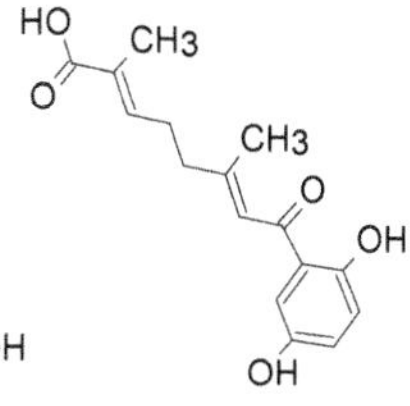

Orirubenone D

Fig. 1 Contd. ...

...Fig. 1 Contd.

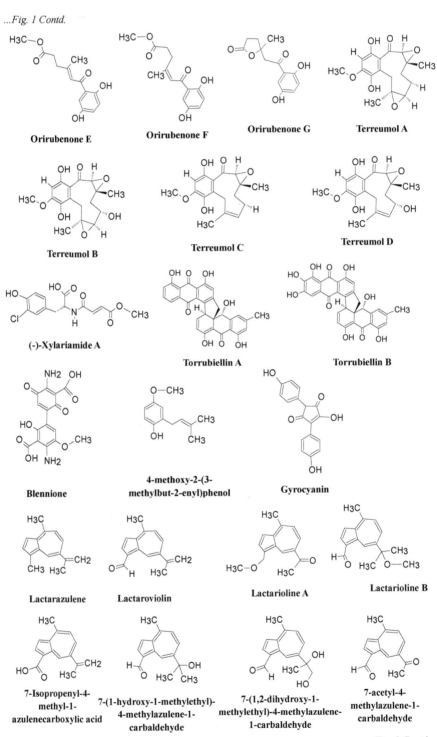

Orirubenone E

Orirubenone F

Orirubenone G

Terreumol A

Terreumol B

Terreumol C

Terreumol D

(-)-Xylariamide A

Torrubiellin A

Torrubiellin B

Blennione

4-methoxy-2-(3-methylbut-2-enyl)phenol

Gyrocyanin

Lactarazulene

Lactaroviolin

Lactarioline A

Lactarioline B

7-Isopropenyl-4-methyl-1-azulenecarboxylic acid

7-(1-hydroxy-1-methylethyl)-4-methylazulene-1-carbaldehyde

7-(1,2-dihydroxy-1-methylethyl)-4-methylazulene-1-carbaldehyde

7-acetyl-4-methylazulene-1-carbaldehyde

Fig. 1 Contd. ...

...Fig. 1 Contd.

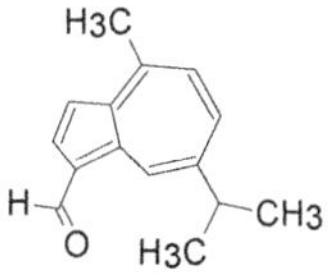

4-methyl-7-(1-methylethyl)azulene-1-carbaldehyde

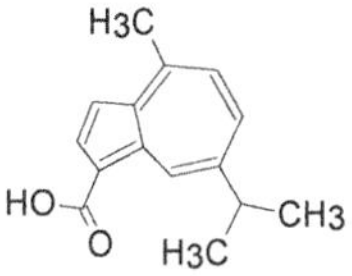

4-methyl-7-(1-methylethyl)azulene-1-carboxylic acid

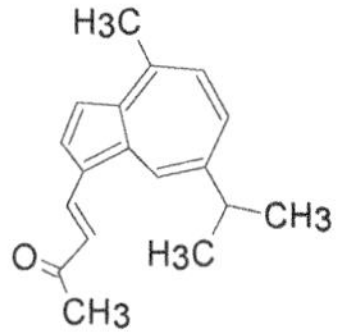

1-[(15E)-buten-17-one]-4-methyl-7-isopropylazulene

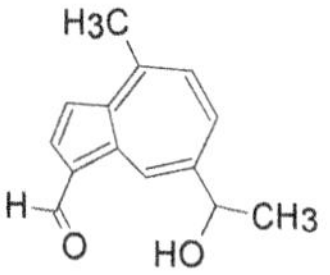

7-(1-hydroxyethyl)-4-methyl-1-azulenecarboxaldehyde

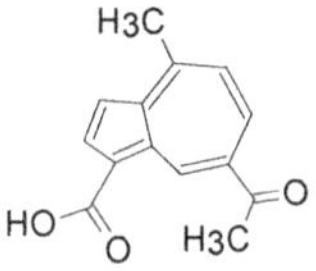

7-acetyl-4-methylazulene-1-carboxylic acid

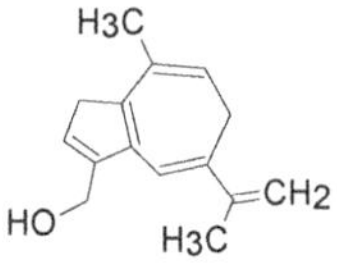

15-hydroxy-3,6-dihydrolactarazulene

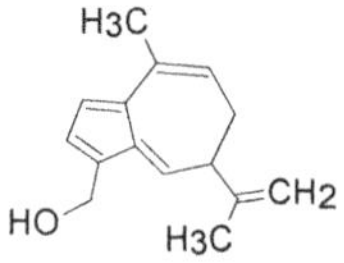

15-hydroxy-6,7-dihydrolactarazulene

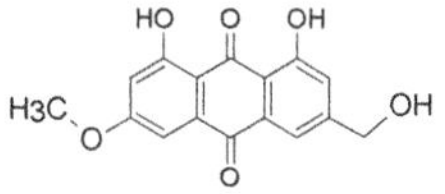

Fallacinol/Phallacinol

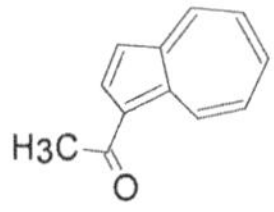

Acetylazulene

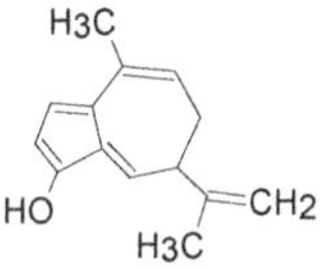

Dihydroazulen-1-ol

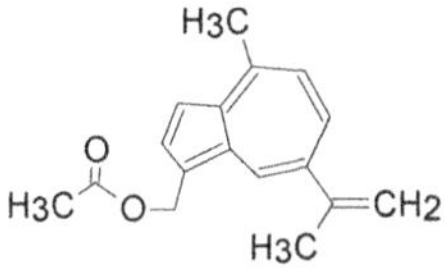

Stearoyldeterrol

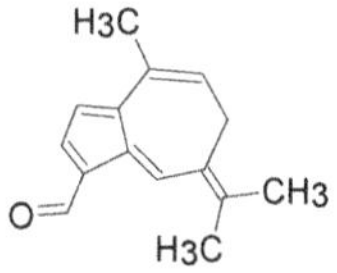

1,3,5,7(11),9-pentaenyl-14-guaianal

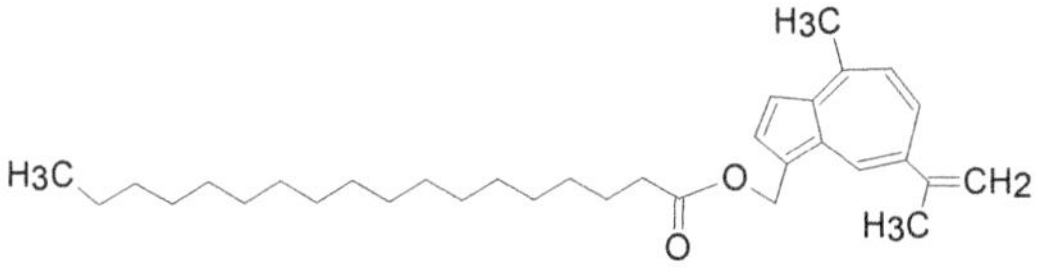

1-hydroxymethyl-4-methyl-7-(1-methylethenyl)azulene stearate

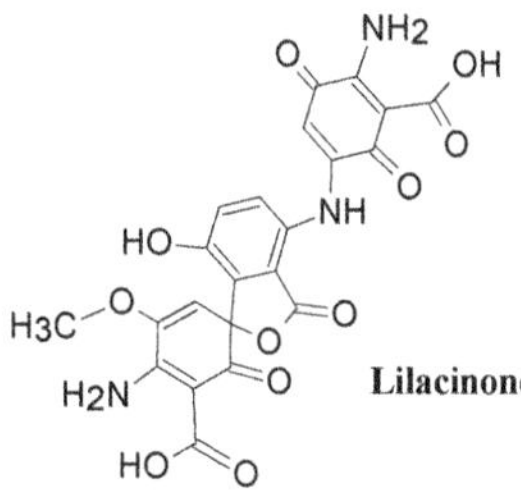

Lilacinone

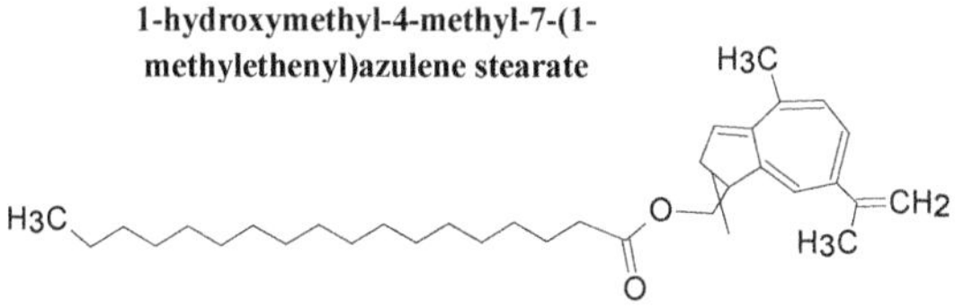

1-stearoyloxymethylene-4-methyl-7-isopropenylazulene

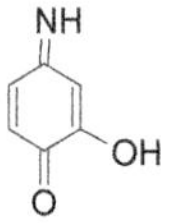

2-hydroxy-4-imino cyclohexa-2,5-dienone

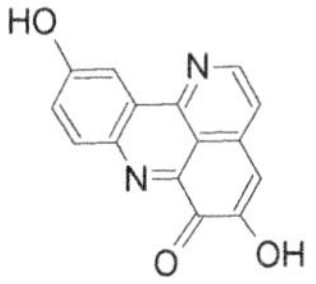

Necatarone

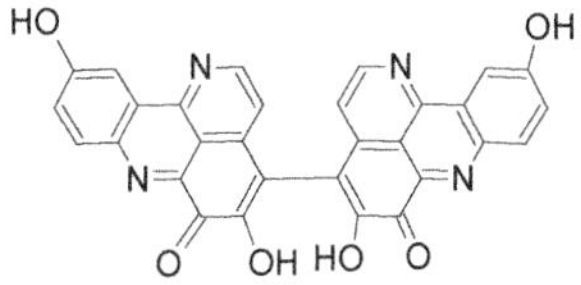

Necataronedehydrodimer

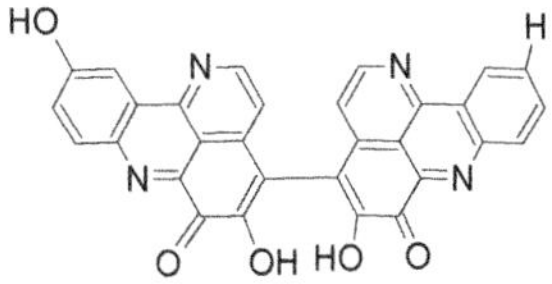

10-deoxydehydrodimer

Fig. 1 Contd. ...

...*Fig. 1 Contd.*

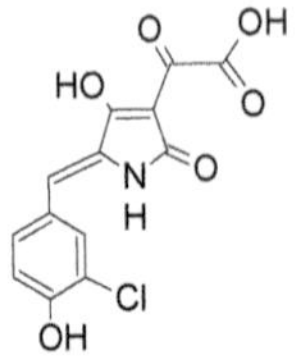

Pachydermin

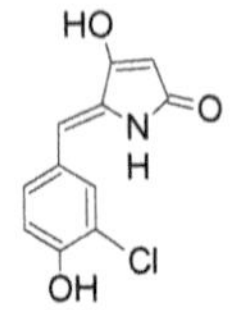

5-(3-chloro-4-hydroxy benzylidene) tetramic acid

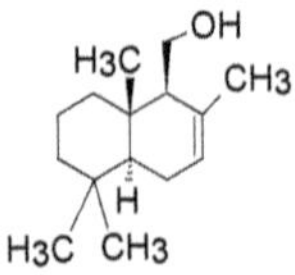

Drimenol

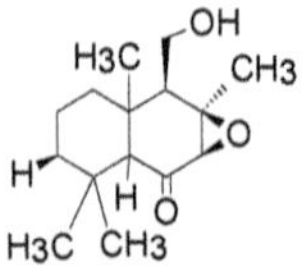

Uvidin A

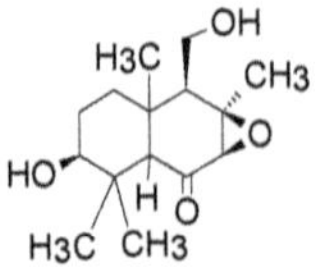

Uvidin B

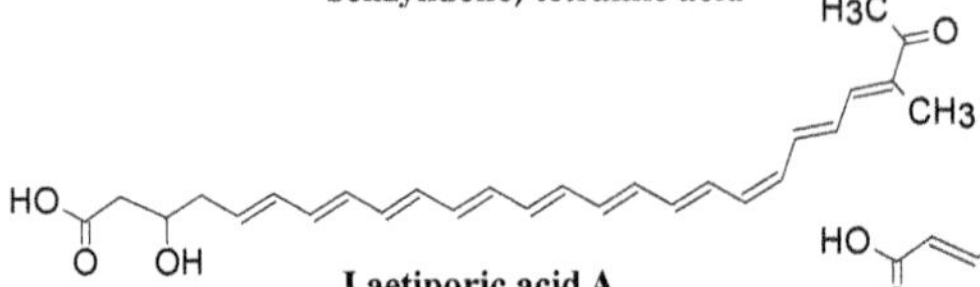

Laetiporic acid A

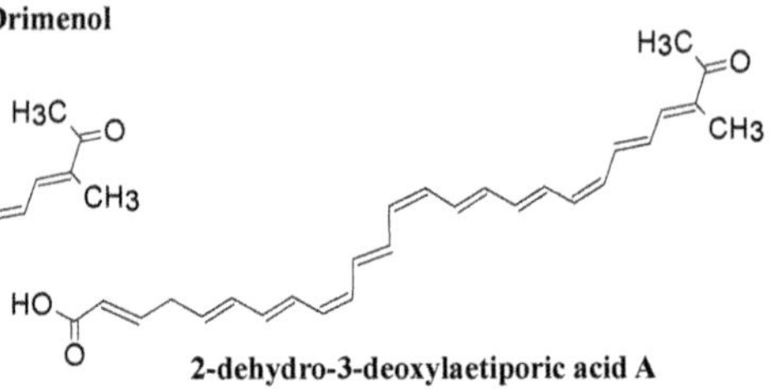

2-dehydro-3-deoxylaetiporic acid A

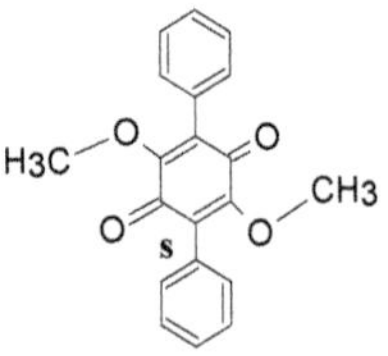

Betulinan A

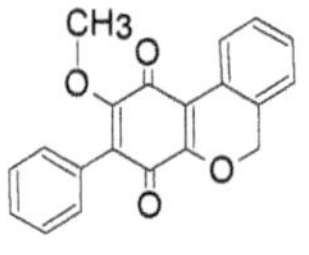

Betulinan B

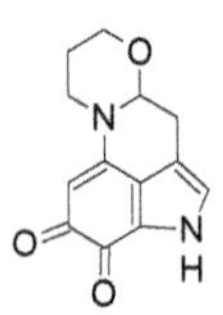

Haematopodin A

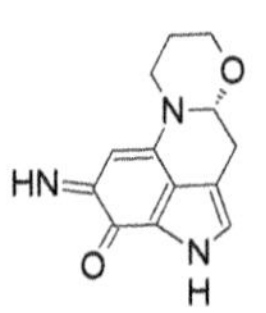

Haematopodin B

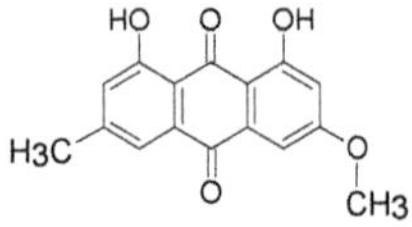

Physcion

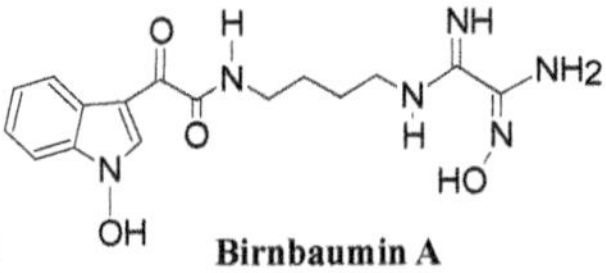

Birnbaumin A

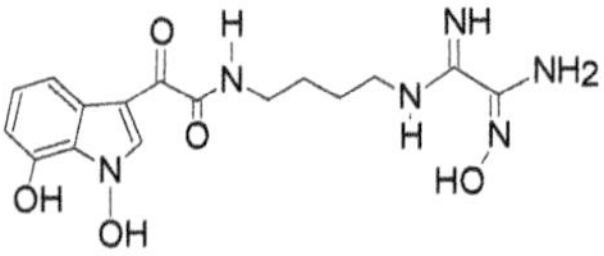

Birnbaumin B

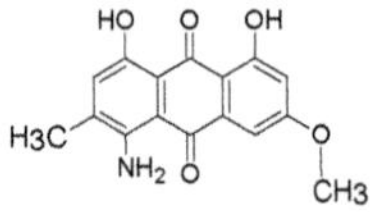

4-aminophyscion

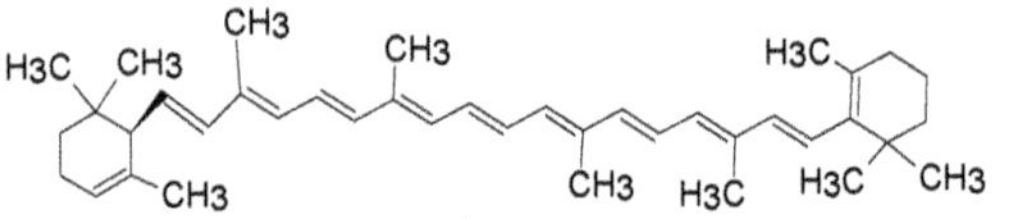

α-carotene

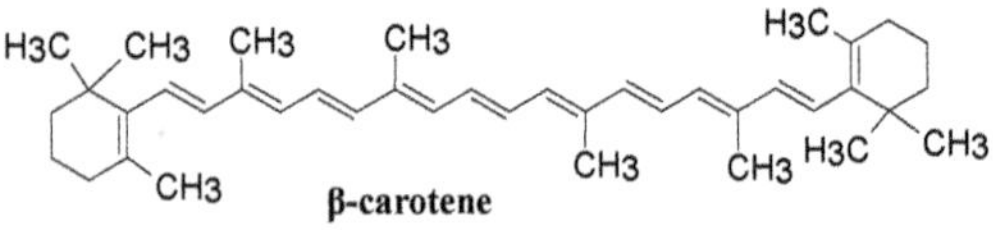

β-carotene

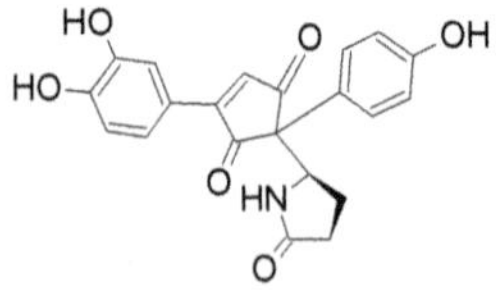

4-(3,4-dihydroxyphenyl)-2-(4-hydroxyphenyl)-2-(2-pyrrolidon-5-yl)-4-cyclopentene-1,3-dione

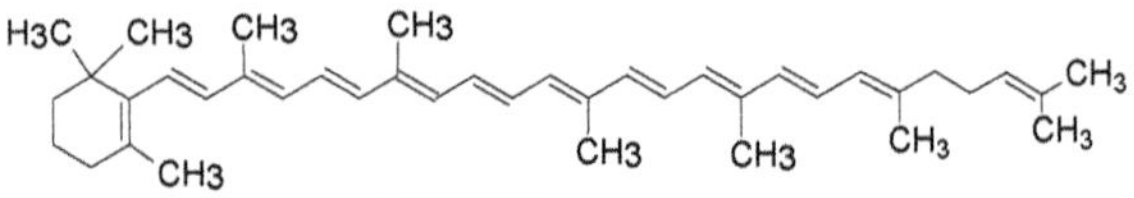

γ-carotene

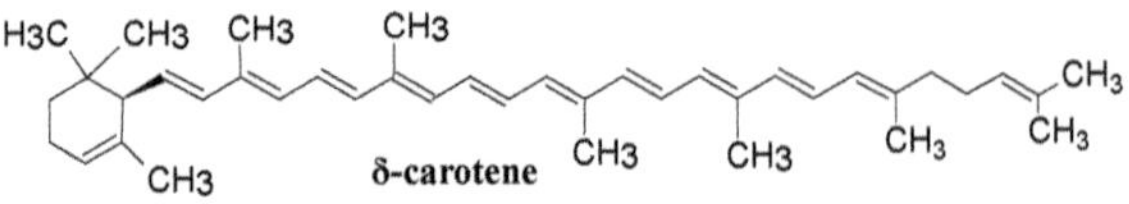

δ-carotene

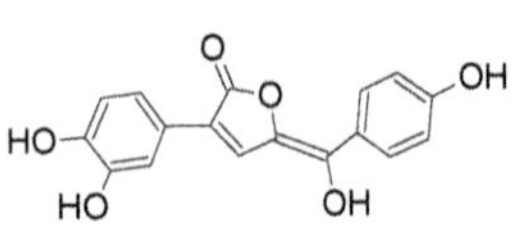

(4Z)-5-hydroxy-2-(3,4-dihydroxyphenyl)-5-(4-hydroxyphenyl)-2,4-pentadien-4-olide

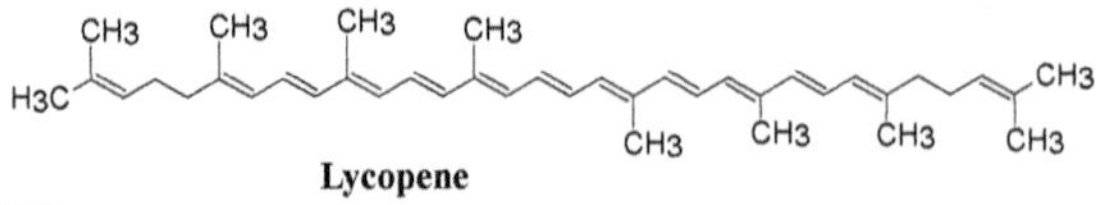

Lycopene

Fig. 1 Contd. ...

...*Fig. 1 Contd.*

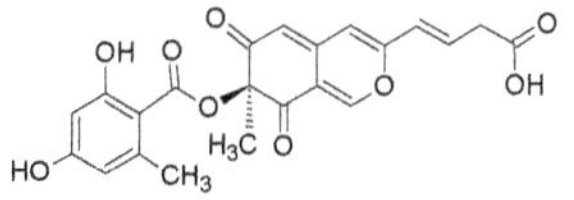

(+)-**Mitorubrin** 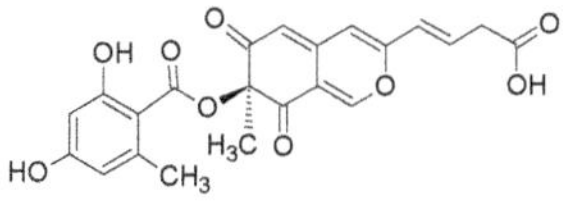   (+)-**Mitorubrinic acid** 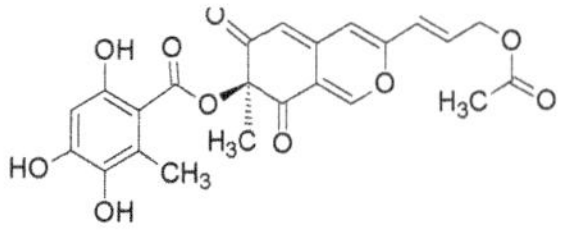   (+)-**6″-Hydroxymitorubrinol acetate**

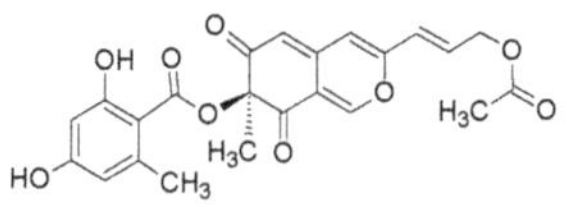

(+)-**Mitorubrinol acetate** 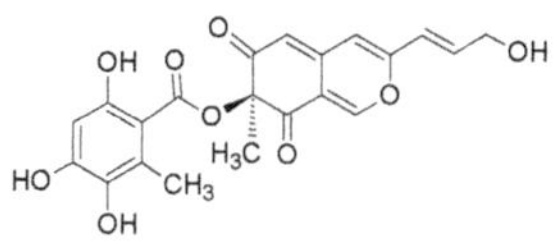   (+)-**6″-Hydroxymitorubrinol** 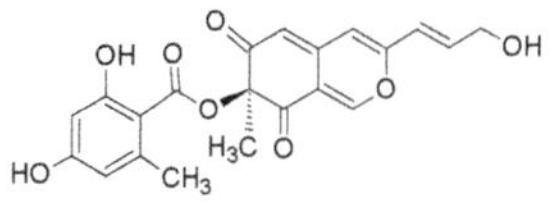   (+)-**Mitorubrinol**

Fig. 1. Pigments of diverse chemical classes from macrofungi.

2.1 *Ascomycota*

2.1.1 *Hypocreales*

Many genera of the order Hypocreales, such as *Hypocrella, Calcarisporium, Gliocephalotrichum, Torrubiella, Cordyceps*, and *Ophiocordyceps* have been reported to be excellent producers of a wide range of pigments. *Hypocrella bambusae*, belonging to the family Clavicipitaceae, has been found to produce pigments, hypocrellins A and B (Ma et al., 2004). Some of the aurovertin-type pigments, such as aurovertins A-E, aurovertin M, aurovertins T-U have been reported from parasitic fungus, *Calcarisporium arbuscula* (Mulheirn et al., 1974; Guo et al., 2013; Zhao et al., 2016). *Gliocephalotrichum simplex* has shown high production of extracellular melanin pigment in the culture media when supplemented with tyrosine, thereby suggesting the possible use of melanin in the field of biotechnology (Jalmi et al., 2012). Four yellow, amorphous pyridone and tetramic acid alkaloids (torrubiellones A-D) have been isolated from the spider-pathogenic fungus *Torrubiella* sp. BCC 2165 and pale yellow torrubiellones A-B and E from *Torrubiella longissima*; whereas torrubiellins A-B, emodin, and chrysophanol have been extracted from leafhopper pathogenic fungus *Torrubiella* spp. BCC 28517 (Isaka et al., 2010; Isaka et al., 2012; Isaka et al., 2014). Species of the genus *Cordyceps* have also been found to produce pigments, for example, yellow pigment cordycepoid A has been reported from *Cordyceps bifusispora*, whereas yellow cordycepin from *Cordyceps* sp. (Lu et al., 2013; Tuli et al., 2014; Ramesh et al., 2019). Researchers have reported two yellow pigments from *Paecilomyces farinosus* (currently known as *Cordyceps farinosa*) and water-soluble red pigments from *Isaria farinosa* (currently known as *Cordyceps farinosa*) (Lang et al., 2005; Velmurugan et al., 2010; Caro et al., 2012; Ramesh et al., 2019). Along with these, pigments, such as erythrostominone, deoxyerythrostominone, 4-*O*-methyl erythrostominone, epierythrostominol, deoxyerythrostominol, and 3,5,8-TMON (3,5,8-trihydroxy-6-methoxy-2-(5-oxohexa-1,3-dienyl)-1,4-naphthoquinone) have been obtained from *Ophiocordyceps unilateralis* (formerly known as *Cordyceps unilateralis*) (Caro et al., 2017). Similarly, a reddish-orange pigment, skyrin, has been discovered from *Epichloe bertonii* (formerly known as *Hyperdermium bertonii*) (Caro et al., 2017).

2.1.2 Helotiales

Macrofungi are used to obtain spalted wood (wood coloured internally by fungi, with a long history of use in West European marquetry and intarsia art), the most prominent being those from the genus *Chlorociboria*. The two species known in Western Europe, *Chlorociboria aeruginascens* and *Chlorociboria aeruginosa* produce xylindein, a highly conjugated quinone pigment (Robinson et al., 2012a; Hinsch, 2015; Hinsch et al., 2015; Hernandez et al., 2016). Besides, microscopic analysis of the pigments extracted from *C. aeruginosa*, and *C. aeruginascens* has also been carried out by some researchers (Vega Gutierrez and Robinson, 2017).

2.1.3 Xylariales

Members of the family Xylariaceae and Hypoxylaceae of the order Xylariales possess an exceptional pigment-production potential. A study reports for the first time the presence of aphlegmacin-type pigment, phlegmacin A 8,8'-di-*O*-methyl ether along with two known yellow pigments [(S)-torosachrysone-8-*O*-methyl ether and emodin-6,8-di-*O*-methyl ether] from an Ascomycete, *Xylaria euglossa* (Wang et al., 2005). A yellow cytotoxic pigment, (–)-xylariamide A, has been isolated from *Xylaria* species (Davis, 2005, Zhou and Liu, 2010). Several researchers have studied the ability of *Xylaria polymorpha* to produce black pigment melanin in different wood samples, such as sugar maple and beech (Tudor, 2013; Tudor et al., 2013; Tudor et al., 2014; Robinson et al., 2014). One of the studies on spalting fungi from the southern Amazon forest of Peru mentions the zone line producing ability of certain Ascomycetes (*Xylaria guianensis, Xylaria hypoxylon*, and *Xylaria curta*) in wood (Vega Gutierrez and Robinson, 2015).

Besides family Xylariaceae, many genera of the family Hypoxylaceae (*Hypoxylon, Daldinia, Jackrogersella*, and *Entonaema*) produce a diverse class of pigments. Members of the genera *Hypoxylon* have been proved to be a rich source of pigments. Species of the family Hypoxylaceae, *H. fragiforme* has been reported to produce pigments, such as hypoxyxylerone, fragiformins A-B, mitorubrine azaphilones [Mitorubrin, (+)-mitorubrinol, mitorubrinol-acetate, and (+)-mitorubrinic acid], rutilins C-D, fragirubrins A-E, and lenormandin F (Steglich et al., 1974; Edwards et al., 1991; Caro et al., 2017; Surup et al., 2018). In addition to this, several other species have also been found to produce a variety of pigments. For instance, pigments rubiginosins A-C, mitorubrin, entonaemin A and hypomiltin have been reported from *H. rubiginosum*, mitorubrine azaphilones [Mitorubrine, (+)-mitorubrinol, mitorubrinol-acetate, and (+)-mitorubrinic acid] from *H. howeanum*, vermelhotin and hypoxyvermelhotins A-C from *H. lechatii*, daldinins A-C, daldinins E-F and BNT (1,1'–binaphthalene–4,4'–5,5'–tetrol) from *H. fuscum*, rickenyls B-D from *H. rickii*, yellow mitorubrinol derivatives [(+)-6″-hydroxymitorubrinol acetate, (+)-mitorubrinol acetate,(+)-6″-hydroxymitorubrinol,(+)-mitorubrinol] from *H. fulvo-sulphureum*, hypoxylone from *H. sclerophaeum*, rutilins A-B from *H. rutilum*, and lenormandins A-G from *H. jaklitschii* (Quang et al., 2004a; Quang et al., 2004b; Quang et al., 2005b; Quang et al., 2006d; Stadler et al., 2007; Stadler et al., 2008; Sir et al., 2015; Caro et al., 2017). Besides, different species of the genus *Daldinia* (*Daldinia bambusicola, D. caldariorum, D. concentrica, D. eschscholzii, D. childiae,*

D. clavata, D. fissa, D. grandis, D. lloydi, D. loculata, D. petriniae, D. singularis) have been reported to produce a wide variety of pigments, such as daldinol, daldinals A-C and BNT (1,1'-binaphthalene-4,4'-5,5'-tetrol), daldinins A-C (Hashimoto et al., 1994; Caro et al., 2017). Moreover, *Daldinia concentrica* has been reported to produce two aromatic steroids (17β,20R,22E,24R)-19-norergosta-1,3,5,7,9,14,22-heptaene, (17β,20R,22E,24R)-1-methyl-19-norergosta-1,3,5,7,9,14,22-heptaene and pale yellow pigment, concentricolide (Qin and Liu, 2004; Qin et al., 2006; Caro et al., 2017). Another member of Hypoxylaceae, *Jackrogersella cohaerens* (previously known as *Annulohypoxylon cohaerens*) has been mentioned to produce multiformin A, as well as different cohaerin variants (cohaerins A-K) (Quang et al., 2005a; Quang et al., 2006c; Caro et al., 2017). A recent study has reported glaziellin A, entonaemins A-B, rubiginosin A, and comazaphilone D from *Entonaema liquescens* (formerly known as *Glaziella splendens*) (Kim et al., 2019).

2.1.4 Pleosporales

Certain species of the family Shiraiaceae, especially *Shiraia bambusicola* and *Rubroshiraia bambusae*, have been reported for their pigment-production potential. *Shiraia bambusicola* has been described to produce red, perylenequinone pigments, such as hypocrellins A-D and red pigment shiraiarin (Kishi et al., 1991; Fang et al., 2006; Cai et al., 2008). Another study has shown that species *Rubroshiraia bambusae* produces a high amount of pigments, hypocrellin A and B compared to *S. bambusicola* (Dai et al., 2019).

2.1.5 Pezizales

A red pigment, aleuriaxanthin, has been isolated from an orange peel fungus *Aleuria aurantia* (Margalith, 1992; Ramesh et al., 2019). Researchers have extracted and purified the melanin pigment from black truffle *Tuber melanosporum* and by using UV and IR spectroscopy, they have characterised the purified melanin (Harki et al., 1997).

2.2 Basidiomycota

2.2.1 Agaricales

Agaricales is one of the diverse and distinct orders of Basidiomycota to show a large number of pigment-producing mushrooms belonging to different families, such as Agaricaceae, Amanitaceae, Hygrophoraceae, Hygrophoropsidaceae, Tricholomataceae, Strophariaceae, Pleurotaceae, Cortinariaceae, Stephanosporaceae, Physalacriaceae, Hymenogasteraceae and Schizophyllaceae.

The family Agaricaceae has been reported to be a good source of colourful fungi, producing a wide variety of pigments. A red iminoquinone, 2-hydroxy-4-iminocyclohexa-2,5-dienone, black pigments [melanin, and 4-glutaminyhydroxylbenzene (GHB) melanin] have been reported from *Agaricus bisporus* and yellow azaquinone agaricone from *Agaricus xanthoderma* (Margalith, 1992; Hanson, 2008a, b; Ramesh et al., 2019). Another member *Leucocoprinus birnbaumii*, commonly known as flowerpot parasol, has been mentioned to produce two yellow 1-hydroxyindole pigments (birnbaumins A-B) (Bartsch et al., 2005). A sulphur-rich melanin pigment has been reported from edible termite mushroom

Macrolepiota albuminosa (formerly known as *Termitomyces albuminosus*) using scanning electron microscopy as well as UV-VIS and FTIR spectroscopy (De Souza et al., 2018).

Pigment-producing mushrooms are also observed in the family, Pleurotaceae. A gilled mushroom, *Pleurotus citrinopileatus* (formerly known as *Pleurotus cornucopiae* var. *citrinopileatus*) has been mentioned to produce yellow pheomelanin-like pigment (Shirasaka et al., 2012). Black pigment melanin has been reported from *Pleurotus djamor*, *Pleurotus citrinopileatus* and *Pleurotus cystidiosus* based on chemical tests, melanisation assay, UV, IR and EPR spectroscopy (Selvakumar et al., 2008; Vallimayil and Eyini, 2013). A cholesterol reducer red pigment, mevinolin (lovastatin), has been reported from *Pleurotus ostreatus* (Parthasarathy and Sathiyabama, 2015; Ramesh et al., 2019). Family Amanitaceae also contains mushrooms possessing attractive, bright and beautiful colours. A number of bright coloured pigments, such as amavadin (blue), muscarufin, muscapurpurin (purple), muscaurins I–VII (orange), muscoflavin (yellow) and an uncharacterised orange pigment have been described from the fly agaric, *Amanita muscaria* (Bayer and Kneifel, 1972; Muso, 1976; Li and Oberlies, 2005; Hanson, 2008b; Duran et al., 2009; Karuppan et al., 2014); whereas hispidin and bisnoryangonin have been reported from *Amanita pantherina* (Repke et al., 1978).

Certain unique and colourful pigments have also been observed in some of the colourful mushrooms belonging to the family Hygrophoraceae and Hygrophoropsidaceae. A witch's hat mushroom, *Hygrocybe conica*, has been reported to have yellow pigments, such as muscaflavin and hygroaurins (Muso, 1976; Li and Oberlies, 2005). Likewise, two yellow β-carboline alkaloids (harmane and norharmane) have been reported from *Hygrophorus eburneus*; whereas yellow-green pigment, brunnein A, was discovered from *Hygrophorus agathosmus* (formerly known as *Hygrophorus hyacinthinus*) (Teichert et al., 2008c). A pale brown clitocybin A and yellow clitocybin D have been extracted from the culture broth of *Hygrophoropsis aurantiaca* (formerly known as *Clitocybe aurantiaca*) (Kim et al., 2008; Kim et al., 2009).

Further, members of the family Strophariaceae and Tricholomataceae have also proven their pigment-production potential. A yellow pigment, (3S)-torosachrysone-8-*O*-methyl ether, has been discovered from *Tricholoma equestre* and *T. sulphureum*, orirubenones A-G (yellow) from *Tricholoma orirubens*, aurantricholone (orange-red), terreumols A-D (yellow) from *Tricholoma terreum*; whereas aurantricholides A-B (yellow) have been observed in *T. aurantium* (Gill and Steglich, 1987; Gill, 1994; Klostermeyer et al., 2000; Hanson, 2008b; Kawagishi et al., 2004; Sakai et al., 2005; Velisek and Cejpek, 2011; Yin et al., 2013). Moreover, the pigment-production ability has also been observed in species of the genera *Hypholoma* and *Pholiota*. Researchers have extracted yellow pigments, hypholomines A-B from the fruit bodies of sulphur tuft mushroom, i.e., *Hypholoma fasciculare* (Fiasson et al., 1977). Along with these, pigments hispidin, bisnoryangonin and β-keto ester have also been reported from *Hypoloma* species (Brady and Benedict, 1972; Gill, 1994; Velisek and Cejpek, 2011). A polyketide pigment, squarrosidine, and two styrylpyrone pigments (hispidin and bisnoryangonin) have been described respectively from *Pholiota squarrosa* and *P. squarroso-adiposa* (Brady and Benedict, 1972; Gill, 1994; Wangun and Hertweck, 2007; Velisek and Cejpek, 2011).

Besides, many other macrofungi of the order Agaricales have shown distinct pigments. Orange-yellow pigments, such as xerulin, dihydroxerulin and xerulinic acid have been reported from *Oudemansiella melanotricha* (family Physalacriaceae) (Kuhnt et al., 1990). Also, pigments stephanosporin and 2-chloro-4-nitrophenol have been isolated from the carrot truffle, *Stephanospora caroticolor* (family Stephanosporaceae) which are responsible for the orange colour of Gasteromycete (Lang et al., 2001). Likewise, the yellow-brown styrylpyrone pigments (hispidin and bisnoryangonin) have been discovered from species of the genus *Gymnopilus* (family Hymenogastraceae), such as *Gymnopilus junonius* (formerly known as *Gymnopilus spectabilis*) and *G. punctifolius* (Repke et al., 1978; Lee et al., 2008). A split gill mushroom, *Schizophyllum commune* (family Schizophyllaceae) has been found to produce pigments of diverse colours, such as melanin (black-brown), indigo (blue), indirubin (red), isatin (orange-red) and tryptanthrin (yellow) (Miles et al., 1956; Epstein and Miles, 1966; Hosoe et al., 1999; Arun et al., 2015; Ramesh et al., 2019). Recently scientists have screened different fungi for the melanin-biosynthesis ability and based on results, they have revealed that Basidiomycete, *Armillaria cepistipes* (family Physalacriaceae) produced the highest amount of melanin (27.98 g/L) (Ribera et al., 2019).

Some of the species of the genus *Mycena* have been found to be potential producers of pigments. A species *Mycena rosea* has been reported to produce red pigments, mycearubins A and B (Peters and Spiteller, 2007a; Zhou and Liu, 2010; Velisek and Cejpek, 2011); while *Mycena sanguinolenta* produce two blue alkaloid pigments (sanguinones A-B), one red indoloquinone alkaloid (sanguinolentaquinone) and yellow pigment decarboxydehydrosanguinone A (Peters and Spiteller, 2007b; Zhou and Liu, 2010; Velisek and Cejpek, 2011). An antibacterial, red polyne pigment, mycenaaurin A, has been reported from *Mycena aurantiomarginata* (Jaeger and Spiteller, 2010; Ramesh et al., 2019). Similarly, a study on *Mycena pelianthina* has revealed red pyrroloquinoline alkaloids (pelianthinarubins A-B, and mycenarubins A and D) from the fruit bodies (Pulte et al., 2016). Many colourful pigments, such as red pigments (haematopodin, haematopodin B, mycearubins A-B and mycearubins D-F), yellow pigments (mycenaflavins A-C) and a purple pigment mycenaflavin D have been discovered from *Mycena haematopus* (Baumann et al., 1993; Hopmann and Steglich, 1996; Peters et al., 2008; Velisek and Cejpek, 2011; Lohmann et al., 2018).

Moreover, certain members of Agaricales belonging to group 'Incertae sedis' have also shown the pigment-production ability. For example, a dung-loving bird nest fungi *Cyathus stercoreus* has reported three yellow antioxidant pigments (cyathusals A-C) along with known yellow pigment, pulvinatal (Kang et al., 2007), whereas a xylotrophic Basidiomycetes, *Fistulina hepatica* has been mentioned to produce carotenoid pigments (Velygodska and Fedotov, 2016).

Members of the genus *Cortinarius*, belonging to the family Cortinariaceae, are rich sources of different pigments. For instance, a dark-orange pigment, fallacinol (6-*O*-methoxycitreorosein) from *Cortinarius cinnabarinus*, bright yellow pigment, flavomannin-6,6'-di-*O*-methyl ether from *C. cinnamomeoluteus* and *C. croceus*, flavomannin-6,6'-di-*O*-methyl ether and (3S)-torosachrysone-8-*O*-

methyl ether from *C. citrinus*, green pigment (3*R*)-atrochrysone from *C. atrovirens* and *C. odoratus*, (3*S*)-torosachrysone-8-*O*-methyl ether from *C. fulmineus* and *C. splendens*, scaurins A and B from *C. scaurus*, canthin-6-one, infractins A and B from *C. infractus* (Gill and Steglich, 1987; Gill, 1994; Gill, 1996; Velisek and Cejpek, 2011). Along with these, many other species of this genus have been reported to produce a variety of pigments, viz., *Cortinarius basirubescens* produces emodin, physcion, erythroglaucin, (1*S*,3*S*)-austrocortirubin, (1*S*,3*S*)-austrocortilutein, (1*S*,3*R*)-austrocortilutein, (*S*)-torosachrysone, 6-methylxanthopurpurin-3-*O*-methyl ether, *Cortinarius odorifer* produces phlegmacin, anhydrophlegmacin, *Cortinarius sinapicolor* produces hydroxyphlegmacinquinone, (3*S*,3'*S*, P)-anhydrophlegmacin-9,10-quinone 8'-*O*-methyl ether, (2'*S*,3*S*,3'*S*, P)-2'-hydroxyanhydrophlegmacin-9,10-quinone 8'-*O*-methyl ether, dermocanarin-4, *Cortinarius subtortus* produces 6-hydroxyquinoline-8-carboxylic acid and 7-hydroxy-1-oxo-1,2-dihydro isoquinoline-5-carboxylic acid, *Cortinarius rufo-olivaceus* produces rufoolivacins A-B; whereas *Cortinarius atrovirens* produces atrovirin B (Steglich and Topper-Petersen, 1972; Antonowitz et al., 1994; Gill, 1995; Gill et al., 1998; Elsworth et al., 1999; Teichert et al., 2008a; Zhang et al., 2009; Beattie et al., 2010; Zhou and Liu, 2010).

Another species *Cortinarius brunneus* has been revealed to produce *β*-carboline alkaloid pigments [brunneins A–C and 3-(7-hydroxy-9H- -carboline-1-yl) propanoic acid] along with *N*-glucosyl-1H-indole derivatives [1-(1-*β*-glucopyranosyl)-3-(methoxymethyl)-1H-indole, 1-(1-*β*-glucopyranosyl)-1H-indole-3-carbaldehyde, N-1-*β*-glucopyranosyl-3-(carboxymethyl)-1H-indole, N-1-*β*-glucopyranosyl-3-(2-methoxy-2-oxoethyl)-1H-indole, 1-(1-*β*-glucopyranosyl)-1H-indole-3-acetic acid and methyl 1-(1-*β*-glucopyranosyl)-1H-indole-3-acetate (Teichert et al., 2007; Teichert et al., 2008b; Zhou and Liu, 2010). An edible fungus, *Cortinarius purpurascens*, has reported nine anthraquinone-related pigments, such as rufoolivacin, rufoolivacins C-D, leucorufoolivacin, physcion, verbindung cr11, verbindung cr60, citreorosein 6,8-dimethyl ether, and 1-hydroxy-3-methyl-2-isopropanyl-6,8-dimethoxyanthraquinone (Bai et al., 2013). Some of the yellow-red pigments, such as emodin, emodin-1-*β*-D-glucopyranoside, physcion, dermocybin, dermorubin, dermoxanthone, and flavomannin have been obtained from *Cortinarius sanguineus* (formerly known as *Dermocybe sanguinea*), and *C. semisanguineus* (Gill, 1999; Räisänen et al., 2001a; Räisänen, 2002; Räisänen et al., 2002; Velisek and Cejpek, 2011; Caro et al., 2012; Gessler et al., 2013; Zalas et al., 2015; Räisänen, 2019; Ramesh et al., 2019). Similarly, pigments like physcion, dermocybin, dermocybin-1-*β*-D glycopyranoside, dermorubin, 5-chlorodermorubin, dermolutein, 5-chlorodermolutein, dermoglaucin, endocrocin, 5,7-dichloroendocrocin have been mentioned from *Cortinarius vitiosus* (Räisänen, 2019). Recently, four new diterpenoid pigments (pyromyxones A-D) possessing nor-guanacastepene skeleton have been obtained from the Chilean mushroom *Cortinarius pyromyxa* (Lam et al., 2019).

Besides these, certain *Cortinarius* species, which were previously included in the genus *Dermocybe*, have also been found to produce a diverse class of pigments, like austrocortilutein, 1-deoxyaustrocortilutein, austrocortirubin, 1-deoxyaustrocortirubin and torosachrysone from *Cortinarius persplendidus* (formerly known as *Dermocybe splendida*), icterinoidins A-C, atrovirin B, hypericin, skyrin from *C. icterinoides* (formerly known as *D. icterinoides*), austrovenetin, protohypericin, hypercin, skyrin and hypercin-like pigment from *C. austrovenetus* (formerly known as *D. austroveneta*), hypercin and skyrin from *C. alienatus* (formerly known as *D. alienata*), dermocanarins 1-3, physcion, 4-aminophyscion and erythroglaucin from *C. canarius* (formerly known as *D. canaria*), cardinalins 2-6 from *C. cardinalis* (formerly known as *D. cardinalis*) (Antonowitz et al., 1994; Gill and Yu, 1994; Gill, 1995; Gill and Gimenez, 1995). Two atropisomeric austrocolorins A$_1$ and B$_1$ belonging to a rare class 10,10'-coupled dihydroanthracenone dimers have been isolated from an Australian toadstool, *Dermocybe* sp. (Beattie et al., 2004). Likewise, an antimicrobial pigment, phallacinol, has also been reported from *Dermocybe* sp. (Gessler et al., 2013; Ramesh et al., 2019).

3. Anthraquinones and their Production by Macrofungi

Anthraquinones and their derivatives constitute a group in a large variety of quinoid compounds with circa seven hundred molecule structures being described so far in nature (Fouillaud et al., 2016); of these, about one-third have been found in plants, two-thirds in fungi and lichen (Diaz-Muñoz et al., 2018) and a minuscule share in bacteria and insects. Lichens are symbiotic organisms comprised of alga and fungal partners and due to this symbiotic origin lichens and fungi contain similar types of colourants.

The colourless 9,10-anthraquinone (Fig. 2) is one of the oldest known dye chromophores and a well-known class of dye structures (Hunger, 2003). When examining the synthetic dye structures, the anthraquinones' tinctorial strength is low as compared to, for example, azo dyes and therefore, anthraquinones have lost their value as synthetic colourants; furthermore, they are expensive to produce (Gordon and Gregory, 1983; Hunger, 2003). Synthetic anthraquinone structures are used in applications, such as acid dyes for wool and polyamides, disperse dyes for polyesters and vat dyes for cotton and cellulose fibres (Hunger, 2003).

In biocolourant perspective, anthraquinones are interesting as they are among the most stable natural secondary metabolites and natural dye structures, producing bright colours and in natural dyes, their tinctorial strength is high. The number and quality of donor groups and their positions in the anthraquinone ring system affect the wide variety of achieved colours. Anthraquinone may be considered as consisting of two isolated benzoyl chromogens in which the substituents, located in different rings, interact much less than those in the same benzoyl chromogen. Therefore, most synthetic anthraquinone dyes are 1,4- or 1,2-di-substituted and almost any shade of a colour can be produced by incorporating two donor groups of different strengths in the 1- and 4-positions of the anthraquinone ring (Gordon and Gregory, 1983). In natural compounds, however, tetra-substituted (1,4,5,8-) anthraquinones are more

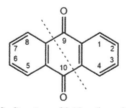

Fig. 2. Structure of 9,10-anthraquinone.

bathochromic than di- (1,4-) or tri-substituted (1,2,4-), as with the greater number of substituents the circuit of electrons expands cause bathochromic shift towards longer wavelengths (Hunger, 2003).

The disadvantages of anthraquinones as colourants are their potential health risks like 9,10-anthraquinone is carcinogenic (PubChem database, 2020). Also, some substituted anthraquinones, but not all, have been found to possess mutagenic and toxic properties (Brown, 1980, Räisänen et al., 2020, PubChem database, 2020). When discussing the safety issues, it is good to keep in mind that anthraquinones can also be found in various foods consumed by humans, such as peas, cabbage, lettuce and beans which can contain them in an amount of 0.0436 mg/kg of fresh weight (Diaz-Muñoz et al., 2018) and emodin and other plant-derived anthraquinones have been well-known laxatives (Brusick and Mengs, 1997), indicating that toxicity is dependent on dosage and length of exposure.

Because the anthraquinone structure has great potential as a biocolourant, it is important to examine closely the chemical structures and select the most promising ones, i.e., choose those with non-toxicity as well as good stability, colour fastness and affinity to the intended substrate and to the application tests.

The studies (Räisänen et al., 2001a, 2002) show that in addition to emodin (Table 2), dermocybin and dermorubin show potential as textile dyes for various fibre types and dyeing techniques. All these anthraquinones have been found in fungi, especially in genus *Cortinarius*, subgroup *Dermocybe*. According to the studies of mutagenicity (Umbuzeiro et al., 2020), emodin showed positive response as before (Dong et al., 2016; Chen et al., 2018) but dermocybin and dermorubin showed negative toxicity response (Umbuzeiro et al., 2020).

There is a comprehensive review article of fungal secondary metabolites covered by Zhou and Liu (2010), and one focusing on anthraquinones from marine-derived fungi written by Fouillaud et al. (2016). Shahid et al. (2013) list a variety of applications of natural colourants, such as in cosmetics, food and textiles, whereas the review by Hyde et al. (2019) shows the wide potential of fungi in applications where the colour applications are mentioned among the 50 other listed options.

3.1 *Anthraquinones in Macrofungi*

Anthraquinones are found broadly in fungal and lichen species. In macrofungi, they are abundantly present in genus *Cortinarius*, especially in subgroup *Dermocybe*. *Cortinarius*, with a cosmopolitan distribution, contains about three thousand described and many yet unexplored species (Niskanen et al., 2012; Royal Botanic Gardens, Victoria, Hyde et al., 2020). In recent years, more effort has been dedicated to the taxonomy of *Cortinarius* sp. and many previously unknown species have been discovered. The DNA-based research has been the key tool for identifying new species (Niskanen et al., 2012; Hyde et al., 2020).

Efficient colour-producing macrofungi, which contain large amounts of colourants, could be considered as sources. In the *Dermocybe* species, the amount of anthraquinones in dry weight of fungal mass can be as high as 6 per cent, whereas for plant sources, viz., in madder root (*Rubia* sp.), the amounts of colourants are typically 1.5–4 per cent but might require several years of growth period (Räisänen, 2019; Bechtold and Mussak, 2009). Of course, when considering large-scale colourant-production, growing, filamentous fungi in liquid medium is the most feasible way but macrofungi can offer alternatives in particular applications since the collection of fungi requires less sophisticated equipment and access to sources is usually easy. In the Northern Hemisphere, there have been development projects to increase utilisation of natural resources in the creation of products and services with added value. Foraging networks have been recognised as an efficient resource for natural material collection (Rutanen, 2014).

Anthraquinone compounds found in macrofungi have great variation in their chemical compositions and structures. The smallest ones are hydroxyl or methyl substituted with molecular weights of 240 g/mol onward and occur in their free form as aglycones. However, very often anthraquinones are bound to sugar residues, resulting in, for example, 1-β-D-glucopyranosides. The larger anthraquinone molecules may have two-ring systems connected in different ways to each other. Such compounds are, for example, icterinoidin B_1, austrocolorin A_1 and hypericin (Table 2) (Zhou and Liu, 2010; Gessler et al., 2013).

Table 2. Examples of the variety of anthraquinone structures existing in macrofungi *Cortinarius* sp. (Räisänen, 2019; Zhou and Liu, 2010).

Compounds name	Molecular Weight [g/ mol] and Formula	Chemical structure	Species
Emodin	270.0528 $C_{15}O_5H_{10}$		*C. sanguineus,* *C. semisanguineus*
Dermocybin-1-β-D-glucopyranoside	478.1111 $C_{22}O_{12}H_{22}$		*C. vitiosus,* *C. sanguineus,* *C. semisanguineus*
Icterinoidin B_1	528.1056 $C_{29}H_{20}O_{10}$		*C. icterinoides*
Austrocolorin A_1	602.2152 $C_{34}H3_4O_{10}$		*Dermocybe* sp. WAT 26641
Hypericin	504.0845 $C_{30}H_{16}O_8$		*C. austrovenetus*

As colourants for textiles, most wanted structures would be the ones of aglycones. Such compounds are less water soluble as, for example, the combination with glycoside increases water solubility and also the size of the compound.

3.2 Hymenochaetales

Pigments with different hues have been reported from the colourful fruit bodies of mushrooms of the genus *Hymenochaete, Inonotus, Inocutis* and *Phellinus* belonging to family Hymenochaetaceae (Lee and Yun, 2006; Velisek and Cejpek, 2011). Many antioxidant pigments, including hispidin and its derivatives, such as inoscavins A-E, methylinoscavins A-D, phelligridins D-F, davallialactone, methydavallialactone and interfungins A-C have been isolated from *Hymenochaete xerantica* (formerly known as *Inonotus xeranticus*) (Kim et al., 1999; Lee et al., 2006a, 2006b; Lee et al., 2007b; Lee and Yun, 2006; Zhou and Liu, 2010; Velisek and Cejpek, 2011).

Different species of the genus *Inonotus* have also been mentioned to be prolific producers of a wide variety of pigments. *Inonotus hispidus*, commonly known as shaggy bracket fungus, was previously noted as producing two important pigments: hispidin (yellow-brown) and hispolone (yellow) (Ali et al., 1996). Researchers have studied the black pigment melanin-production potential of *Inonotus hispidus* in the wood substrate (Tudor, 2013; Tudor et al., 2013; Robinson et al., 2012b). Similarly, some of the investigators have discovered the melanin complex from medicinal mushrooms, *Inonotus obliquus* (Babitskaya et al., 2002). Moreover, many novel antioxidant polyphenols, such as inonoblins A-C (yellow) along with some known compounds (phelligridins D-G) have been extracted from the fruit body of *Inonotus obliquus* (Lee et al., 2007a). Similarly, a number of yellow pigments (rheadinin, hypholomine A, hypholomine B, hispidin and bisnoryangonin) and one orange pigment (phellinin A) have been isolated from mycelium of *Inocutis rheades* (formerly known as *Inonotus rheades*) and identified by UV, IR, MS and NMR spectroscopic analysis (Olennikov et al., 2017).

Plenty of colourful pigments have been discovered from another genus, *Phellinus*. Mushroom *Phellinus linetus* has been known to produce pigments of diverse colours, like brown pigments (meshimakobnols A-B), orange pigments (phellifuropyranone A, phelligridin G), yellow pigments (phelligridimer A, davallialactone, methyldavallialactone, inoscavin A, interfungin A, hypholomine B, ellagic acid), and a yellow-brown pigment hispidin (Nagatsu et al., 2004; Wang et al., 2005; Kojima et al., 2008; Lee et al., 2008; Zhou and Liu, 2010). Similarly, pigments with a wide range of colours, such as yellow pigments (phelligridimer A, phelligridins C-D, J, and hispolon), orange pigments (phelligridins E-I), brown pigments (meshimakobnols A-B) and yellow-brown hispidin, have been reported from the fruit bodies of another medicinal mushroom, *Phellinus igniarius* (Kirk et al., 1975; Mo et al., 2003a; Mo et al., 2003b; Mo et al., 2004; Wang et al., 2005; Wang et al., 2007b; Velisek and Cejpek, 2011; Kim et al., 2016; Thanh et al., 2017). *Phellinus ellipsoideus* (formerly known as *Fomitiporia ellipsoidea*) has also been found to produce pigments, such as hispidin, phelligridin K, inoscavin C and inonoblin B (Zan et al., 2015). Besides this, several other medicinally important species of the genus *Phellinus* have been found to produce pigments, such asmelanin complex by *Phellinus robustus*, a novel phenylpropanoid-derived polyketide pinillidine by *Phellinus pini*, and novel spiroindene pigment (phelliribsin A) by *Phellinus ribis* (Bisko et al., 2007; Wangun and Hertweck, 2007; Kubo et al., 2014). Certain yellow and orange styrylpyrone pigments (phellinins A-C) along with yellow-brown pigment, hispidin, have been reported from the culture broth of *Phellinus* sp. KACC93057P (Lee et al., 2009a; 2009b, Lee et al., 2010).

3.3 Polyporales

Several members of different families of the order Polyporales, such as Polyporaceae, Oxyporaceae, Laetiporaceae and Phanerochaetaceae have exhibited pigment-production potential. Members of the family Polyporaceae mainly contain pigments which are derivatives of polyporic acids and terphenylquinones. Some of the investigators have reported certain phenoxazinone-type pigments (pycnoporin, cinnabarinic acid, cinnabarin and tramesanguin) from the *Pycnoporus* spp. (*P. cinnabarinus*, *P. coccineus* and *P. sanguineus*) (Sullivan and Henry, 1971; Smania Jr et al., 2003; Dias and Urban, 2009; Velisek and Cejpek, 2011; Ramesh et al., 2019) and betulinans A and B from the plant pathogen, *Lenzites betulina* (Lee et al., 1996). Different carotenoid pigments as well as one anti-diabetic pigment, fomentariol, have been discovered from a plant pathogenic fungus, *Fomes fomentarius*, which is commonly known as 'hoof fungus' (Velisek and Cejpek, 2011; Velygodska and Fedotov, 2016). Some researchers have studied different fungi, such as *Fomes fomentarius* (family Polyporaceae), *Trametes versicolor* (family Polyporaceae) and *Oxyporus populinus* (family Oxyporaceae) for their ability to produce black pigment melanin in the wood substrate (Tudor, 2013; Tudor et al., 2013). Later on, they did a comparative study on *in vivo* melanin synthesis by *Oxyporus populinus* in sugar maple and *Fomes fomentarius* in birch, along with the study of *in vitro* pigment production by *Trametes versicolor* and *Inonotus hispidus* in sugar maple and beech (Tudor et al., 2014). Diverse carotenoid pigments, as well as certain non-carotenoid polyene pigments (laetiporic acids A-C and 2-dehydro-3-deoxylaetiporic acid A) have been isolated from wood-rotting bracket fungus, *Laetiporus sulphureus*, belonging to family Laetiporaceae (Weber et al., 2004; Davoli et al., 2005; Popa et al., 2016; Velygodska and Fedotov, 2016). Species of the family Phanerochaetaceae also produce coloured pigments, for example, *Hapalopilus rutilans* (formerly known as *Hapalopilus nidulans/Polyporus nidulans*) has been mentioned to produce a dark red pigment, polyporic acid and blue pigments (corticins A-B) by *Terana coerulea* (formerly known as *Corticium coeruleum*) (Hanson, 2008b; Velisek and Cejpek, 2011). Species of *Ganoderma*, such as *Ganoderma applanatum* and *G. lucidum* have also been reported for their pigment-production ability (Karuppan et al., 2014).

3.4 Russulales

Mushrooms of the genus *Lactarius* are commonly known as milk-caps, as they secrete milky exude when they are cut or damaged. Along with the milky exude, many species have been reported for their pigment-production potential. *Lactarius deliciosus* has been found to contain sesquiterpenoids, such as blue lactarazulene [1,4-dimethyl-7-(1-methylethyl)azulene], red-violet lactaroviolin [4-methyl-7-(1-methylethyl)azulene-1-carbaldehyde], red azulenes [acetylazulene, dihydroxyazulene,

7-acetyl-4-methylazulene-1-carbaldehyde,   7-(1,2-dihydroxy-1-methylethyl)-4-methylazulene-1-carbaldehyde and 7-acetyl-4-methylazulene-1-carboxylic acid], orange   pigments   (dihydroazulen-1-ol,15-hydroxy-3,6-dihydrolactarazulene,15-hydroxy-6,7-dihydrolactarazulene), purple pigments [7-(1-hydroxyethyl)-4-methyl-1-azulenecarboxaldehyde and 7-isopropenyl-4-methyl-1-azulenecarboxylic acid] and red-orange-brown quinizarine (Yang et al., 2006; Rai, 2009; Zhou and Liu, 2010; Velisek and Cejpek, 2011; Tala et al., 2017; Tian et al., 2019; Ramesh et al., 2019). Another species, *Lactarius hatsudake* has also been revealed to produce pigments, such as lactarioline A (blue), lactarioline B (red), lactaroviolin (red-violet), 7-(1-hydroxy-1-methylethyl)-4-methylazulene-1-carbaldehyde   (red-purple),   4-methyl-7-(1-methylethyl)azulene-1-carboxylic acid (purple), 4-methyl-7-(1-methylethyl)azulene-1-carbaldehyde   (red-purple),1-[(15E)-buten-17-one]-4-methyl-7-isopropylazulene (green), 4-methyl-7-isopropylazulene-1-carboxylic acid (red), 1-formyl-4-methyl-7-isopropyl azulene (red),1-formyl-4-methyl-7-(1-hydroxy-1-methylethyl) azulene (red) (Fang et al., 2006; Fang et al., 2007; Xu et al., 2010; Zhou and Liu, 2010). Besides this, several other species of the genus *Lactarius* have been described to produce a wide range of pigments, especially red pigment 3,5,7(11),9-pentaenyl-14-guaianal by *Lactarius sanguifluus*, blue pigments [1-hydroxymethyl-4-methyl-7-(1-methylethenyl) azulene   stearate   and   1-stearoyloxymethylene-4-methyl-7-isopropenylazulene] by *Lactarius indigo*, uvidin A, uvidin B, and drimenol by *Lactarius uvidus*, lactarane and secolactarane sesquiterpenoids by *Lactarius scrobiculatus*, 4-methoxy-2-(3-methylbut-2-enyl) phenol by *Lactarius fuliginosus* and *Lactarius picinus*, aminobenzoquinoneblennione (green) by *Lactarius blennius*, lilacinone (red) by *Lactarius lilacinus*, necatarone, necataronedehydrodimer, and 10-deoxydehydrodimer by *Lactarius turpis* and blue azulenes by *Lactarius* sp. (Harmon et al., 1980; Spiteller and Steglich, 2002; Spiteller et al., 2003; Velisek and Cejpek, 2011; Caro et al., 2017).

Mushrooms of the order *Russulales* have been reported to be a rich source of meroterpenoid pigments. An ochre brittle-gill mushroom, *Russula ochroleuca* has been reported to produce red pigment, ochroleucin A, and yellow pigment, ochroleucin B (Sontag et al., 2006; Velisek and Cejpek, 2011). Similarly, meroterpenoid pigments have also been reported from several species of the genus *Albatrellus* belonging to family Albatrellaceae, e.g., pigments albatrelins D-F have been discovered from the *Albatrellus ovinus*, grifolinones B-C and albatrellin from *A. confluens*, grifolinone B, albatrellin and 16-hydroxyalbatrellin from *A. flettii*, purple pigment grifolinone B from *A. caeruleoporus*; whereas green pigment cristatomentin from *A. cristatus* (Quang et al., 2006b; Koch and Steglich, 2007; Yang et al., 2008; Koch et al., 2010; Liu et al., 2013). *Albatrellus confluens* has also been found to produce yellow pigments, such as aurovertins B-C, E and aurovertins I-S (Wang et al., 2005; Zhou et al., 2009; Guo et al., 2013).

Other families of the order *Russulales* (Phanerochaeraceae and Peniophoraceae) have also been revealed to possess unique pigment-producing fungi. A saprophytic fungus, *Atheliachaete sanguinea* (formerly known as *Phanerochaete sanguinea*) has been found to produce pigments like xylerythrin (black crystal with green lustre), 5-*O*-methylxylerythrin (black), peniophorinin (black), penioflavin (yellow) (Abrahamsson and Innes, 1966; Gripenberg, 1970; Girpenberg, 1978; Velisek and Cejpek, 2011), whereas *Phanerochaete affinis* (formerly known as

Peniophora affinis) produces peniophorin A (pink-red) and peniophorin B (brown-black) (Gerber et al., 1980). A species of *Peniophora* collected from the southern Amazon forest of Peru has been described to produce the zone lines in wood (Vega Gutierrez and Robinson, 2015). Besides this, one unidentified stereaceous Basidiomycete has been reported to produce two yellow polyene pigments [(3Z,5E,7E,9E,11E,13Z,15E,17E)-18-methyl-19-oxoicosa-3,5,7,9,11,13,15,17-octaenoic acid and (3E,5Z,7E,9E,11E,13E,15Z,17E,19E)-20-methyl-21-oxodocosa-3,5,7,9,11,13,15,17,19-nonaenoic acid] upon physical injury to the mycelia (Schwenk et al., 2014).

3.5 Boletales

Mushrooms of the order Boletales contain a very diverse class of pigments. Many families of the order Boletales, such as Boletaceae, Gomphidiaceae, Paxillaceae, Serpulaceae, Rhizophogaceae, Suillaceae, Tapinellaceae, Sclerodermataceae and Calostomataceae comprise colourful mushrooms possessing bright coloured pigments. Major pigment-producing genera include *Boletus*, *Caloboletus*, *Austroboletus*, *Gastroboletus*, *Pulveroboletus*, *Imleria*, *Chamonixia*, *Leccinum*, *Tapinella*, *Suillus*, *Gomphidius*, *Chalciporus*, *Paxillus*, *Melanogaster*, *Serpula*, *Rhizophogon*, *Calostoma*, *Scleroderma*, *Pisolithus* and *Chroogomphus*.

Several macrofungi of the family Boletaceae have been reported for their beautiful colours. *Boletus subalpinus* (formerly known as *Gastroboletus subalpinus*) has been found to produce an orange pigment, variegatic acid (Besl and Bresinsky, 1997). Two polyene pigments, such as boletocrocins A and B, have been isolated from fruit bodies of *Boletus laetissimus* (Kahner et al., 1998); whereas pigments curtisin and 9-deoxycurtisin have been reported from *Boletus curtisii*, which was responsible for the bright yellow colour of the mushroom (Brockelmann et al., 2004). Three pulvinic acid derivatives [atromentic acid (orange-red), xerocomic acid (red-orange), variegatic acid (orange)] have been extracted from *Caloboletus calopus* (formerly known as *Boletus calopus*) (Huang et al., 2009; Zhou and Liu, 2010). Genus *Pulveroboletus* also contains a variety of pigments such as vulpinic acid from *Pulveroboletus cramesinus*, variegatic acid and variegatorubin, and xerocomic acid from *Pulveroboletus lignicola*, pulverolide, pulveravens A-B and vulpinic acid from *Pulveroboletus ravenelii* (Duncan et al., 2003, Zhang et al., 2006). Some of the investigators have detected 2-naphthoic acid derivatives (austrogracilins A-B) from air-dried fruit bodies of *Austroboletus gracilis* by HPLC with diode array detection (Bartsch et al., 2004). *Imleria badia* (formerly known as *Xerocomus badius*) has been found to possess pigments badione A and norbadione A (Steffan and Steglich, 1984; Aumann et al., 1989; Velisek and Cejpek, 2011). Pigments sclerocitrin, chalcitrin, norbadione A, variegatic acid and variegatorubin have been discovered from the peppery bolete, *Chalciporus piperatus* (Winner *et al.*, 2004; Zhao and Liu, 2010; Velisek and Cejpek, 2011). Species of the genus *Chamonixia* have shown a distinct class of pigments. Some of the investigations have revealed three blue 2,5-diarylcyclopentane-l,3-diones (chamonixin, gyrocyanin and gyroporin) from sporophores of macrofungus *Chamonixia caespitosa* which are responsible for the blue colour of the fruit body (Steglich et al., 1977; Feling et al., 2001). Similarly, two yellow pigments [pachydermin, and 5-(3-chloro-4-hydroxybenzylidene) tetramic acid] have been discovered from the New Zealand Basidiomycete *Leccinum pachyderme* (formerly known as *Chamonixia pachydermis*) (Lang et al., 2006).

Numerous species of *Gomphidius* have been found to produce a variety of pigments, such as gomphidic acid, xerocomic acid, atromentic acid, gomphilactone and bolegrevilol have been produced by *Gomphidius glutinosus*, gomphidic acid, xerocomic acid, atromentic acid, gomphilactone, bolegrevilol by *G. maculatus*, atromentic acid, bolegrevilol from *G. roseus* and xerocomic acid, variegatic acid, variegatorubin by *G. subroseus* (Steglich et al., 1969; Knight and Pattenden, 1976; Besl and Bresinsky, 1997). Also, certain species of the genus *Chroogomphus* possess derivatives of pulvinic acid and boviquinones, for instance, xerocomic acid and boviquinones from *Chroogomphus tomentosus*, xerocomic acid, boviquinone-3, boviquinone-4 from *C. helveticus*, xerocomic acid, atromentic acid, boviquinone-3, boviquinone-4, diboviquinone-3,4, diboviquinone-4,4, methylenediboviquinone-3,3, methylenediboviquinone-3,4 and methylenediboviquinone-4,4 from *C. rutilus* (Besl and Bresinsky, 1997; Ramesh et al., 2019).

Members of the genus *Suillus* have emerged as an excellent source of pigments. A variety of pigments, such as grevillins A-D, atromentin, atromentic acid, xerocomic acid, variegatic acid, variegatorubin, amitenone, methyl bovinate, boviquinone-4, 3',4',4-trihydroxypulvinone, cyclovariegatin, thelephoric acid, suillin, bovilactone-4,4, gyroporin, chamonixin, involutin, tridentoquinone, flazin, aurantricholide B, pyrandione and furanones have been reported from different species of the genus *Suillus*, such as *Suillus acidus, S. aeruginascens, S. amabilis, S. albidipes, S. americanus, S. bellinii, S. brevipes, S. bovinus, S. bresadolae, S. clintonianus, S. collinitus, S. granulatus, S. grisellus, S. grevillei, S. hirtellus, S. flavidus, S. lakei, S. leptopus, S. luteus, S. neoalbidipes, S. placidus, S. plorans, S. pictus, S. pseudobrevipes, S. punctatipes, S. punctipes, S. riparius, S. serotinus, S. sibiricus, S. spectabilis, S. spraguei, S. tridenticus, S. tomentosus, S. umbonatus, S. variegatus* and *S. viscidus* (Besl and Bresinsky, 1997; Muhlbauer et al., 1998; Dong et al., 2007; Besl et al., 2008; Lang et al., 2008; Huang et al., 2009; Lang et al., 2009; Zhou and Liu, 2010; Velisek and Cejpek, 2011; Ramesh et al., 2019).

Certain genera of the family Paxillaceae (e.g., *Paxillus* and *Melanogaster*) and Sclerodermataceae (e.g., *Scleroderma* and *Pisolithus*) have the ability to produce pigments. Researchers have discovered diarylcyclopentenone pigments, such as involutin, chamonixin and gyroporin from *Paxillus involutus*, commonly known as brown roll-rim (Feling et al., 2001; Braesel et al., 2015). Using biochemical, genetic and transcriptomic analyses in addition to stable-isotope labelling with synthetic precursors, researchers have revealed that atromentin is the key intermediate in the biosynthesis pathway of diarylcyclopentenone (Braesel et al., 2015). Besides the known pigment, involutin, some new yellow pigments [involutone, (3,4-dihydroxyphenyl)-2-(4-hydroxyphenyl)-2-(2-pyrrolidon-5-yl)-4-cyclopentene-1,3-dione and (4Z)-5-hydroxy-2-(3,4-dihydroxyphenyl)-5-(4-hydroxyphenyl)-2,4-pentadien-4-olide] have been reported from *Paxillus involutus* (Antkowiak et al., 2003; Mikołajczyk and Antkowiak, 2009; Zhou and Liu, 2010). Another member of this family, i.e., a subterranean fungus *Melanogaster broomeianus* has been revealed to synthesise a polyene pigment melanocrocin (Aulinger et al., 2001). Different pulvinic acid dimers (sclerocitrin, badione A, norbadione A and xerocomic acid) have been extracted from the fruit body of common earthball mushroom, *Scleroderma citrinum* (Winner et al., 2004; Zhao and Liu, 2010; Velisek and Cejpek, 2011). Similarly, *Pisolithus arrhizus*, commonly known as a dead man's foot, has been found to produce pigments, like badione A (chocolate brown) and norbadione A (golden yellow) (Winner et al., 2004; Velisek and Cejpek, 2011).

Along with these, many other investigations have reported violet spiromentins A-D from *Tapinella atrotomentosa* and *Tapinella panuoides*. Moreover, *Tapinella atrotomentosa* has also been reported to produce orange-yellow flavomentins A-D, osmundalactone (purple), 5-hydroxy-hex-2-en-4-olide (purple), xerocomic acid (red-orange) and atromentic acid (orange-red) (Gill and Steglich, 1987; Liu, 2006, Velisek and Cejpek, 2011; Beni et al., 2018). *Serpula lacrymans*, a dry-rot fungus is reported to possess red pigment, xerocomorubin (Hanson, 2008b). Researchers have shown that bacteria induce pigment formation in Basidiomycete *Serpula lacrymans* upon co-culturing with the bacterium (*Bacillus subtilis* or *Pseudomonas putida*, or *Streptomyces iranensis*). They have observed increased secretion of atromentin-derived pigments, i.e., variegatic, xerocomic, isoxerocomic and atromentic acid upon co-culturing (Tauber et al., 2016). A major pigment, rhizopogone (red), along with minor pigment, 2-acetoxyrhizopogone, has been extracted from *Rhizopogon pumilionum* (Lang et al., 2009; Zhou and Liu, 2010). A stalked puffball *Calostoma cinnabarinum* has been reported to produce a polyene pigment, calolstomal, which is responsible for the red-orange color of the puffball (Gruber and Steglich, 2007; Zhou and Liu, 2010).

3.6 Thelephorales

A diverse class of pigments has been discovered by several investigators from a number of species of the genus *Thelephora* (family Thelephoraceae). An orange pigment, ganbajunin A, has been reported from *Thelephora ganbajun* (Hu et al., 2001), whereas a grey antioxidant pigment, thelephorin A, has been obtained from the mushroom, *Thelephora vialis* (Tsukamoto et al., 2002). Many benzoyl p-terphenyl derivatives, such as thelephantins A-D (grey), thelephantin E (red-brown), thelephantin F (grey), thelephantin G (green-brown), thelephantin H (grey) and thelephorin A (grey) along with atromentin (brown) have been isolated from fruit bodies of the *Thelephora aurantiotincta* and their structural elucidation has been done by UV, IR, MS and NMR spectroscopic analysis (Quang et al., 2003a; 2003b; Quang et al., 2006a). Similarly, p-terphenyl derivative pigments, such as terrestrin B (reddish-violet), terrestrin C (red), terrestrin D (greyish violet), terrestrin E (greyish grey), terrestrin F (blue), terrestrin G (grey), thelephantin F (grey), thelephantin H (grey) have been reported from *Thelephora terrestris* (Radulovic et al., 2005). Mushrooms of the family Bankeraceae have also been found to produce a unique class of pigments, e.g., *Boletopsis leucomelaena* produces brown pigment, cycloleucomelone (Jagers et al., 1987; Steglich et al., 2000), *Sarcodon leucopus* produces nitrogenous terphenylquinols, sarcoviolins, i.e., sarcoviolon α (violet); whereas species of *Hydnellum* produce red pigment, aurantiacin (Cali et al., 2004; Velisek and Cejpek, 2011).

3.7 Auriculariales

In 2015, Prados-Rosales and colleagues isolated the melanin pigment from the edible mushroom, *Auricularia auricula*, and performed the structural characterisation, using sophisticated spectroscopic as well as physical/imaging techniques. A similar study by Zou and Ma in 2018 revealed *Auricularia auricula* RF201 as the highest producer of melanin among different strains of *A. auricula*, showing a maximum yield of 493.9 mg/L under submerged fermentation. Another species, *Auricularia nigricans* obtained as spalting fungus from the southern Amazon forest of Peru, has been reported to produce zone lines in wood (Vega Gutierrez and Robinson, 2015). In a recent study, natural melanin was discovered from *Exidia nigricans* and its melanin nature was confirmed by different chemical tests as well as FTIR and Raman spectroscopy analysis (Lopusiewicz, 2018).

3.8 Cantharellales

A variety of carotenoid pigments have been reported from the order Cantharellales. For example, α-carotene, β-carotene, lycopene and γ- and δ-isomers of carotene have been discovered from golden chantharelle, *Cantharellus cibarius* (Velisek and Cejpek, 2011); whereas a xanthophyll, canthaxanthin, and β-carotene have been reported from pink to red-orange cinnabar chanterelle, *C. cinnabarinus* (Haxo, 1950; Velisek and Cejpek, 2011).

3.9 Corticiales

A red pigment, phlebiarubrone, and three violet terphenylquinone pigments (4'-hydroxyphlebiarubrone, 4',4"-dihydroxyphlebiarubrone and 3/,4',4"-trihydroxyphlebiarubrone) have been reported from *Punctularia atropurpurascens* (Anke, 1984).

3.10 Microstromatales

A Basidiomycete, *Quambalaria cyanescens*, belonging to family Quambalariaceae, has been reported to produce three biologically active pigments (quambalarine A, quambalarine B, and mompain) (Stodulkova et al., 2015).

4. Toxicity of Anthraquinones

Anthraquinones show potential as colourants, especially in long-term artifacts, due to their stability and colour properties. Many fungal species produce them, but concerns have been raised on fungi's safety as future microbial cell factories. Comprehensive knowledge of already recognised fungal metabolites, pathogenic strains and toxin producers have been called for (Goyal et al., 2017). Especially when used for food and cosmetic purposes, the knowledge of safety of colourants and co-production of mycotoxins along with the target compound is crucial. Currently, the EU and USA have banned some fungal colourants for use in food due to the risk of possible contamination by toxic metabolites. Therefore, it is important to explore new mycotoxin free strains of fungi (Goyal et al., 2017). Such anthraquinone sources are, for example, macrofungi *Cortinarius,* subgroup *Dermocybe* (Räisänen et al., 2020).

The metabolic engineering of fungal strains gives new possibilities to eliminate toxins and produce more targeted end products, meaning even a single compound dye production and modification of core structure in a way which enables enhanced performance on the intended base material. Such kinds of start-up companies already exist (Pili, 2020).

Also, concerns about anthraquinones' safety have been discussed and it has been proposed that the amount and position of substituent affects the toxicity of an anthraquinone. The chemical structures of anthraquinones and the functional groups, for example, hydroxyl, alkyl, alkoxy or chlorine, attached at specific positions, can confer different biological activities and genotoxicity profiles (Brown, 1980; Inoue et al., 2009; Fotia et al., 2012) and anthraquinones are also known for their ability to interact with DNA and inhibit enzyme activity (Diaz-Muñoz et al., 2018).

There are an increasing number of reports regarding the adverse effects of emodin. It has been reported to induce genotoxicity, reproductive toxicity and also hepatotoxicity and nephrotoxicity, presumably connected to high doses and long-term use (Dong et al., 2016; Chen et al., 2018), but on the other hand, emodin-related compounds, emodin-1-β-D-glucopyranoside and dermocybin, have not reported these properties (Brown, 1980; Umbuzeiro et al., 2020).

For natural dyes it is essential to test the pure compounds and their possible toxic effects; even that natural dyes in applications are usually mixtures of several compounds. Testing mixtures leaves too much room for speculation on the actual crucial factor and it is known that single compounds may act differently compared to mixtures (Räisänen et al., 2020). No less important is the dosage and the way the colourant is in contact with the human body. The concentration of colourant in long-term artefacts, such as textiles, is low (maximum being 4 per cent on the weight of the fibre) and the skin contact in use is short. Hence, biocolourants in textile pose infinitesimal, if any, health risk to the user (Räisänen et al., 2020). More risks may occur when colourants end up in large quantities in the aquatic environment during the textile manufacturing process or in cases where workers are not properly protected, for example, from dye-dust when dosing.

5. Conclusion

Colours of macrofungi not only attract the fauna, but also humans. Bright and unique pigments of these macrofungi attract attention across the world due to their usage for a variety of value-added purposes. This has ultimately increased the demand of macrofungal pigments in the market. Additional beneficial attributes of macrofungal pigments may lay down new options and open up new areas for their application in different areas. Even though numerous pigments have been reported from several macrofungi, still availability of an unexplored abundant diversity of macrofungi provides scope for meticulous exploration of novel, valuable and useful pigments.

Because of the great potential in a variety of applications, it is important to examine closely the chemical structures and characteristics of compounds and to select the most promising ones, especially those that are non-toxic and stable during processing.

Acknowledgments

AL and SKS thank Director, ARI for providing the facility. AL also thank CSIR, New Delhi (India) for providing research fellowship (SRF) and SP Phule, Pune University, Pune (India) for allowing to register for Ph.D degree. LD was supported by grants from the *Conseil Régional de La Réunion*, Reunion Island, France (grant # 'COLORMAR RUN110356' and '*Arômes et colorants microbiens* RUN003261'). RR was supported by the Strategic Research Council at the Academy of Finland (grand # 327178, 2019).

References

Abrahamsson, S. and Innes, M. (1966). Molecular structure of xylerythrin, a fungus pigment. Acta Crystallogr., 21(6): 948–956.

Ainsworth, G.C. (2008). Ainsworth & Bisby's Dictionary of the Fungi, CABI. Abingdon, UK, p. 771.

Ali, N.A., Jansen, R., Pilgrim, H., Liberra, K. and Lindequist, U. (1996). Hispolon, a yellow pigment from *Inonotus hispidus.* Phytochem., 41(3): 927–929.

Anke, H., Casser, I., Herrfnann, R. and Steglich, W. (1984). New terphenylquinones from mycelial cultures of *Punctularia atropurpurascens* (Basidiomycetes). Zeitschrift für Naturforschung C, 39(7-8): 695–698.

Antkowiak, R., Antkowiak, W.Z., Banczyk, I. and Mikolajczyk, L. (2003). A new phenolic metabolite, involutone, isolated from the mushroom *Paxillus involutus.* Can. J. Chem., 81(1): 118–124.

Antonowitz, A., Gill, M., Morgan, P.M. and Yu, J. (1994). Coupled anthraquinones from the toadstool *Dermocybe icterinoides.* Phytochem., 37(6): 1679–1683.

Arun, G., Eyini, M. and Gunasekaran, P. (2015). Characterisation and biological activities of extracellular melanin produced by *Schizophyllum commune* (Fries). Ind. J. Exp. Biol., 53(6): 380–387.

Aulinger, K., Besl, H., Spiteller, P., Spiteller, M. and Steglich, W. (2001). Melanocrocin, a polyene pigment from *Melanogaster broomeanus* (Basidiomycetes). Zeitschrift für Naturforschung C, 56(7-8): 495–498.

Aumann, D.C., Clooth, G., Steffan, B. and Steglich, W. (1989). Complexation of cesium 137 by the cap pigments of the Bay Boletus (*Xerocomus badius*). Angew. Chem., 28(4): 453–454.

Babitskaya, V.G., Scherba, V.V., Ikonnikova, N.V., Bisko, N.A. and Mitropolskaya, N.Y. (2002). Melanin complex from medicinal mushroom *Inonotus obliquus* (Pers.: Fr.) Pilat (Chaga) (Aphyllophoro mycetidae). Int. J. Med. Mush., 4: 39–145.

Badalyan, S.M. and Zambonelli, A. (2019). Biotechnological exploitation of macrofungi for the production of food, pharmaceuticals and cosmeceuticals. pp. 199–230. *In*: Sridhar, K.R. and Deshmukh, S.K. (eds.). Advances in macrofungi: Diversity, Ecology and Biotechnology. CRC Press, Boca Raton, USA.

Bai, M.S., Wang, C., Zong, S.C., Lei, M. and Gao, J.M. (2013). Antioxidant polyketide phenolic metabolites from the edible mushroom *Cortinarius purpurascens.* Food Chem., 141(4): 3424–3427.

Bartsch, A., Bröckelmann, M.G., Steffan, B. and Steglich, W. (2004). Austrogracilin A and B, two 2-naphthoic acid derivatives from the mushroom *Austroboletus gracilis.* ARKIVOC, 10: 13–19.

Bartsch, A., Bross, M., Spiteller, P., Spiteller, M. and Steglich, W. (2005). Birnbaumin A and B: Two unusual 1-hydroxyindole pigments from the 'flower pot parasol' *Leucocoprinus birnbaumii.* Angew. Chem., 44(19): 2957–2959.

Baumann, C., Bröckelmann, M., Fugmann, B., Steffan, B., Steglich, W. and Sheldrick, W.S. (1993). Haematopodin, an unusual pyrroloquinoline derivative isolated from the fungus *Mycena haematopus*, Agaricales. Angew. Chem., 32(7): 1087–1089.

Bayer, E. and Kneifel, H. (1972). Isolation of amavadin, a vanadium compound occuring in *Amanita muscaria*. Zeitschrift für Naturforschung B, 27(2): 207.

Beattie, K., Elsworth, C., Gill, M., Milanovic, N.M., Prima-Putra, D. and Raudies, E. (2004). Austrocolorins A₁ and B₁: Atropisomeric 10, 10′-linked dihydroanthracenones from an Australian *Dermocybe* sp. Phytochem., 65(8): 1033–1038.

Beattie, K.D., Rouf, R., Gander, L., May, T.W., Ratkowsky, D. et al. (2010). Antibacterial metabolites from Australian macrofungi from the genus *Cortinarius*. Phytochem., 71(8-9): 948–955.

Bechtold, T. and Mussak, R. (2009). Handbook of Natural Colourants. Wiley series in Renewable Resources, John Wiley & Sons, Chichester, p. 434.

Beni, Z., Dekany, M., Kovacs, B., Csupor-Loffler, B., Zomborszki, Z.P. et al. (2018). Bioactivity-guided isolation of antimicrobial and antioxidant metabolites from the mushroom *Tapinella atrotomentosa*. Molecules, 23(5): 1082.

Besl, H., Michler, I., Preuss, R. and Steglich, W. (1974). Pigments of fungi, XXII Grevillin D, the main pigment of *Suillus granulatus*, *S. luteus* and *S. placidus* (Boletales) (author's transl). Zeitschrift fur Naturforschung (Section C-Biosciences), 29(11-12): 784–786.

Besl, H. and Bresinsky, A. (1997). Chemosystematics of Suillaceae and Gomphidiaceae (suborder Suillineae). Pl. Syst. Evol., 206(1-4): 223–242.

Besl, H., Bresinsky, A., Kilpert, C., Marschner, W., Schmidt, H.M. and Steglich, W. (2008). Isolation and synthesis of methyl bovinate, an unusual pulvinic acid derivative from *Suillus bovinus* (Basidiomycetes). Zeitschrift für Naturforschung B, 63(7): 887–893.

Bhandari, B. and Jha, S.K. (2017). Comparative study of macrofungi in different patches of Boshan community forest in Kathmandu, Central Nepal, *Botanica Orientalis*. J. Pl. Sci., 11: 43–48.

Bisko, N.A., Shcherba, V.V. and Mitropolskaya, N.Y. (2007). Study of melanin complex from medicinal mushroom *Phellinus robustus* (P. Karst.) Bourd. et Galz. (Aphyllophoro mycetideae). Int. J. Med. Mush., 9(2): 177–184.

Borovicka, J., Kotrba, P., Gryndler, M., Mihaljevic, M., Randa, Z. et al. (2010). Bioaccumulation of silver in ectomycorrhizal and saprobic macrofungi from pristine and polluted areas. Sci. Tot. Environ., 408(13): 2733–2744.

Brady, L.R. and Benedict, R.G. (1972). Occurrence of bisnoryangonin in *Pholiota squarroso-adiposa*. J. Pharmaceut. Sci., 61(2): 318. 10.1002/jps.2600610251.

Braesel, J., Gotze, S., Shah, F., Heine, D., Tauber, J. et al. (2015). Three redundant synthetases secure redox-active pigment production in the Basidiomycete *Paxillus involutus*. Chem. Biol., 22(10): 1325–1334.

Brockelmann, M.G., Dasenbrock, J., Steffan, B., Steglich, W., Wang, Y. et al. (2004). An unusual series of thiomethylated canthin-6-ones from the North American mushroom *Boletus curtisii*. Eur. J. Org. Chem., 2004(23): 4856–4863.

Brown, J.P. (1980). A review of the genetic effects of naturally occurring flavonoids, anthraquinones and related compounds. Mutation Research, 75(3): 243–277.

Brusick, D. and Mengs, U. (1997). Assessment of the genotoxic risk from laxative Senna products. Environ. Mol. Mutagen., 29(1): 1–9.

Butler, M.J. and Day, A.W. (1998). Fungal melanins: A review. Can. J. Microbiol., 44(12): 1115–1136.

Cai, Y., Din, Y., Ta, G. and Lia, X. (2008). Production of 1, 5-dihydroxy-3-methoxy-7-methylanthracene-9, 10-dione by submerged culture of *Shiraia bambusicola*. J. Microbiol. Biotechnol., 18(2): 322–327.

Calì, V., Spatafora, C. and Tringali, C. (2004). Sarcodonins and sarcoviolins, bioactive polyhydroxy-p-terphenyl pyrazinediol dioxide conjugates from fruiting bodies of the Basidiomycete *Sarcodon leucopus*. Eur. J. Org. Chem., 2004(3): 592–599.

Camassola, M. (2013). Mushrooms—The incredible factory for enzymes and metabolites productions. Ferment. Technol., 2(1): 1000e117.10.4172/2167-7972.1000e117.

Caro, Y., Anamale, L., Fouillaud, M., Laurent, P., Petit, T. and Dufosse, L. (2012). Natural hydroxyanthraquinoid pigments as potent food-grade colourants: An overview. Nat. Prod. Bioprospect., 2(5): 174–193.

Caro, Y., Venkatachalam, M., Lebeau, J., Fouillaud, M. and Dufossé, L. (2017). Pigments and colorants from filamentous fungi. pp. 499–568. *In*: Merillon, J.-M. and Ramawat, K.G. (eds.). Fungal Metabolites. Springer International Publishing, Cham, Switzerland.

Chang, S.T. and Miles, P.G. (1992). Mushroom biology—A new discipline. Mycologist, 6(2): 64–65.

Chen, C., Gao, J., Wang, T.-S., Guo, C., Yan, Y.-J. et al. (2018). NMR-based metabolomic techniques identify the toxicity of emodin in HepG2 cells. Sci. Rep., 8: 9379.10.1038/s41598-018-27359-4.

Cheong, P.C.H., Tan, C.S. and Fung, S.Y. (2018). Medicinal mushrooms: Cultivation and pharmaceutical impact. pp. 287–304. *In*: Singh, B.P., Passari, A.K. and Lallawmsanga, C. (eds.). Biology of Macrofungi. Springer, Cham, Switzerland.

Conceicao, A.A., Cunha, J.R.B., Vieira, V.O., Pelaez, R.D.R., Mendonça, S. et al. (2018). Bioconversion and biotransformation efficiencies of wild macrofungi. pp. 361–377. *In*: Singh, B., Lallawmsanga, Passari A. (eds.). Biology of Macrofungi, Fungal Biology. Springer, Cham, Switzerland.

Dai, D.Q., Wijayawardene, N.N., Tang, L.Z., Liu, C., Han, L.H. et al. (2019). *Rubroshiraia* gen. nov., a second hypocrellin-producing genus in Shiraiaceae (Pleosporales). MycoKeys, 58: 1–26.

Davis, R.A. (2005). Isolation and structure elucidatio of the new fungal metabolite (–)-sylariamide A. J. Nat. Prod., 68(5): 769–772.

Davoli, P., Mucci, A., Schenetti, L. and Weber, R.W.S. (2005). Laetiporic acids, a family of non-carotenoid polyene pigments from fruit bodies and liquid cultures of *Laetiporus sulphureus* (Polyporales, Fungi). Phytochem., 66(7): 817–823.

De Bernardi, M., Mellerio, G., Vidari, G., Vita-Finzi, P. and Fronza, G. (1980). Fungal metabolites, Part 5, Uvidins, new drimane sesquiterpenes from *Lactarius uvidus* Fries. J. Chem. Soc., Perkin Trans., 1: 221–226.

De Silva, D.D., Rapior, S., Sudarman, E., Stadler, M., Xu, J. et al. (2013). Bioactive metabolites from macrofungi: Ethnopharmacology, biological activities and chemistry. Fungal Diversity, 62(1): 1–40.

De Souza, R.A., Kamat, N.M. and Nadkarni, V.S. (2018). Purification and characterisation of a sulphur rich melanin from edible mushroom *Termitomyces albuminosus* Heim. Mycology, 9(4): 296–306.

Dias, D.A. and Urban, S. (2009). HPLC and NMR studies of phenoxazone alkaloids from *Pycnoporus cinnabarinus*. Nat. Prod. Comm., 4(4): 489–498.

Diaz-Muñoz, G., Miranda, I.L., Sartori, S.K., De Rezende, D.C. and Diaz, M.A.N. (2018). Anthraquinones: An overview. Stud. Nat, Prod. Chem., 58: 313–338.

Dighton, J. (2019). The roles of macrofungi (Basidiomycotina) in terrestrial ecosystems. pp. 70–104. *In*: Sridhar, K.R. and Deshmukh, S.K. (eds.). Advances in Macrofungi: Diversity, Ecology and Biotechnology. CRC-Press, Boca Raton, USA.

Dong, X., Fu, J., Yin, X., Cao, S., Li, X. et al. (2016). Emodin: A review of its pharmacology, toxicity and pharmacokinetics. Phytother. Res., 30(8): 1207–1218.

Dong, Z.J., Wang, F., Wang, R.R., Yang, L.M., Zheng, Y.T. and Liu, J.K. (2007). Chemical constituents of the fruiting bodies from the Basidiomycete *Suillus granulatus*. Chin. Trad. Herbal Drugs, 38: 17–19.

Duarte, A.W.F., de Menezes, G.C.A., Silva, T.R., Bicas, J.L., Oliveira, V.M. and Rosa, L.H. (2019). Antarctic fungi as producers of pigments. pp. 305–318. *In*: Rosa, L.H. (ed.). Fungi of Antarctica. Springer Nature, Switzerland.

Duncan, C.J., Cuendet, M., Fronczek, F.R., Pezzuto, J.M., Mehta, R.G. et al. (2003). Chemical and biological investigation of the fungus *Pulveroboletus ravenelii*. J. Nat. Prod., 66(1): 103–107.

Duran, N., De Conti, R. and Teixeira, M.F.S. (2009). Pigments from fungi: Industrial perspective. pp. 185–225. *In*: Rai, M. (ed.). Advances in Fungal Biotechnology. IK International Publishing House Pvt. Ltd., New Delhi.

Edwards, R.L., Fawcett, V., Maitland, D.J., Nettleton, R., Shields, L. and Whalley, A.J. (1991). Hypoxyxylerone: A novel green pigment from the fungus *Hypoxylon fragiforme* (Pers.: Fries) Kickx. J. Chem. Soc. Chem. Com., 1991(15): 1009–1010.

Elsworth, C., Gill, M., Gimenez, A., Milanovic, N.M. and Raudies, E. (1999). Pigments of fungi, Part 50, 1 Structure, biosynthesis and stereochemistry of new dimeric dihydroanthracenones of the phlegmacin type from *Cortinarius sinapicolor* Cleland. J. Chem. Soc. Perkin Trans., 1(2): 119–126.

Epstein, E. and Miles, P.G. (1966). Identification of indirubin as a pigment produced by mutant cultures of fungus *Schizophyllum commune*. Bot. Mag., 79(940): 566–571.

Fang, L.Z., Shao, H.J., Yang, W.Q. and Liu, J.K. (2006). Two new azulene pigments from the fruiting bodies of the Basidiomycete *Lactarius hatsudake*. Helv. Chim. Acta, 89(7): 1463–1466.

Fang, L.Z., Yang, W.Q., Dong, Z.J., Shao, H.J. and Liu, J.K. (2007). A new azulene pigment from the fruiting bodies of the Basidiomycete *Lactarius hatsudake* (Russulaceae). Acta Botanica Yunnanica, 29(1): 122–124.

Feling, R., Polborn, K., Steglich, W., Muhlbacher, J. and Bringmann, G. (2001). The absolute configuration of the mushroom metabolites involutin and chamonixin. Tetrahedron, 57(37): 7857–7863.

Fernando, D., Senathilake, K., Nanayakkara, C., de Silva, E.D., Wijesundera, R.L. et al. (2018). An insight into isolation of natural products derived from macrofungi as antineoplastic agents: A review. Int. J. Collab. Res. Int. Med. Pub. Health, 10(2): 816–828.

Fiasson, J.L., Gluchoff-Fiasson, K. and Steglich, W. (1977). Über die farbundfluoreszenzstoffe des grünblättrigenschwefelkopfes (*Hypholoma fasciculare*, Agaricales). Chemische Berichte, 110(3): 1047–1057.

Fotia, C., Avnet, S., Granchi, D. and Baldini, N. (2012). The natural compound alizarin as an osteotropic drug for the treatment of bone tumors. J. Ortho. Res., 30(9): 1486–1492.

Fouillaud, M., Venkatachalam, M., Girard-Valenciennes, E., Caro, Y. and Dufossé, L. (2016). Anthraquinones and derivatives from marine-derived fungi: Structural diversity and selected biological activities. Marine Drugs, 14(4): 64.

Ganeshpurkar, A., Rai, G. and Jain, A.P. (2010). Medicinal mushrooms: Towards a new horizon. Pharmacog. Rev., 4(8): 127–135.

Gates, G.M., Mohammed, C., Wardlaw, T., Ratkowsky, D.A. and Davidson, N.J. (2011). The ecology and diversity of wood-inhabiting macrofungi in a native *Eucalyptus obliqua* forest of southern Tasmania, Australia. Fungal Ecology, 4(1): 56–67.

Gedminas, A., Lynikienė, J. and Povilaitienė, A. (2015). Entomopathogenic fungus *Cordyceps militaris*: distribution in South Lithuania, *in vitro* cultivation and pathogenicity tests. Baltic Forestry, 21(2): 359–368.

Gerber, N.N., Shaw, S.A. and Lechevalier, H.A. (1980). Structures and antimicrobial activity of peniophorin A and B, two polyacetylenic antibiotics from *Peniophora affinis* Burt. Antimicrob. Ag. Chemother., 17(4): 636–641.

Gessler, N.N., Egorova, A.S. and Belozerskaya, T.A. (2013). Fungal anthraquinones. Appl. Biochem. Microbiol., 49(2): 85–99.

Gill, M. and Steglich, W. (1987). Pigments of fungi (Macromycetes). Fortschritte der Chemieorganischer Naturstoffe, Springer, Vienna, p. 297.

Gill, M. (1994). Pigments of fungi (Macromycetes). Nat. Prod. Rep., 11: 67–90.

Gill, M. and Yu, J. (1994). The cardinalins, the first pyranonaphthoquinones from the higher fungi. Nat. Prod. Lett., 5(3): 211–216.

Gill, M. (1995). New pigments of *Cortinarius* Fr. and *Dermocybe* (Fr.) *Wünsche* (Agaricales) from Australia and New Zealand. Beihefte Zur Sydowia, 10: 73–87.

Gill, M. and Gimenez, A. (1995). Pigments of fungi, Part 40, Dermocanarins 1-3, unique naphthol-and naphthoquinone-anthraquinone dimers that contain a macrocyclic lactone ring from the fungus *Dermocybe canaria* Horak. J. Chem. Soc. Perkin Trans., 1(6): 645–651.

Gill, M. (1996). Pigments of fungi (Macromycetes). Nat. Prod. Rep., 13: 513–528.

Gill, M., Gimenez, A., Jhingran, A.G., Milanovic, N.M. and Palfreyman, A.R. (1998). Pigments of fungi, Part 49, 1 Structure and biosynthesis of dermocanarin 4, a naphthoquinone-dihydroanthracenone dimer from the fungus *Cortinarius sinapicolor* Cleland. J. Chem. Soc. Perkin Trans., 1, 998(20): 3431–3436.

Gill, M. (1999). Pigments of fungi (Macromycetes). Nat. Prod. Rep., 16: 301–317.

Glamoclija, J., Ciric, A., Nikolic, M., Fernandes, A., Barros, L. et al. (2015). Chemical characterization and biological activity of Chaga (*Inonotus obliquus*), a medicinal mushroom. J. Ethnopharmacol., 162: 323–332.

Gordon, P.F. and Gregory, P. (1983). Organic Chemistry in Colour. Springer-Verlag, Heidelberg, p. 322.

Goyal, S., Ramawat, K.G. and Merillon, J.M. (2017). Different shades of fungal metabolites: An overview. pp. 1–29. *In*: Jean-Michel, M. and Ramawat, K.G. (eds.). Fungal Metabolites, Springer, Cham, Switzerland.

Gripenberg, J. (1970). Fungus pigments, XXI, peniophorinin, a further pigment produced by *Peniophora sanguinea* Bres. Acta Chem. Scand., 24(10): 3449–3454.

Gripenberg, J. (1978). Fungus pigments, XXV, Penioflavin. Acta Chem. Scand., 32(1): 75–76.

Gruber, G. and Steglich, W. (2007). Calostomal, a polyene pigment from the Gasteromycete *Calostoma cinnabarinum* (Boletales). Zeitschrift für Naturforschung B, 62(1): 129–131.

Guo, H., Feng, T., Li, Z.H. and Liu, J.K. (2013). Ten new aurovertins from cultures of the Basidiomycete *Albatrellus confluens*. Nat. Prod. Bioprospect., 3(1): 8–13.

Hanson, J.R. (2008a). Fungal metabolites derived from amino acids. pp. 32–46. *In*: Hanson, J.R. (ed.). The Chemistry of Fungi. The Royal Society of Chemistry, Thomas Graham House, Cambridge.

Hanson, J.R. (2008b). Pigments and odours of fungi. pp. 127–142. *In*: Hanson, J.R. (ed.). The Chemistry of Fungi. The Royal Society of Chemistry, Thomas Graham House, Cambridge.

Harki, E., Talou, T. and Dargent, R. (1997). Purification, characterisation and analysis of melanin extracted from *Tuber melanosporum* Vitt. Food Chem., 58(1-2): 69–73.

Harmon, A.D., Weisgraber, K.H. and Weiss, U. (1980). Preformed azulene pigments of *Lactarius indigo* (Schw.) Fries (Russulaceae, Basidiomycetes). Experientia, 36(1): 54–56.

Hashimoto, T., Tahara, S., Takaoka, S., Tori, M. and Asakawa, Y. (1994). Structures of daldinin A-C, three novel azaphilone derivatives from ascomycetous fungus *Daldinia concentrica*. Chem. Pharmaceut. Bull., 42(11): 2397–2399.

Hawksworth, D.L. (2019). The macrofungal resource. pp. 1–9. *In*: Sridhar, K.R. and Deshmukh, S.K. (eds.). Advances in Macrofungi: Diversity, Ecology and Biotechnology. CRC Press, Florida, USA.

Haxo, F. (1950). Carotenoids of the mushroom *Cantharellus cinnabarinus*. Bot. Gaz., 112(2): 228–232.

Heo, Y.M., Kim, K., Kwon, S.L., Na, J., Lee, H. et al. (2018). Investigation of filamentous fungi producing safe, functional water-soluble pigments. Mycobiology, 46(3): 269–277.

Hernandez, V.A., Galleguillos, F. and Robinson, S. (2016). Fungal pigments from spalting fungi attenuating blue stain in *Pinus* spp. Int. J. Biodeter. Biodegr., 107: 154–157.

Hinsch, E.M. (2015). A comparative analysis of extracted fungal pigments and commercially available dyes for colourising textiles, M.Sc. Thesis, Oregon State University, Oregon, p. 69.

Hinsch, E.M., Chen, H.L., Weber, G. and Robinson, S.C. (2015). Colorfastness of extracted wood-staining fungal pigments on fabrics: A new potential for textile dyes. J. Tex. Apparel Technol. Manag., 9(3): 1–11.

Hopmann, C. and Steglich, W. (1996). Pigments of fungi, 65! Synthesis of Haematopodin—A pigment from the mushroom *Mycena haematopus* (Basidiomycetes). Liebigs Annalen, 1996(7): 1117–1120.

Hosoe, T., Nozawa, K., Kawahara, N., Fukushima, K., Nishimura, K. et al. (1999). Isolation of a new potent cytotoxic pigment along with indigotin from the pathogenic basidiomycetous fungus *Schizophyllum commune*. Mycopathologia, 146(1): 9–12.

Hou, W., Lian, B., Dong, H., Jiang, H. and Wu, X. (2012). Distinguishing ectomycorrhizal and saprophytic fungi using carbon and nitrogen isotopic compositions. Geosci. Front., 3(3): 351–356.

Hu, L., Gao, J.M. and Liu, J.K. (2001). Unusual poly (phenylacetyloxy)-substituted 1, 1′: 4′, 1 ″-terphenyl derivative from fruiting bodies of the Basidiomycete *Thelephora ganbajun*. Helvetica Chimica Acta, 84(11): 3342–3349.

Huang, Y.T., Onose, J.I., Abe, N. and Yoshikawa, K. (2009). *In vitro* inhibitory effects of pulvinic acid derivatives isolated from Chinese edible mushrooms, *Boletus calopus* and *Suillus bovinus*, on cytochrome P450 activity. Biosci. Biotechnol. Biochem., 73(4): 855–860.

Hunger, K. (ed.). (2003). Industrial Dyes—Chemistry, Properties, Applications. Wiley-VCH, Weinheim, p. 648.

Hyde, K.D., Dong, Y., Phookamsak, R., Jeewon, R., Bhat, D.J. et al. (2020). Fungal diversity notes. pp. 1151–1276. Taxonomic and phylogenetic contributions on genera and species of fungal taxa. Fungal Diversity, 100: 5–277. https://doi.org/10.1007/s13225-020-00439-5.

Hyde, K.D., Xu, J., Rapior, S., Jeewon, R., Lumyong, S. et al. (2019). The amazing potential of fungi: 50 ways we can exploit fungi industrially. Fungal Diversity, 97: 1–136.

Inoue, K., Yoshida, M., Takahashi, M., Fujimoto, H., Shibutani, M. et al. (2009). Carcinogenic potential of alizarin and rubiadin, component of madder colour, in a rat medium-term multi-organ bioassay. Cancer Science, 100(12): 2261–2267.

Isaka, M., Chinthanom, P., Supothina, S., Tobwor, P. and Hywel-Jones, N.L. (2010). Pyridone and tetramic acid alkaloids from the spider pathogenic fungus *Torrubiella* sp. BCC 2165. J. Nat. Prod., 73(12): 2057–2060.

Isaka, M., Palasarn, S., Tobwor, P., Boonruangprapa, T. and Tasanathai, K. (2012). Bioactive anthraquinone dimers from the leafhopper pathogenic fungus *Torrubiella* sp. BCC 28517. J. Antibiot., 65(11): 571–574.

Isaka, M., Haritakun, R., Intereya, K., Thanakitpipattana, D. and Hywel-Jones, N.L. (2014). Torrubiellone E, an antimalarial *N*-hydroxypyridone alkaloid from the spider pathogenic fungus *Torrubiella longissima* BCC 2022. Nat. Prod. Comm., 9(5): 627–628.

Jaeger, R.J. and Spiteller, P. (2010). Mycenaaurin A, an antibacterial polyene pigment from the fruiting bodies of *Mycena aurantiomarginata*. J. Nat. Prod., 73(8): 1350–1354.

Jagers, E., Hillen-Maske, E. and Steglich, W. (1987). Metabolites of *Boletopsis leucomelaena* (Basidiomycetes): Clarification of the chemical nature of leucomelone and protoleucomelone. Z. Naturforsch, 42b(10): 1349–1353.

Jalmi, P., Bodke, P., Wahidullah, S. and Raghukumar, S. (2012). The fungus *Gliocephalotrichum simplex* as a source of abundant, extracellular melanin for biotechnological applications. World J. Microbiol. Biotechnol., 28(2): 505–512.

Jin-Ming, G. (2006). New biologically active metabolites from Chinese higher fungi. Curr. Org. Chem., 10(8): 849–871.

Kahner, L., Dasenbrock, J., Spiteller, P., Steglich, W., Marumoto, R. and Spiteller, M. (1998). Polyene pigments from fruit-bodies of *Boletus laetissimus* and *B. rufo-aureus* (Basidiomycetes). Phytochemistry, 49(6): 1693–1697.

Kang, H.S., Jun, E.M., Park, S.H., Heo, S.J., Lee, T.S. et al. (2007). Cyathusals A, B, and C, antioxidants from the fermented mushroom *Cyathus stercoreus*. J. Nat. Prod., 70(6): 1043–1045.

Karuppan, P., Subramanian, C.S., Velusamy, K., Sadasivam, M., Mahalingam, P. et al. (2014). Prospective aspects of myco-chrome as promising future textiles. pp. 412–416. *In*: Proceedings of 8th International Conference on Mushroom Biology and Mushroom Products, New Delhi, 19–22 November 2014, vol. I and II, ICAR- Directorate of Mushroom Research, New Delhi.

Kawagishi, H., Tonomura, Y., Yoshida, H., Sakai, S. and Inoue, S. (2004). Orirubenones A, B, and C, novel hyaluronan-degradation inhibitors from the mushroom *Tricholoma orirubens*. Tetrahedron, 60(33): 7049–7052.

Kim, J.P., Yun, B.S., Shim, Y.K. and Yoo, I.D. (1999). Inoscavin A, a new free radical scavenger from the mushroom *Inonotus xeranticus*. Tetrahed. Lett., 40(36): 6643–6644.

Kim, J.Y., Kim, D.W., Hwang, B.S., Woo, E.E., Lee, Y.J. et al. (2016). Neuraminidase inhibitors from the fruiting body of *Phellinus igniarius*. Mycobiology, 44(2): 117–120.

Kim, J.Y., Woo, E.E., Ha, L.S., Ki, D.W., Lee, I.K. and Yun, B.S. (2019). Neuraminidase inhibitors from the fruiting body of *Glaziella splendens*. Mycobiology, 47(2): 256–260.

Kim, Y.H., Cho, S.M., Hyun, J.W., Ryoo, I.J., Choo, S.J. et al. (2008). A new antioxidant, clitocybin A, from the culture broth of *Clitocybe aurantiaca*. J. Antibiot., 61(9): 573–576.

Kim, Y.H., Ryoo, I.J., Choo, S.J., Xu, G.H., Lee, S. et al. (2009). Clitocybin D, a novel human neutrophil elastase inhibitor from the culture broth of *Clitocybe aurantiaca*. J. Microbiol. Biotechnol., 19(10): 1139–1141.

Kinge, T.R., Apalah, N.A., Nji, T.M., Acha, A.N. and Mih, A.M. (2017). Species richness and traditional knowledge of macrofungi (mushrooms) in the Awing forest reserve and communities, northwest region, Cameroon. J. Mycol., 2017: 2809239. 10.1155/2017/2809239.

Kirk, T.K., Lorenz, L.F. and Larsen, M.J. (1975). Partial characterisation of a phenolic pigment from sporocarps of *Phellinus igniarius*. Phytochemistry, 14(1): 281–284.

Kishi, T., Tahara, S., Taniguchi, N., Tsuda, M., Tanaka, C. and Takahashi, S. (1991). New perylenequinones from *Shiraia bambusicola*. Planta Medica, 57(4): 376–379.

Klostermeyer, D., Knops, L., Sindlinger, T., Polborn, K. and Steglich, W. (2000). Novel benzotropolone and 2H-furo [3, 2-b] benzopyran-2-one pigments from *Tricholoma aurantium* (Agaricales). Eur. J. Org. Chem., 2000(4): 603–609.

Knight, D.W. and Pattenden, G. (1976). Synthesis of the pulvinic acid pigments of lichen and fungi. J. Chem. Soc. Chem. Comm., (16): 660–661.

Koch, B. and Steglich, W. (2007). Meroterpenoid pigments from *Albatrellus flettii* (Basidiomycetes). Eur. J. Org. Chem., 2007(10): 1631–1635.

Koch, B., Kilpert, C. and Steglich, W. (2010). Cristatomentin, a green pigment of mixed biogenetic origin from *Albatrellus cristatus* (Basidiomycetes). Eur. J. Org. Chem., 2010(2): 359–362.

Kojima, K., Ohno, T., Inoue, M., Mizukami, H. and Nagatsu, A. (2008). Phellifuropyranone A: A new furopyranone compound isolated from fruit bodies of wild *Phellinus linteus*. Chem. Pharmaceut. Bull., 56(2): 173–175.

Kubo, M., Liu, Y., Ishida, M., Harada, K. and Fukuyama, Y. (2014). A new spiroindene pigment from the medicinal fungus *Phellinus ribis*. Chem. Pharmaceut. Bull., 62(1): 122–124.

Kuhnert, E., Heitkämper, S., Fournier, J., Surup, F. and Stadler, M. (2014). Hypoxyvermelhotins A-C, new pigments from *Hypoxylon lechatii* sp. nov. Fungal Biology, 118(2): 242–252.

Kuhnt, D., Anke, T., Besl, H., Bross, M., Herrmann, R. et al. (1990). Antibiotics from Basidiomycetes, XXXVII, New inhibitors of cholesterol biosynthesis from cultures of *Xerula melanotricha* Doerfelt. J. Antibiot., 43(11): 1413–1420.

Lagashetti, A.C., Dufossé, L., Singh, S.K. and Singh, P.N. (2019). Fungal pigments and their prospects in different industries. Microorganisms, 7(12): 604.

Lallawmsanga, C., Passari, A.K. and Singh, B.P. (2018). Exploration of macrofungi in sub-tropical semi-evergreen Indian forest ecosystems. pp. 1–14. *In*: Singh, B.P., Passari, A.K. and Lallawmsanga, C. (eds.). Biology of Macrofungi. Springer, Cham, Switzerland.

Lam, Y.T., Palfner, G., Lima, C., Porzel, A., Brandt, W. et al. (2019). Nor-guanacastepene pigments from the Chilean mushroom *Cortinarius pyromyxa*. Phytochemistry, 165: 112048.

Lang, G., Blunt, J.W., Cummings, N.J., Cole, A.L. and Munro, M.H. (2005). Paecilosetin, a new bioactive fungal metabolite from a New Zealand isolate of *Paecilomyces farinosus*. J. Nat. Prod., 68(5): 810–811.

Lang, G., Cole, A.L., Blunt, J.W. and Munro, M.H. (2006). An unusual oxalylatedtetramic acid from the New Zealand Basidiomycete *Chamonixia pachydermis*. J. Nat. Prod., 69(1): 151–153.

Lang, M., Spiteller, P., Hellwig, V. and Steglich, W. (2001). Stephanosporin, a 'traceless' precursor of 2-chloro-4-nitrophenol in the Gasteromycete *Stephanospora caroticolor*. Angew. Chem., 40(9): 1704–1705.

Lang, M., Muhlbauer, A., Graf, C., Beyer, J., Lang-Fugmann, S. et al. (2008). Studies on the structure and biosynthesis of tridentoquinone and related meroterpenoids from the mushroom *Suillus tridentinus* (Boletales). Eur. J. Org. Chem., 2008(5): 816–825.

Lang, M., Jaagers, E., Polborn, K. and Steglich, W. (2009). Structure and absolute configuration of the fungal ansabenzoquinone *Rhizopogone*. J. Nat. Prod., 72(2): 214–217.

Lee, I.K., Yun, B.S., Cho, S.M., Kim, W.G., Kim, J.P. et al. (1996). Betulinans A and B, two benzoquinone compounds from *Lenzites betulina*. J. Nat. Prod., 59(11): 1090–1092.

Lee, I.K. and Yun, B.S. (2006). Hispidinanalogs from the mushroom *Inonotus xeranticus* and their free radical scavenging activity. Bioorg. Med. Chem. Lett., 16(9): 2376–2379.

Lee, I.K., Seok, S.J., Kim, W.K. and Yun, B.S. (2006a). Hispidin derivatives from the mushroom *Inonotus xeranticus* and their antioxidant activity. J. Nat. Prod., 69(2): 299–301.

Lee, I.K., Jung, J.Y., Seok, S.J., Kim, W.G. and Yun, B.S. (2006b). Free radical scavengers from the medicinal mushroom *Inonotus xeranticus* and their proposed biogenesis. Bioorg. Med. Chem. Lett., 16(21): 5621–5624.

Lee, I.K., Kim, Y.S., Jang, Y.W., Jung, J.Y. and Yun, B.S. (2007a). New antioxidant polyphenols from the medicinal mushroom *Inonotus obliquus*. Bioorg. Med. Chem. Lett., 17(24): 6678–6681.

Lee, I.K., Kim, Y.S., Seok, S.J. and Yun, B.S. (2007b). Inoscavin E, a free radical scavenger from the fruiting bodies of *Inonotus xeranticus*. J. Antibiot., 60(12): 745–747.

Lee, I.K., Cho, S.M., Seok, S.J. and Yun, B.S. (2008). Chemical constituents of *Gymnopilus spectabilis* and their antioxidant activity. Mycobiology, 36(1): 55–59.

Lee, I.K., Seo, G.S., Jeon, N.B., Kang, H.W. and Yun, B.S. (2009a). Phellinins A1 and A2, new styrylpyrones from the culture broth of *Phellinus* sp. KACC93057P: I. Fermentation, taxonomy, isolation and biological properties. J. Antibiot., 62(11): 631–634.

Lee, I.K., Jung, J.Y., Kim, Y.H. and Yun, B.S. (2009b). Phellinins A1 and A2, new styrylpyrones from the culture broth of *Phellinus* sp. KACC93057P: II. Physicochemical properties and structure elucidation. J. Antibiot., 62(11): 635–637.

Lee, I.K., Jung, J.Y., Kim, Y.H. and Yun, B.S. (2010). Phellinins B and C, new styrylpyrones from the culture broth of *Phellinus* sp. J. Antibiot., 63(5): 263–266.

Lee, Y.S., Kang, Y.H., Jung, J.Y., Kang, I.J., Han, S.N. et al. (2008). Inhibitory constituents of aldose reductase in the fruiting body of *Phellinus linteus*. Biol. Pharmaceut. Bull., 31(4): 765–768.

Li, C. and Oberlies, N.H. (2005). The most widely recognised mushroom: Chemistry of the genus *Amanita*. Life Sciences, 78(5): 532–538.

Liu, J.K. (2006). Natural terphenyls: Developments since 1877. Chem. Rev., 106(6): 2209–2223.

Liu, L.Y., Li, Z.H., Ding, Z.H., Dong, Z.J., Li, G.T. et al. (2013). Meroterpenoid pigments from the Basidiomycete *Albatrellus ovinus*. J. Nat. Prod., 76(1): 79–84.

Lohmann, J.S., Wagner, S., von Nussbaum, M., Pulte, A., Steglich, W. and Spiteller, P. (2018). Mycenaflavin A, B, C, and D: pyrroloquinoline alkaloids from the fruiting bodies of the mushroom *Mycena haematopus*. Chemistry-A Eur. J., 24(34): 8609–8614.

Lopusiewicz, L. (2018). Isolation, characterisation, and biological activity of melanin from *Exidia nigricans*. World Sci. News, 91: 111–129.

Lu, R.L., Luo, F.F., Hu, F.L., Huang, B., Li, C.R. and Bao, G.H. (2013). Identification and production of a novel natural pigment, cordycepoid A, from *Cordyceps bifusispora*. Appl. Microbiol. Biotechnol., 97(14): 6241–6249.

Ma, G., Khan, S.I., Jacob, M.R., Tekwani, B.L., Li, Z. et al. (2004). Antimicrobial and antileishmanial activities of hypocrellins A and B. Antimicrob. Ag. Chemother., 48(11): 4450–4452.

Margalith, P.Z. (1992). Pigment Microbiology. Chapman and Hall, London, p. 156.

Mikolajczyk, L. and Antkowiak, W.Z. (2009). Structure studies of the metabolites of *Paxillus involutus*. Heterocycles, 79(1): 423–426.

Miles, P.G., Lund, H. and Raper, J.R. (1956). The identification of indigo as a pigment produced by a mutant culture of *Schizophyllum commune*. Arch. Biochem. Biophys., 62(1): 1–5.

Mo, S., Yang, Y.C., He, W. and Shi, J.G. (2003a). Two pyrone derivatives from fungus *Phellinus igniarius*. Chin. Chem. Lett., 14(7): 704–706.

Mo, S., He, W., Yang, Y.C. and Shi, J.G. (2003b). Two benzyl dihydroflavones from *Phellinus igniarius*. Chin. Chem. Lett., 14(8): 810–813.

Mo, S., Wang, S., Zhou, G., Yang, Y., Li, Y. et al. (2005). Phelligridins C-F: Cytotoxic pyrano [4, 3-c] [2] benzopyran-1, 6-dione, and Furo [3, 2-c] pyran-4-one derivatives from the fungus *Phellinus igniarius*. J. Nat. Prod., 67(5): 823–828.

Mueller, G.M., Schmit, J.P., Leacock, P.R., Buyck, B., Cifuentes, J. et al. 2007. Global diversity and distribution of macrofungi. Biodivers. Conser., 16(1): 37–48.

Muhlbauer, A., Beyer, J. and Steglich, W. (1998). The biosynthesis of the fungal meroterpenoids boviquinone-3 and-4 follows two different pathways. Tetrahed. Lett., 39(29): 5167–5170.

Mukherjee, G., Mishra, T. and Deshmukh, S.K. (2017). Fungal pigments: An overview. pp. 525–541. *In*: Satyanarayana, T., Deshmukh, S.K. and Johri, B.N. (eds.). Developments in Fungal Biology and Applied Mycology. Springer, Singapore.

Mulheirn, L.J., Beechey, R.B., Leworthy, D.P. and Osselton, M.D. (1974). Aurovertin B, a metabolite of *Calcarisporium arbuscula*. Journal of the Chemical Society, Chemical Communications, (21): 874–876.

Muso, H. (1976). The pigments of fly agaric, *Amanita muscaria*. Tetrahedron, 35(24): 2843–2853.

Nagatsu, A., Itoh, S., Tanaka, R., Kato, S., Haruna, M. et al. (2004). Identification of novel substituted fused aromatic compounds, meshimakobnol A and B, from natural *Phellinus linteus* fruit body. Tetrahed. Lett., 45(30): 5931–5933.

Nelsen, S.F. (2010). Bluing components and other pigments of boletes. Fungi, 3(4): 11–14.

Niskanen, T., Laine, S. and Liimatainen, K. (2012). *Cortinarius sanguineus* and equally red species in Europe with an emphasis on northern European material. Mycologia, 104(1): 242–253.

Olennikov, D.N., Gornostai, T.G. and Penzina, T.A. (2017). Rheadinin, a new bis (styrylpyrone) from mycelium of *Inonotus rheades*. Chem. Nat. Comp., 53(4): 629–631.

Parthasarathy, R. and Sathiyabama, M. (2015). Lovastatin-producing endophytic fungus isolated from a medicinal plant *Solanum xanthocarpum*. Nat. Prod. Res., 29(24): 2282–2286.

Peters, S. and Spiteller, P. (2007a). Mycenarubins A and B, red pyrroloquinoline alkaloids from the mushroom *Mycena rosea*. Eur. J. Org. Chem., 2007(10): 1571–1576.

Peters, S. and Spiteller, P. (2007b). Sanguinones A and B, blue pyrroloquinoline alkaloids from the fruiting bodies of the mushroom *Mycena sanguinolenta*. J. Nat. Prod., 70(8): 1274–1277.

Peters, S., Jaeger, R.J. and Spiteller, P. (2008). Red pyrroloquinoline alkaloids from the mushroom *Mycena haematopus*. Eur. J. Org. Chem., 2008(2): 319–323.

Piątek, M. (1999). Parasitic macrofungi (Basidiomycetes) on fruit shrubs and trees in the Tarnów town (S Poland). Acta Mycol., 34(2): 329–344.

Pili. (2020). https://www.pili.bio/ (accessed on June 25, 2020).

Popa, G., Cornea, C.P., Luta, G., Gherghina, E., Israel-Roming, F. et al. (2016). Antioxidant and antimicrobial properties of *Laetiporus sulphureus* (Bull.) Murrill. AgroLife Sci. J., 5(1): 168–173.

Prados-Rosales, R., Toriola, S., Nakouzi, A., Chatterjee, S., Stark, R. et al. (2015). Structural characterization of melanin pigments from commercial preparations of the edible mushroom *Auricularia auricula*. J. Agric. Food Chem., 63(33): 7326–7332.

PubChem database. (2020). *Anthraquinone*, National Centre for Biotechnology Information, US National Library of Medicine, Bethesda, Maryland.https://pubchem.ncbi.nlm.nih.gov/compound/ Anthraquinone (accessed on June 24, 2020).

Pulte, A., Wagner, S., Kogler, H. and Spiteller, P. (2016). Pelianthinarubins A and B, red pyrroloquinoline alkaloids from the fruiting bodies of the mushroom *Mycena pelianthina*. J. Nat. Prod., 79(4): 873–878.

Qin, X.D. and Liu, J.K. (2004). Natural aromatic steroids as potential molecular fossils from the fruiting bodies of the Ascomycete *Daldinia concentrica*. J. Nat. Prod., 67(12): 2133–2135.

Qin, X.D., Dong, Z.J., Liu, J.K., Yang, L.M., Wang, R.R. et al. (2006). Concentricolide, an anti-HIV agent from the Ascomycete *Daldinia concentrica*. Helv. Chim. Acta, 89(1): 127–133.

Quang, D.N., Hashimoto, T., Nukada, M., Yamamoto, I., Hitaka, Y. et al. (2003a). Thelephantins A, B, and C: three benzoyl p-terphenyl derivatives from the inedible mushroom *Thelephora aurantiotincta*. Phytochemistry, 62(1): 109–113.

Quang, D.N., Hashimoto, T., Hitaka, Y., Tanaka, M., Nukada, M. et al. (2003b). Thelephantins D–H: five p-terphenyl derivatives from the inedible mushroom *Thelephora aurantiotincta*. Phytochemistry, 63(8): 919–924.

Quang, D.N., Hashimoto, T., Tanaka, M., Stadler, M. and Asakawa, Y. (2004a). Cyclic azaphilonesdaldinins E and F from the Ascomycete fungus *Hypoxylon fuscum* (Xylariaceae). Phytochemistry, 65(4): 469-473.

Quang, D.N., Hashimoto, T., Stadler, M. and Asakawa, Y. (2004b). New azaphilones from the inedible mushroom *Hypoxylon rubiginosum*. J. Nat. Prod., 67(7): 1152–1155.

Quang, D.N., Hashimoto, T., Nomura, Y., Wollweber, H., Hellwig, V. et al. (2005a). Cohaerins A and B, azaphilones from the fungus *Hypoxylon cohaerens*, and comparison of HPLC-based metabolite profiles in *Hypoxylon* sect, *Annulata*. Phytochemistry, 66(7): 797–809.

Quang, D.N., Hashimoto, T., Stadler, M. and Asakawa, Y. (2005b). Dimeric azaphilones from the xylariaceous Ascomycete *Hypoxylon rutilum*, Tetrahedron, 61(35): 8451–8455.

Quang, D.N., Hashimoto, T. and Asakawa, Y. (2006a). Inedible mushrooms: A good source of biologically active substances. The Chem. Rec., 6(2): 79–99.

Quang, D.N., Hashimoto, T., Arakawa, Y., Kohchi, C., Nishizawa, T. et al. (2006b). Grifolin derivatives from *Albatrellus caeruleoporus*, new inhibitors of nitric oxide production in RAW 264.7 cells. Bioorg. Med. Chem., 14(1): 164–168.

Quang, D.N., Stadler, M., Fournier, J., Tomita, A. and Hashimoto, T. (2006c). Cohaerins C-F, four azaphilones from the xylariaceous fungus *Annulohypoxylon cohaerens*. Tetrahedron, 62(26): 6349–6354.

Quang, D.N., Harinantenaina, L., Nishizawa, T., Hashimoto, T., Kohchi, C. et al. (2006d). Inhibition of nitric oxide production in RAW 264.7 cells by azaphilones from xylariaceous fungi. Biol. Pharmaceut. Bull., 29(1): 34–37.

Radulovic, N., Quang, D.N., Hashimoto, T., Nukada, M. and Asakawa, Y. (2005). Terrestrins A-G: p-terphenyl derivatives from the inedible mushroom *Thelephora terrestris*. Phytochemistry, 66(9): 1052–1059.

Rai, M. (2009). Advances in Fungal Biotechnology. New Delhi, India: I. K. International Pvt Ltd., p. 545.

Ramesh, C., Vinithkumar, N.V., Kirubagaran, R., Venil, C.K. and Dufossé, L. (2019). Multifaceted applications of microbial pigments: Current knowledge, challenges and future directions for public health implications. Microorganisms, 7(7): 186.

Redhead, S.A. (1997). Macrofungi of British Columbia: Requirements for Inventory (vol. 28). British Columbia, Ministry of Forests Research Programme, p. 119.

Repke, D.B., Leslie, D.T. and Kish, N.G. (1978). GLC-Mass spectral analysis of fungal metabolites. J. Pharmaceut. Sci., 67(4): 485–487.

Ribera, J., Panzarasa, G., Stobbe, A., Osypova, A., Rupper, P. et al. (2019). Scalable biosynthesis of melanin by the Basidiomycete *Armillaria cepistipes*. J. Agric. Food Chem., 67(1): 132–139.

Robinson, S.C., Tudor, D., Snider, H. and Cooper, P.A. (2012a). Stimulating growth and xylindein production of *Chlorociboria aeruginascens* in agar-based systems. AMB Express, 2(1): 15.

Robinson, S.C., Tudor, D. and Cooper, P.A. (2012b). Utilising pigment-producing fungi to add commercial value to American beech (*Fagus grandifolia*). Appl. Microbiol. Biotechnol., 93(3): 1041–1048.

Robinson, S.C., Tudor, D., Zhang, W.R., Ng, S. and Cooper, P.A. (2014). Ability of three yellow pigment-producing fungi to colour wood under controlled conditions. Int. Wood Prod. J., 5(2): 103–107.

Royal Botanic Gardens Victoria. Systematics and Taxonomy of Australian *Cortinarius*. https://www.rbg.vic.gov.au/science/projects/mycology/systematics-and-taxonomy-of-australian-cortinarius (accessed on June 24, 2020).

Rutanen, J. (2014). Luonnontuotteetmonipuolistuvissaarvoverkoissa – luonnontuotealantoimintaohjelma (2020). [Natural products in versatile value nets – plan of action 2020 for natural products' sector], Mikkeli: University of Helsinki Ruralia institute. http://hdl.handle.net/10138/229380 (accessed on June 26, 2020).

Räisänen, R., Nousiainen, P. and Hynninen, P.H. (2001a). Emodin and dermocybin natural anthraquinones as high-temperature disperse dyes for polyester and polyamide. Tex. Res. J., 71(10): 922–927.

Räisänen, R. (2002). Anthraquinones from the Fungus *Dermocybe sanguinea* as Textile Dyes, Ph.D. thesis. University of Helsinki, Finland, p. 107.

Räisänen, R., Nousiainen, P. and Hynninen, P.H. (2002). Dermorubin and 5-chlorodermorubin natural anthraquinone carboxylic acids as dyes for wool. Tex. Res. J., 72(11): 973–976.

Räisänen, R. (2019). Fungal colorants in applications—Focus on *Cortinarius* species. Colour Technol., 135(1): 22–31.

Räisänen, R., Primetta, A., Nikunen, S., Honkalampi, U., Nygren, H. et al. (2020). Examining safety of biocolourants from fungal and plant sources—Examples from *Cortinarius* and *Tapinella*, *Salix* and *Tanacetum* spp. and dyed woollen fabrics. Antibiotics, 9(5): 266.

Sakai, S., Tomomura, Y., Yoshida, H., Inoue, S. and Kawagishi, H. (2005). Orirubenones D to G, novel phenones from the *Tricholoma orirubens* mushroom. Biosci. Biotechnol. Biochem., 69(8): 1630–1632.

Schwenk, D., Nett, M., Dahse, H.M., Horn, U., Blanchette, R.A. and Hoffmeister, D. (2014). Injury-induced biosynthesis of methyl-branched polyene pigments in a white-rotting Basidiomycete. J. Nat. Prod., 77(12): 2658–2663.

Selvakumar, P., Rajasekar, S., Periasamy, K. and Raaman, N. (2008). Isolation and characterization of melanin pigment from *Pleurotus cystidiosus* (telomorph of *Antromycopsis macrocarpa*). World J. Microbiol. Biotechnol., 24(10): 2125–2131.

Sen, T., Barrow, C.J. and Deshmukh, S.K. (2019). Microbial pigments in the food industry—Challenges and the way forward. Front. Nutr., 6: 7.10.3389/fnut.2019.00007.

Shahid, M. and Shahid-ul-Islam Mohammad, F. (2013). Recent advancements in natural dye applications: A review. J. Clean. Prod., 53: 310–331.

Shang, Y., Feng, P. and Wang, C. (2015). Fungi that infect insects: Altering host behaviour and beyond. PLoS Pathog., 11(8): e1005037.org/10.1371/journal.ppat.1005037.

Shirasaka, N., Nagashima, C., Yamaguchi, Y. and Terashita, T. (2012). Evaluation of yellow pigment from *Pleurotus cornucopiae* var. *Citrino pileatus*. Jap. Soc. Mush. Sci. Biotechnol., 19(4): 175–180.

Sinclair, W.A. and Lyon, H.H. (2005). Diseases of Trees and Shrubs (2nd ed.). Comstock Publishing Associates, Ithaca, New York, p. 660.

Singh, R. (2017). A review on different benefits of mushroom. IOSR J. Pharm. Biol. Sci., 12(1): 107–111.

Singh, S.M., Singh, P.N., Singh, S.K. and Sharma, P.K. (2014). Pigment, fatty acid and extracellular enzyme analysis of a fungal strain *Thelebolus microsporus* from Larsemann Hills, Antarctica. The Polar Record, 50(1): 31. 10.1017/S0032247412000563.

Sir, E.B., Kuhnert, E., Surup, F., Hyde, K.D. and Stadler, M. (2015). Discovery of new mitorubrin derivatives from *Hypoxylon fulvo-sulphureum* sp. nov. (Ascomycota, Xylariales). Mycol. Prog., 14(5): 28. 10.1007/s11557-015-1043-1.

Smania, A., Marques, C.J.S., Smania, E.F.A., Zanetti, C.R., Carobrez, S.G. et al. (2003). Toxicity and antiviral activity of cinnabarin obtained from *Pycnoporus sanguineus* (Fr.). Murr. Phytother. Res., 17(9): 1069–1072.

Sontag, B., Rüth, M., Spiteller, P., Arnold, N., Steglich, W. et al. (2006). Chromogenic meroterpenoids from the mushrooms *Russula ochroleuca* and *R. viscida*. Eur. J. Org. Chem., 2006(4): 1023–1033.

Spiteller, P. and Steglich, W. (2002). Blennione, green aminobenzoquinone derivative from *Lactarius blennius*. J. Nat. Prod., 65(5): 725–727.

Spiteller, P., Arnold, N., Spiteller, M. and Steglich, W. (2003). Lilacinone, a red aminobenzoquinone pigment from *Lactarius lilacinus*. J. Nat. Prod., 66(10): 1402–1403.

Stadler, M., Fournier, J., Quang, D.N. and Akulov, A.Y. (2007). Metabolomic studies on the chemical ecology of the Xylariaceae (Ascomycota). Nat. Prod. Comm., 2(3): 287–304.

Stadler, M., Fournier, J., Granmo, A. and Beltran-Tejera, E. (2008). The red hypoxylons of the temperate and subtropical Northern Hemisphere. North Am. Fungi, 3: 73–125.

Stamets, P. and Zwickey, H. (2014). Medicinal mushrooms: Ancient remedies meet modern science. Integra. Med., 13(1): 46–47.

Steffan, B. and Steglich, W. (1984). Pigments from the cap cuticle of the Bay boletus (*Xerocomus badius*). Angew. Chem., 23(6): 445–447.

Steglich, W., Furtner, W. and Prox, A. (1969). Fungus pigments, 3, Xerocomic acid and gomphidic acid, 2 chemotaxonomically interesting pulvinic acid derivatives from *Gomphidius glutinosus* (schff) fr. Zeitschrift fur Naturforschung C, 24(7): 941–942.

Steglich, W. and Töpper-Petersen, E. (1972). Fungal pigments XII, Phlegmacin and anhydrophlegmacin, new pigments from *Cortinarius odorifer* (Agaricales). Zeitschrift fur Naturforschung, 27(10): 1286–1287.

Steglich, W., Klaar, M. and Furtner, W. (1974). (+)-mitorubrin derivatives from *Hypoxylon fragiforme*. Phytochemistry, 13: 2874–2875.

Steglich, W., Thilmann, A., Besl, H. and Bresinsky, A. (1977). Pigments of fungi, 29, 2, 5-diarylcyclopentane-1, 3-diones from *Chamonixia caespitosa* (Basidiomycetes). Zeitschrift für Naturforschung C, 32(1-2): 46–48.

Steglich, W., Fugmann, B. and Lang-Fugmann, S. (2000). ROMPP Encyclopedia Natural Products. Georg Thieme Verlag, New York, p. 730.

Stodulkova, E., Cisarova, I., Kolarik, M., Chudickova, M., Novak, P. et al. (2015). Biologically active metabolites produced by the Basidiomycete *Quambalaria cyanescens*. PLoS ONE, 10(2): e0118913.10.1371/journal.pone.0118913.

Subramanian, C.S., Velusamy, K. and Karuppan, P. (2014). Orange dye from *Pycnoporus* sp. for textile industries. pp. 456–460. *In*: Proceedings of 8th International Conference on Mushroom Biology and Mushroom Products, New Delhi, 19–22 November 2014, vol. I and II, ICAR—Directorate of Mushroom Research, New Delhi.

Sullivan, G. and Henry, E.D. (1971). Occurrence and distribution of phenoxazinone pigments in the genus *Pycnoporus*. J. Pharmaceut. Sci., 60(7): 1097–1098.

Surup, F., Narmani, A., Wendt, L., Pfutze, S., Kretz, R. et al. (2018). Identification of fungal fossils and novel azaphilone pigments in ancient carbonised specimens of *Hypoxylon fragiforme* from forest soils of Chatillon-sur-Seine (Burgundy). Fungal Diversity, 92(1): 345–356.

Tala, M.F., Qin, J., Ndongo, J.T. and Laatsch, H. (2017). New azulene-type sesquiterpenoids from the fruiting bodies of *Lactarius deliciosus*. Nat. Prod. Bioprospect., 7(3): 269–273.

Tang, X., Mi, F., Zhang, Y., He, X., Cao, Y. et al. (2015). Diversity, population genetics, and evolution of macrofungi associated with animals. Mycology, 6(2): 94–109.

Tauber, J.P., Schroeckh, V., Shelest, E., Brakhage, A.A. and Hoffmeister, D. (2016). Bacteria induce pigment formation in the Basidiomycete *Serpula lacrymans*. Environ. Microbiol., 18(12): 5218–5227.

Teichert, A., Schmidt, J., Porzel, A., Arnold, N. and Wessjohann, L. (2007). Brunneins A-C, β-carboline alkaloids from *Cortinarius brunneus*. J. Nat. Prod., 70(9): 1529–1531.

Teichert, A., Schmidt, J., Porzel, A., Arnold, N. and Wessjohann, L. (2008a). (Iso)-quinoline alkaloids from fungal fruiting bodies of *Cortinarius subtortus*. J. Nat. Prod., 71(6): 1092–1094.

Teichert, A., Schmidt, J., Porzel, A., Arnold, N. and Wessjohann, L. (2008b). N-glucosyl-1H-indole derivatives from *Cortinarius brunneus* (Basidiomycetes). Chem. Biodiver., 5(4): 664–669.

Teichert, A., Lübken, T., Schmidt, J., Kuhnt, C., Huth, M. et al. (2008c). Determination of β-carboline alkaloids in fruiting bodies of *Hygrophorus* spp. by liquid chromatography/electrospray ionisation tandem mass spectrometry. Phytochem. Anal., 19(4): 335–341.

Thanh, N.T., Tuan, N.N., Kuo, P.C., Dung, D.M., Phuong, D.L. et al. (2017). Chemical constituents from the fruiting bodies of *Phellinus igniarius*. Nat. Prod. Res., 32(20): 2392–2397.

Tian, C.K., Yuan, R.Y., Wang, Y.X., Chen, L., Wu, Z. et al. (2019). Two new guaiane sesquiterpenes from the fruiting bodies of *Lactarius deliciosus*. J. Asian Nat. Prod. Res., 10.1080/10286020.2019.1695781.

Tsukamoto, S., Macabalang, A.D., Abe, T., Hirota, H. and Ohta, T. (2002). Thelephorin A: A new radical scavenger from the mushroom *Thelephora vialis*. Tetrahedron, 58(6): 1103–1105.

Tudor, D. (2013). Fungal Pigment Formation in Wood Substrate, Ph.D. thesis. University of Toronto, Toronto, ON, Canada, p. 189.

Tudor, D., Robinson, S.C. and Cooper, P.A. (2013). The influence of pH on pigment formation by lignicolous fungi. Int. Biodeter. Biodegr., 80: 22–28.

Tudor, D., Robinson, S.C., Krigstin, T.L. and Cooper, P.A. (2014). Microscopic investigation on fungal pigment formation and its morphology in wood substrates. The Open Mycol. J., 8(1): 174–186.

Tuli, H.S., Sandhu, S.S. and Sharma, A.K. (2014). Pharmacological and therapeutic potential of *Cordyceps* with special reference to Cordycepin. 3 Biotech., 4(1): 1–12.

Umbuzeiro, G., Albuquerque, A., Morales, D., Szymczyk, M. and Freeman, H.S. (2020). Strategy for the evaluation of mutagenicity of natural dyes: Application to the BioColour project. *In*: SETAC Sci. Con. 2020, SETAC Europe, Dublin, p. 337.

Vallimayil, J. and Eyini, M. (2013). Effect of metal salts on extracellular melanin and laccase production by *Pleurotus djamor* (fr.) Boedijn and *Pleurotus citrinopileatus* Singer. Int. J. Adv. Biol. Res., 3(1): 46–50.

Valverde, M.E., Hernández-Pérez, T. and Paredes-López, O. (2015). Edible mushrooms: Improving human health and promoting quality life. Int. J. Microbiol., 2015: 376387. 10.1155/2015/376387.

Van der Sar, S.A., Blunt, J.W., Cole, A.L., Din, L.B. and Munro, M.H. (2005). Dichlorinated pulvinic acid derivative from a Malaysian *Scleroderma* sp. J. Nat. Prod., 68(12): 1799–1801.

Vega Gutierrez, S.M. and Robinson, S.C. (2015). Initial studies on the diversity of spalting fungi in the southern amazon forest of Peru. pp. 59–60. *In*: The Challenge of Complexity; Società Italiana di Microbiologia Agraria-Alimentare e Ambientale, Florence, Italy.

Vega Gutierrez, S.M. and Robinson, S.C. (2017). Microscopic analysis of pigments extracted from spalting fungi. Journal of Fungi, 3(1): 15.10.3390/jof3010015.

Velisek, J. and Cejpek, K. (2011). Pigments of higher fungi—A review. Czech J. Food Sci., 29(2): 87–102.

Velmurugan, P., Lee, Y.H., Nanthakumar, K., Kamala-Kannan, S., Dufossé, L. et al. (2010). Water-soluble red pigments from *Isaria farinosa* and structural characterisation of the main colored component. J. Basic Microbiol., 50(6): 581–590.

Velygodska, A.K. and Fedotov, O.V. (2016). The production and analysis of carotenoid preparations from some strains of xylotrophic Basidiomycetes, Visnyk of Dnipropetrovsk University. Biology, Ecology, 24(2): 290–294.

Wang, F., Luo, D.Q. and Liu, J.K. (2005). Aurovertin E, a new polyene pyrone from the Basidiomycete *Albatrellus confluens*. J. Antibiot., 58(6): 412–415.

Wang, X.N., Tan, R.X., Wang, F., Steglich, W. and Liu, J.K. (2005). The first isolation of a phlegmacin type pigment from the Ascomycete *Xylaria euglossa*. Zeitschrift fur Naturforschung B, 60(3): 333–336.

Wang, Y., Wang, S.J., Mo, S.Y., Li, S., Yang, Y.C. and Shi, J.G. (2005). Phelligridimer A, a highly oxygenated and unsaturated 26-membered macrocyclic metabolite with antioxidant activity from the fungus *Phellinus igniarius*. Org. Lett., 7(21): 4733–4736.

Wang, Y., Shang, X.Y., Wang, S.J., Mo, S.Y., Li, S. et al. (2007b). Structures, biogenesis, and biological activities of pyrano [4, 3-c] isochromen-4-one derivatives from the fungus *Phellinus igniarius*. J. Nat. Prod., 70(2): 296–299.

Wangun, H.V.K. and Hertweck, C. (2007). Squarrosidine and pinillidine: 3, 3′-fused bis (styrylpyrones) from *Pholiota squarrosa* and *Phellinus pini*. Eur. J. Org. Chem., 2007(20): 3292–3295.

Weber, R.W.S., Mucci, A. and Davoli, P. (2004). Laetiporic acid, a new polyene pigment from the wood-rotting Basidiomycete *Laetiporus sulphureus* (Polyporales, fungi). Tetrahed. Lett., 45(5): 1075–1078.

Winner, M., Giménez, A., Schmidt, H., Sontag, B., Steffan, B. and Steglich, W. (2004). Unusual pulvinic acid dimers from the common fungi *Scleroderma citrinum* (common earthball) and *Chalciporus piperatus* (peppery bolete). Angew. Chem., 43(14): 1883–1886.

Xu, G.H., Kim, J.W., Ryoo, I.J., Choo, S.J., Kim, Y.H. et al. (2010). Lactariolines A and B: New guaiane sesquiterpenes with a modulatory effect on interferon-γ production from the fruiting bodies of *Lactarius hatsudake*. J. Antibiot., 63(6): 335–337.

Yang, X.L., Luo, D.Q., Dong, Z.J. and Liu, J.K. (2006). Two new pigments from the fruiting bodies of the Basidiomycete *Lactarius deliciosus*. Helv. Chim. Acta, 89(5): 988–990.

Yang, X.L., Qin, C., Wang, F., Dong, Z.J. and Liu, J.K. (2008). A new meroterpenoid pigment from the Basidiomycete *Albatrellus confluens*. Chem. Biodiver., 5(3): 484–489.

Yarlagadda, R. (2013). A systematic review on some medicinal mushrooms showing antioxidant and anticancer activities. Medical Data, 5(3): 253–260.

Ye, L., Li, H., Mortimer, P.E., Xu, J., Gui, H. et al. (2019). Substrate preference determines macrofungal biogeography in the greater Mekong sub-region. Forests, 10(10): 824.

Yin, X., Feng, T., Li, Z.H., Dong, Z.J., Li, Y. and Liu, J.K. (2013). Highly oxygenated meroterpenoids from fruiting bodies of the mushroom *Tricholoma terreum*. J. Nat. Prod., 76(7): 1365–1368.

Zalas, M., Gierczyk, B., Bogacki, H. and Schroeder, G. (2015). The *Cortinarius* fungi dyes as sensitisers in dye-sensitised solar cells. Int. J. Photoener., 2015: 653740.10.1155/2015/653740.

Zan, L.F., Bao, H.Y., Bau, T. and Li, Y. (2015). A new antioxidant pyrano [4, 3-c][2] benzopyran-1, 6-dione derivative from the medicinal mushroom *Fomitiporia ellipsoidea*. Nat. Prod. Comm., 10(2): 315–316.

Zhang, A.L., Qin, J.C., Bai, M.S., Gao, J.M., Zhang, Y.M. et al. (2009). Rufoolivacin B, a novel polyketide pigment from the fruiting bodies of the fungus *Cortinarius rufo-olivaceus* (Basidiomycetes). Chin. Chem. Lett., 20(11): 1324–1326.

Zhang, L., Wang, F., Dong, Z.J., Steglich, W. and Liu, J.K. (2006). A new butenolide-type fungal pigment from the mushroom *Pulveroboletus ravenelii*. Heterocycles, 68(7): 1455–1458.

Zhao, H., Wu, R., Ma, L.F., Wo, L.K., Hu, Y.Y. et al. (2016). Aurovertin-type polyketides from *Calcarisporium arbuscula* with potent cytotoxic activities against triple-negative breast cancer. Helv. Chim. Acta, 99(7): 543–546.

Zhou, Z.Y., Liu, R., Jiang, M.Y., Zhang, L., Niu, Y. et al. (2009). Two new cleistanthane diterpenes and a new isocoumarine from cultures of the Basidiomycete *Albatrellus confluens*. Chem. Pharmaceut. Bull., 57(9): 975–978.

Zhou, Z.Y. and Liu, J.K. (2010). Pigments of fungi (Macromycetes). Nat. Prod. Rep., 27(11): 1531–1570.

Zou, Y. and Ma, K. (2018). Screening of *Auricularia auricula* strains for strong production ability of melanin pigments. Food Sci. Technol., 38(1): 41–44.

15

Insights on the Structural and Functional Attributes of Melanin from Termitophilic Agaricales[Ψ]

Rosy Agnes De Souza* and Nandkumar Mukund Kamat*

1. INTRODUCTION

Melanins (*Melanos* = Greek word for black) are most stable negatively-charged amorphous compounds and resistant biochemical moieties, insoluble in water having high molecular weight polymers of phenolic compounds with strong antioxidant properties, which enhance the survival and competitive ability of the organism in certain environmental conditions (Riley, 1997; Casadevall et al., 2000). Natural melanin shows many biochemical activities in terms of environmental survival (Eisenman et al., 2020), such as photoprotection, antianemia, antioxidation, antiradiation (Casadevall et al., 2017), antibiofilm (Bin et al., 2012) and so on, which suggests that natural melanin has the potential to be used in food products and medicines (Fig. 1). Melanins are also found to have importance in several non-biological contexts due to their insulating and semiconducting properties and applications as ion-exchange resins and redox polymers (Adhyaru et al., 2003).

Melanins are structurally very diverse and carry three types of polymers, such as aseumelanin $(DOPA)_n$, pheomelanin $(Cysteinyl\ DOPA)_n$ and allomelanin $(DHN)_n$. Melanins are often complexed with protein and less often with carbohydrates

Mycological Laboratory, Department of Botany, Goa University, Taleigao Plateau, Goa-403206, India.
* Corresponding authors: rosyagnesdsouza@gmail.com; nkamat@unigoa.ac.in
[Ψ] This chapter is dedicated to the research on structural and functional aspects of *Termitomyces melanin*.

Fig. 1. Possible future application *Termitomyces* melanin (modified from Cordero and Casadevall, 2017; Agustinho and Nosanchuk, 2017).

(Cheng et al., 2004). Production of microbial pigments is one of the emerging fields of research due to the growing interest of industries for safer, easily degradable and ecofriendly products (Pombeiro-Sponchiado et al., 2017). In fungi, melanins are produced either through the l-3–4 dihydroxyphenylalanine (L-DOPA) melanin pathway, mostly in Basidiomycetes or through the 1,8-dihydroxynaphthalene (DHN) pathway, usually found in Ascomycetes (Pal et al., 2013).

2. Melanins in Macrofungi

Fungal melanins have been identified from many fungi present during their developmental stages, such as sclerotia, spores and hyphal pigmentation due to wounding or as result of light exposure and extreme environmental conditions. In most cases, spores, fruit bodies or cultured mycelium and some submerged cultured mycelium have been used for isolation of macrofungal melanins and are characterised by applying various techniques (Table 1). Some macrofungal species reported for melanins are *Agaricus bisporus* (Zaidi et al., 2014), *Auricularia auricula* (Sun et al., 2016a, b), *Chroogomphus rutilus* (Hu et al., 2015), *Cenococcum geophilum* (Fernandez and Koide, 2013), *Hypoxylon archer* (Wu et al., 2008), *Inonotus obliquus* (Babitskaya et al., 2000), *Lachnum* (Ye et al., 2014), *Ophiocordyceps sinensis* (Dong and Yao, 2012), *Pleurotus cystidiosus* (Selvakumar et al., 2008), *Schizophyllum commune* (Arun et al., 2015) and *Tuber melanosporum* (Harki et al., 1997). Action of tyrosinase leading to formation of DOPA from tyrosine is reported in *A. bisporus*, *T. melanosporum*, whereas there is involvement of laccase in *Cryptococcus*

Table 1. Type of macrofungal melanin and characterisation techniques.

Macrofungal species	Melanin type	Localisation	Techniques	References
Agaricus bisporus	GHB, PAP,	Fruiting bodies	cDNA based real time -qPCR	Weijn et al., 2013
Auricularia auricula	Pheomelanin, DHN, eumelanin	Fruiting bodies, extracellular medium	Elemental analysis, UV-VIS, FTIR, Liquid NMR, SS-NMR, HPLC, SEM, TEM, EPR	Zou et al., 2015; Prados-Rosales et al., 2015; Sun et al., 2016 a, b
Exidia nigricans	DOPA	Fruit bodies	UV, IR, Raman	Lopusiewicz, 2018a
Lachnum sp.	Pheomelanin	Mycelial	UV-VIS, FTIR, Elemental analysis, ESI-MS, NMR, pyrolysis GC/MS	Zong et al., 2017
Scleroderma citrinum	DOPA	Gleba of Fruiting bodies	UV, IR, Raman	Lopusiewicz, 2018b
Tuber melanosporum	DOPA	Fruit body	Elemental analysis, FTIR	Harki et al., 1997
Pleurotus cystidiosus	DOPA	Mycelial	UV-VIS, FTIR, EPR	Selvakumar et al., 2008
Inonotus hispidus	DOPA	solid state fermentation	SEM, DLS, Elemental UV–visible, FTIR, EPR	Hou et al., 2019
Inonotus obliquus		Mycelial	UV-VIS, FTIR	Babitskaya et al., 2000

neoformans, Lentinula edodes (Belozerskaya et al., 2017). Thus, fungi can be used to produce a high yield of substance in inexpensive culture medium and by making the bioprocess economically viable on the industrial scale.

2.1 Melanin in Termitomyces Fruit Bodies (Teleomorphs)

Termitomyces species have high nutraceutical value and provide a good source of protein, fibre, fatty acids, essentials and non-essential amino acids and minerals with low lipid content (Botha and Eicker, 1992; Aletor, 1995; Kansci et al., 2003; Adejumo and Awosanya, 2005; Tsai et al., 2006; Baraza et al., 2007; Nabubuya et al., 2010; Earanna et al., 2013; Nakalembe and Kabasa, 2013; Devi et al., 2014). *Termitomyces* are a rich source of useful cerebrosides—termitomycesphins A-E with potential in the field of medicine, specially to cure human neuronal disease, such as Parkinson's, Alzheimer's and Prion disease (Qi et al., 2000, 2001; Qu et al., 2012). *Termitomyces* saponins and polysaccharides exhibit neuritogenic, analgesic and anti-inflammatory activities (Lu et al., 2008; Mitra et al., 2014). A very interesting feature of these species is a centrally dark fruit body. In natural habitat of *Termitomyces* the melanogenesis could be associated only with the teleomorphic epigeal stage. Many species sport a distinctly dark melanised perforatorium/umbo at the epigeal fruiting stage. The presence of black to grey, brownish pigmentation in fruit bodies of various *Termitomyces* species is reported in numerous taxonomic descriptions (Table 2) without establishing chemical identity of this pigmentation (Otieno, 1968;

Table 2. Pigmentation and main morphological characters of melanin in *Termitomyces* spp.

Termitomyces taxon	Pigmentation and main morphological characters
T. albiceps	Pileus reddish-brown, perforatorium conical, stipe cylindrical, terminating as a sclerotic disk, pseudorrhiza with black or brown surface
T. albuminosus	Pileus white, surface covered with wrinkles, sub-conical to expanded, obtusely umbonate, whitish with greyish-brown scales, perforatorium black, stipe white with greyish tint, smooth, fibrous, annulus present, pseudorrhiza dark brown
T. badius	Pileus buffy brown, striated texture from apex to the edge, with sharply pointed umbo, stipe cream, with slightly bulbous base
T. bulborhizus	Pileus pale-brown or darker, rugose with obtuse perforatorium, stipe with white to yellowish-brown floccules, swollen at base just below ground level, pseudorrhiza pale coloured, long, slender
T. clypeatus	Pileus silky greyish-brown to ash grey with pale margin, smooth to slightly fibrillose, globose, umbo low or absent, perforatorium long, spiniform, smooth, brown to blackish brown, annulus absent
T. cylindricus	Pileus greyish-brown, smooth, umbo bluntly pointed or poorly developed, stipe bulbous near the soil surface with yellowish-green pseudorrhiza, no annulus
T. entolomoides	Pileus rimose, blackish-grey with bluish tinges, with obtusely conical umbo, stipe grey swollen at ground level, annulus absent
T. eurrhizus	Pileus greyish-brown to fuliginous, viscid, rugulose, without velar remnants, pointed blackish-brown perforatorium, very dark-blackish pseudorrhiza, annulus inconstant, stipe bulbous near soil surface
T. floccosus	Pileus surface brownish orange at centre, then pale-brownish grey, gradually slightly paler towards margin, stipe off-white to pale-grey, pseudorrhiza surface whitish to greyish-white
T. fragilis	Pileus surface brownish-gray or greyish-brown at centre, mid of centre is darker: brownish-grey, grey, reddish-grey, greyish-white, light-grey elsewhere
T. gilvus	Pileus surface brownish orange- to dark-brown at the centre, brownish-yellow to orange white towards margin, perforatorium usually dark brown, stipe pale-brown on the bulb, paeudorrhiza white to pale brown
T. globulus	Pileus dull-orange or tawny-brown with pale margin, smooth to slightly fibrillose, globose, large 15–20 cm, perforatorium low, small and poorly defined, stipe usually without annulus, pseudorrhiza brownish
T. heimii	Pileus whitish, smooth with greyish-brown, broad umbo, annulus present
T. indicus	Pileus white to creamy, umbo brown papillate, stipe with very short stout rooting base, without pseudorrhiza
T. lanatus	Pileus covered by thick, grey, woolly veil hiding the umbo beneath, annulus present
T. le-testui	Pileus brownish, tomentose, large (not more than 30 cm), covered by small brown granular squamules at disc and with pronounced, cylindric-clavate perforatorium, dark brown, annulus persistent
T. macrocarpus	Pileus fuliginous with a conical perforatorium, stipe blackish-brown, swollen fusoid, but not sub-bulbous
T. mammiformis	Pileus white, pale-brown or whitish grey (3–7 cm), rimose with few velar remnants, perforatorium mammiform, scrobiculate, blackish-brown, annulus present

Table 2 contd... .

... Table 2 contd.

Termitomyces taxon	Pigmentation and main morphological characters
T. medius	Pileus buffy-brown and smooth without bluish colours, umbo sharply pronounced, stipe white, not swollen at ground level, annulus absent
T. radicatus	Pileus greyish-brown or pale-orange, smooth, with dark spiniform umbo, annulus present
T. reticulatus	Pileus creamy white with appressed or upturned brown scales and with unpronounced to underdeveloped umbo, annulus present
T. robustus	Pileus brownish or dark-brown, concentrically scrobiculate or radially ridged, glabrous, perforatorium differentiated sub mammilate, numerous rhizomorphs, stipe cream to pale ochraceous, almost cylindrical to narrowly fusoid, no annulus, pseudorrhiza partly blackish
T. sagittiformis	Pileus small to medium greyish sepia with dull margins, smooth, umbo obtuse, stipe creamy-white, fusiform extending to brown fusoid pseudorrhiza, no annulus
T. singidensis	Widespread in East Africa, pileus brownish, with greyish fluffy veil, perforatorium pointed, broad conical
T. spiniformis	Pileus grey, smooth, perforatorium long, strongly spiniform, warty, dark fuliginose, annulus present
T. striatus	Pileus brownish, large up to 12 cm, stipe annulate, annulus present
T. striatus var *striatus*	Pileus ochraceous brown to grey brown (up to 12 cm), not concentrically scrobiculate, at most radially striate
T. striatus f. annulatus	Pileus whitish on surface, (up to 12 cm), brownish around the disc showing tiny concolourous squamules, not concentrically scrobiculate, at most radially striate, perforatorium conical to subcylindric-mammiform, annulus conspicuous, persistent and downwards oriented
T. striatus f. ochraceus	Pileus cream to ochraceous brown, annulus absent
T. striatus f. griseus	Pileus grey to greyish-brown, annulus absent
T. striatus f. striatus	Pileus surface greyish-brown, (up to 12 cm), ochraceous brown to orange, darkish at centre, showing tiny concolourous squamules, not concentrically scrobiculate, at most radially striate, perforatorium conspicuous, annulus inconspicuous or as membranous squamules on stipe
T. striatus f. griseiumboides	Pileus beige-greyish to greyish-dark (2–5 cm), contrasting grey-blackish perforatorium, stipe with swelling at the base before tapering to form pseudorrhiza
T. striatus f. subumbonatus	Pileus whitish to pale-greyish, perforatorium inconspicuous, subumbonate with a greyish-brown to violaceous-brown umbo showing tiny squamules, stipe showing no swelling at the base
T. striatus f. brunneus	Pileus uniformly and constantly chocolate brown to dark-brown from origin, plicate to sulcate striate, perforatorium nipple-shaped (mammiform), stipe showing no swelling at the base
T. striatus f. pileatus	Pileus changing from greyish-orange with squamules when young, to cocoa-brown or leather-brown when mature, glabrous and usually subinfundibiliform at maturity, perforatorium obtuse to obtusely conical, stipe showing no swelling at the base
T. titanicus	Pileus ash-grey to very large (more than 30 cm), sub-squamulose without pronounced broad umbo, stipe with annulus, pseudorrhiza pale, not entirely blackish

Table 2 contd... .

... Table 2 contd.

Termitomyces taxon	Pigmentation and main morphological characters
T. tylerianus	Pileus smoky white or greyish white, smooth to striate, umbo brown, pointed to conical, stipe bulbous near to soil surface, with long, pale creamish pseudorrhiza, annulus absent
T. umkowaan	Pileus yellowish-brown or greyish-yellow, smooth, radially wrinkled, umbo spiniform to sharply pointed, stipe pale, smooth, lacking flocculus, bulbous near the soil surface, with long rust-brown pseudorrhiza, annulus absent
T. upsilocystidiatus	Pileus surface pale-brownish grey, gradually slightly paler toward margin, stipe off-white, pseudorrhiza white

Note: Taxa appear alphabetically; only species for which sufficient information is available are included; species not included due to insufficient information on pigmentation are *T. albidolaevis, T. albidus, T. albus, T. arghakhenchensis, T. biyi, T. brunneopileatus, T. cartilagineus, T. citriophyllus, T. congolensis, T. dominicalensis, T. epipolius, T. griseiumbo, T. infundibuliformis, T. intermedius, T. longiradicatus, T. magoyensis, T. mboudaeina, T. meipengianus, T. narobiensis, T. orientalis, T. palpensis, T. perforans, T. poliomphax, T. poonensis, T. quilonensis, T. subclypeatus, T. subhyalinus, T. subumkowaan.*

(*Source*: Heim, 1977; Piearce, 1987; Pegler and Vanhaecke, 1994; Wei et al., 2004; Mossebo et al., 2009; de Kesel, 2011; Tibuhwa, 2012; Karun and Sridhar, 2013; Aryal et al., 2016; Aryal and Budathoki, 2016; Ye et al., 2019; Sathiya Seelan et al., 2020; Tang et al., 2020).

Pegler and Rayner, 1969; Pegler, 1969; Natarajan, 1979; Van der Westhuizen and Eicker, 1990; Pegler and Vanhaecke, 1994; Abdullah and Rusea, 2009; Srivastava et al., 2011; de Kesel, 2011; Tibuhwa, 2012; Karun and Sridhar, 2013; Aryal et al., 2016). As described in Table 2, among the 61 confirmed species, we could identify about 30–39 distinct species showing black to grey, brown pigmentation in their fruit body parts. During our study, we could observe that the *Termitomyces* fruit bodies show black to grey pigmentation at the sight of injury, specially perforatorium/umbo and pseudorrhiza, but occasionally stipe is also pigmented due to injury. Recently de Souza et al. (2018) produced, characterised and confirmed this black pigment from *Termitomyces* fruit bodies or cultures as sulphur-rich pheomelanin.

2.2 *Termitomyces Melanin from Mycelial Cultures*

In some fungi, melanins are found in the cell wall or outermost layer intracellularly and are likely to be cross-linked to polysaccharides. *Termitomyces* melanin was found to be extracted from culture filtrate and biomass. The melanin found outside the fungal cells is known to be distinct from intracellular melanin and is secreted into the medium. This can be formed via two ways: (a) release of laccase enzyme by the fungi into the culture medium, which oxidises some medium components into melanins; (b) secretion of phenolic compounds into the culture medium which may slowly auto-oxidise to form melanins (Pal et al., 2013). Laccases are among the three important ligninolytic enzymes of Basidiomycetes fungi and they are involved in several physiological activities of the Basidiomycetes fungi, such as mycelial growth, melanin production (Arun et al., 2014). Babitskaya et al. (2000, 2002) reported that melanin production in *Inonotus obliquus* is correlated with phenol oxidase synthesis, whose precursor is copper-dependent.

Melanin granules in the fibrillar matrix are broken off into shake-culture media and are sometimes considered as extracellular melanins but Bell and Wheeler (1986) preferred to designate these as wall-bound melanins, because the fibrils are an extension of the wall. But when phenol oxidases are secreted into the external environment to oxidise the phenolic compound or when phenols are released into the external environment and later get oxidised to produce melanin may be considered to be extracellular melanins. Melanin deposition in fungal cell wall involves interactions with chitin molecules and destruction in chitin metabolism leading to release of pigment extracellularly (Cordero and Casadevall, 2017). Chrissian et al. (2020) have reported that the cell-wall balance between chitin and chitosan directly influences the retention of melanin pigments. Melanin granules are known to be localised in the cell wall and our microscopic observations indicated the same. They are found in cross-link to polysaccharide components, including β-linked glucan, chitin, mannan and galactofuran. Therefore, the extracted melanins are known to contain cell wall polysaccharides, chitin, proteins and lipids, which makes it difficult to analyse 'pure' melanin (De la Rosa et al., 2017). Our microscopic analysis of *Termitomyces* melanin showed that the melanin was cell-wall-bound. Based on literature, we could also see that some *Termitomyces* cultures, like *T. heimii, T. aurantiacus* growing on PDA were reported for black pigment formation after seven days and similar observations were made in our study (Siddiquee et al., 2012, 2015).

3. Physical Properties of Melanin

Melanin in the current study was isolated by acid precipitation for characterisation. The pigment obtained was insoluble in water, ethanol, chloroform and acetone, while soluble in 1 N NaOH. The solubility of melanin in alkali solutions is believed to come from the deprotonation of the various acid/base moieties found in melanin, such as carboxylic acid and phenolic groups (Bronze-Uhle et al., 2013). The pigment was bleached by NaOCl and H_2O_2 as also reported earlier, indicating resistance to solvents, bleaching when subjected to the action of various oxidants and the capacity to reduce directly ammonical solutions of silver nitrate. Its high insolubility, opacity of melanin, amorphous nature, heterogenous nature due to undefined chemical entities and harsh extraction procedures are major obstacles to the progress in melanin research. Isolation of melanin from natural sources is quite difficult as it is tightly bound to cellular components (d'Ischia et al., 2013). During the melanin purification process, it is known that acid hydrolysis treatment is alone insufficient to remove macromolecules; thus further purification can be achieved by chloroform treatments (Harki et al., 1997). The structure of melanin polymers is little understood and an accurate definition of melanin does not exist. However, the following criteria, such as black colour, insolubility in water and organic solvents, resistance to degradation by hot or cold concentrated acids, bleaching by oxidising agents, solubilisation by hot alkali solution, indicate that the isolated pigment match with those of the fungal melanin pigment isolated from other mushrooms or fungi (Arun et al., 2014). DOPA-melanin formation occurs by oxidation of tyrosine by tyrosinase (o-diphenol oxygen oxidoreductase) or laccase (p-diphenol oxygen oxidoreductase), thus leading to formation of red or brown pigments. Red pigment is formed due to polymerisation of cysteinly DOPA and brown pigments are formed when mixed polymers are formed with DOPA and cysteinly DOPA. These melanins are considered to be wall-bound melanins (Bell and wheeler, 1986).

4. Techniques Involved in Melanin Characterisation

Prota (1993) stated that due to line-broadening and the inordinate number of resonance signals in eumelanin and pheomelanin, interpretation of the spectra becomes rather challenging, although some promising results have been obtained by the solid phase ^{13}C and ^{14}N NMR spectroscopy. Indeed, he suggested, the only reliable approach at present available for classifying natural melanins rests on elemental analysis and degradation experiments to define the nature and origin of the main structural units of the pigment polymer. Analytical and degradative studies suggest that sulphur-containing pheomelanin (cys-dopa-derived) pigments are in fact structural variants of eumelanins arising by partial peroxidative cleavage of 5,6-dihydroxyindole units. Pheomelanin are formed due to involvement of sulphur-containing compounds, which liberate cysteins through the action of a glutamyl transpeptidase. Later DOPA quinines connect with cysteins to form 5-S-cysteinly-DOPA and 2-S-cysteinly-DOPA, which give benzothiazin intermediates that polymerise to produce pheomelanins (Belozerskaya et al., 2017). Synthesis of melanin requires the presence of L-tyrosine and L-cysteine amino acids (Bilinska, 2001).

4.1 *UV Spectra of Melanin from Fruit Bodies and Mycelial Cultures*

Melanin UV spectra of *Termitomyces* shows the characteristic absorption peaks in the UV region and melanins in general are known for absorption in UV region. This is mainly due to the presence of many complex conjugated structures of aromatic regions in the melanin molecule (Ou-Yang et al., 2004). There is a decrease in the absorption with increase in wavelength almost linearly in the case of melanins (Zhang et al., 2015) and same was observed for *Termitomyces* melanins. Among the biological pigments, only melanins are known to absorb all visible wavelengths and this characteristic feature is responsible for the dark color of the pigment (Bell and Wheeler, 1986; Goncalves et al., 2012). In UV spectra of *Termitomyces* melanin, the log of optical density when plotted against the wavelength (400–600 nm) produced a linear curve with negative slopes from –0.002 to –0.003. Such characteristic straight lines with negative slopes are usually employed to identify melanins. Some fungi, such as *Phyllosticta capitalensis* and *Auricularia auricula*, showed a slope ranging from –0.0015 to –0.0030 (Ellis and Griffiths, 1974; Suryanarayanan et al., 2004; Bin et al., 2012; Zhang et al., 2015). These observations were consistent with our spectral data obtained for all fruit body parts as well as pellet culture of *Termitomyces*. Li et al. (2015) reported that melanin profiles show spectral shifts from low to high due to increase in the molecular size of the melanin clusters. There were also no absorption peaks at 280 and 260 nm, which meant there were no impurities of nucleic acid, lipids and other proteins in melanin isolated from *Termitomyces*.

4.2 Scanning Electron Microscopy

Electron images of melanin were 30–150 nm in diameter and found to be deposited in outer layers of the cell wall (Alviano et al., 1991). Under acidic conditions, melanins are known to aggregate and at high pH, smaller oligomers with low degree of polymerisation are observed. Melanin behaves like a polyelectrolyte and this property depends on the ionisation state of the carboxylic, phenolic and amine groups present in it. The solubility of melanin is high in alkaline conditions and this favours the optimal synthesis of nanostructures (Apte et al., 2013). Sepia melanin showed under SEM that at lower magnification, it appeared aggregates, whereas when magnification was increased to 20–50 KX, it was seen that these aggregates were composed of small granules with sizes ranging from 1 μm to 200 nm (Mbonyiryivuze et al., 2015a). Beltran-Gracia et al. (2014) showed SEM results that indicated that the melanin pigment extracted from *Mycospharella fijiensis* mycelium showed spherical granules tightly aggregated with varying diameter size 100–300 nm and melanin extracted from culture media indicated amorphous material without definable structures. Melanin purification steps lead to dehydration, thus making the polymer more aggregated and resulting in loss of capacity for physiological interactions (Nicolaus, 1968; Prota, 1992). The aggregated structure of melanin is postulated to prevent reactive oxygen species formation because photoactive residues are less exposed (Beltran-Garcia et al., 2014). *Termitomyces* SEM micrographs of melanin were in agreement with the ones reported in literature. Cell-wall-bound melanin granules are known to have larger granules which help to decrease porosity and increase the absorptive properties to toxic molecules in polymer. Pure melanin indicated fractal dimensions (D) in the range of 1.195–1.733 as consistent with Eom et al. (2017). Cordero and Casadevall (2017) proposed a local-order-global-disorder model for melanin structure where they suggested that when indolic/phenolic monomers are polymerised into a series of planar arrangements of regularly interspaced stacked layers, they can cross-link into more heterogeneous and disordered macromolecule configuration. This involves a combination of π-stacking, hydrogen and ionic bonded nanostructures of unclear dimensions, which then aggregate to form disordered particles with spherical dimensions, known as melanin granules. Fungal cell-wall melanin consists of small granules and to keep them in place, chitin may play an important role in scaffolding (Eisenman and Casadevall, 2012).

4.3 Elemental Composition

Elemental analysis of *Termitomyces* melanin mainly indicated the composition percentage as C = 54.679 per cent, H = 3.544 per cent, N = 2.492 per cent, O = 26.924 per cent and S = 12.361 per cent as reported by de Souza et al. (2018). Sava et al. (2001) reported that the multiple acid hydrolysis reduces the percent nitrogen content to a constant level. Repeated acid hydrolysis helps to remove contaminated proteins. But, acid hydrolysis along with washing with chloroform reduces the aliphatic chains, hydrocarbons and proteins (Harki et al., 1997). The researchers also concluded that most European truffles are rich in the sulphur-containing amino acids, 1.2–5.9 g/100 g protein and these are also known to produce melanin (Sawaya et al., 1985; Wang and Marcone, 2011). As truffles are edible mycorrhizal hypogeous mushrooms living in symbiosis with host plant roots to accomplish their complex life cycle (Harley and Smith, 1983; Saltarelli et al., 2008), these results could be extended to symbiotic *Termitomyces* species. In a recent study of the biosynthetic pathways leading to the formation of melanin in mature *Tuber melanosporum*, it was shown that cysteine and tyrosine act as precursors (Harki et al., 1997, 2006). This could explain the high sulphur content of *Termitomyces* melanin like that of *T. melanosporum*. Sun et al. (2016a) also reported the presence of eumelanin along with pheomelanin in edible mushroom *Auricularia auricula* as the elemental composition showed the presence of sulphur content. According to Ye et al. (2014), about 14.83 per cent sulphur content was found by elemental analysis from *Lachnum* YM404 strain. Costa et al. (2015) reported C = 45.8 per cent, H = 4.3 per cent, N = 7.7 per cent, S = 16 per cent and O = 26.2 per cent. Sulphur content (8–12 per cent) indicates proximity to pheomelanin, which mainly consists of sulphur-containing benzothiazine and benzathiozol derivatives (Ito and Fujita, 1985; Belozerskaya et al., 2017). According to Bull (1970), in *Aspergillus nidulans*, the melanin pigment varied in elemental composition with response to the growth medium. Atomic C/H and O/C ratios are used to elucidate structural formulae of melanins. The C/H measures the degree of aromaticity of melanins, while the O/C ratio is an indicator of the carbohydrate and carboxylic acid contents (De la Rosa et al., 2017). Ito and Fujita (1985) reported the elemental composition of pheomelanin with C = 46.24 per cent, H = 4.46 per cent, N = 9.36 per cent, S = 9.78 per cent and O = 30.16 per cent. In *Termitomyces*, several sulphur containing amino acids (methionine and cysteine) have been reported from different species, such as *T. robustus, T. microcarpus, T. umkowaani, T. sagittaeformis* and *T. reticulatus* (Alofe, 1991; Botha and Eicker, 1992; Ijeh et al., 2016; Sun et al., 2017). Rahmad et al. (2014) also identified sulphite reductase enzyme from *T. heimii*, which plays a role in sulphur assimilation. The possible role of sulphur content in *Termitomyces* species has been described by de Souza and Kamat (2019).

4.4 *FTIR Spectral Characteristics of Melanin*

Infrared spectra of melanins from several biological sources show general similarity and there are detailed differences between the pigments from different species. Infrared absorption spectra contain important spectral characteristics for identification of melanin pigments. Previously such characteristic IR spectral bands were reported from several fungal species, such as *Auricularia auricula* (Zhang et al., 2015), *Phyllosticta capitalensis* (Suryanarayanan et al., 2004) and *Pleurotus cystidiosus* (Selvakumar et al., 2008). *Termitomyces* species showed absorption band at 2964 cm⁻¹ and 2891 cm⁻¹, indicating the presence of CH_3 and CH_2 aliphatic groups. The other bands, like 1724 cm⁻¹, 1585 cm⁻¹ and 1442 cm⁻¹ indicated C=O, C=C and C=N/N–H groups, whereas 1263 cm⁻¹ indicated phenolic C–O–H band. Besides, other characteristic bands indicated the presence of aromatic rings and sulphur at 800 cm⁻¹ and 678 cm⁻¹. After acid hydrolysis, spectra of DOPA and pure melanin showed reduced or absence of certain spectral absorptions. Strong band was observed at 2930–2920 cm⁻¹, medium band at 2860–2850 cm⁻¹ assigned to stretching vibration of aliphatic C-H and 1460–1450 cm⁻¹ for bending vibration of aliphatic

C-H (Harik et al., 1997). The most intense band in the region of 1710–1580 cm^{-1} was typical of melanins and resulted from the valence vibrations of C=O-groups (Azarko et al., 2001). The two bands at 1720 and 1503 cm^{-1}, were attributed to carboxylic acid groups and quinoneimine-containing adjacent sulphur groups, respectively in the cys-DOPA spectrum (Costa et al., 2015). The absorbance bands below 1650 cm^{-1} are due to the quinone intermediate and extensive conjugation generated during phenolic radical polymerisation. Absorbance at the vicinity of 1200 cm^{-1} is due to the polymerisation of a variety of ethers. The band at 800 cm^{-1} may be due to alkene as observed in native melanin (Pierce and Rast, 1995).

4.5 *Electron Paramagnetic Resonance Properties*

The EPR spectroscopy is widely used to confirm the presence of melanin, because it determines the presence of stable populations of free radicals trapped in the pigment. Melanin polymers show paramagnetic character and o-semiquinone free radicals with spin S = ½ and thus can be identified by EPR. Free radicals present in melanin absorb microwaves in the magnetic field and are responsible for EPR spectra as absorption is the base of EPR spectroscopy. The EPR spectra of eumelanin appear to be simple, whereas single line and EPR of pheomelanin appear to be complex shaped with unresolved hyperfine structures (Pilawa et al., 2017). *Termitomyces* melanin reported EPR peak at 2.00968 (de Souza et al., 2018). The EPR spectra obtained from biomass melanin of *Aspergillus flavus*, *A. niger*, *A. tubingensis* and *A. tamarii* showed an EPR signal in a single line with g factors of 2.00585, 2.00574, 2.00577 and 2.00614, respectively (Pal et al., 2013). The quinone residues present in melanin are associated with the five-member ring structure that can alternate between the fully-reduced form (phenol form) and the two-electron oxidation product (quinone form) via semi-quinone state. Such oxidation-reduction reactions involving phenolic compounds are known to use melanin as a precursor-induced pigment to mediate nano-particle synthesis in other biological systems and this has been exploited in nanobiotechnology (Apte et al., 2013). Melanin is also used as nano-carrier for pH responsive drug release, thus indicating potential use in the pharmaceutical and biomedical fields (Araujo et al., 2014). Blois et al. (1964) concluded that observed unpaired electrons are associated with trapped free radicals in the melanin polymer. Chodurek et al. (2012) reported o-semiquinone free radicals with the characteristic EPR spectra and g-factors 2.0045–2.0060 in DOPA-melanin and melanin isolated from A-375 cells. At X-band (9 GHz), the EPR spectra of pheomelanin shows broader signal with total width approximately at 3 mT and g = 2.005, whereas eumelanin represented a single asymmetric line 0.4–0.6 mT wide with a g factor 2.004 (d'Ischia et al., 2013). The EPR spectra were consistent with the cysteinyl DOPA melanin observed by d'Ischia et al. (2013). The presence of paramagnetic centres due to quinoid groups explains the ability of melanin pigments to deactivate free radicals and peroxides and absorb heavy metals and toxic electrophilic metabolites (Gessler et al., 2014). Melanin from *Asperigillus nidulans* MEL1 and MEL2 mutants showed g value at 2.007 in the EPR studies, which was similar to that of synthetic DOPA melanin (Goncalves et al., 2012). The high g values found for the melanins suggest that some element other than carbon and hydrogen plays a role in the free-radical species, such as nitrogen ($g \sim 2.0031$) and sulphur ($g \sim 2.0080$) (Saiz-jimenez and Shafizaded, 1985).

4.6 XRD Features

The X-ray analysis suggests that they are planar structures that differ in their stacking differences. It is observed that melanins show 0.34 nm spacing, which corresponds to the adventitious parallel stacking of aromatic units in randomly oriented local domains (Duff et al., 1988). The X-ray studies could only reveal lack of crystallinity. The X-ray diffraction patterns of the observed sample indicated that it is amorphous in nature as it produced broad features in the diffraction spectrum, known as non-Braggs features and resulting from the absence of coherent scattering from regular and repeating structures (Casadevall et al., 2012). From the peaks position of the XRD spectra of *Termitomyces* melanin, it indicated peaks at 2θ = 27.56, 31.92, 45.63, 56.65, 66.42, 75.49 and 83.19, which have the corresponding reflections (111), (200), (220), (222), (400), (420) and (422), respectively as reported in sepia melanin crystallographic patterns. A broad diffraction peak (2θ = 10–90°) was observed on the XRD spectra of *Termitomyces* melanin. The X-ray diffraction technique for analysis of amorphous material is limited (Mbonyiryivuze et al., 2015b). In amorphous material, incoherent component cannot be calculated and subtracted because the orientations of structural elements are random and the resulting spectrum represents an integral spatial average (Casadevall et al., 2012).

4.7 NMR Studies

Melanin structural elements include substituted hydroxyindoles, indolequinones, pyrroles, free carboxylic acid groups, phenolic hydroxyls and carbon-based free radicals. Presence of uncyclised aliphatic chains, originating from monomeric precursors, has been confirmed by the cross-polymerisation/magic angle spin NMR technique (Duff et al., 1988).

4.8 ss¹H NMR Studies

The ss¹H NMR spectrum of *Termitomyces* melanin showed signals at 6.111, 3.550 ppm, which occurred due to protons on carbons attached to nitrogen and/or oxygen atoms, 1.684, 1.315, 0.198 ppm for CH_3 groups of alkyl fragments and 5.363 ppm indicating pyrrole –CH group of a carboxyl-substituted indole. Characteristic signal was obtained at 2.089, 2.832 ppm, indicating proton attached to the sulphur. The values were compared with sepia melanin ¹H NMR spectra, which indicated chemical shifts at 3.70–4.20 ppm protons on carbons attached to nitrogen or oxygen; 0.8–1.0 ppm for CH_3 groups of alkyl fragments; 8.4, 6.45, 6.88, 7.2 ppm for pyrrole –CH group of a carboxyl-substituted indole. There are a few reports on NMR and mass spectral analysis of fungal melanin. Arun et al. (2015) reported doublets at 7.8 and 7.9 ppm, which were assigned to C (7) H and C (6) H protons of 4, 5 di-substituted indole and the substitution may be due to CH_3-COO group as the signal at 2.421 or 2.612 or 2.341 ppm was seen. This confirmed the pheomelanin nature of pigment in *Schizophyllum commune*.

4.9 *¹³C NMR Studies*

Herve et al., (1994) indicated that CP/MAS spectra are identical to MAS, thus concluding that proton concentration in melanin are large enough to permit a complete transfer of magnetisation with contact time of 2 ms and no spectral distortions are introduced by high spinning rate. Thus, cross-polarisation technique can be used for melanin. *Termitomyces* melanin showed carbonyl peaks at 200–170 ppm and aromatic regions at 160–110 ppm. Chemical shift at 40–30 ppm appeared due to open-chain aliphatic carbons, specifically 45–40 ppm indicated = C-S and others like 71, 56, 52, 33, 30 were due to aliphatic carbons in cysteine/DOPA. The ¹³C NMR spectral-band assignments of *Termitomyces* melanin were compared with other fungal melanin, such as *Oidiodendron tenuissimum, Trichoderma harzianum, Ulocladium atrum, Henderson ulatoruloidea, Eurotium echinulatum* (Knicker et al., 1995) and also compared with DOPA (Duff et al., 1988) and sepia melanins (Duff et al., 1988; Adhyaru et al., 2003; Herve et al., 1994). Chemical shifts at δ 95–130 to protonated carbon atoms and δ 130–145 to quaternary carbon atoms of polydopamine present in melanin backbones are indicated (Adhyaru et al., 2003). The ¹³C spectra displays aromatic regions (110–160 ppm), indicating a common heterogenous, amorphous, aromatic core structure in fungal melanins. Additional spectral features evident from carboxylic groups (168–174 ppm) and open chain aliphatics (30–60 ppm) were reported for solid fungal melanins. Open-chain aliphatic carbons could arise from unreacted L-DOPA and aliphatic amine structural elements could arise from coupling of dopamine and/or quinone structural units (Chatterjee et al., 2014). Tian et al. (2003) reported that carbon near sulphur shows signal at 45–40 ppm and CH_2 tyrosine/DOPA-based signal at 40–35 ppm in ¹³C spectrum, consistent with our results.

4.10 *MALDI-TOF Analysis*

In recent years, matrix-assisted laser desorption/ionisation-time of flight mass spectrometry (MALDI-TOF) method using laser beam (Laser Desorption Ionisation) on solid compound mixed with suitable matrix has become useful for understanding high molecular weight substances, such as protein, oligonucleotides and synthetic polymers (Seraglia and Traldi, 1993). Thus, it was of interest to test for understanding the fragmentation patters in melanin, which was a natural polymer isolated from *Termitomyces*. The MALDI-TOF spectrum of isolated melanin pigment from *Termitomyces* demonstrated the presence of several types of molecular ions in the spectrum having m/z values of 372.801, 388.950, 402.563, 411.938, 428.495, 434.684, 438.342, 450.340, 511.156, 533.192, 544.193, 551.172, 557.226, 563.693, 567.196, 579.518, 586.294, 597.273, 603.580, 611.258, 625.532, 633.255, 647.312, 659.283, 682.293, 695.272, 709.267, 715.570, 727.242, 749.280, 761.307, 778.966, which in turn indicate the presence of a mixture of degradation products of melanin. The MALDI-TOF mass spectrum of *Termitomyces* melanin indicates a complex melanin characteristic of polydisperse nature. The structure of natural melanin is quite irregular and it is difficult to assign the observed MS signals to predict specific fragment structure. The structure of isolated melanin is also expected to differ to some extent, based on the source. However, if one notes the possible fragments reported by the previous researchers (Banerjee et al., 2014, Arun et al., 2015), it is observed that there is a drop of 2–3 units in molecular weight (551.172 to 550.139; 579.518 to 576.12; and 715.570 to 713.170). This may be due to the removal of hydrogen at a point of junctions to the next part of the repeating unit. Hence, the observed molecular weights of fragments in case of our sample are in agreement with what is reported by others.

4.11 *Effect of Inhibitor on Melanin Synthesis*

Inhibition of melanin production in *Termitomyces* was observed in colonies exposed to 100 µg/ml concentration Kojic acid (KA) when compared to 10, 1 µg/ml concentrations and control without KA. There was no effect on *Termitomyces* melanin production, when exposed to tricyclazole inhibitor of varied concentrations 1, 10 to 100 µg/ml. The aim of this study was to determine which pathway (DHN or DOPA) produces melanin in the *Termitomyces* species, using kojic acid, which is a DOPA pathway inhibitor and tricyclazole, which is DHN pathway inhibitor. Pal et al. (2013) reported a faint-brown extracellular pigment surrounding the *Aspergillus tubingensis* colony even in the presence of inhibitors and no melanin was detected in the culture filtrate of liquid cultures grown in the presence of tricyclazole, in spite of the filtrate being faintly pigmented. Kojic acid inhibits the enzyme tyrosinase, which catalyzes the first two steps, i.e., tyrosine oxidation to DOPA to dopaquinone of the DOPA pathway, whereas tricyclazole inhibits the enzymes tetra- and tri-hydroxynaphthalene reductases (4HNR and 3HNR) of pentaketide melanin pathway, which catalyses the reduction of T4HN to scytalone and T3HN to vermelone, respectively in many fungi. Besides these, synthetic inhibitors and several other natural inhibitors obtained from various sources, inhibit melanin synthesis (Fernandes and Kerkar, 2017). The minimum concentration required of tricyclazole to inhibit reductase activity varies from < 0.1 to 10 µg/ml for different species (Bell and Wheeler, 1986). However, our results showed that *Termitomyces* melanin is produced by DOPA melanin pathway.

5. Constraints in Studying *Termitomyces* Melanin

In *Termitomyces*, melanin extraction created several constraints while conducting termitomelanin research, such as the complete extraction of melanin from mycelium was not possible by NaOH extraction method as the mycelium after melanin extraction procedure still appeared to be black/brown in colour. Even repeated extraction with NaOH could not help. Thus, the use of enzymatic method can be applied, which is described by Prados-Rosales et al. (2015) and a comparative study can be performed for large-scale production of melanin. Industrial fermenters could be used to produce large-scale *Termitomyces* mycelial biomass which could be used for extraction of melanin. Since the enzymatic method was not cost-effective in present study, we did not use this technique for melanin extraction as our aim was to characterise the melanin pigment from *Termitomyces* and not estimate its yield. In terms of increasing yield of melanin for future the studied enzymatic method could be adapted.

6. *Termitomyces* Melanin Structural and Functional Significance

Early Cretaceous period is known to have a large quantity of highly melanised fungal spores (Dadachova and Casadevall, 2008). Fungus growing termites originated in African rainforest about 31 Mya (Nobre et al., 2011). The melanins are known to provide structural strength and protection because the phenolic or indolic polymers are recalcitrant to degradation (Henson et al., 1999). This nature of melanin is required to protect the fruit bodies as they emerge out from the underground termite mound soil to the surface by passing through different layers of soil. During emergence, friction is caused by obstacles, such as rocks, roots, sharp mineral crystals, pebbles and hard soil. Melanin production in *Termitomyces* fruit bodies is initially seen in perforatium or driller, which later becomes part of umbo. For initiation of this type of melanin, there might be involvement of several epigenetic factors. After complete differentiation of mushrooms the scale (remnants of universal veil) and ring/annulus on stipe (remnants of partial veil) remain melanised, but rest of the mushroom parts show selective melanisation. Mechanical pressure is one of the important factors to initiate melanin synthesis in genus *Termitomyces*. As in response to grazing damage done by snails, beetles and other mycophagous, insects cause melanisation. In *in vitro*, this was experimentally proved by cutting culture plugs and incubation, which showed intense zones of melanin-pigment production around the injury. Melaninsation was also noticed during the course of mechanical injury, while mushrooms emerged from the soil layers to the surface.

Since melanins are known to give tensile strength, this may play an important role to bore through hard soils as melanins confer greater mechanical invasiveness to its hyphae along with the influence of other factors, such as osmolyte accumulation and turgor pressure (Henson et al., 1999). In addition, melanin might help to prevent heavy metal toxicity as melanins will bind to metals present in the fungus comb or soil during its growth. Schweitzer et al. (2009) reported that photoelectric effect and Compton scattering increases with the presence of sulphur content in melanin as compared to non-sulphur-containing melanins; hence it could be used as radioprotectant. Besides just melanin, there is also need for biolubricant, such as extracellular polysaccharide material, which will aid in a smooth drilling process of mushroom to the surface.

6.1 *Proposed Biosynthesis Pathway for Termitomyces Melanin*

Various characterisation, analysis and inhibitor studies could help to postulate possible melanin production pathway in *Termitomyces*, which need further attention to solve experimental studies as has been depicted in Fig. 2. Characteristic observation noticed in *Termitomyces* melanin is that it is rich in sulphur content, which can be similar to that reported in *Tuber* spp., thus indicating formation of Pheomelanin type. Besides, previous reports indicate that *Termitomyces* contain the sulphur source from amino acids also which may be from regoliths in natural condition and which might be enriching their habitat and helping them to produce sulphur-rich melanin. Inhibitors study also proved it to be DOPA type of melanin.

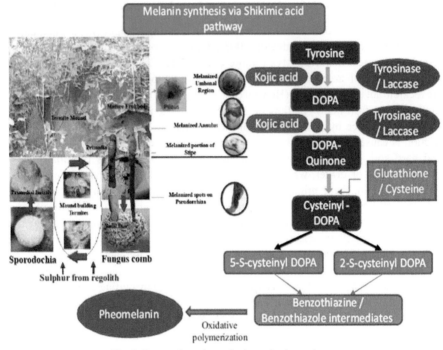

Fig. 2. Proposed termito-melanin production pathway.

7. Conclusion, Prospects and Future Challenges

Edible melanin has gained importance in medical industries, thus one can use *Termitomyces* melanin, which is found to be similar to human melanin in properties of injury-based induction phenomena. It can be also used as a model organism for understanding the melanin-based mechanisms. Exact mechanism of melanin synthesis needs to be elucidated by using available species of *Termitomyces*. Nevertheless, several aspects remain diffuse and need more insight in understanding the localisation of melanin from mycelial cultures by using other techniques, such as Transmission Electron Microscopy (TEM). Besides, several structural questions could be answered using computational techniques to follow the structural details of extracted purified *Termitomyces* melanin. Various problems were also encountered while purifying *Termitomyces* melanin. There is need to identify if there are any techniques/protocols to get rid of bound proteinaceous residues to purified melanin. Will the nature and activity of melanin will still remain the same after removing such bound proteins? Such questions are required to be answered in future research to develop or answer the hypotheses. Proteomics studies could be performed to understand various protein expressions for understanding radiation effects, which in turn will confirm the radiation effect on *Termitomyces* species with or without melanin. In future studies, there is need to improve the *Termitomyces* melanin-producing strain and standardisation of melanin yield affecting factors. Melanin research has to focus towards drug delivery/material for implantology and also in cancer therapy to prevent side effects by absorbing radiations. In future, one should aim at application of *Termitomyces* melanin, such as synthesis of clinical probes may find application in pharmaceutical as well as cosmetic industries as a component of photo-protective creams that protect the skin from possible oxidative damage and can be also used as oral radioprotectors, suitable for treatment before acute radiotherapy or to alleviate the effects of radiation during catastrophic events. Edible melanin can also be used as a food additive to prevent rancidity by quenching the bacterial quorum sensing (Zhu et al., 2011; Sun et al., 2016a). In future, various studies can be performed to assess usefulness of this non-toxic melanin for clinical and other technological exploitation by understanding the melanin structure-property-function relationship.

Acknowledgements

Authors would like to acknowledge all those who contributed in the field of fungal melanin and *Termitomyces* research in the last few decades. We acknowledge the Goa University Fungus Culture Collection and Research Unit (GUFCCRU), Department of Botany, Goa University, Scanning Electron Microscope facility and University Science Instrumentation Centre (USIC), Goa University, SAIF-IIT-Bombay, CSIR-CSMCRI and Rajiv Gandhi Centre for Biotechnology, Kerala. Rosy Agnes De Souza acknowledges DST-Inspire fellowship, Government of India, New Delhi.

References

Abdullah, F. and Rusea, G. (2009). Documentation of inherited knowledge on wild edible fungi from Malaysia. Blumea., 54(1-3): 35–38.

Adejumo, T.O. and Awosanya, O.B. (2005). Proximate and mineral composition of four edible mushroom species from south-western Nigeria. Afr. J. Biotechnol., 4(10): 1084–1088.

Adhyaru, B.B., Akhmedov, N.G., Katritzky, A.R. and Bowers, C.R. (2003). Solid-state cross-polarisation magic angle spinning ^{13}C and ^{15}N NMR characterisation of sepia melanin, sepia melanin free acid and human hair melanin in comparison with several model compounds. Magn. Reson. Chem., 41(6): 466–474.

Agustinho, D.P. and Nosanchuk, J.D. (2017). Functions of fungal melanin. *In*: Reference Module in Life Sciences, ISBN: 978-0-12-809633-8.

Aletor, V.A. (1995). Compositional studies on edible tropical species of mushrooms. Food Chem., 54(3): 265–268.

Alofe, F.V. (1991). Amino acids and trace minerals of three edible wild mushrooms from Nigeria. J. Food Compost. Anal., 4(2): 167–174.

Alviano, C.S., Farbiarz, S.R., De Souza, W., Angluster, J. and Travassos, L.R. (1991). Characterisation of *Fonsecaea pedrosoi* melanin. Microbios., 137(4): 837–844.

Apte, M., Girme, G., Bankar, A., Ravi Kumar, A. and Zinjarde, S. (2013). 3, 4-dihydroxy-L-phenylalanine-derived melanin from *Yarrowiali polytica* mediates the synthesis of silver and gold nanostructures. J. Nanobiotechnol., 11(1): 1–9.

Araujo, M., Viveiros, R., Correia, T.R., Correia, I.J., Bonifácio, V.D. et al. (2014). Natural melanin: A potential pH-responsive drug release device. Int. J. Pharma., 469(1): 140–145.

Arun, G., Angeetha, M., Eyini, M. and Gunasekaran, P. (2014). Effect of copper sulphate and resorcinol on the extracellular production of melanin and laccase by *Schizophyllum commune* Fr. and *Pleurotus cystidiosus* var. *Formosensis*. Ind. J. Adv. Plant Res., 1(5): 55–61.

Arun, G., Eyini, M. and Gunasekaran, P. (2015). Characterisation and biological activities of extracellular melanin produced by *Schizophyllum commune* (Fries). Indian J. Exp. Biol., 53(6): 380–387.

Aryal, H.P. and Budathoki, U. (2016). Ethnomycology of *Termitomyces* R. Heim in Nepal. J. Yeast Fungal Res., 7(4): 28–38.

Aryal, H.P., Ghimire, S.K. and Budhathoki, U. (2016). *Termitomyces*: New to the science. J. Plant Sci. Res., 3(1): 150–157.

Azarko, I.I., Karpovich, I.A., Kukulianskaja, T.A., Kurchenko, V.P., Novikov, D.A. et al. (2001, September). Super-high-frequency study of biologically active compounds. *In*: 11th International Conference 'Microwave and Telecommunication Technology'. Conference Proceedings (IEEE Cat. No. 01EX487), pp. 95-96.

Babitskaya, V.G., Shcherba, V.V. and Ikonnikova, N.V. (2000). Melanin complex of the fungus *Inonotus obliquus*. Appl. Biochem. Microbiol., 36(4): 377–381.

Banerjee, A., Supakar, S. and Banerjee, R. (2014). Melanin from the nitrogen-fixing bacterium *Azotobacter chroococcum*: A spectroscopic characterisation. PLoS ONE, 9(1): e84574.

Baraza, L.D., Joseph, C.C., Moshi, M.J. and Nkunya, M.H.H. (2007). Chemical constituents and biological activity of three Tanzanian wild mushroom species. Tanz. J. Sci., 33: 1–7.

Bell, A.A. and Wheeler, M.H. (1986). Biosynthesis and functions of fungal melanins. Ann. Rev. Phytopathol., 24(1): 411–451.

Belozerskaya, T.A., Gessler, N.N. and Aver'yanov, A.A. (2017). Melanin pigments of fungi. pp. 1–29. *In*: Merillon, J.M. and Ramawat, K. (eds.). Fungal Metabolites. Reference Series in Phytochemistry, Springer Cham.

Beltran-Garcia, M.J., Prado, F.M., Oliveira, M.S., Ortiz-Mendoza, D., Scalfo, A.C., Pessoa Jr, A. et al. (2014). Singlet molecular oxygen generationby light-activated DHN-melanin of the fungal pathogen *Mycosphaerella fijiensis* in black Sigatoka disease of bananas. PloS ONE, 9(3): e91616.

Bilinska, B. (2001). On the structure of human hair melanins from an infrared spectroscopy analysis of their interactions with Cu^{2+} ions. Spectrochim. Acta A., 57(12): 2525–2533.

Bin, L., Wei, L., Xiaohong, C., Mei, J. and Mingsheng, D. (2012). *In vitro* antibiofilm activity of the melanin from *Auricularia auricula*, an edible jelly mushroom. Ann. Microbiol., 62(4): 1523–1530.

Blois, M.S., Zahlan, A.B. and Maling, J.E. (1964). Electron spin resonance studies on melanin. Biophys. J., 4(6): 471–490.

Botha, W.J. and Eicker, A. (1992). Nutritional value of *Termitomyces* mycelial protein and growth of mycelium on natural substrates. Mycol. Res., 96(5): 350–354.

Bronze-Uhle, E.S., Batagin-Neto, A., Xavier, P.H., Fernandes, N.I., De Azevedo, E.R. and Graeff, C.F. (2013). Synthesis and characterisation of melanin in DMSO. J. Mol. Struct., 1047: 102–108.

Bull, A.T. (1970). Chemical composition of wild-type and mutant *Aspergillus nidulans* cell walls. The nature of polysaccharide and melanin constituents. Microbios, 63(1): 75–94.

Casadevall, A., Rosas, A.L. and Nosanchuk, J.D. (2000). Melanin and virulence in *Cryptococcus neoformans*. Curr. Opin. Microbiol., 3(4): 354–358.

Casadevall, A., Cordero, R.J., Bryan, R., Nosanchuk, J. and Dadachova, E. (2017). Melanin, radiation, and energy transduction in fungi. Microbiol. Spectr., 5(2): FUNK-0037-2016.

Chrissian, C., Camacho, E., Fu, M.S., Prados-Rosales, R., Chatterjee, S. et al. (2020). Melanin deposition in two *Cryptococcus* species depends on cell-wall composition and flexibility. J. Biol. Chem., 295(7): 1815–1828.

Chatterjee, S., Prados-Rosales, R., Tan, S., Itin, B., Casadevall, A. and Stark, R.E. (2014). Demonstration of a common indole-based aromatic core in natural and synthetic eumelanins by solid-state NMR. Org. Biomol. Chem., 12(34): 6730–6736.

Cheng, Q., Kinney, K.A., Whitman, C.P. and Szaniszlo, P.J. (2004). Characterisation of two polyketide synthase genes in *Exophiala lecanii-corni*, a melanised fungus with bioremediation potential. Bioorg. Chem., 32(2): 92–108.

Chodurek, E., Zdybel, M., Pilawa, B. and Dzierzewicz, Z. (2012). Examination by EPR spectroscopy of free radicals in melanins isolated from A-375 cells exposed on valproic acid and cisplatin. Acta Pol. Pharma., 69(6): 1334–1341.

Cordero, R.J. and Casadevall, A. (2017). Functions of fungal melanin beyond virulence. Fungal Biol. Rev., 31(2): 99–112.

Costa, T.G., Feldhaus, M.J., Vilhena, F.S., Heller, M., Micke, G.A. et al. (2015). Preparation, characterisation, cytotoxicity and antioxidant activity of DOPA melanin modified by amino acids: Melanin-like oligomeric aggregates. J. Brazil. Chem. Soc., 26(2): 273–281.

d'Ischia, M., Wakamatsu, K., Napolitano, A., Briganti, S., Garcia-Borron, J.C. et al. (2013). Melanins and melanogenesis: Methods, standards, protocols. Pig. Cell Melan. Res., 26(5): 616–633.

Dadachova, E. and Casadevall, A. (2008). Ionising radiation: How fungi cope, adapt, and exploit with the help of melanin. Curr. Opin. Microbiol., 11(6): 525–531.

de Kesel, A. (2011). Provisional macroscopic key to the edible mushrooms of tropical Africa: 100+ taxa from the Zambezian and Sudanian region. MycoAfrica. AMA., 4(1): 1–9.

De la Rosa, J.M., Martin-Sanchez, P.M., Sanchez-Cortes, S., Hermosin, B., Knicker, H. and Saiz-Jimenez, C. (2017). Structure of melanins from the fungi *Ochroconis lascauxensis* and *Ochroconis anomala* contaminating rock art in the Lascaux Cave. Sci. Rep., 7(1): 13441.

de Souza, R.A., Kamat, N.M. and Nadkarni, V.S. (2018). Purification and characterisation of a Sulphur rich melanin from edible mushroom *Termitomyces albuminosus* Heim. Mycology, 9(4): 296–306.

de Souza, R.A. and Kamat, N.M. (2019). *Termitomyces* holomorph benefits from anomalous sulphur content in teleomorph. IJLSR, 7(1): 186–192.

Devi, M.B., Singh, S.M. and Singh, N.I. (2014). Nutrient analysis of indigenous *Termitomyces eurrhizus* (Berk.) Heim of Manipur, India. Int. J. Curr. Microbiol. Appl. Sci., 3(6): 491–496.

Dong, C. and Yao, Y. (2012). Isolation, characterisation of melanin derived from *Ophiocordyceps sinensis*, an entomogenous fungus endemic to the Tibetan Plateau. J. Biosci. Bioeng., 113(4): 474–479.

Duff, G.A., Roberts, J.E. and Foster, N. (1988). Analysis of the structure of synthetic and natural melanins by solid-phase NMR. Biochemistry, 27(18): 7112–7116.

Earanna, N., Nandini, K., Ganavi, M.C., Sajeevan, R.S. and Nataraja, K.N. (2013). Molecular and nutritional characterisation of an edible mushroom from the Western Ghats of Karnataka, India—A new report. Mushroom Res., 22(1): 1–7.

Eisenman, H.C. and Casadevall, A. (2012). Synthesis and assembly of fungal melanin. Appl. Microbiol. Biotechnol., 93(3): 931–940.

Eisenman, H.C., Greer, E.M. and McGrail, C.W. (2020). The role of melanins in melanotic fungi for pathogenesis and environmental survival. Appl. Microbiol. Biotechnol., 104: 4247–4257.

Ellis, D.H. and Griffiths, D.A. (1974). The location and analysis of melanins in the cell walls of some soil fungi. Can. J. Microbiol., 20(10): 1379–1386.

Eom, T., Woo, K., Cho, W., Heo, J.E., Jang, D. et al. (2017). Nanoarchitecturing of natural melanin nanospheres by layer-by-layer assembly: Mmacroscale anti-inflammatory conductive coatings with optoelectronic tenability. Biomacromolecules, 18(6): 1908–1917.

Fernandes, M.S. and Kerkar, S. (2017). Microorganisms as a source of tyrosinase inhibitors: Review. Ann. Microbiol., 67(4): 343–358.

Fernandez, C.W. and Koide, R.T. (2013). The function of melanin in the ectomycorrhizal fungus *Cenococcum geophilum* under water stress. Fungal Ecol., 6(6): 479–486.

Gessler, N.N., Egorova, A.S. and Belozerskaia, T.A. (2014). Melanin pigments of fungi under extreme environmental conditions. Prikl. Biokhim. Mikrobiol., 50(2): 125–134.

Goncalves, R.C.R., Lisboa, H.C.F. and Pombeiro-Sponchiado, S.R. (2012). Characterisation of melanin pigment produced by *Aspergillus nidulans*. World J. Microbiol. Biotechnol., 28(4): 1467–1474.

Harki, E., Talou, T. and Dargent, R. (1997). Purification, characterisation and analysis of melanin extracted from *Tuber melanosporum* Vitt. Food Chem., 58(1): 69–73.

Harki, E., Bouya, D. and Dargent, R. (2006). Maturation-associated alterations of the biochemical characteristics of the black truffle *Tuber melanosporum* Vitt. Food Chem., 99(2): 394–400.

Harley, J.L. and Smith, S.E. (1983). Mycorrhizal Symbiosis. Academic Press, London.

Heim, R. (1977). Termites et Champignons; les champignons termitophiles d'Afrique Noire etd'Asiemeridionale, Paris: Societe Nouvelle Des Editions Boubee.

Henson, J.M., Butler, M.J. and Day, A.W. (1999). The dark side of the mycelium: Melanins of phytopathogenic fungi. Annu. Rev. Phytopathol., 37(1): 447–471.

Hu, W.L., Dai, D.H., Huang, G.R. and Zhang, Z.D. (2015). Isolation and characterisation of extracellular melanin produced by *Chroogomphus rutilus* D447. Am. J. Food Technol., 10(2): 68–77.

Hou, R., Liu, X., Xiang, K., Chen, L., Wu, X. et al. (2019). Characterisation of the physicochemical properties and extraction optimisation of natural melanin from *Inonotus hispidus* mushroom. Food Chem., 277: 533–542.

Ijeh, I.I., Eke, I.N., Ugwu, C.C. and Ejike, E.C.C.C. (2016). Myco-nourishment from the Wild: Chemical analyses of the nutritional and amino acid profile of *Termitomyces robustus* harvested from Uzuakoli, Nigeria. Nat. Prod. Chem. Res., 4(4): 225.

Ito, S. and Fujita, K. (1985). Microanalysis of eumelanin and pheomelanin in hair and melanomas by chemical degradation and liquid chromatography. Anal. Biochem., 144(2): 527–536.

Kansci, G., Mossebo, D.C., Selatsa, A.B. and Fotso, M. (2003). Nutrient content of some mushroom species of the genus *Termitomyces* consumed in Cameroon. Food/Nahrung., 47(3): 213–216.

Karun, N.C. and Sridhar, K.R. (2013). Occurrence and distribution of *Termitomyces* (Basidiomycota, Agaricales) in the Western Ghats and on the west coast of India. Czech Mycol., 65(2): 233–254.

Knicker, H., Almendros, G., González-Vila, F.J., Lüdemann, H.D. and Martin, F. (1995). ^{13}C and ^{15}NNMR analysis of some fungal melanins in comparison with soil organic matter. Org. Geochem., 23(11): 1023–1028.

Li, Y., Liu, J., Wang, Y., Chan, H.W., Wang, L. and Chan, W. (2015). Mass spectrometric and spectrophotometric analyses reveal an alternative structure and a new formation mechanism for melanin. Anal. Chem., 87(15): 7958–7963.

Lopusiewicz, L. (2018a). Isolation, characterisation and biological activity of melanin from *Exidia nigricans*. World Sci. News, 91: 111–129.

Lopusiewicz, L. (2018b). *Scleroderma citrinum* melanin: Isolation, purification, spectroscopic studies with characterisation of antioxidant, antibacterial and light barrier properties. World Sci. News, 94(2): 115–130.

Lu, Y.Y., Ao, Z.H., Lu, Z.M., Xu, H.Y., Zhang, X.M. et al. (2008). Analgesic and anti-inflammatory effects of the dry matter of culture broth of *Termitomyces albuminosus* and its extracts. J. Ethnopharmacol., 120(3): 432–436.

Mbonyiryivuze, A., Nuru, Z.Y., Ngom, B.D., Mwakikunga, B., Dhlamini, S.M. et al. (2015a). Morphological and chemical composition characterisation of commercial *Sepia* melanin. Phys. Mater. Chem., 3(1): 22–27.

Mbonyiryivuze, A., Omollo, I., Ngom, B.D., Mwakikunga, B., Dhlamini, S.M. et al. (2015b). Natural dye sensitizer for Grätzel cells: *Sepia* melanin. Phys. Mater. Chem., 3(1): 1–6.

Mitra, P., Mandal, N.C. and Acharya, K. (2014). Phytochemical characteristics and free radical scavenging activity of ethanolic extract of *Termitomyces microcarpus* R Heim. Der Pharm. Lett., 6(5): 92–98.

Mossebo, D.C., Njounkou, A.L., Piatek, M., Kengni, B. and Diasbe, M.D. (2009). *Termitomyces striatus* f. *Pileatus* f. nov. and f. *brunneus* f. nov. from Cameroon with a key to central African species. Mycotaxon, 107(1): 315–329.

Nabubuya, A., Muyonga, J.H. and Kabasa, J.D. (2010). Nutritional and hypocholesterolemic properties of *Termitomyces microcarpus* mushrooms. AJFAND, 10(3): 2235–2257.

Nakalembe, I. and Kabasa, J.D. (2013). Fatty and amino acids composition of selected wild edible mushrooms of Bunyoro Sub-region, Uganda. AJFAND, 13(1): 7225–7241.

Natarajan, K. (1979). South Indian Agaricales V: *Termitomyces heimii*. Mycologia, 71(4): 853–855.

Nicolaus, R.A. (1968). Melanins. Hermann, Paris.

Nobre, T., Fernandes, C., Boomsma, J.J., Korb, J. and Aanen, D.K. (2011). Farming termites determine the genetic population structure of *Termitomyces* fungal symbionts. Mol. Ecol., 20(9): 2023–2033.

Otieno, N.C. (1968). Further contributions to a knowledge of termite fungi in east Africa: The genus *Termitomyces* Heim. Sydowia, 22: 160–165.

Ou-Yang, H., Stamatas, G. and Kollias, N. (2004). Spectral responses of melanin to ultraviolet A irradiation. J. Investig. Dermatol., 122(2): 492–496.

Pal, A.K., Gajjar, D.U. and Vasavada, A.R. (2013). DOPA and DHN pathway orchestrate melanin synthesis in *Aspergillus* species. Med. Mycol., 52(1): 10–18.

Pegler, D.N. (1969). Studies on African Agaricales: II. Kew Bull., 23(2): 219–249.

Pegler, D.N. and Rayner, R.W. (1969). A contribution to the Agaric flora of Kenya. Kew Bull., 23(3): 347–412.

Pegler, D.N. and Vanhaecke M. (1994). *Termitomyces* of southeast Asia. Kew Bull., 49(4): 717–736.

Piearce, G.D. (1987). The genus *Termitomyces* in Zambia. Mycologist, 1(3): 111–116.

Pierce, J.A. and Rast, D.M. (1995). A comparison of native and synthetic mushroom melanins by Fourier-transform infrared spectroscopy. Phytochemistry, 39(1): 49–55.

Pilawa, B., Zdybel, M. and Chodurek, E. (2017). Application of electron paramagnetic resonance spectroscopy to examine free radicals in melanin polymers and the human melanoma malignum cells. pp. 79–103. *In*: Blumenberg, M. (ed.). Melanin, Rijeka: InTech Open.

Pombeiro-Sponchiado, S.R., Sousa, G.S., Andrade, J.C., Lisboa, H.F. and Gonçalves, R.C. (2017). Production of melanin pigment by fungi and its biotechnological applications. pp. 47–73. *In*: Blumenberg, M, (ed.). Melanin, Intech Open. ISBN 978-953-51-5472-3.

Prados-Rosales, R., Toriola, S., Nakouzi, A., Chatterjee, S., Stark, R. et al. (2015). Structural characterisation of melanin pigments from commercial preparations of the edible mushroom *Auricularia auricula*. J. Agric. Food Chem., 63(33): 7326–7332.

Prota, G. (1992). Melanins and Melanogenesis. San Diego, CA: Academic Press, New York.

Prota, G. (1993). Regulatory mechanisms of melanogenesis: Beyond the tyrosinase concept. J. Investig. Dermatol., 100(2): S156–S161.

Qi, J., Ojika, M. and Sakagami, Y. (2000). Termitomycesphins A-D, novel neuritogenic cerebrosides from the edible Chinese mushroom *Termitomyces albuminosus*. Tetrahedron, 56(32): 5835–5841.

Qi, J., Ojika, M. and Sakagami, Y. (2001). Neuritogenic cerebrosides from an edible Chinesemushroom. Part 2: Structures of two additional termitomycesphins and activity enhancement of an inactive cerebroside by hydroxylation. Bioorg. Med. Chem., 9(8): 2171–2177.

Qu, Y., Sun, K., Gao, L., Sakagami, Y., Kawagishi, H. et al. (2012). Termitomycesphins G and H, additional cerebrosides from the edible Chinese mushroom *Termitomyces albuminosus*. Biosci. Biotechnol. Biochem., 76(4): 791–793.

Rahmad, N., Al-Obaidi, J.R., Rashid, N.M.N., Zean, N.B., Yusoff, M.H.Y.M. et al. (2014). Comparative proteomic analysis of different developmental stages of the edible mushroom *Termitomyces heimii*. Biol. Res., 47(1): 30.10.1186/0717-6287-47-30.

Riley, P.A. (1997). Melanin. Int. J. Biochem. Cell Biol., 29(11): 1235–1239.

Saiz-Jimenez, C. and Shafizadeh, F. (1985). Electron spin resonance spectrometry of fungal melanins. Soil Sci., 139(4): 319–325.

Saltarelli, R., Ceccaroli, P., Cesari, P., Barbieri, E. and Stocchi, V. (2008). Effect of storage on biochemical and microbiological parameters of edible truffle species. Food Chem., 109(1): 8–16.

Sathiya Seelan, J.S., Shu Yee, C., She Fui, F., Dawood, M., Tan, Y.S. et al. (2020). New Species of *Termitomyces* (Lyophyllaceae, Basidiomycota) from Sabah (Northern Borneo), Malaysia. Mycobiology, 48(2): 95–103.

Sava, V.M., Galkin, B.N., Hong, M.Y., Yang, P.C. and Huang, G.S. (2001). A novel melanin-like pigment derived from black tea leaves with immuno-stimulating activity. Food Res. Int., 34(4): 337–343.

Sawaya, W.N., Al-Shalhat, A., Al-Sogair, A. and Al-Mohammad, M. (1985). Chemical composition and nutritive value of truffles of Saudi Arabia. J. Food Sci., 50(2): 450–453.

Schweitzer, A.D., Howell, R.C., Jiang, Z., Bryan, R.A., Gerfen, G. et al. (2009). Physico-chemical evaluation of rationally designed melanins as novel nature-inspired radioprotectors. PloS ONE, 4(9): e7229.

Selvakumar, P., Rajasekar, S., Periasamy, K. and Raaman, N. (2008). Isolation and characterisation of melanin pigment from *Pleurotus cystidiosus* (telomorph of *Antromycopsis macrocarpa*). World J. Microbiol. Biotechnol., 24(10): 2125–2131.

Seraglia, R., Traldi, P., Elli, G., Bertazzo, A., Costa, C. and Allegri, G. (1993). Laser desorptionionisation mass spectrometry in the study of natural and synthetic melanins I -Tyrosine melanins. Biol. Mass Spectrom., 22(12): 687–697.

Siddiquee, S., Yee, W.Y., Taslima, K., Fatihah, N.H.N., Kumar, S.V. and Hasan, M.M. (2012). Sequence analysis of the ribosomal DNA internal transcribed spacer regions in *Termitomyces heimii* species. Ann. Microbiol., 62(2): 797–803.

Siddiquee, S., Rovina, K., Naher, L., Rodrigues, K.F. and Uzzaman, M.A. (2015). Phylogenetic relationships of *Termitomyces aurantiacus* inferred from internal transcribed spacers DNA sequences. Adv. Biosci. Biotechnol., 6(5): 358–367.

Srivastava, B., Dwivedi, A.K. and Pandley, V.N. (2011). Morphological characterisation and yield potential of *Termitomyces* spp. mushroom in Gorakhpur forest division. BEPLS., 1(1): 54–56.

Sun, L., Liu, Q., Bao, C. and Fan, J. (2017). Comparison of free total amino acid compositions and their functional classifications in 13 wild edible mushrooms. Molecules, 22(3): 350, 10 pp.

Sun, S., Zhang, X., Sun, S., Zhang, L., Shan, S. and Zhu, H. (2016a). Production of natural melanin by *Auricularia auricula* and study on its molecular structure. Food Chem., 190: 801–807.

Sun, S., Zhang, X., Chen, W., Zhang, L. and Zhu, H. (2016b). Production of natural edible melanin by *Auricularia auricula* and its physicochemical properties. Food Chem., 196: 486–492.

Suryanarayanan, T.S., Ravishankar, J.P., Venkatesan, G. and Murali, T.S. (2004). Characterisation of the melanin pigment of a cosmopolitan fungal endophyte. Mycol. Res., 108(08): 974–978.

Tang, S.M., He, M.Q., Raspe, O., Luo, X., Zhang, X.L. et al. (2020). Two new species of *Termitomyces* (Agaricales, Lyophyllaceae) from China and Thailand. Phytotaxa, 439(3): 231–242.

Tian, S., Garcia-Rivera, J., Yan, B., Casadevall, A. and Stark, R.E. (2003). Unlocking the molecular structure of fungal melanin using 13C biosynthetic labelling and solid-state NMR. Biochemistry, 42(27): 8105–8109.

Tibuhwa, D.D. (2012). *Termitomyces* species from Tanzania, their cultural properties and unequalled Basidiospores. JBLS., 3(1): 140–159.

Tsai, S.Y., Weng, C.C., Huang, S.J., Chen, C.C. and Mau, J.L. (2006). Non-volatile taste components of *Grifola frondosa, Morchella esculenta* and *Termitomyces albuminosus* mycelia. LWT-Food Sci. Technol., 39(10): 1066–1071.

Van der Westhuizen, G.C.A. and Eicker, A. (1990). Species of *Termitomyces* occurring in South Africa. Mycol. Res., 94(7): 923–937.

Wang, S. and Marcone, M.F. (2011). The biochemistry and biological properties of the world's most expensive underground edible mushroom: Truffles. Food Res. Int., 44(9): 2567–2581.

Wei, T.-Z., Yao, Y.-J., Wang, B. and Pegler, D.N. (2004). *Termitomyces bulborhizus* sp. nov. from China, with a key to allied species. Mycol. Res., 108(12): 1458–1462.

Weijn, A., Bastiaan-Net, S., Wichers, H.J. and Mes, J.J. (2013). Melanin biosynthesis pathway in *Agaricus bisporus* mushrooms. Fungal Genet. Biol., 55: 42–53.

Wu, Y., Shan, L., Yang, S. and Ma, A. (2008). Identification and antioxidant activity of melanin isolated from *Hypoxylon archeri*, a companion fungus of *Tremella fuciformis*. J. Basic Microbiol., 48(3): 217–221.

Ye, M., Guo, G.Y., Lu, Y., Song, S., Wang, H.Y. and Yang, L. (2014). Purification, structure and antiradiation activity of melanin from *Lachnum* YM 404. Int. J. Biol. Macromol., 63: 170–176.

Ye, L., Karunarathna, S.C., Li, H., Xu, J., Hyde, K.D. and Mortimer, P.E. (2019). A survey of *Termitomyces* (Lyophyllaceae, Agaricales), including a new species, from a subtropical forest in Xishuangbanna, China. Mycobiology, 47(4): 391–400.

Zaidi, K.U., Ali, A.S. and Ali, S.A. (2014). Purification and characterisation of melanogenic enzyme Tyrosinase from button mushroom. Enzyme Res., 2014: ID 120739.

Zhang, W., Du, F., Wang, L., Zhao, L., Wang, H. and Ng, T.B. (2015). Hydrolysis of oligosaccharides by a thermostable α-galactosidase from *Termitomyces eurrhizus*. Int. J. Mol. Sci., 16(12): 29226–29235.

Zhu, H., He, C.C. and Chu, Q.H. (2011). Inhibition of quorum sensing in *Chromobacterium violaceum* by pigments extracted from *Auricularia auricula*. Lett. Appl. Microbiol., 52(3): 269–274.

Zong, S., Li, L., Li, J., Shaikh, F., Yang, L. and Ye, M. (2017). Structure characterisation and lead detoxification effect of carboxymethylated melanin derived from *Lachnum* sp. Appl. Biochem. Biotechnol., 182(2): 669–686.

Zou, Y., Zhao, Y. and Hu, W. (2015). Chemical composition and radical scavenging activity of melanin from *Auricularia auricula* fruiting bodies. LWT-Food Sci. Technol., 35(2): 253–258.

Bioremediation

16

Macrofungal Mycelia as Sources of Biopesticides

Leopold M Nyochembeng

1. INTRODUCTION

Current chemical agriculture characterised by application of synthetic chemical substances to control pests and diseases poses severe environmental and human health hazards. These pesticides (herbicides, insecticides, fungicides, nematicides and others) contribute to environmental pollutants and often exhibit high and acute residual toxicity, so much so that some can have carcinogenic, teratogenic and other side effects (Srivastava and Sharma, 2011). In the United States, for example, streptomycin had been permissible to manage fire blight of apple and pear as a standard control measure in conventional and organic orchards (McManus and Stockwell, 2000). However, residual streptomycin detected in apple fruits (Mayerhofer et al., 2009) and streptomycin-resistant genes, discovered in *E. amylovora* isolates from shoots, blossom and rootstock in apple orchards (Tancos et al., 2016), indicated the possibility of horizontal gene transfer to non-target micro-organisms. Hence, the development of alternatives to streptomycin and other synthetic chemicals used for pest management in agriculture has become a top priority. In recent years, the role of biopesticides has received more attention primarily due to increased environmental awareness and safety that have ushered in new ways of farming and pest management, leading to a paradigm shift in environmental stewardship. In this new sustainable production system, biopesticides play a key role since the use of conventional synthetic pesticides, that are known to be very unspecific in action and exhibit general and high levels of toxicity, is increasingly being discouraged or banned (Barseghyan et al., 2016).

Department of Biological and Environmental Sciences, CCS-Bonner Wing Room 218 B, Alabama A&M University, PO Box 1208 Normal, AL 35762, United States.
Email: leopold.nyochembeng@aamu.edu

Biopesticides are naturally occurring substances from various biosources, including plants, animals, insects and microorganisms that control pests—biological competitors of higher plants that include weeds, insects and plant pathogens. 'Biopesticide', therefore, is a collective term for valuable naturally-produced and biologically-active compounds with specific toxicities against plant pests, thus exhibiting fungicidal, herbicidal, bactericidal, insecticidal, nematicidal and antiviral properties. Biopesticides have been classified as (1) products and byproducts of naturally occurring substances that control pests; (2) microorganisms and/or their products and byproducts that control pests; (3) substances produced by plants that contain added genetic material plant-incorporated protectants called PIPs. Research on biopesticides has intensified partly due to the increasing demand for synthetic pesticide alternatives applicable in the organic farming industry, which is currently the fastest growing sector of sustainable agriculture worldwide. Interest in biopesticides is spurred by their widespread availability, easy degradability, ability to manifest various modes of action, low toxicity to humans and non-target organisms and low cost (Lengai and Muthomi, 2018).

Several sources of biopesticides now exist for pest management in sustainable agricultural systems as significant research efforts focus on developing alternatives to synthetic pesticides. These include the use of non-pathogenic bacteria, macrofungi and microfungi and their metabolites, plant or compost extracts, essential oils, bacteriophages and nanomaterials (Griffin et al., 2017). This chapter discusses the role of macrofungal mycelia as sources of biopesticides. Macrofungi are fleshy fungi within two major phyla (Basidiomycota and Ascomycota of the kingdom fungi) and include the mushrooms defined as fruit bodies found either underground or above ground and visible to the unaided eye and harvestable by hand (Chang and Miles, 2004). The literature abounds in studies relating the uses of macrofungi and their metabolites in human health, with little emphasis on their biopesticidal applications in agriculture.

2. Macrofungal Characteristics and Biopesticide Potential

Macrofungi with biopesticidal potential exhibit many desirable and requisite characteristics, which include prolific growth rate for substrate colonisation and biomass production. This attribute is significant for metabolite production, biodegradability and cost effectiveness. Their ecological adaptability and ubiquitous nature enables them to grow and thrive under various temperate, tropical/subtropical warm and humid climates and they are widely available ecologically acceptable for application. They are a rich source of natural antibiotics and secondary metabolites that have a variety of targeted functions and actions and are less harmful to non-target organisms (Jo et al., 2014; Sivanandhan et al., 2017). An important characteristic feature is that macrofungal organisms act synergistically with other biocontrol systems. For example, studies have shown that the culture filtrate of *Lentinula edodes* enhanced the biocontrol activity of *Cryptococcus laurentii* and *Pichia membranifaciens* against *Penicillium expansum* (blue mould pathogen in post-harvest apples) in apple fruits and induced resistance in the host (Tolaini et al., 2010; Wang et al., 2013, Kaur et al., 2016).

3. Metabolites of Macrofungi

Macrofungi retain the potential and have a remarkable capacity to synthesise secondary metabolites with diverse properties and functions. The secondary metabolites are usually not necessary for normal growth and reproduction of fungus (Agyare and Agana, 2016) but play important roles in competition, defence and survival in unfavourable environments (Xiao and Zhong, 2016). The production of these metabolites is often under the regulation of key environmental factors (Shi et al., 2010), including chemical, biological or biochemical signals, nutritional status (C, N and C/N ratio) and other physical elicitors (culture condition, pH, heat stress) (Ren et al., 2019). Numerous compounds have been isolated from macrofungal species and many more are yet to be determined or characterised. The functional metabolites of macrofungi have been classified as antimicrobial sesquiterpenoids, diterpenoids, terpenes, steroids, anthraquinones, benzoic acid derivatives, quinolines and as primary metabolites such as oxalic acid and as high-molecular weight compounds, mainly peptides and proteins (Nedelkoska et al., 2019). Other less known metabolites have been determined to be a series of acetylenic acids thought to be ubiquitous in nature and especially common in Basidiomycetes (Jiang et al., 2008). These acids have a photosensitising activity, which makes them toxic to bacteria, viruses and insects (Huang et al., 2016).

In vitro production of macrofungal biopesticides is often governed by some factors that affect their biopesticidal content and activity or efficacy. Submerged cultivation tends to increase the amount and perhaps content of the biopesticidal compounds (Rathore et al., 2019; Lu et al., 2020). Medium composition, culture temperature and exogenous additives also affect the content of the secondary metabolites (Lu et al., 2020). Kaur et al. (2016) observed phytotoxicity of metabolites containing shiitake mycelia culture filtrate after fermentation in sucrose medium for 30 days that was attributable to the presence of oxalic acid. However, the oxalic acid content was significantly reduced when fermentation of shiitake mycelia occurred in glucose medium for 15 days.

3.1 Mechanism of Action

Macrofungal biopesticides may inhibit fungal growth by preventing respiration or inducing cell deformations (Weber et al., 1990) and cytostatic activities. Several modes of action are exhibited by macrofungal biopesticides, including hyperparasitism (Priya et al., 2019), antibiosis, secretion of toxic metabolites, deployment of lytic enzymes and parasitism. Some macrofungal species exhibiting nematicidal activity deploy specialised structures—acanthocysts, stephanocysts and echinocysts (Karasinski, 2013) that participate as attractants and for cuticle penetration or produce paralysing toxins (Balaes and Tanase, 2016).

Agricultural production worldwide is increasingly seeking effective alternative products to current synthetic pesticides that are environmentally friendly and can be used solely or incorporated into integrated pest management (IPM) strategies for mitigating pest problems. Sivanandhan et al. (2017) stated that such products must be cost effective, biodegradable, target-specific and have new modes of action. Macrofungi belonging to Basidiomycetes synthesise a plethora of secondary metabolites that exhibit potent antibacterial, antifungal, antiviral, nematicidal, insecticidal and herbicidal activities. These biopesticides continue to be discovered and evaluated for controlling agricultural pests and diseases.

3.2 Anti-plant Pathogenic Bacteria

Most plant pathogenic bacteria are Gram-negative; however, there are a few agriculturally important phytopathogenic bacteria that are Gram-positive. Some examples include *Clavibacter michiganensis*, which cause bacterial canker in tomato and other crops; *Streptomyces scabies*, which cause potato common scab disease. Extracts of many mushroom species, especially those belonging to the orders Ganodermatales, Poriales, Agaricales and Stereales possess potent antibacterial activity and the metabolites could be good sources of natural antibiotics for controlling plant pathogenic bacteria (Alves et al., 2012).

3.2.1 Gram-positive Bacteria

The Gram-positive plant pathogenic bacterium, *Clavibacter michiganensis* subsp. *sepedonicus* causes bacterial wilt and ring rot of potato tubers and was inhibited by extracts from *Clytocybe geotropa*, which contained an active protein clitocypin (Dreo et al., 2007; Sivanandhan et al., 2017).

3.2.2 Gram-negative Bacteria

Extracts of many members of macrofungal Basidiomycetes have been shown to exhibit antibacterial activity and some have been suggested to be potential substitutes or replacements for antibiotics used in managing bacterial diseases in crop production. Kaur et al. (2016) demonstrated that shiitake (*Lentinula edodes*) mycelial culture filtrate (Le_{mcf}) from several strains inhibited *Xanthomonas campestris* pv. *vesicatoria* cause of bacterial spot of tomato. The authors further showed that the extracts had antibacterial activity against *Erwinia amylovora* which is the cause of fire blight diseases in pears and apples. The potency of the extracts was equivalent to that of 100 ppm (100 µg/ml) streptomycin sulphate (Kaur et al., 2016; Kaur et al., 2019). Other workers have described a wide range of antibacterial activities of extracts from *Clytocybe geotropa* against *Ralstonia solanacearum*, *Erwinia carotovora* subsp. *carotovora*, *Pseudomonas syringae* pv. *syringae*, and *Xanthomonas campestris* pv. *vesicatoria* (Sivanandhan et al., 2017).

3.3 Anti-plant Pathogenic Fungi

In recent years, research to find natural alternatives to conventional synthetic fungicides for crop production has been intensified and significant consideration directed to macrofungal sources for novel biofungicides. Culture filtrates of the higher Basidiomycete, *Clitocybe nuda* containing a hydrophilic compound of molecular weight between 500 and 1000 Da, were used to reduce the incidence of *Phytophthora* blight of pepper, caused by *Phytophthora capsici* and the incidence of bacterial leaf spot on pepper, caused by *Xanthomonas axonopodis* pv. *vesicatoria*

(Chen and Huang, 2009). The compound was further characterised as pH insensitive and stable at high temperature. Extracts from basidiocarps of *Coprinus comatus* and *Ganoderma carnosum* had inhibitory effects against *Fusarium oxysporum* and *Alternaria brassicae in vitro* (Srivastava and Sharma, 2011). Priya et al. (2019) also demonstrated that organic solvent extracts of many macrofungal species, including *Coprinus comatus, Ganoderma lucidum, Lentinus edodes* and *Trametes versicolor*, exhibited more than 50 per cent growth inhibition in mycelia of *Colletotrichum capsici* causative agent of fruit rot of chili peppers. The metabolites of *G. lucidum* inhibited about 80 per cent spore germination in *C. capsici*. Ethanolic extract of *Pleurotus florida* showed highest inhibition activity against *Fusarium oxysporum* while *P. pulmonarius* displayed maximum zone of inhibition against *A. solani* (Dahima et al., 2020).

3.4 Anti-plant Parasitic Nematodes

There is increasing research effort to replace the synthetic nematicides with more natural sources for controlling plant parasitic nematodes. In China, extracts of wild macrofungal strains of Basidiomycetes and mycelia fermentation filtrate showed significant selective toxicity against phytopathogenic nematodes. *Mutinus caninus* (strain F149) exhibited 97.7 per cent mortality against *Meloidogyne incognita* and *Heterodera glycines* (Chen et al., 2010), while *Amanita excels* (strain F155) extracts and fermentation filtrate displayed 90 per cent toxicity against *M. incognita* (Chen et al., 2010). Specific strains of *Pleurotus cornucopiae, P. eryngii, P. ostreatus*, and *Lentinula edodes* exhibited potent nematicidal activity, as much as 82–99 per cent mortality (Sivanandhan et al., 2017). Other toxin-producing nematophagous of higher Basidiomycetes including *Coprinus, Hericium coralloides* and *Pleurotus pulmonarius* exhibit potent nematicidal effect attributable to the production of p-Anisaldehyde and other aromatic metabolites and fatty acids (Hyde et al., 2019). Members of the family Pleurotaceae include nematophagous species, such as the oyster mushroom *Pleurotus ostreatus* known to release toxin-laden droplets that can paralyse the nematode on contact followed by penetration of nematode body by the fungal feeding hyphae and complete digestion. Several compounds have been isolated from mycelial extracts of species of *Pleurotus* with potent nematicidal activity identified as (*E*)-2-decenedioic acid, *S*-coriolic acid and linoleic acid (Degenkolb and Vilcinskas, 2016). Direct application of the *Pleurotus* spp. in soil has been proposed as a cost-effective approach for management of plant parasitic nematodes (Palizi et al., 2009; Degenkolb and Vilcinskas, 2016).

3.5 Bioherbicidal Activity

Very limited research has been conducted to develop macrofungal sources of bioherbicides. Allelochemicals of fungal species have been touted as potential sources for development of novel bioherbicidal compounds as alternatives to synthetic herbicides. Methanol extracts of the edible Basidiomycete, *Ramaria flava*, collected at the Amanos mountains of Turkey at 25 mg/ml inhibited the germination, radicle and plumule growth and total chlorophyll content in *Cucumis sativus* (Bozdogan et al., 2016).

3.6 Anti-plant Viruses

Few studies have been conducted on the antiviral activity of macrofungi against plant viruses. Strains of the potent species *Ganoderma lucidum* and mycelial culture filtrate of the polypore *Fomes fomentarius* were found to be effective in inhibiting tobacco mosaic virus (TMV) and the mechanical transmission of TMV, respectively (Sivanandhan et al., 2017). Similarly, the aqueous extracts of *Lentinula edodes* displayed antiviral activity against cowpea aphid-borne mosaic virus (Di Piero et al., 2010). Wang et al. (2013) purified and characterised a novel lectin with antiphytoviral activity from the wild mushroom, *Paxillus involutus*. Table 1 summarises macrofungi with known biopesticidal activities in agriculture.

Table 1. Metabolites of some macrofungal species exhibiting biopesticidal activities against crop pests.

	Metabolite	Pest control	References
Armita phalloides	180 KDa protein Extract	Bacterial wilt (*Ralstonia solanacearum*)	Erjavec et al. (2016)
Clitocyte geotropa	Clitocypin	*Clavibacter michiganensis* sub sp. *sepedonicus*	Dreo et al. (2007)
Clitocyte nuda	MW 500–1000	*Phytophthora capsici* and bacterial leaf spot	Chen and Huang (2009)
Coprinus comatus		*Fusarium oxysporum* and *Alternaria brassicae*	Srivastava and Sharma (2011)
Ganoderma lucidum		*Colletotrichum capsici*	Priya et al. (2019)
Coprinus, Hericium coralloides and *Pleurotus pulmonarius*	p-Anisaldehyde, fatty acids	Plant parasitic nematodes	Hyde et al. (2019)
Pleurotus spp.	(E)-2-decanedioic acid, S-coriolic acid and linoleic acid	Plant parasitic nematodes	Degenkolb and Vilcinskas (2016)
Ramaria flava	MeOH extract	Inhibition of germination in *Cucumis sativa*	Bozdogan et al. (2016)
Paxillus involutus	Lectins	Antiphytoviral activity	Wang et al. (2013)
Lentinula edodes		Cowpea aphid-borne mosaic virus	Di Piero et al. (2010)
Lentinula edodes	Culture filtrate	Bacterial spot	Kaur et al. (2016)

3.7 Control of Post-harvest Diseases

The biocontrol potential of higher Basidiomycetes has been harnessed for control and prevention of post-harvest losses to moulds and other decaying fungi that can have tremendous impact on the fruit market. Barneche et al. (2016) extracted and tested metabolites from several higher Basidiomycete mushrooms (*Bjerkandera adusta, Dictyopanus pusillus, Ganoderma resinaceum* and *Laetiporus sulphureus*) using malt extract broth, at pH 5, at 20°C and static culture conditions and observed antimicrobial activity against *Aspergillus oryzae, Botrytis cinerea, Laetiporus sulphureus, Penicillium expansum, Rhizopus stolonifer*, and *Xanthomonas vesicatoria*, displaying the strongest activity against *Xanthomonas campestris* pv.

vesicatoria. Extracts of the edible medicinal mushroom were also shown to bolster the biocontrol activity of the yeast *Cryptoccocus laurentii* against *Penicillium expansum*, a post-harvest pathogen of apples, which produces the hazardous mycotoxin patulin (Zjalica et al., 2010). *Lentinula edodes* culture filtrates were also found to enhance the growth of *C. laurentii* and its catalase, superoxide dismutase and glutathione peroxidase activities as antioxidants.

4. Conclusion and Prospects

Macrofungi are prominent sources of biopesticides and continue to evoke significant research attention in exploiting their rich metabolites for sustainable pest management in agriculture, thus alleviating the risks posed by current synthetic pesticide application. They are widely available, versatile and easily cultivable. Although many macrofungal biopesticides have been commercialised, there are surmountable challenges to the beneficial exploitation of these macrofungi. These challenges are mainly associated with metabolite compound formulation, registration, commercialisation and large-scale adoption and application. Various formulations for commercialisation already exist and include powders, emulsions, oil solutions, granular formulations and microencapsulations (Liu and Li, 2004; Patel et al., 2011; Hyde et al., 2019). These challenges deserve priority consideration in future research in addition to sustaining efforts at continued development of new promising macrofungal sources of biopesticides.

References

Agyare, C. and Agana, T.A. (2019). Bioactive metabolites from Basidiomycetes. pp. 230–252. *In:* Sridhar, K.R. and Deshmukh, S.K. (eds.). Advances of Macrofungi: Diversity, Ecology and Biotechnology. CRC Press, Boca Raton, USA.

Alves, M.J., Ferreira, I.C., Dias, J., Teixeira, V., Martins, A. and Pintado, M. (2012). A review on antimicrobial activity of mushroom (Basidiomycetes) extracts and isolated compounds. Planta Med., 78: 1707–1718.

Balaes, T. and Tanase, C. (2016). Basidiomycetes as potential biocontrol agents against nematodes. Rom. Biotechnol. Lett., 21(1): 11185–11192.

Barneche, S., Jorcin, G., Cecchetto, G., Cerdeiras, M.P. Vázquez, A. and Alborés, S. (2016). Screening for antimicrobial activity of wood rotting higher Basidiomycetes mushrooms from Uruguay against phytopathogens. Int. J. Med. Mushrooms, 18: 261–267.

Barseghyan, G.S., Barazaniand, A. and Wasser, S.P. (2016). Medicinal mushrooms with anti-phytopathogenic and insecticidal properties. pp. 137–153. *In:* Petre, M. (ed.). Mushroom Biotechnology, Development and Applications. Academic Press.

Bozdogan, A., Eker, T., Bozok, F., Ulukanli, Z., Dogan, H.H. and Buyukalaca, S. (2016). Multiple antioxidant and bioherbicidal assays of the edible mushroom species 'Ramaria flava' in Amanos mountains. Biointerface Res. Appl. Chem., 6: 1681–1685.

Chang, S.-T. and Miles, P.G. (2004). Mushrooms: Cultivation, Nutritional Value, Medicinal Effect and Environmental Impact, 2nd ed. CRC Press, Boca Raton.

Chen, J.-T. and Huang, J.-W. (2009). Control of plant diseases with secondary metabolite of *Clitocybe nuda*. New Biotechnol., 26: 193–198.

Chen, L., Chen, Y., Zhang, G. and Duam, Y. (2010). Nematicidal activity of extraction and fermentation filtrate of Basidiomycetes collected in Liaoning Province, China. Chin. J. Biol. Con., 26: 467–473.

Dahima, V., Doshi, A. and Singh, H. (2020). Screening of antifungal activity of *Pleurotus pulmonarius*, *Pleurotus florida* and *Shizophyllum commune*. Int. J. Curr. Microbiol. App. Sci., 9(04): 997–1004.

Degenkolb, T. and Vilcinskas, A. (2016). Metabolites from nematophagous fungi and nematicidal natural products from fungi as alternatives for biological control. Part II: Metabolites from nematophagous Basidiomycetes and non-nematophagous fungi. Appl. Microbiol. Biotechnol., 100: 3813–3824.

Di Piero, R.M., Novaes, Q.S. and Pascholati, S.F. (2010). Effect of *Agaricus brasiliensis* and *Lentinula edodes* mushrooms on the infection of passionflower with cowpea aphid-borne mosaic virus. Braz. Arch. Biol. Technol., 53: 269–278.

Dreo, T., Zelko, M., Skubic, J., Teixeira, V. and Ravnikar, M. (2007). Antibacterial activity of proteinaceous extracts of higher Basidiomycetes mushrooms against plant pathogenic bacteria. Int. J. Med. Mushrooms, 9: 226–237.

Erjavec, J., Ravnikar, M., Brzin, J., Grebenc, T., Blejec, A. et al. (2016). Antibacterial activity of wild mushroom extracts on bacterial wilt pathogen *Ralstonia solanacearum*. Plant Dis., 100: 453–464.

Griffin, K., Gambley, C., Brown, P. and Li, Y. (2017). Copper-tolerance in *Pseudomonas syringae*, pv. *tomato* and *Xanthomonas* spp. and the control of diseases associated with these pathogens in tomato and pepper. A systematic literature review. Crop Prot., 96: 144–150.

Huang, Y., Zhang, S.-B., Chen, H.-P., Zhao, Z.-Z., Li, Z.-H. et al. (2016). New acetylenic acids and derivatives from the Basidiomycete *Craterellus lutescens* (Cantharellaceae). Fitoterapia, 115: 177–181.

Hyde, K.D., Xu, J., Rapior, S., Jeewon, R., Lumyong, S. et al. (2019). The amazing potential of fungi: 50 ways we can exploit fungi industrially. Fungal Diversity 97: 1–136.

Jiang, M.Y., Wang, F., Dong, Z.J., Zhang, Y., Zhu, H.J. and Liu, J.K. (2008). A new hydroxyl acetylenic fatty acid from the Basidiomycete *Craterellus aureus* (Cantharellaceae). Acta Bot. Yunnanica, 30: 614–616.

Jo, W.S., Hossain, M.A. and Park, S.C. (2014). Toxicological profiles of poisonous, edible, and medicinal mushrooms. Mycobiology, 42: 215–220.

Karasinski, D. (2013). *Lawrynomyces*, a new genus of corticioid fungi in the Hymenochaetales. Acta Mycologica, 48(1): 5–11.

Kaur, H., Nyochembeng, L.M., Mentreddy, S.R., Banerjee, P. and Cebert, E. (2016). Assessment of the antimicrobial activity of *Lentinula edodes* against *Xanthomonas campestris* pv *vesicatoria*. Crop Prot., 89: 284–288.

Kaur, H., Nyochembeng, L.M., Mentreddy, S.R., Banerjee, P. and Cebert, E. (2019). Optimisation of fermentation conditions of *Lentinula edodes* (Berk) Pegler (Shiitake Mushroom) mycelia as a potential biopesticide. J. Agric. Sci., 11(13): 10.5539/jas.v11n13p1.

Lengai, G.M.W. and Muthomi, J.W. (2018). Biopesticides and their role in sustainable agricultural production. J. Biosci. Med., 6: 7–41.

Liu, X.Z. and Li, S. (2004). Fungal secondary metabolites in biological control of crop pests. pp. 723–747. *In*: An, Z.Q. (ed.). Handbook of Industrial Mycology. Marcel Dekker, New York.

Lu, H., Lou, H., Hu, J., Liu, Z. and Chen, Q. (2020). Macrofungi: A review of cultivation strategies, bioactivity, and application of mushrooms. Comp. Rev. Food Sci. Food Saf., 19: 2333–2356.

Mayerhofer, G., Schwaiger-Nemirova, I., Kuhn, T., Girsch, L. and Allerberger, F. (2009). Detecting streptomycin in apples from orchards treated for fire blight. J. Antimicrob. Chemother., 63: 1076–1077.

McManus, P. and Stockwell, V. (2000). Antibiotics for plant disease control: Silver bullets or rusty sabers. APSnet Features. https://doi.org/10.1094/APSnetFeature-2000-0600.

Nedelkoska, D.N., Kalevska, T., Atanasova-Pancevska, N., Karadelev, M., Uzunoska, Z. and Kungulovski, D. (2019). Evaluation of bactericidal activity of selected wild macrofungi extracts against *Escherichia coli*. J. Agric. Pl. Sci., 17(2): 53–57.

Palizi, P., Goltapeh, E.M., Pourjam, E. and Safaie, N. (2009). Potential of oyster mushrooms for the biocontrol of sugar beet nematode (*Heterodera schachtii*). J. Plant Prot. Res., 49: 27–33.

Patel, A.V., Jakobs-Schonwandt, D., Rose, T. and Vorlop, K.D. (2011). Fermentation and microencapsulation of the nematophagous fungus *Hirsutella rhossiliensis* in a novel type of hollow beads. Appl. Microbiol. Biotechnol., 89: 1751–1760.

Priya, K., Thiribhuvanamala, G., Kamalakannan, A. and Krishnamoorthy, A.S. (2019). Antimicrobial activities of biomolecules from mushroom fungi against *Colletotrichum capsici* (Syd.) Butler and Bisby, the fruit rot pathogen of chilli. Int. J. Curr. Microbiol. Appl. Sci., 8(6): 1172–1186.

Rathore, H., Prasad, S., Kapri, M., Tiwari, A. and Sharma, S. (2019). Medicinal importance of mushroom mycelium: Mechanisms and applications. J. Func. Foods, 56: 182–193.

Ren, A., Shi, L., Zhu, J., Yu, H., Jiang, A. et al. (2019). Shedding light on the mechanisms underlying the environmental regulation of secondary metabolite ganoderic acid in *Ganoderma lucidum* using physiological and genetic methods. Fungal Gen. Biol., 128: 43–48.

Shi, L., Ren, A., Mu, D. and Zhao, M. (2010). Current progress in the study on biosynthesis and regulation of ganoderic acids. Appl. Microbiol. Biotechnol., 88: 1243–1251.

Sivanandhan, S., Khusro, A., Paulraj, M.G., Ignacimuthu, S. and Al-Dhabi, N.A. (2017). Biocontrol properties of Basidiomycetes: An overview. J. Fungi, 3(1): 2. 10.3390/jof3010002.

Srivastava, M.P. and Sharma, N. (2011). Antimicrobial activities of basidiocarp of some basidiomycetes strains against bacteria and fungi. J. Mycol Plant Pathol., 41: 332–334.

Tancos, K., Villani, S., Kuehne, S., Borejsza-Wysocka, E., Breth, D. et al. (2016). Prevalence of streptomycin-resistant *Erwinia amylovora* in New York apple orchards. Pl. Dis., 100: 802–809. 10.1094/PDIS-09-15-0960-RE.

Tolaini, V., Zjalic, S., Reverberi, M., Fanelli, C., Fabbri, A. et al. (2010). *Lentinula edodes* enhances the biocontrol activity of *Cryptococcus laurentii* against *Penicillium expansum* contamination and patulin production in apple fruits. Int. J. Food Microbiol., 138: 243–249.

Wang, J., Wang, H.-Y., Xia, X.-M., Li, P.-P. and Wang, K.-Y. (2013). Synergistic effect of *Lentinula edodes* and *Pichia membranifaciens* on inhibition of *Penicillium expansum* infections. Post-harvest Biol. Technol., 81: 7–12.

Wang, S.X., Zhang, G.Q., Zhao, S., Xu, F. and Zhou, Y. (2013). Purification and characterization of a novel lectin with antiphytovirus activities from the wild mushroom *Paxillus involutus*. Prot. Pept. Lett., 20(7): 767–774.

Weber, W., Anke, T., Steffan, B. and Steglich, W. (1990). Antibiotics from Basidiomycetes. XXXII. Strobilurin E: A new cytostatic and antifungal (E)-BETA-methoxyacrylate antibiotic from *Crepidotus fulvotomentosus* Peck. J. Antibiot., 43: 207–212.

Xiao, H. and Zhong, J.-J. (2016). Production of useful terpenoids by higher-fungus cell factory and synthetic biology approaches. Tr. Biotechnol., 34: 242–255.

Zjalica, T.S., Reverberi, M., Fanelli, C., Fabbri, A.A., Del Fiore, A. et al. (2010). *Lentinula edodes* enhances the biocontrol activity of *Cryptococcus laurentii* against *Penicillium expansum* contamination and patulin production in apple fruits. Int. J. Food Microbiol., 138(3): 243–249.

Index

Printed and bound by CPI Group (UK) Ltd, Croydon, CR0 4YY

24/10/2024

01778304-0013